Probability Theory and Elements of Measure Theory

SERIES IN QUANTITATIVE METHODS FOR DECISION MAKING

ROBERT L. WINKLER, Consulting Editor

Probability and Statistics for Decision Making, Ya-lun Chou
Statistics: Probability, Inference, and Decision, Volumes I and II and Combined ed., W. L. Hays and R. L. Winkler
Decision Making and the Theory of the Firm, I. Horowitz
An Introduction to Probability, Decision, and Inference, I. H. LaValle
Elements of Probability and Statistics, S. A. Lippman
Modern Mathematical Methods for Economics and Business, R. E. Miller
Applied Multivariate Analysis, S. J. Press
Linear Optimization, W. A. Spivey and R. M. Thrall
An Introduction to Bayesian Inference and Decision, R. L. Winkler

INTERNATIONAL SERIES IN DECISION PROCESSES

INGRAM OLKIN, Consulting Editor

Probability Theory and Elements of Measure Theory, H. Bauer
Probability and Statistics for Decision Making, Ya-lun Chou
A Basic Course in Statistics with Sociological Application, 2d ed., T. R. Anderson and M. Zelditch, Jr.
Statistics: Probability, Inference, and Decision, Volumes I and II and Combined ed., W. L. Hays and R. L. Winkler
Decision Making and the Theory of the Firm, I. Horowitz
Introduction to Statistics, R. A. Hultquist
Reliability Handbook, B. A. Kozlov and I. A. Ushakov (edited by J. T. Rosenblatt and L. H. Koopmans)
An Introduction to Probability, Decision, and Inference, I. H. LaValle
Elements of Probability and Statistics, S. A. Lippman
Modern Mathematical Methods for Economics and Business, R. E. Miller
Applied Multivariate Analysis, S. J. Press
Fundamental Research Statistics for the Behavioral Sciences, J. T. Roscoe
Linear Optimization, W. A. Spivey and R. M. Thrall
Applied Probability, W. A. Thompson, Jr.
Elementary Statistical Methods, 3d ed., H. M. Walker and J. Lev
An Introduction to Bayesian Inference and Decision, R. L. Winkler

Probability Theory and Elements of Measure Theory

Heinz Bauer

University of Erlangen-Nürnberg

Translated by

Lisa Rosenblatt

HOLT, RINEHART AND WINSTON, INC.

New York · Chicago · San Francisco · Atlanta
Dallas · Montreal · Toronto · London · Sydney

Library of Congress Catalog Card Number: 78-156000
ISBN: 0-03-081621-1
AMS 1970 Subject Classification Numbers: 6001, 6002, 2801, 2802
Printed in the United States of America
2 3 4 5 038 1 2 3 4 5 6 7 8 9

Gratefully Dedicated
to My Honored Teacher
Professor Otto Haupt

PREFACE

This book is intended to serve as a guide to the student of probability theory. It is meant to provide the reader with a knowledge of the most important ideas, methods, and results of this mathematical theory which, at present, is developing and branching out rapidly. The introductory nature and restricted scope of the book permits only treatment of those aspects of the theory which are today considered "classical." Nevertheless, the rudiments of stochastic processes and, in particular, of Markov processes, which are central to recent investigations and also important in nonmathematical applications, are discussed.

Since probability theory is intimately connected with measure and integration theory, the book also has a second goal: to provide the reader with a thorough familiarity with the essentials of measure theory. But this is developed only to the extent necessary to deal with problems in probability theory. Many important topics in measure theory, for example, the theory of Radon measures on locally compact spaces with no countability condition or the theory of Haar measures, are not discussed. On the other hand, measure theory is developed independently of probabilistic considerations. Thus, the reader who is interested only in measure theory can study Chapters 1–3 and 7–8 independently, since the remaining parts of the book are concerned exclusively with probability theory.

The book arose from a two-semester, first-year graduate course in probability theory that I gave at the University of Hamburg. Mrs. Lisa Rosenblatt undertook the difficult task of translating this book, originally published in German. The English version differs from the German one mainly in the additional problems. I have also tried to eliminate errors and improve some passages.

Jörn Lembcke, Klaus Janssen, and Clifford C. Brown were very helpful with advice and criticism in revising the manuscript. I offer my gratitude to them. I am particularly grateful to Mrs. Lisa Rosenblatt for the translation and to Edwin Hewitt and Victor Klee for their interest in having this book published in English. Finally, I wish to express my appreciation to Holt, Rinehart and Winston, Inc. for the efficient and understanding role in bringing this translation to its completion.

Erlangen, Germany
December 1971

Heinz Bauer

CONTENTS

Probability Theory and Elements of Measure Theory

INTRODUCTION

Probability theory owes its existence to the desire to develop mathematical models for studying random phenomena. The search for mathematical models that permit the investigation of the space surrounding us is responsible for the development of geometry. But geometry has long since abandoned the direction suggested by the initial problem and has, in the course of its development, constantly refined concepts and methods, leading to the investigation of new problems of algebraic and topological nature. However, probability theory did not go through a similar development until more recent times. Even though the development of the theory was motivated by the many possibilities for application, it nevertheless achieved recognition within mathematics as a discipline in its own right. This is mainly due to its strong connections with other areas of mathematics such as number theory, ergodic theory, function theory, and second-order partial differential equations, which have been developed in recent years. Decisive for its recognition as an independent mathematical discipline, however, was the realization that in its foundations and structure the theory satisfies the standards of rigor customary for a discipline of present day mathematics.

The evidence for this was presented by A. N. Kolmogorov [38] in his axiomatic approach to probability theory, proposed in 1933 and now generally accepted. Through his work probability theory became established within general measure and integration theory.

The first part of this book presents an introduction to measure and integration theory. The probabilistic motivation here is set aside in favor of a more geometric one. Only afterwards, and hence in far reaching agreement with the historical development, it is shown that the most significant measure-theoretic concepts also allow a probabilistic interpretation. The second part of the book introduces probability theory, which reaches its first climax in the discussion of the law of large numbers. The third part is a continuation of measure and integration theory. First the functional-analytic aspect of integration theory is treated via a discussion of the Daniell–Stone theorem. Then measures on topological spaces, particularly on locally compact and Polish spaces, are handled. Only the most elementary knowledge of general topology is assumed throughout. A short chapter on Fourier analysis, which establishes only the most important results for the case of Euclidean space $\mathbf{R}^n$, concludes this part.

In the fourth part the probabilistic considerations are continued. After a chapter on limit theorems, the notion of conditional expectation is introduced and thus the door to the theory of stochastic processes is opened. Of this theory, chiefly martingales (mostly with discrete parameter set) and the interplay between Markov processes and Markov semigroups are treated.

The exposition largely follows the accepted paths. Nevertheless, I also hope to provide the specialist with some new detail. In particular, the consistent reference to Dynkin systems in place of the usual but less tractable and less productive monotone systems might be new. After careful consideration I followed the classical route in the presentation of integration theory, going first through measure and then through integration theory. The requirements of probability theory seem to me still to be best satisfied in this way. I have nevertheless tried to present the discussion of the concept of integral in such a fashion that the reader can learn the functional-analytic approach to the theory without difficulty in the third part.

The examples interspersed generously in the exposition often have the character of exercises, and the reader is strongly urged to work these out carefully, as well as the additional exercise material. Not only for this, but also for reading the text itself, the reader will have need of paper and pencil in order to follow through and complete in detail the ideas which are sometimes only outlined.

NOTATION

The field of real numbers with the usual topology, that is, the *real line*, is always denoted by $\mathbf{R}$. When $(+)\infty$ and $-\infty$ are included, $\mathbf{R}$ becomes the *extended* (or *compactified*) *real line* $\bar{\mathbf{R}}$. Computations with the numbers $\pm\infty$ are done in the usual way. The convention

$$0 \cdot (+\infty) = 0 \cdot (-\infty) = (+\infty) \cdot 0 = (-\infty) \cdot 0 = 0,$$

which we use from now on, is not universally common. It proves to be useful for all measure- and probability-theoretic considerations. The set of all numbers $x \in \mathbf{R}$ ($x \in \bar{\mathbf{R}}$) with $x \geqq 0$ is denoted by $\mathbf{R}_+$ ($\bar{\mathbf{R}}_+$).

As usual, the symbols $\cup$, $\cap$, $\bigcup$, $\bigcap$, and $\complement$ are used for the set-theoretic operations, in particular, $\complement B$ is the complement of B with respect to the underlying space. $A \subset B$ means that A is a subset of B; $A = B$ is allowed here. We write $A \setminus B$ for the *difference set* $A \cap \complement B$ of two sets A and B, that is, the set of all $x \in A$ such that $x \notin B$. The empty set is denoted by $\varnothing$.

The symbol $f: A \to B$ means that f is a mapping of the set A into the set B. In the special case $B = \mathbf{R}$ ($B = \bar{\mathbf{R}}$) we speak of a *real* (*numerical*) *function*. The *restriction* of a mapping $f: A \to B$ to a subset $A' \subset A$ is denoted by $\text{rest}_{A'}f$.

A *sequence* $a_1, a_2, \ldots$ of elements of a set A is a mapping of the set of natural numbers $1, 2, \ldots$, denoted by $\mathbf{N}$, into the set A. Hence we write sequences in the form $(a_n)_{n \in \mathbf{N}}$ or $(a_n)_{n=1,2,\ldots}$ or also simply (a_n).

If the indexing is different, for example, $(a_n)_{n=0,1,\ldots}$ this will always be mentioned. Similarly, finite sequences appear in the form $(a_i)_{i=1,\ldots,n}$, and so on.

If f and g are real functions on a set A, then $f + g$, fg, and so on, denote the functions defined by $x \to f(x) + g(x)$, $x \to f(x)g(x)$, and so on. Numerical functions are handled analogously; but then $f(x) + g(x)$, for example, must be defined for all $x \in A$. Correspondingly, $\Sigma f_n = \Sigma_{n=1}^{\infty} f_n$ for a sequence (f_n) of numerical functions denotes the function $x \to \Sigma_{n=1}^{\infty} f_n(x)$. A numerical function f defined on a subset of $\bar{\mathbf{R}}$ is said to be *isotone* (*antitone*) if it is monotone increasing (monotone decreasing) in the usual sense. More generally, we also use this terminology when the domain of definition is a set A partially ordered by a relation $\leqq$ and $x \leqq y$ ($x, y \in A$) implies $f(x) \leqq f(y)$ [or $f(y) \leqq f(x)$]. If $x < y$ ($x, y \in A$) implies $f(x) < f(y)$ [or $f(y) < f(x)$], then we say that f is *strictly* isotone (or antitone).

Topological concepts are always used in the sense of N. Bourbaki. This applies in particular to the notion of compactness, that is, the compact spaces considered in this book will all be compact Hausdoff spaces. Where it seems useful we use the notation $A \equiv B$ to signify that the object A is equal to the object B *by definition*. The symbol $\lrcorner$ indicates *end of proof.*

part I

Measure and Integration Theory

1

MEASURE THEORY

1.1 σ-ALGEBRAS AND THEIR GENERATORS

Let Ω be an arbitrary set and $\mathfrak{P}(\Omega)$ its power set, that is, the system of all subsets of Ω. Then for every family $(A_i)_{i \in I}$ of elements of $\mathfrak{P}(\Omega)$, the union $\bigcup_{i \in I} A_i$ and the intersection $\bigcap_{i \in I} A_i$ are also in $\mathfrak{P}(\Omega)$. Further, for every set A, $\mathfrak{P}(\Omega)$ also contains its complement $\complement A$. In what follows we shall be interested in systems of sets $\mathfrak{A} \subset \mathfrak{P}(\Omega)$ which share these properties with $\mathfrak{P}(\Omega)$ for every countable index set I.

1.1.1. Definition. A system $\mathfrak{A}$ of subsets of a set Ω is called a *σ-algebra* (in Ω) if it has the following properties:

$$\Omega \in \mathfrak{A}; \tag{1.1.1}$$

$$A \in \mathfrak{A} \Rightarrow \complement A \in \mathfrak{A}; \tag{1.1.2}$$

$$\text{for every sequence } (A_n) \text{ of sets of } \mathfrak{A}, \bigcup_{n=1}^{\infty} A_n \text{ lies in } \mathfrak{A}. \tag{1.1.3}$$

Examples

1. $\complement\, \mathfrak{P}(\Omega)$ is always a σ-algebra.

2. For every set Ω, the system of all sets $A \subset \Omega$ for which either A or A is countable[1] is a σ-algebra. Property (1.3) is obtained here as follows: If all sets A_n are countable, then $\bigcup A_n$ is also countable; if one A_{n_0} is not

[1] Countable means finite or countably infinite.

countable, then $\complement \bigcup A_n = \bigcap \complement A_n$ is contained in $\complement A_{n_0}$ and therefore countable.

3. If $\mathfrak{A}$ is a σ-algebra in a set Ω and Ω' is a subset of Ω, then

$$\Omega' \cap \mathfrak{A} \equiv \{\Omega' \cap A : A \in \mathfrak{A}\} \tag{1.1.4}$$

is a σ-algebra $\mathfrak{A}'$ in Ω'. It is called the *trace* of $\mathfrak{A}$ in Ω'. If Ω' lies in $\mathfrak{A}$, then $\mathfrak{A}'$ consists of all subsets of Ω' belonging to $\mathfrak{A}$.

4. Suppose that Ω and Ω' are sets, $\mathfrak{A}'$ is a σ-algebra in Ω', and $T: \Omega \to \Omega'$ is a mapping of Ω into Ω'. Then the system of sets

$$T^{-1}(\mathfrak{A}') \equiv \{T^{-1}(A') : A' \in \mathfrak{A}'\} \tag{1.1.5}$$

is a σ-algebra in Ω. This follows from the known properties of set-theoretic operations for inverse mappings (here T^{-1}).

Every σ-algebra $\mathfrak{A}$ has the following properties, dual to (1.1.1) and (1.1.3):

$$\emptyset \in \mathfrak{A}; \tag{1.1.6}$$

$$\text{for every sequence } (A_n) \text{ of sets of } \mathfrak{A}, \bigcap_{n=1}^{\infty} A_n \text{ lies in } \mathfrak{A}. \tag{1.1.7}$$

Since $\emptyset = \complement \Omega$ and $\bigcap A_n = \complement \bigcup A'_n$ where $A'_n \equiv \complement A_n$, this follows from (1.1.1)–(1.1.3). Note that

$$A_1 \cup \cdots \cup A_n = A_1 \cup \cdots \cup A_n \cup \emptyset \cup \emptyset \cup \cdots$$

and

$$A_1 \cap \cdots \cap A_n = A_1 \cap \cdots \cap A_n \cap \Omega \cap \Omega \cap \cdots.$$

Thus $\mathfrak{A}$ contains the union and intersection of every finite collection of sets in $\mathfrak{A}$. From this and (1.1.2) it follows that

$$A, B \in \mathfrak{A} \Rightarrow A \setminus B = A \cap \complement B \in \mathfrak{A}. \tag{1.1.8}$$

The following theorem is important for the construction of σ-algebras.

1.1.2. Theorem. Every intersection of (finitely or infinitely many) σ-algebras in a set Ω is itself again a σ-algebra in Ω.

The proof is obtained by simple verification of properties (1.1.1)–(1.1.3).

It now follows that for every system $\mathfrak{E}$ of subsets of Ω there exists a *smallest σ-algebra* $\mathfrak{A}(\mathfrak{E})$ *containing* $\mathfrak{E}$; that is, $\mathfrak{A}(\mathfrak{E})$ is a σ-algebra with $\mathfrak{E} \subset \mathfrak{A}(\mathfrak{E})$, and for every σ-algebra $\mathfrak{A}'$ with $\mathfrak{E} \subset \mathfrak{A}'$, we have $\mathfrak{A}(\mathfrak{E}) \subset \mathfrak{A}'$. For the proof, consider the system Σ of all σ-algebras $\mathfrak{A}'$ in Ω with $\mathfrak{E} \subset \mathfrak{A}'$; $\mathfrak{P}(\Omega)$, for example, is an element of Σ. Then $\mathfrak{A}(\mathfrak{E})$ is the intersection of all $\mathfrak{A}' \in \Sigma$ which according to Theorem 1.1.2 has all the desired properties. We call $\mathfrak{A}(\mathfrak{E})$ the *σ-algebra generated by* $\mathfrak{E}$ (in Ω) and $\mathfrak{E}$ a *generator of* $\mathfrak{A}(\mathfrak{E})$.

Examples

5. If $\mathfrak{E}$ itself is a σ-algebra in Ω, then $\mathfrak{E} = \mathfrak{A}(\mathfrak{E})$.

6. If $\mathfrak{E}$ consists of only one set $A \subset \Omega$, then $\mathfrak{A}(\mathfrak{E})$ is the σ-algebra of the four sets $\emptyset$, A, $\complement A$, and Ω.

As generator of a σ-algebra $\mathfrak{A}$ we often will encounter systems $\mathfrak{E}$ of subsets of Ω that already possess some of the properties of a σ-algebra. Of particular interest are rings of subsets of Ω.

1.1.3. Definition. A system $\mathfrak{R}$ of subsets of a set Ω is called a *ring* (in Ω) if it has the following properties:

$$\emptyset \in \mathfrak{R}; \tag{1.1.9}$$
$$A, B \in \mathfrak{R} \Rightarrow A \setminus B \in \mathfrak{R}; \tag{1.1.10}$$
$$A, B \in \mathfrak{R} \Rightarrow A \cup B \in \mathfrak{R}. \tag{1.1.11}$$

If in addition we impose the requirement

$$\Omega \in \mathfrak{R}, \tag{1.1.12}$$

then $\mathfrak{R}$ is called an *algebra* (in Ω).

Every ring $\mathfrak{R}$ contains, along with two (and hence finitely many) sets, not only their union, but also their *intersection*: for any two sets $A, B \in \mathfrak{R}$ we have $A \cap B = A \setminus (A \setminus B)$.

1.1.4. Theorem. A system $\mathfrak{R}$ of subsets of a set Ω is an algebra if and only if it has properties (1.1.1), (1.1.2), and (1.1.11).

Proof. By definition every algebra has properties (1.1.1) and (1.1.11). From (1.1.10) we obtain (1.1.2). The converse follows from the identity

$$A \setminus B = A \cap \complement B = \complement(\complement A \cup B),$$

which holds for arbitrary sets A, B, and from $\emptyset = \complement\,\Omega$. ⌟

Examples

7. Every σ-algebra is an algebra.

8. For every infinite set Ω, the system of all sets $A \subset \Omega$ for which either A or $\complement A$ is finite is an algebra, although not a σ-algebra.

9. The system of all finite subsets of a set Ω is a ring, and for finite Ω it is an algebra.

10. The smallest ring existing in a set Ω consists of only $\emptyset$.

PROBLEMS

1. For every system $\mathfrak{E}$ of subsets of a set Ω, there exists a smallest ring $\mathfrak{R}(\mathfrak{E})$ in Ω containing $\mathfrak{E}$. The ring $\mathfrak{R}(\mathfrak{E})$ is said to be *generated* by $\mathfrak{E}$ in Ω. Prove the existence of $\mathfrak{R}(\mathfrak{E})$. Determine $\mathfrak{R}(\mathfrak{E})$ and $\mathfrak{A}(\mathfrak{E})$ in the case where $\mathfrak{E}$ consists of two arbitrary sets $A, B \subset \Omega$. When do $\mathfrak{R}(\mathfrak{E})$ and $\mathfrak{A}(\mathfrak{E})$ coincide in this case (in general)?
2. For arbitrary sets A and B, the set

$$A \triangle B \equiv (A \setminus B) \cup (B \setminus A)$$

is called the *symmetric difference* of A and B. Prove:
(a) $A \triangle B = B \triangle A$;
(b) $(A \triangle B) \triangle C = A \triangle (B \triangle C)$;
(c) $A \triangle A = \emptyset$;
(d) $A \triangle \emptyset = A$;
(e) $(A \triangle B) \cap C = (A \cap C) \triangle (B \cap C)$.
3. Deduce from Problem 2: A set $\mathfrak{R} \subset \mathfrak{P}(\Omega)$ is a ring in Ω if and only if it is a commutative ring in the sense of algebra with respect to the operations $\triangle$ (as addition) and $\cap$ (as multiplication).
4. A subset $\mathfrak{N}$ of a ring $\mathfrak{R}$ in Ω is called an *ideal* (in $\mathfrak{R}$) if it has the following three properties:
(a) $\emptyset \in \mathfrak{N}$;
(b) $N \in \mathfrak{N}, M \subset N, M \in \mathfrak{R} \Rightarrow M \in \mathfrak{N}$.
(c) $M, N \in \mathfrak{N} \Rightarrow M \cup N \in \mathfrak{N}$.
Prove: $\mathfrak{N} \subset \mathfrak{R}$ is an ideal in $\mathfrak{R}$ if and only if $\mathfrak{N}$ is an ideal (in the usual algebraic sense) in the algebraic ring $\mathfrak{R}$ (compare Problem 3). Every ideal in $\mathfrak{R}$ is itself a ring in Ω.

1.2 DYNKIN SYSTEMS

It is often difficult to verify directly whether a given system of sets is a σ-algebra. The following concept,[2] due to E. B. Dynkin [31], provides a way to overcome this difficulty.

[2] The term "λ-system" is used in [31].

1.2.1. Definition. A system $\mathfrak{D}$ of subsets of a set Ω is called a *Dynkin system* (in Ω) if it has the following properties:

$$\Omega \in \mathfrak{D}; \tag{1.2.1}$$

$$D, E \in \mathfrak{D}, D \subset E \Rightarrow E \setminus D \in \mathfrak{D}; \tag{1.2.2}$$

for every sequence (D_n) of pairwise disjoint sets of $\mathfrak{D}$,

$$\bigcup_{n=1}^{\infty} D_n \text{ lies in } \mathfrak{D}. \tag{1.2.3}$$

Examples

1. Every σ-algebra is obviously a Dynkin system.

2. Let Ω be a finite set with exactly $2p$ elements ($p \in \mathbf{N}$). Then the system $\mathfrak{D}$ of all sets $D \subset \Omega$ with an even number $2q$ of elements ($q = 0, 1, \ldots, p$) is a Dynkin system. If $p > 1$, then $\mathfrak{D}$ is not an algebra and therefore also not a σ-algebra.

The precise relationship between the concepts of σ-algebra and Dynkin system is the following:

1.2.2. Theorem. A Dynkin system $\mathfrak{D}$ is a σ-algebra if and only if the intersection of any two sets in $\mathfrak{D}$ is also in $\mathfrak{D}$.

Proof. We need to show only that every Dynkin system $\mathfrak{D}$ which contains the intersection of any two sets in $\mathfrak{D}$ is a σ-algebra. Properties (1.1.1) and (1.1.2) then follow directly from (1.2.1) and (1.2.2). Property (1.1.3) is obtained as follows: We have

$$A \cup B = A \cup (B \setminus A \cap B) \quad \text{and} \quad A \cap (B \setminus A \cap B) = \varnothing.$$

Hence $\mathfrak{D}$ contains the union of any two of its elements, and therefore also the union of any finite number of its elements. For a sequence (D_n) of sets of $\mathfrak{D}$ we have

$$\bigcup_{n=1}^{\infty} D_n = \bigcup_{n=0}^{\infty} (D'_{n+1} \setminus D'_n)$$

if we set

$$D'_0 \equiv \varnothing \text{ and } D'_n \equiv D_1 \cup \cdots \cup D_n \qquad (n = 1, 2, \ldots).$$

The sets $D'_{n+1} \setminus D'_n$ are pairwise disjoint and by the above argument and (1.2.2) they lie in $\mathfrak{D}$. By (1.2.3), $\bigcup D_n$ then also lies in $\mathfrak{D}$. ┘

Just as for σ-algebras (or rings and algebras, respectively), we find that for every system of sets $\mathfrak{E} \subset \mathfrak{P}(\Omega)$, there exists a smallest Dynkin system $\mathfrak{D}(\mathfrak{E})$ containing $\mathfrak{E}$. It is called *the Dynkin system generated by* $\mathfrak{E}$.

The significance of Dynkin systems is given primarily by the following theorem.

1.2.3. Theorem. For every system of sets $\mathfrak{E} \subset \mathfrak{P}(\Omega)$ which contains the intersection of every pair of its sets, we have

$$\mathfrak{D}(\mathfrak{E}) = \mathfrak{A}(\mathfrak{E}). \tag{1.2.4}$$

Proof. Since every σ-algebra is a Dynkin system, $\mathfrak{A}(\mathfrak{E})$ is a Dynkin system containing $\mathfrak{E}$ and thus $\mathfrak{D}(\mathfrak{E}) \subset \mathfrak{A}(\mathfrak{E})$. Conversely, if $\mathfrak{D}(\mathfrak{E})$ is a σ-algebra, then the dual relation $\mathfrak{A}(\mathfrak{E}) \subset \mathfrak{D}(\mathfrak{E})$ follows. By Theorem 1.2.2, we need to show only that $\mathfrak{D}(\mathfrak{E})$ contains the intersection of every pair of its sets. To prove this property we define for each $D \in \mathfrak{D}(\mathfrak{E})$

$$\mathfrak{D}_D \equiv \{Q \in \mathfrak{P}(\Omega): Q \cap D \in \mathfrak{D}(\mathfrak{E})\}.$$

Simple computations show that $\mathfrak{D}_D$ is a Dynkin system. For every $E \in \mathfrak{E}$, by the above hypothesis on $\mathfrak{E}$ we have $\mathfrak{E} \subset \mathfrak{D}_E$ and thus $\mathfrak{D}(\mathfrak{E}) \subset \mathfrak{D}_E$. For every $D \in \mathfrak{D}(\mathfrak{E})$ and every $E \in \mathfrak{E}$, we then have $E \cap D \in \mathfrak{D}(\mathfrak{E})$; but then $\mathfrak{E} \subset \mathfrak{D}_D$ and hence $\mathfrak{D}(\mathfrak{E}) \subset \mathfrak{D}_D$ for every $D \in \mathfrak{D}(\mathfrak{E})$. This is precisely the desired property of $\mathfrak{D}(\mathfrak{E})$. ⌟

Systems of sets $\mathfrak{E}$ that contain the intersection (union) of every pair of their sets will be called *$\cap$-stable* (*$\cup$-stable*).

PROBLEMS

1. Prove: A set $\mathfrak{D} \subset \mathfrak{P}(\Omega)$ is a Dynkin system in Ω if and only if it satisfies the following four properties:
   (a) $\Omega \in \mathfrak{D}$;
   (b) $D, E \in \mathfrak{D}, D \subset E \Rightarrow E \setminus D \in \mathfrak{D}$;
   (c) $D, E \in \mathfrak{D}, D \cap E = \varnothing \Rightarrow D \cup E \in \mathfrak{D}$;
   (d) $\bigcup_{n=1}^{\infty} D_n \in \mathfrak{D}$ for each increasing sequence $(D_n)_{n \in \mathbf{N}}$ in $\mathfrak{D}$.
2. Determine $\mathfrak{D}(\mathfrak{E})$ in the case where $\mathfrak{E}$ consists of only two subsets A and B of Ω. Prove: $\mathfrak{D}(\mathfrak{E}) = \mathfrak{A}(\mathfrak{E})$ if and only if $A \cap B = \varnothing$ or $A \cap \complement B = \varnothing$ or $\complement A \cap B = \varnothing$ or $\complement A \cap \complement B = \varnothing$.

1.3 CONTENTS AND PREMEASURES

If we pose the problem of defining a measure (p-dimensional volume) for certain subsets of Euclidean space $\mathbf{R}^p$, then we will try to solve this problem first for "geometrically simple" subsets. Only then we will consider more complicated sets and attempt to introduce their measure with the help of the previously defined measures of simple sets. This is the underlying idea in the following discussion of the general notion of measure. Again let Ω be an arbitrary set.

1.3.1. Definition. Let $\mathfrak{R}$ be a ring in Ω and μ a numerical function defined on $\mathfrak{R}$. We call μ a *premeasure* (on $\mathfrak{R}$) if

$$\mu(\emptyset) = 0; \tag{1.3.1}$$

$$\mu \geqq 0; \tag{1.3.2}$$

for every sequence (A_n) of pairwise disjoint sets of $\mathfrak{R}$ with $\bigcup_{n=1}^{\infty} A_n \in \mathfrak{R}$, we have

$$\mu\left(\bigcup_{n=1}^{\infty} A_n\right) = \sum_{n=1}^{\infty} \mu(A_n) \qquad (\sigma\text{-}additivity).^3 \tag{1.3.3}$$

We call μ a *content* if in place of (1.3.3) we have only

for every two and hence for every finite number of pairwise disjoint sets $A_1, \ldots, A_n \in \mathfrak{R}$,

$$\mu\left(\bigcup_{i=1}^{n} A_i\right) = \sum_{i=1}^{n} \mu(A_i) \qquad (\textit{finite additivity}). \tag{1.3.3'}$$

By (1.3.1), every premeasure is obviously a content. We need only set $A_{n+1} = A_{n+2} = \cdots = \emptyset$ in (1.3.3). (For the converse see Example 3 below.)

Examples

1. For every ring $\mathfrak{R}$ in Ω and every point $\omega \in \Omega$, the following function ϵ_ω is a premeasure on $\mathfrak{R}$:

$$\epsilon_\omega(A) \equiv \begin{cases} 1, & \omega \in A; \\ 0, & \omega \notin A. \end{cases} \tag{1.3.4}$$

We call ϵ_ω the premeasure on $\mathfrak{R}$ defined by the *unit mass* at ω.

2. Let $\mathfrak{A}$ be the σ-algebra defined in Section 1.1, Example 2 for the case of an *uncountable* set Ω. If we set $\mu(A) = 0$ or 1, depending on whether A or $\complement A$ is countable, then μ is a premeasure on $\mathfrak{A}$. We note that for any two disjoint subsets of Ω, at most one can have a countable complement.

[3] The term *total additivity* is also used.

3. For a countably infinite set Ω, let $\mathfrak{A}$ be the algebra defined in Section 1.1, Example 8. If we set $\mu(A) = 0$ or $+\infty$, depending on whether A or $\complement A$ is finite, then μ is a content but not a premeasure.

4. Let $\mu_1, \mu_2, \ldots$ be a sequence of contents (premeasures) on a ring $\mathfrak{R}$; further, let $\alpha_1, \alpha_2, \ldots$ be a sequence of real numbers $\geqq 0$. Then

$$\mu \equiv \sum_{n=1}^{\infty} \alpha_n \mu_n$$

is again a content (premeasure) on $\mathfrak{R}$.

Every *content* μ on a ring $\mathfrak{R}$ also has the following properties (here $A, B, A_1, A_2, \ldots$ are arbitrary sets in $\mathfrak{R}$):

$$\mu(A \cup B) + \mu(A \cap B) = \mu(A) + \mu(B); \tag{1.3.5}$$

$$A \subset B \Rightarrow \mu(A) \leqq \mu(B) \qquad (\textit{isotonicity}); \tag{1.3.6}$$

$$A \subset B,\ \mu(A) < +\infty \Rightarrow \mu(B \setminus A) = \mu(B) - \mu(A) \qquad (\textit{subtractivity}); \tag{1.3.7}$$

$$\mu\Big(\bigcup_{i=1}^{n} A_i\Big) \leqq \sum_{i=1}^{n} \mu(A_i) \qquad (\textit{subadditivity}); \tag{1.3.8}$$

for every sequence (A_n) of pairwise disjoint sets in $\mathfrak{R}$ with $\bigcup A_n \in \mathfrak{R}$, we have

$$\sum_{n=1}^{\infty} \mu(A_n) \leqq \mu\Big(\bigcup_{n=1}^{\infty} A_n\Big). \tag{1.3.9}$$

Proof. For arbitrary sets $A, B \in \mathfrak{R}$,

$$A \cup B = A \cup (B \setminus A) \qquad \text{and} \qquad B = (A \cap B) \cup (B \setminus A).$$

By finite additivity it follows that

$$\mu(A \cup B) = \mu(A) + \mu(B \setminus A) \qquad \text{and} \qquad \mu(B) = \mu(A \cap B) + \mu(B \setminus A)$$

and further, by addition,

$$\mu(A \cup B) + \mu(A \cap B) + \mu(B \setminus A) = \mu(A) + \mu(B) + \mu(B \setminus A).$$

Now (1.3.5) follows if $\mu(B \setminus A)$ is finite. If $\mu(B \setminus A) = \infty$, then the formulas just derived for $\mu(A \cup B)$ and $\mu(B)$ show that $\mu(A \cup B) = \mu(B) = \infty$ and hence (1.3.5) also holds. In the case $A \subset B$, the formula derived for $\mu(B)$ yields

$$\mu(B) = \mu(A) + \mu(B \setminus A).$$

Since $\mu \geqq 0$, this yields (1.3.6) as well as (1.3.7). If we set $B_1 \equiv A_1$, $B_2 \equiv A_2 \setminus A_1, \ldots, B_n \equiv A_n \setminus A_1 \cup \cdots \cup A_{n-1}$, then $B_1, \ldots, B_n$

are pairwise disjoint sets in $\mathfrak{R}$ and hence $\mu\left(\bigcup B_i\right) = \Sigma\mu(B_i)$. Since $\bigcup B_i = \bigcup A_i$, $B_i \subset A_i$ $(i = 1, \ldots, n)$, and because of the isotonicity of μ, we have (1.3.8). To prove (1.3.9) it suffices to note that for every sequence (A_n) of pairwise disjoint sets in $\mathfrak{R}$ with $A \equiv \bigcup A_n \in \mathfrak{R}$,

$$\mu(A_1) + \cdots + \mu(A_m) = \mu(A_1 \cup \cdots \cup A_m) \leqq \mu(A) \qquad (m = 1, 2, \ldots).$$

It follows that

$$\sum_{n=1}^{\infty} \mu(A_n) \leqq \mu(A). \quad \lrcorner$$

Finally if μ is a *premeasure* on $\mathfrak{R}$, then for arbitrary sets $A_0, A_1, \ldots \in \mathfrak{R}$,

$$A_0 \subset \bigcup_{n=1}^{\infty} A_n \Rightarrow \mu(A_0) \leqq \sum_{n=1}^{\infty} \mu(A_n). \tag{1.3.10}$$

Since $A_0 = \bigcup (A_0 \cap A_n)$, by (1.3.6) we can assume $A_0 = \bigcup A_n$. Then we can define $B_1 \equiv A_1$, $B_2 \equiv A_2 \setminus A_1, \ldots, B_n \equiv A_n \setminus (A_1 \cup \cdots \cup A_{n-1}), \ldots$ and proceed as in the proof of (1.3.8).

The next theorem characterizes premeasures by other properties, related to σ-additivity. Its formulation is simplified by the following:

Notation. Let $E, E_1, E_2, \ldots$ be sets. By $E_n \uparrow E$ $(E_n \downarrow E)$ we mean that $E_1 \subset E_2 \subset \cdots$ and $E = \bigcup E_n$ $(E_1 \supset E_2 \supset \cdots$ and $E = \bigcap E_n$. The sequence (E_n) thus "converges" to E isotonely (antitonely).

1.3.2. Theorem. For a content μ on a ring $\mathfrak{R}$ we consider the following properties:

(a) μ is a premeasure.

(b) For every sequence (A_n) of sets of $\mathfrak{R}$ with $A_n \uparrow A \in \mathfrak{R}$,

$$\lim_{n\to\infty} \mu(A_n) = \mu(A) \qquad \textit{(continuity from below)}.$$

(c) For every sequence (A_n) of sets of $\mathfrak{R}$ with $A_n \downarrow A \in \mathfrak{R}$ and $\mu(A_n) < \infty$ for all n,

$$\lim_{n\to\infty} \mu(A_n) = \mu(A) \qquad \textit{(continuity from above)}.$$

(d) For every sequence (A_n) of sets of $\mathfrak{R}$ with $A_n \downarrow \varnothing$ and $\mu(A_n) < \infty$ for all n,

$$\lim_{n\to\infty} \mu(A_n) = 0 \qquad (\varnothing\textit{-continuity}).$$

Then the following implications hold:

$$\text{(a)} \Leftrightarrow \text{(b)} \Rightarrow \text{(c)} \Rightarrow \text{(d)}.$$

If μ is finite on $\mathfrak{R}$, that is, $\mu(A) < \infty$ for all $A \in \mathfrak{R}$, then properties (a)–(d) are equivalent.

Proof. (a) $\Rightarrow$ (b): We can assume $A_1 = \varnothing$. Then $B_n \equiv A_n \setminus A_{n-1}$, $n = 2, 3, \ldots$ are pairwise disjoint sets in $\mathfrak{R}$ with

$$A = \bigcup_{n=2}^{\infty} B_n \qquad \text{and} \qquad A_n = B_2 \cup \cdots \cup B_n.$$

Then by the σ-additivity of μ,

$$\mu(A) = \sum_{n=2}^{\infty} \mu(B_n) = \lim_{n\to\infty} \sum_{i=2}^{n} \mu(B_i) = \lim_{n\to\infty} \mu(A_n).$$

(b) $\Rightarrow$ (a): Let (A_n) be a sequence of pairwise disjoint sets of $\mathfrak{R}$ with $A \equiv \bigcup A_n \in \mathfrak{R}$. If we set $B_n \equiv A_1 \cup \cdots \cup A_n$, then $B_n \uparrow A$ and hence $\mu(A) = \lim \mu(B_n)$. Since μ is finitely additive we have

$$\mu(B_n) = \mu(A_1) + \cdots + \mu(A_n)$$

and thus $\mu(A) = \Sigma\mu(A_n)$. Hence μ is σ-additive and therefore a premeasure.

(b) $\Rightarrow$ (c): By (1.3.7), $\mu(A_1 \setminus A_n) = \mu(A_1) - \mu(A_n)$ hold for all $n = 1, 2, \ldots$. From $A_n \downarrow A$ it follows that $A_1 \setminus A_n \uparrow A_1 \setminus A$, where all the sets involved are in $\mathfrak{R}$. Then from (b),

$$\mu(A_1 \setminus A) = \lim \mu(A_1 \setminus A_n) = \mu(A_1) - \lim \mu(A_n).$$

Thus (c) follows, since $A \subset A_n$ implies $\mu(A) < \infty$ and hence $\mu(A_1 \setminus A) = \mu(A_1) - \mu(A)$.

(c) $\Rightarrow$ (d): This is immediate.

Finally assume μ is finite. Then we show that:

(d) $\Rightarrow$ (b): If (A_n) is a sequence of sets of $\mathfrak{R}$ with $A_n \uparrow A \in \mathfrak{R}$, then $A \setminus A_n \downarrow \varnothing$, and hence (noting the finiteness of μ) $\lim \mu(A \setminus A_n) = 0$. Since μ is a finite content, we have further that $\mu(A \setminus A_n) = \mu(A) - \mu(A_n)$. Now (b) follows. ⅃

Remark. Example 3 of this section shows that a $\varnothing$-continuous content need not be a premeasure. Therefore we cannot in general drop the hypothesis of finiteness in the assertion "(d) $\Rightarrow$ (a)."

PROBLEMS

1. Consider for a finite set Ω the premeasure $\mu \equiv \sum_{\omega \in \Omega} \epsilon_\omega$ on $\mathfrak{P}(\Omega)$. Check that $\mu(A)$ is the number of elements of an arbitrary set $A \subset \Omega$. Prove that any premeasure μ on $\mathfrak{P}(\Omega)$ is of the form $\mu = \sum_{\omega \in \Omega} \alpha_\omega \epsilon_\omega$, where $\alpha_\omega = \mu(\{\omega\})$.
2. Let Ω be a set. Define for a subset $A \subset \Omega$: $\mu(A)$ = number of points in A, if A is finite; $\mu(A) = +\infty$, if A is infinite. Then μ is a premeasure on $\mathfrak{P}(\Omega)$.
3. Let μ be a premeasure on a ring $\mathfrak{R}$ in Ω. Denote the $\mathfrak{N}_\mu$ the set of all $N \in \mathfrak{R}$ satisfying $\mu(N) = 0$ (that is, the set of the so-called *μ-null sets*). Prove the $\mathfrak{N}_\mu$ has the following properties:
   (a) $\emptyset \in \mathfrak{N}_\mu$;
   (b) $N \in \mathfrak{N}_\mu$, $M \subset N$, $M \in \mathfrak{R} \Rightarrow M \in \mathfrak{N}_\mu$;
   (c) $\bigcup_{n=1}^{\infty} N_n \in \mathfrak{N}_\mu$ for each sequence $(N_n)_{n\in\mathbf{N}}$ in $\mathfrak{N}_\mu$ for which $\bigcup_{n=1}^{\infty} N_n \in \mathfrak{R}$.

   Subsets of $\mathfrak{R}$ with these properties are called *σ-ideals* (compare Section 1.1, Problem 4). If μ is only a content, then $\mathfrak{N}_\mu$ is an ideal in $\mathfrak{R}$.
4. Let $\mathfrak{N}$ be a σ-ideal in a ring $\mathfrak{R}$. Then

$$\mu(A) \equiv \begin{cases} 0, & A \in \mathfrak{N} \\ +\infty, & A \in \mathfrak{R} \setminus \mathfrak{N} \end{cases}$$

   defines a premeasure on $\mathfrak{R}$ satisfying $\mathfrak{N}_\mu = \mathfrak{N}$.
5. Let μ be a finite content on a ring $\mathfrak{R}$. Define

$$d(A,B) := \mu(A \triangle B) \qquad (A,B \in \mathfrak{R}).$$

   Then d is a pseudometric on $\mathfrak{R}$, that is, d has all properties of a metric with the only exception that $d(A,B) = 0$ does not necessarily imply $A = B$.
6. Let μ be a premeasure on a ring $\mathfrak{R}$ in Ω. Define

$$\tilde{\mathfrak{R}} \equiv \{A \in \mathfrak{P}(\Omega) : A \cap R \in \mathfrak{R}, \qquad \text{for all } R \in \mathfrak{R}\}$$

   and

$$\tilde{\mu}(A) \equiv \sup\{\mu(R) : R \subset A,\ R \in \mathfrak{R}\} \qquad (A \in \tilde{\mathfrak{R}}).$$

   Then $\tilde{\mathfrak{R}}$ is an algebra in Ω satisfying $\mathfrak{R} \subset \tilde{\mathfrak{R}}$, and $\tilde{\mu}$ is a premeasure on $\tilde{\mathfrak{R}}$ which extends μ.

7. Let μ be a finite content on a ring $\mathfrak{R}$. Prove the following generalization of formula (1.3.5):

$$\mu\Big(\bigcup_{i=1}^{n} A_i\Big) = \sum_{1 \leqq i \leqq n} \mu(A_i) - \sum_{1 \leqq i < j \leqq n} \mu(A_i \cap A_j) + \sum_{1 \leqq i < j < k \leqq n} \mu(A_i \cap A_j \cap A_k) - + \cdots + (-1)^{n-1}\mu(A_1 \cap \cdots \cap A_n)$$

for all $A_1, \ldots, A_n \in \mathfrak{R}$ and all $n \in \mathbf{N}$.

8. Let $(\mu_n)_{n \in \mathbf{N}}$ be a sequence of premeasures on a ring $\mathfrak{R}$ which is increasing in the sense: $\mu_n(A) \leqq \mu_{n+1}(A)$ for all $n \in \mathbf{N}$ and all $A \in \mathfrak{R}$. Then $\mu(A) \equiv \lim_{n \to \infty} \mu_n(A)$ exists for all $A \in \mathfrak{R}$. Prove that μ is again a premeasure.

1.4 LEBESGUE PREMEASURE

Now let Ω be the Euclidean space $\mathbf{R}^p$ ($p = 1, 2, \ldots$). For any two points $a = (\alpha_1, \ldots, \alpha_p)$ and $b = (\beta_1, \ldots, \beta_p)$ with coordinates α_i, β_i we write $a \leqq b$ ($a \lhd b$) if $\alpha_i \leqq \beta_i$ ($\alpha_i < \beta_i$) for all $i = 1, \ldots, p$. A *right half-open interval* in $\mathbf{R}^p$ is a set of the form

$$[a, b[\; \equiv \{x \in \mathbf{R}^p \colon a \leqq x \lhd b\}, \tag{1.4.1}$$

provided a and b are points of $\mathbf{R}^p$ with $a \leqq b$. Geometrically we are dealing with an "open toward the right" parallelotope parallel to the axes. Obviously $[a,b[$ is nonempty if and only if $a \lhd b$; in this case $[a,b[$ determines the points a and b uniquely.

For every such interval $[a,b[$ we call the real number

$$(\beta_1 - \alpha_1) \cdot \ldots \cdot (\beta_n - \alpha_n) \tag{1.4.2}$$

its p-dimensional *elementary content*. It is equal to zero if and only if $[a,b[\; = \varnothing$, and thus if $a \lhd b$ does not hold (although $a \leqq b$).

From now on $\mathfrak{J}^p$ will always denote the set of all right half-open intervals in $\mathbf{R}^p$, $\mathfrak{F}^p$ the system of all finite unions of sets in $\mathfrak{J}^p$. The elements of $\mathfrak{F}^p$ are called p-dimensional *figures*. We have $\mathfrak{J}^p \subset \mathfrak{F}^p$.

1.4.1. Lemma. For any two intervals $I, J \in \mathfrak{J}^p$, we have $I \cap J \in \mathfrak{J}^p$ and $J \setminus I \in \mathfrak{F}^p$. Every figure is the union of finitely many pairwise disjoint intervals of $\mathfrak{J}^p$.

Proof. Let $I = [a,b[$ and $J = [c,d[$ with $a \leqq b$ and $c \leqq d$; the respective coordinates of these points are $\alpha_i, \beta_i, \gamma_i, \delta_i$ ($i = 1, \ldots, p$). If we now set $e \equiv (\max (\alpha_1,\gamma_1), \ldots, \max (\alpha_p,\gamma_p))$ and $f \equiv (\min (\beta_1,\delta_1), \ldots, \min (\beta_p,\delta_p))$, then $I \cap J = [e,f[$ if $e \leq f$, or $I \cap J = \emptyset$ otherwise. Thus, $I \cap J$ always lies in $\mathfrak{I}^p$. Since $J \setminus I = J \setminus I \cap J$ and $I \cap J \in \mathfrak{I}^p$, we can assume $I \subset J$ and $I \neq \emptyset$ for the second part of the proof. Then I and J uniquely determine the points a, b, c, d, and $c \leq a \triangleleft b \leq d$.

Now in $[a,b[$ with $a = (\alpha_1, \ldots, \alpha_p)$ and $b = (\beta_1, \ldots, \beta_p)$ let us replace α_i by γ_i and at the same time β_i by α_i, or, α_i by β_i and at the same time β_i by δ_i in all possible ways, that is, more precisely, let us make this kind of replacement for the elements i of each nonempty subset of $\{1, \ldots, p\}$. In this way we obtain $3^p - 1$ pairwise disjoint intervals in $\mathfrak{I}^p$ with $J \setminus I$ as union. Thus, $J \setminus I$ is a figure that can be represented as the union of finitely many pairwise disjoint sets of $\mathfrak{I}^p$. That this is true for every figure $F = I_1 \cup \cdots \cup I_n \in \mathfrak{F}^p$ with $I_1, \ldots, I_n \in \mathfrak{I}^p$ is obtained as follows:

$$F = I_1 \cup (I_2 \setminus I_1) \cup (I_3 \setminus I_1 \cup I_2) \cup \cdots \cup (I_n \setminus I_1 \cup \cdots \cup I_{n-1})$$

is the union of n pairwise disjoint sets each of which is of the form $I \setminus J_1 \cup \cdots \cup J_m$, where $I, J_1, \ldots, J_m$ are intervals of $\mathfrak{I}^p$. But this follows from

$$I \setminus J_1 \cup \cdots \cup J_m = \bigcap_{i=1}^{m} (I \setminus J_i),$$

when we write every $I \setminus J_i$ as the union of finitely many pairwise disjoint intervals of $\mathfrak{I}^p$ and apply the distributive law and the first part of the assertion. ⌟

1.4.2. Theorem. $\mathfrak{F}^p$ is a ring in $\mathbf{R}^p$.

Proof. The only nonobvious property of a ring is (1.1.10), according to which for every two sets F, $G \in \mathfrak{F}^p$, $F \setminus G$ also lies in $\mathfrak{F}^p$. By definition there exist intervals

$$I'_1, \ldots, I'_m, I''_1, \ldots, I''_n \in \mathfrak{I}^p$$

with

$$F = \bigcup_{i=1}^{m} I'_i \quad \text{and} \quad G = \bigcup_{j=1}^{n} I''_j.$$

But then

$$F \setminus G = \bigcup_{i=1}^{m} \left(\bigcap_{j=1}^{n} (I'_i \setminus I''_j) \right)$$

and thus it remains to show only that each of the sets $\bigcap_{j=1}^{n} (I'_i \setminus I''_j)$ is a figure.

By Lemma 1.4.1, $I'_i \setminus I''_j$ is always a figure. Thus it suffices to show that the intersection of two and therefore of finitely many figures is again a figure. But if $F = \bigcup_{i=1}^{m} I'_i$ and $G = \bigcup_{j=1}^{n} I''_j$ are two figures, then by the distributive law, $F \cap G$ is the union of the sets $I'_i \cap I''_j$ $(i = 1, \ldots, m;\ j = 1, \ldots, n)$, and this is again a figure by Lemma 1.4.1. ┘

Every figure is by definition the union of finitely many intervals of $\mathfrak{J}^p$. Hence, $\mathfrak{F}^p \subset \mathfrak{R}$ for every ring $\mathfrak{R}$ in $\mathbf{R}^p$ with $\mathfrak{J}^p \subset \mathfrak{R}$. Thus, Theorem 1.4.2 shows that $\mathfrak{F}^p$ is the *ring generated by* $\mathfrak{J}^p$ in $\mathbf{R}^p$.

Our geometric interpretation now makes the validity of the following theorem evident.

1.4.3. Theorem. There exists precisely one content λ on $\mathfrak{F}^p$ such that for every $I \in \mathfrak{J}^p$, $\lambda(I)$ is equal to the p-dimensional elementary content. This content is real-valued.

Proof. By Lemma 1.4.1, every figure $F \in \mathfrak{F}^p$ has a representation $F = I_1 \cup \cdots \cup I_n$ as the union of finitely many disjoint intervals $I_1, \ldots, I_n \in \mathfrak{J}^p$. For every real-valued content λ on $\mathfrak{F}^p$ we then have

$$\lambda(F) = \lambda(I_1) + \cdots + \lambda(I_n),$$

that is, λ is uniquely determined by its value on $\mathfrak{J}^p$ and real-valued. It remains to show only the existence of λ. To this end we first define λ only on $\mathfrak{J}^p$: For each $I \in \mathfrak{J}^p$, let λ be the p-dimensional elementary content of I. Then:

(a) Let $I = [a,b[\ \in \mathfrak{J}^p$, $a = (\alpha_1, \ldots, \alpha_p)$ and $b = (\beta_1, \ldots, \beta_p)$ with $a \leqq b$ and let γ be a real number with $\alpha_i \leqq \gamma \leqq \beta_i$ for some $i = 1, \ldots, p$. The hyperplane with the equation $\xi_i = \gamma$ decomposes I into two disjoint intervals $I_1 = [a',b[$, $I_2 = [a,b'[\ \in \mathfrak{J}^p$, where $a'(b')$ arises from a (b) by changing the ith coordinate of a (b) into γ. Then (1.4.2) yields $\lambda(I) = \lambda(I_1) + \lambda(I_2)$. By induction we obtain:

(b) If we decompose some $I = [a,b[\ \in \mathfrak{J}_p$ by finitely many hyperplanes of the type described in (a) into pairwise disjoint intervals $I_1, \ldots, I_n \in \mathfrak{J}^p$, then $\lambda(I) = \lambda(I_1) + \cdots + \lambda(I_n)$. More generally we have:

(c) For finitely many pairwise disjoint $I_1, \ldots, I_n \in \mathfrak{J}^p$ with $I_0 \equiv I_1 \cup \cdots \cup I_n \in \mathfrak{J}^p$, we have $\lambda(I_0) = \lambda(I_1) + \cdots + \lambda(I_n)$. We can obviously assume that each I_j is nonempty; then there exist points $a_j = (\alpha_{j1}, \ldots, \alpha_{jp})$ and $b_j = (\beta_{j1}, \ldots, \beta_{jp})$ in $\mathbf{R}^p$ with $a_j \lhd b_j$ and $I_j = [a_j,b_j[$ $(j = 0, 1, \ldots, n)$. By intersecting I_0 with all hyperplanes with the equations $\xi_i = \alpha_{ji}$ and $\xi_i = \beta_{ji}$ $(i = 1, \ldots, p;\ j = 1, \ldots, n)$, I_0 is decomposed into pairwise disjoint intervals $I'_1, \ldots, I'_m \in \mathfrak{J}^p$. Each of

$I_1, \ldots, I_n$ is in the union of certain of these $I'_1, \ldots, I'_m$. If we now apply (b) $n + 1$ times, the asserted equality follows.

(d) Now if

$$F = I_1 \cup \cdots \cup I_n = J_1 \cup \cdots \cup J_m$$

are two representations of a figure $F \in \mathfrak{F}^p$ as the union of pairwise disjoint intervals, then

$$\lambda(I_1) + \cdots + \lambda(I_n) = \lambda(J_1) + \cdots + \lambda(J_m).$$

Indeed $I_j = I_j \cap F = \bigcup_{i=1}^{m} (I_j \cap J_i)$ is a representation of I_j as the union of pairwise disjoint intervals $I_j \cap J_1, \ldots, I_j \cap J_m$. Consequently, by (c),

$$\lambda(I_j) = \sum_{j=1}^{m} \lambda(I_j \cap J_i) \qquad (j = 1, \ldots, n).$$

If we interchange the roles of i and j, we obtain analogously

$$\lambda(J_i) = \sum_{j=1}^{n} \lambda(I_j \cap J_i) \qquad (i = 1, \ldots, m).$$

Both equations together yield $\Sigma\lambda(I_j) = \Sigma\lambda(J_i)$.

(e) Thus for every $F \in \mathfrak{F}^p$, the number $\Sigma\lambda(I_j)$ is independent of the particular representation

$$F = I_1 \cup \cdots \cup I_n$$

of F as the union of finitely many pairwise disjoint $I_1, \ldots, I_n \in \mathfrak{J}^p$.

By setting

$$\lambda(F) \equiv \lambda(I_1) + \cdots + \lambda(I_n),$$

we extend λ to a real function defined on $\mathfrak{F}^p$ which we again denote by λ. This function is nonnegative and finitely additive by (d). Moreover, since $\emptyset \in \mathfrak{J}^p$ and $\lambda(\emptyset) = 0$, we have a content with the desired properties. ┘

1.4.4. Theorem. The content λ on $\mathfrak{F}^p$ is a premeasure.

Proof. Because of the finiteness of λ, according to Theorem 1.3.2 it suffices to verify the $\emptyset$-continuity of λ. Thus, let (F_n) be an antitone sequence of figures of $\mathfrak{F}^p$. We show that the assumption

$$\delta \equiv \lim \lambda(F_n) = \inf \lambda(F_n) > 0$$

implies

$$\bigcap_{n=1}^{\infty} F_n \neq \emptyset.$$

F_n is the union of finitely many pairwise disjoint intervals $I_1, \ldots, I_m \in \mathfrak{J}^p$. By a small shrinking of these intervals we can obtain a figure $G_n \in \mathfrak{F}^p$ such that $\bar{G}_n$ (= closure of G_n) $\subset F_n$ and

$$\lambda(F_n) - \lambda(G_n) \leqq 2^{-n}\delta.$$

If we set $H_n \equiv G_1 \cap \cdots \cap G_n$, then (H_n) is a sequence of sets of $\mathfrak{F}^p$ with $H_n \supset H_{n+1}$ and $\bar{H}_n \subset \bar{G}_n \subset F_n$. By the boundedness of F_n, $(\bar{H}_n)$ is a sequence of compact subsets of $\mathbf{R}^p$ with $F_n \supset \bar{H}_n \supset \bar{H}_{n+1}$. Consequently,[4] $\bigcap \bar{H}_n \neq \emptyset$ and thus $\bigcap F_n \neq \emptyset$ as asserted, provided every H_n is nonempty. But this can be seen as follows: For every $n = 1, 2, \ldots,$

$$\lambda(H_n) \geqq \lambda(F_n) - \delta(1 - 2^{-n}), \tag{*}$$

which will be shown by induction. (*) holds for $n = 1$ since $H_1 = G_1$ and $\lambda(F_1) - \lambda(G_1) \leqq 2^{-1}\delta$. When (*) is true for n, it is true for $n + 1$. Here we note that $H_{n+1} = G_{n+1} \cap H_n$ and hence, by (1.3.5),

$$\lambda(H_{n+1}) = \lambda(G_{n+1}) + \lambda(H_n) - \lambda(G_{n+1} \cup H_n).$$

By the induction hypothesis we have $\lambda(H_n) \geqq \lambda(F_n) - \delta(1 - 2^{-n})$. By the choice of G_{n+1} we have $\lambda(G_{n+1}) \geqq \lambda(F_{n+1}) - 2^{-n-1}\delta$ and $G_{n+1} \cup H_n \subset F_{n+1} \cup F_n = F_n$, and hence $\lambda(G_{n+1} \cup H_n) \leqq \lambda(F_n)$. Together this yields precisely

$$\lambda(H_{n+1}) \geqq \lambda(F_{n+1}) - 2^{-n-1}\delta - \delta(1 - 2^{-n}) = \lambda(F_{n+1}) - \delta(1 - 2^{-n-1}).$$

If we further note that $\lambda(F_n) \geqq \delta$ by the definition of δ, then (*) yields the bound $\lambda(H_n) \geqq 2^{-n}\delta > 0$ and hence $H_n \neq \emptyset$. ⌟

1.4.5. Definition. The premeasure λ defined on the ring $\mathfrak{F}^p$ of p-dimensional figures in $\mathbf{R}^p$ is called the *Lebesgue premeasure* in $\mathbf{R}^p$ or p-dimensional Lebesgue premeasure. It is henceforth denoted by λ^p.

PROBLEM

Prove: There exists a uniquely determined content μ on $\mathfrak{J}^1$ satisfying

$$\mu([\alpha,\beta[) = \begin{cases} 1, & \text{for } \alpha < 0 \text{ and } \beta \geqq 0 \\ 0, & \text{in all other cases.} \end{cases}$$

Is μ σ-additive?

[4] Compare, for example, Franz [32], Theorem 1.23.2.

1.5 EXTENSION OF A PREMEASURE TO A MEASURE

In the following let Ω be an arbitrary set.

1.5.1. Definition. A premeasure μ defined on a σ-algebra $\mathfrak{A}$ in Ω is called a *measure* on $\mathfrak{A}$.

For every set $A \in \mathfrak{A}$ the number $\mu(A)$ is said to be the *measure of A*.

Examples

1. The premeasure defined in Section 1.3, Example 2 is a measure.

2. If we choose a σ-algebra $\mathfrak{A}$ in Ω for $\mathfrak{R}$ in Section 1.3, Example 1, then ϵ_ω is a measure on $\mathfrak{A}$; ϵ_ω is said to be the measure on $\mathfrak{A}$ defined by a unit mass at ω, or simply the *unit mass at* ω. This notation derives from the interpretation of a measure μ on a σ-algebra in Ω as *mass-* or *charge-distribution* on Ω. Then, correspondingly, for every set $A \in \mathfrak{A}$, $\mu(A)$ is to be considered as the mass "smeared" across A.

3. Let Ω be an arbitrary set. For every set $A \subset \Omega$ let $|A|$ be the number of elements in A. Then $A \to |A|$ is a measure on $\mathfrak{P}(\Omega)$, called the *counting measure* on A (compare Section 1.3, Problem 2).

4. The Lebesgue premeasure λ^p is *not* a measure, since its domain of definition $\mathfrak{F}^p$ is not a σ-algebra. For example, $\mathbf{R}^p$ does not lie in $\mathfrak{F}^p$, since obviously every p-dimensional figure is a bounded subset of $\mathbf{R}^p$.

If $\tilde{\mu}$ is a measure on a σ-algebra $\mathfrak{A}$ in Ω and $\mathfrak{R}$ is a ring in Ω with $\mathfrak{R} \subset \mathfrak{A}$, then the restriction μ of $\tilde{\mu}$ to $\mathfrak{R}$ is a premeasure. The designation "premeasure" is justified subsequently by showing, conversely: For every premeasure μ on a ring $\mathfrak{R}$ there exists a σ-algebra $\mathfrak{A}$ in Ω with $\mathfrak{R} \subset \mathfrak{A}$ and a measure $\tilde{\mu}$ on $\mathfrak{A}$ with $\mu = \operatorname{rest}_{\mathfrak{R}} \tilde{\mu}$. It obviously suffices to choose, for $\mathfrak{A}$, the σ-algebra $\mathfrak{A}(\mathfrak{R})$ generated by $\mathfrak{R}$ in Ω.

1.5.2. Theorem (Extension Theorem). Every premeasure μ on a ring $\mathfrak{R}$ in Ω can be extended in at least one way to a measure $\tilde{\mu}$ on the σ-algebra $\mathfrak{A}(\mathfrak{R})$ generated by $\mathfrak{R}$ in Ω.

Proof. For every set $Q \subset \Omega$, $\mathfrak{U}(Q)$ denote the set of all sequences $(A_n)_{n \in \mathbf{N}}$ of sets of $\mathfrak{R}$ with $Q \subset \bigcup_{n=1}^{\infty} A_n$. Then we can define the following function μ^* on $\mathfrak{P}(\Omega)$:

$$\mu^*(Q) \equiv \begin{cases} \inf\left\{\sum_{n=1}^{\infty} \mu(A_n): (A_n) \in \mathfrak{U}(Q)\right\}, & \text{if } \mathfrak{U}(Q) \neq \varnothing; \\ +\infty, & \text{if } \mathfrak{U}(Q) = \varnothing. \end{cases} \tag{1.5.1}$$

This function has the following properties:

$$\mu^*(\varnothing) = 0; \tag{1.5.2}$$

$$\mu^* \geqq 0; \tag{1.5.3}$$

$$Q_1 \subset Q_2 \Rightarrow \mu^*(Q_1) \leqq \mu^*(Q_2); \tag{1.5.4}$$

$$\mu^*\left(\bigcup_{n=1}^{\infty} Q_n\right) \leqq \sum_{m=1}^{\infty} \mu^*(Q_n); \tag{1.5.5}$$

for every sequence (Q_n) of sets of $\mathfrak{P}(\Omega)$.

Here (1.5.3) follows directly from (1.5.1); for the proof of (1.5.2) we observe that the identical sequence $\varnothing, \varnothing, \ldots$ lies in $\mathfrak{U}(\varnothing)$. We obtain (1.5.4) from the fact that $Q_1 \subset Q_2$ implies $\mathfrak{U}(Q_2) \subset \mathfrak{U}(Q_1)$. For the proof of (1.5.5), we can assume that $\mu^*(Q_n) < \infty$, and thus in particular $\mathfrak{U}(Q_n) \neq \varnothing$ for $n = 1, 2, \ldots$. Then for arbitrary $\epsilon > 0$ and every $n \in \mathbf{N}$ there exists a sequence $(A_{nm})_{m \in \mathbf{N}}$ from $\mathfrak{U}(Q_n)$ with

$$\sum_{m=1}^{\infty} \mu(A_{nm}) \leqq \mu^*(Q_n) + 2^{-n}\epsilon.$$

The double sequence $(A_{nm})_{n,m=1,2,\ldots}$ (considered as a sequence) lies in $\mathfrak{U}(\bigcup Q_n)$, and hence

$$\mu^*(\bigcup Q_n) \leqq \sum_{m,n} \mu(A_{nm}) \leqq \sum \mu^*(Q_n) + \epsilon,$$

whence follows (1.5.5). ┘

Of importance in what follows is the remark that for all $A \in \mathfrak{R}$,

$$\mu^*(Q) \geqq \mu^*(Q \cap A) + \mu^*(Q \cap \complement A), \qquad \text{for all } Q \in \mathfrak{P}(\Omega) \tag{1.5.6}$$

and

$$\mu^*(A) = \mu(A). \tag{1.5.7}$$

Here we can assume that $\mu^*(Q) < \infty$, that is, $\mathfrak{U}(Q) \neq \varnothing$. Now we have

$$\Sigma\mu(A_n) = \Sigma\mu(A_n \cap A) + \Sigma\mu(A_n \setminus A)$$

for every sequence (A_n) of $\mathfrak{U}(Q)$, due to the finite additivity of μ. Further, the sequence $(A_n \cap A)$ lies in $\mathfrak{U}(Q \cap A)$ and the sequence $(A_n \setminus A)$ in $\mathfrak{U}(A \setminus A)$; consequently,

$$\Sigma\mu(A_n) \geqq \mu^*(Q \cap A) + \mu^*(Q \setminus A)$$

for every such sequence (A_n). Hence (1.5.6) follows. Equation (1.5.7) follows from (1.3.10), according to which $\mu(A) \leqq \mu^*(A)$, and from the remark that the sequence $A, \emptyset, \emptyset, \ldots$ lies in $\mathfrak{U}(A)$.

The significance of what has just been proved lies in the claim that the system $\mathfrak{A}^*$ of all sets $A \in \mathfrak{P}(\Omega)$ with property (1.5.6) is a σ-algebra in Ω and $\text{rest}_{\mathfrak{A}^*}\,\mu^*$ is a measure. By (1.5.6), $\mathfrak{R} \subset \mathfrak{A}^*$ and thus $\mathfrak{A}(\mathfrak{R}) \subset \mathfrak{A}^*$. Then by (1.5.7), $\tilde{\mu} \equiv \text{rest}_{\mathfrak{A}(\mathfrak{R})}\mu^*$ is an extension of μ to a measure on $\mathfrak{A}(\mathfrak{R})$. But this is what we are seeking. Thus the proof will be completed by the following theorem and the accompanying definition.

1.5.3. Definition. An *outer measure* on a set Ω is any numerical function μ^* on the power set $\mathfrak{P}(\Omega)$ with properties (1.5.2)–(1.5.5). A subset A of Ω is said to be *μ^*-measurable* if it has property (1.5.6).

The idea of the proof of the extension theorem, due to C. Carathéodory, consists of associating to a premeasure μ on a ring $\mathfrak{R}$ an outer measure by means of Definition 1.5.1 and applying the following theorem.

1.5.4. Theorem. Let μ^* be an outer measure on Ω. Then the system $\mathfrak{A}^*$ of all μ^*-measurable sets $\mathfrak{A} \subset \Omega$ is a σ-algebra in Ω and the restriction of μ^* to $\mathfrak{A}^*$ is a measure.

Proof. First we note that (1.5.6) and hence the requirement that $A \in \mathfrak{A}^*$ for a set $A \in \Omega$ is equivalent to

$$\mu^*(Q) = \mu^*(Q \cap A) + \mu^*(Q \setminus A), \qquad \text{for all } Q \in \mathfrak{P}(\Omega). \quad (1.5.6')$$

Property (1.5.5) applied to the sequence $Q \cap A$, $Q \setminus A$, $\emptyset$, $\emptyset$, ... implies the validity of the inequality dual to (1.5.6) for all $Q \in \mathfrak{P}(\Omega)$. From (1.5.6) or (1.5.6′) it follows immediately that $\Omega \in \mathfrak{A}^*$ and that due to symmetry in A and $\complement A$, the system $\mathfrak{A}^*$ contains the set $\complement A$ whenever it contains A. The following argument shows that $\mathfrak{A}^*$ contains $A \cup B$ whenever $A \in \mathfrak{A}^*$ and $B \in \mathfrak{A}^*$, and thus that $\mathfrak{A}^*$ is an algebra. For every $Q \in \mathfrak{P}(\Omega)$ we have

$$\mu^*(Q) = \mu^*(Q \cap B) + \mu^*(Q \setminus B).$$

If we replace Q here, once by $Q \cap A$ and then by $Q \setminus A = Q \cap \complement A$, we obtain two new equations [valid for all $Q \in \mathfrak{P}(\Omega)$] which, when substituted in (1.5.6′), yield

$$\mu^*(Q) = \mu^*(Q \cap A \cap B) + \mu^*(Q \cap A \cap \complement B) + \mu^*(Q \cap \complement A \cap B) + \mu^*(Q \cap \complement A \cap \complement B). \quad (1.5.8)$$

If we now replace Q by $Q \cap (A \cup B)$, we obtain

$$\mu^*(Q \cap (A \cup B)) = \mu^*(Q \cap A \cap B) + \mu^*(Q \cap A \cap \complement B) + \mu^*(Q \cap \complement A \cap B), \quad (1.5.9)$$

which, together with the preceding equation, yields

$$\begin{aligned} \mu^*(Q) &= \mu^*(Q \cap (A \cup B)) + \mu^*(Q \cap \complement (A \cup B)) \\ &= \mu^*(Q \cap (A \cup B)) + \mu^*(Q \setminus A \cup B). \end{aligned}$$

Since $Q \in \mathfrak{P}(\Omega)$ was arbitrary, we have $A \cup B \in \mathfrak{A}^*$.

Now let (A_n) be a sequence of pairwise disjoint sets of $\mathfrak{A}^*$ and A their union. If we choose $A = A_1$ and $B = A_2$ in (1.5.9), we obtain

$$\mu^*(Q \cap (A_1 \cup A_2)) = \mu^*(Q \cap A_1) + \mu^*(Q \cap A_2).$$

Hence by induction it follows that

$$\mu^*\left(Q \cap \bigcup_{i=1}^{n} A_i\right) = \sum_{i=1}^{n} \mu^*(Q \cap A_i)$$

for all $Q \in \mathfrak{P}(\Omega)$ and $n \in \mathbf{N}$. If we note that, according to what has been proved, $B_n \equiv \bigcup_{i=1}^{n} A_i$ lies in $\mathfrak{A}^*$ and $Q \setminus B_n \supset Q \setminus A$, that is, $\mu^*(Q \setminus B_n) \geqq \mu^*(Q \setminus A)$, then we obtain

$$\mu^*(Q) = \mu^*(Q \cap B_n) + \mu^*(Q \setminus B_n) \geqq \sum_{i=1}^{n} \mu^*(Q \cap A_i) + \mu^*(Q \setminus A)$$

for $n = 1, 2, \ldots$. Finally, by applying (1.5.5) we get

$$\mu^*(Q) \geqq \sum_{n=1}^{\infty} \mu^*(Q \cap A_n) + \mu^*(Q \setminus A) \geqq \mu^*(Q \cap A) + \mu^*(Q \setminus A)$$

and thus by the introductory remark

$$\mu^*(Q) = \sum_{n=1}^{\infty} \mu^*(Q \cap A_n) + \mu^*(Q \setminus A) = \mu^*(Q \cap A) + \mu^*(Q \setminus A)$$

for all $Q \in \mathfrak{P}(\Omega)$. Now A lies in $\mathfrak{A}^*$, and $\mathfrak{A}^*$ is a σ-algebra by Theorem 1.2.2. If in particular we set $Q = A$ in the last double equality, it follows that

$$\mu^*(A) = \sum_{n=1}^{\infty} \mu^*(A_n).$$

Hence $\operatorname{rest}_{\mathfrak{A}^*} \mu^*$ is a measure. ⌋

Now we shall show that in many important cases the measure $\tilde{\mu}$ of Theorem 1.5.2 is uniquely determined.

1.5.5. Theorem (**Uniqueness Theorem**). Let $\mathfrak{E}$ be a $\cap$-stable generator of a σ-algebra $\mathfrak{A}$ in Ω in which a sequence $(E_n)_{n \in \mathbf{N}}$ of sets exists with $E_n \uparrow \Omega$. If μ_1 and μ_2 are measures on $\mathfrak{A}$ with

$$\mu_1(E) = \mu_2(E), \qquad \text{for all } E \in \mathfrak{E}$$

and with

$$\mu_1(E_n) = \mu_2(E_n) < +\infty, \qquad \text{for all } n \in \mathbf{N},$$

then $\mu_1 = \mu_2$ on $\mathfrak{A}$.

Proof. Let E be in $\mathfrak{E}$ with $\mu_1(E) = \mu_2(E) < \infty$. Now consider the system

$$\mathfrak{D}_E \equiv \{D \in \mathfrak{A}: \mu_1(E \cap D) = \mu_2(E \cap D)\}.$$

Obviously $\Omega \in \mathfrak{D}_E$. By (1.3.7), $D_1, D_2 \in \mathfrak{D}_E$, and $D_2 \subset D_1$ imply

$$\begin{aligned}\mu_1(E \cap (D_1 \setminus D_2)) &= \mu_1(E \cap D_1 \setminus E \cap D_2) = \mu_1(E \cap D_1) - \mu_1(E \cap D_2)\\ &= \mu_2(E \cap D_1) - \mu_2(E \cap D_2) = \mu_2(E \cap (D_1 \setminus D_2)),\end{aligned}$$

and thus $D_1 \setminus D_2 \in \mathfrak{D}_E$. Here we note that when $\mu_1(E) = \mu_2(E) < \infty$ we have $\mu_1(E \cap D) < \infty$ and $\mu_2(E \cap D) < \infty$ for all $D \in \mathfrak{A}$ since μ_1 and μ_2 are isotonic. Using the σ-additivity of μ_1 and μ_2, we show analogously that $\mathfrak{D}_E$ also has the third and still missing property (1.2.3) of a Dynkin system. Since $\mathfrak{E}$ is $\cap$-stable, we have $\mathfrak{E} \subset \mathfrak{D}_E$ and hence $\mathfrak{D}(\mathfrak{E}) \subset \mathfrak{D}_E$. For the Dynkin system $\mathfrak{D}(\mathfrak{E})$ generated by $\mathfrak{E}$ we also have, by Theorem 1.2.3, $\mathfrak{D}(\mathfrak{E}) = \mathfrak{A}(\mathfrak{E}) = \mathfrak{A}$. Since $\mathfrak{D}(\mathfrak{E}) \subset \mathfrak{D}_E \subset \mathfrak{A}$, we see that $\mathfrak{D}_E = \mathfrak{A}$. Hence

$$\mu_1(E \cap D) = \mu_2(E \cap D), \qquad \text{for all } D \in \mathfrak{A}.$$

By assumption we have $\mu_1(E_n) = \mu_2(E_n) < +\infty$ for all $n \in \mathbf{N}$. Consequently, $\mu_1(E_n \cap D) = \mu_2(E_n \cap D)$ for $n = 1, 2, \ldots$ and all $D \in \mathfrak{A}$. By Theorem 1.3.2, μ_1 and μ_2 are continuous from below. From $E_n \uparrow \Omega$ it follows that $E_n \cap D \uparrow D$ and hence

$$\mu_1(D) = \lim_{n\to\infty} \mu_1(E_n \cap D) = \lim_{n\to\infty} \mu_2(E_n \cap D) = \mu_2(D)$$

for all $D \in \mathfrak{A}$. Thus $\mu_1 = \mu_2$. ┘

In order to be able to give a sufficient condition for uniqueness of the measure $\tilde{\mu}$ of Theorem 1.5.2, we state:

1.5.6. Definition. A content μ on a ring $\mathfrak{R}$ in Ω is said to be *σ-finite* if there exists a sequence $(A_n)_{n \in \mathbf{N}}$ of sets of $\mathfrak{R}$ with $A_n \uparrow \Omega$ and $\mu(A_n) < +\infty$ for all $n \in \mathbf{N}$.

Examples

5. Suppose a content μ on a ring $\mathfrak{R}$ in Ω is *finite*, that is, it satisfies the condition $\mu(A) < \infty$ for all $A \in \mathfrak{R}$. The σ-finiteness of μ is then equivalent to the existence of a sequence (A_n) in $\mathfrak{R}$ with $A_n \uparrow \Omega$. This condition is not automatically satisfied; for example, choose $\Omega \neq \emptyset$ and $\mathfrak{R} = \{\emptyset\}$.

6. The *Lebesgue premeasure* in $\mathbf{R}^p$ is σ-finite (and finite). If for every $n \in \mathbf{N}$, x_n denotes that point in $\mathbf{R}^p$ all of whose coordinates are equal to n, then $I_n \equiv [-x_n, x_n[$ is an interval in $\mathfrak{J}^p$; we have $I_n \uparrow \mathbf{R}^p$.

7. The counting measure defined in Section 1.5, Example 3 is σ-finite if and only if Ω is countable.

Combining our results, we have:

1.5.7. Theorem. Every σ-finite premeasure μ on a ring $\mathfrak{R}$ in Ω can be extended to a measure $\tilde{\mu}$ on $\mathfrak{A}(\mathfrak{R})$ in exactly one way.

Proof. Only the uniqueness of $\tilde{\mu}$ requires proof. But this follows directly from Theorem 1.5.5 since by virtue of the σ-finiteness of μ, the ring $\mathfrak{R}$ obviously has all the properties of the generator $\mathfrak{E}$ considered there. ┘

Remark. We cannot drop the hypothesis of σ-finiteness of μ in Theorem 1.5.7. As in Example 5, it suffices to choose a nonempty set Ω, and the ring $\mathfrak{R}$ consisting of only the empty set. Two distinct measures μ and ν with the same restriction on $\mathfrak{R}$ are defined on $\mathfrak{A}(\mathfrak{R}) = \{\emptyset, \Omega\}$ by $\mu(\emptyset) = \nu(\emptyset) \equiv 0$ and $\mu(\Omega) \equiv 0$, $\nu(\Omega) \equiv 1$.

Finally, we give another condition equivalent to σ-finiteness of a content:

1.5.8. Lemma. A content μ on a ring $\mathfrak{R}$ in Ω is σ-finite if and only if there exists a sequence $(B_n)_{n \in \mathbf{N}}$ of pairwise disjoint sets in $\mathfrak{R}$ with the properties:

$$\Omega = \bigcup_{n=1}^{\infty} B_n \tag{1.5.10}$$

and

$$\mu(B_n) < +\infty \qquad \text{for all } n \in \mathbf{N}. \tag{1.5.11}$$

Proof. Let μ be σ-finite and (A_n) a sequence in $\mathfrak{R}$ with $A_n \uparrow \Omega$ and $\mu(A_n) < \infty$ for all n. Then the sequence defined by

$$B_n \equiv \begin{cases} A_n \setminus A_{n-1}, & n \geqq 2 \\ A_1, & n = 1 \end{cases} \qquad (n \in \mathbf{N})$$

has all the desired properties; in particular, the sets B_n are pairwise disjoint. Conversely if (B_n) is a sequence in $\mathfrak{R}$ with the properties (1.5.10) and (1.5.11), then

$$A_n \equiv B_1 \cup \cdots \cup B_n \qquad (n \in \mathbf{N})$$

is a sequence in $\mathfrak{R}$ which by (1.3.8) satisfies the conditions of Definition 1.5.6. ┘

PROBLEMS

1. Let $\mathfrak{R}$ be a ring in Ω, and let μ be the premeasure ϵ_ω on $\mathfrak{R}$ for a point $\omega \in \Omega$. Assume that $\{\omega\}$ is the intersection of a sequence of sets in $\mathfrak{R}$. Prove: The exterior measure μ^* derived from μ [in the sense of (1.5.1)] satisfies

$$\mu^*(A) = \begin{cases} 1, & \omega \in A \\ 0, & \omega \in \complement A \end{cases} \qquad (A \in \mathfrak{P}(\Omega)).$$

All subsets of Ω are μ^*-measurable. Hence the extension $\tilde{\mu}$ constructed in Theorem 1.5.2 is the measure ϵ_ω on $\mathfrak{P}(\Omega)$.

2. Consider the premeasure μ of Section 1.3, Example 2. Prove: The exterior measure μ^* derived from μ satisfies

$$\mu^*(A) = \begin{cases} 1, & \text{if } A \text{ is uncountable} \\ 0, & \text{if } A \text{ is countable} \end{cases} \qquad (A \in \mathfrak{P}(\Omega)).$$

μ^* is not a measure (even not a content) on $\mathfrak{P}(\Omega)$. Only the sets of the original σ-algebra $\mathfrak{A}$ are μ^*-measurable.

3. Let μ be a σ-finite measure on a σ-algebra $\mathfrak{A}$ in Ω, and μ^* the exterior measure derived from it. Prove: For every set $Q \in \mathfrak{P}(\Omega)$ there exists a set $A \in \mathfrak{A}$ (called a *measurable hull* of Q) satisfying $Q \subset A$, $\mu^*(Q) = \mu(A)$, and $\mu(B) = 0$ for all $B \in \mathfrak{A}$ satisfying $B \subset A \setminus Q$. [*Hint:* In the case $\mu^*(Q) < +\infty$, prove the existence of a sequence (A_n) in $\mathfrak{A}$ satisfying $Q \subset A_n$ and $\mu(A_n) \leqq \mu^*(Q) + 1/n$ for all $n \in \mathbf{N}$. Then $A \equiv \bigcap_{n=1}^{\infty} A_n$ has the required properties.]

4. A measure μ on a σ-algebra $\mathfrak{A}$ [or the corresponding measure space $(\Omega,\mathfrak{A},\mu)$] is called *complete* if every subset of a μ-null set is in $\mathfrak{A}$; hence, is again a μ-null set. Prove:
   (a) The measure $\operatorname{rest}_{\mathfrak{A}^*}\mu^*$ of Theorem 1.5.4 is complete.
   (b) The measure of Section 1.3, Example 2 is complete.
   (c) Let $\mathfrak{A}$ be a σ-algebra in Ω, and let $\omega \in \Omega$ satisfy $\{\omega\} \in \mathfrak{A}$. The unit mass ϵ_ω is a complete measure on $\mathfrak{A}$ if and only if $\mathfrak{A} = \mathfrak{P}(\Omega)$.

5. (a) Prove that every measure μ on a σ-algebra $\mathfrak{A}$ in Ω can be *completed*, that is, μ can be extended to a complete measure μ_0

defined on a σ-algebra $\mathfrak{A}_0 \supset \mathfrak{A}$ in Ω such that every complete measure μ_∞ on a σ-algebra $\mathfrak{A}_\infty \supset \mathfrak{A}$ which extends μ is an extension of μ_0. $\mathfrak{A}_0$ is called the *μ-completion* of the σ-algebra $\mathfrak{A}$; μ_0 is called the *completion* of μ.

(b) Determine the completion of ϵ_ω in the situation of Problem 4(c).

(c) Prove that $\mathfrak{A}_0$ consists of all sets $A \cup N$ where $A \in \mathfrak{A}$ and N is a subset of a μ-null set, and that $\mu_0(A \cup N) = \mu(A)$ for all these sets.

(d) Prove that the sets A_0 of $\mathfrak{A}_0$ can be characterized by the following condition:

There exist sets $A_1, A_2 \in \mathfrak{A}$ such that $A_1 \subset A_0 \subset A_2$ and $\mu(A_1) = \mu(A_2)$.

6. Let μ be a σ-finite measure on a σ-algebra $\mathfrak{A}$ in Ω, μ^* the outer measure derived from it, and let $\mathfrak{A}^*$ be the σ-algebra of all μ^*-measurable sets in Ω. Prove that $\tilde{\mu}{:}= \text{rest}_{\mathfrak{A}^*}\mu^*$ is the completion of μ, that is, $\mathfrak{A}_0 = \mathfrak{A}^*$ and $\mu_0 = \tilde{\mu}$. [*Hint:* Use Problems 3 and 5.]

1.6 BOREL SETS AND LEBESGUE MEASURE

We continue the investigations of Section 1.4. Again let $\mathfrak{I}^p$ be the set of all right half-open intervals in $\mathbf{R}^p$, $\mathfrak{F}^p$ the ring of all p-dimensional figures, and λ^p the Lebesgue premeasure on $\mathfrak{F}^p$. We noted already that λ^p is σ-finite. Thus by Theorem 1.5.7, λ^p can be extended in exactly one way to a measure on $\mathfrak{A}(\mathfrak{F}^p)$, which henceforth will also be denoted by λ^p. Since every figure is the union of finitely many $I \in \mathfrak{I}^p$, we have

$$\mathfrak{A}(\mathfrak{F}^p) = \mathfrak{A}(\mathfrak{I}^p). \tag{1.6.1}$$

1.6.1. Definition. The elements of the σ-algebra generated by the system $\mathfrak{I}^p$ of half-open intervals in $\mathbf{R}^p$ are called the *Borel sets* of $\mathbf{R}^p$. Correspondingly, $\mathfrak{A}(\mathfrak{I}^p)$ is called the σ-algebra of Borel sets of $\mathbf{R}^p$ and is denoted by $\mathfrak{B}^p$.

The result stated above can now be expressed as follows, using Theorem 1.4.3.

1.6.2. Theorem. There is exactly one measure λ^p on $\mathfrak{B}^p$ which associates with every right half-open interval in $\mathbf{R}^p$ its p-dimensional elementary content.

1.6.3. Definition. The measure λ^p of Theorem 1.6.2 is called the *Lebesgue-Borel measure* (abbreviated L-B-measure) in $\mathbf{R}^p$. For every Borel set $B \in \mathfrak{B}^p$ we call $\lambda^p(B)$ the Lebesgue measure of B.

It is useful to further enlarge this definition as follows:

For every set $C \in \mathfrak{B}^p$, $C \cap \mathfrak{B}^p$ consists of all Borel subsets of C [compare (1.1.4)]. The restriction λ_C^p of λ^p to $C \cap \mathfrak{B}^p$ is a measure. This is called the L-B-measure *on* C.

As an extension of the Lebesgue premeasure, λ^p is *σ-finite* (see Section 1.5, Example 6). More generally, we have

$$\lambda^p(B) < +\infty$$

for every *bounded* set $B \in \mathfrak{B}^p$, since this is contained in a sufficiently large interval $I \in \mathfrak{J}^p$; for example, one of the intervals I_n of Section 1.5, and hence $\lambda^p(B) \leqq \lambda^p(I) < \infty$.

It seems desirable at this point to get a deeper insight into the σ-algebra $\mathfrak{B}^p$ of Borel sets. In particular the question arises of whether topologically interesting subsets (such as open, closed, and compact sets) are Borel sets. The relevant characterization contained in the following theorem is often used to define $\mathfrak{B}^p$.

1.6.4. Theorem. Let $\mathfrak{O}^p$, $\mathfrak{C}^p$, $\mathfrak{K}^p$ denote respectively the system of all open, closed, and compact subsets of $\mathbf{R}^p$. Then

$$\mathfrak{B}^p = \mathfrak{A}(\mathfrak{O}^p) = \mathfrak{A}(\mathfrak{C}^p) = \mathfrak{A}(\mathfrak{K}^p). \tag{1.6.2}$$

Proof. We have $\mathfrak{K}^p \subset \mathfrak{C}^p \subset \mathfrak{A}(\mathfrak{C}^p)$ and hence $\mathfrak{A}(\mathfrak{K}^p) \subset \mathfrak{A}(\mathfrak{C}^p)$. Every set $C \in \mathfrak{C}^p$ is the union of a sequence of sets $C_n \in \mathfrak{K}^p$; if we let K_n denote the compact sphere of radius n with a fixed point $a \in \mathbf{R}^p$ as center, then we may choose $C_n = C \cap K_n$. Then by (1.1.3) we have $\mathfrak{C}^p \subset \mathfrak{A}(\mathfrak{K}^p)$ and hence $\mathfrak{A}(\mathfrak{C}^p) \subset \mathfrak{A}(\mathfrak{K}^p)$, that is, $\mathfrak{A}(\mathfrak{C}^p) = \mathfrak{A}(\mathfrak{K}^p)$. Since the open sets are the complements of the closed sets and conversely, we further have $\mathfrak{A}(\mathfrak{O}^p) = \mathfrak{A}(\mathfrak{C}^p) = \mathfrak{A}(\mathfrak{K}^p)$. Finally we show that $\mathfrak{A}(\mathfrak{O}^p) = \mathfrak{B}^p$. To this end we define (in the usual way) open bounded intervals in $\mathbf{R}^p$ as the sets

$$]a,b[\equiv \{x \in \mathbf{R}^p : a \lhd x \lhd b\}, \tag{1.6.3}$$

where a and b are points in $\mathbf{R}^p$ with $a \leqq b$.[5] Every right half-open interval $[a,b[\in \mathfrak{J}^p$ is then the intersection of a sequence of open, bounded intervals:

$$[a,b[= \bigcap_{n=1}^{\infty}]a_n,b[,$$

where

$$a_n = \left(\alpha_1 - \frac{1}{n}, \ . \ . \ . \ , \alpha_p - \frac{1}{n}\right) \qquad \text{and} \qquad a = (\alpha_1, \ . \ . \ . \ , \alpha_p).$$

[5] See Section 1.4 for the definition of $\lhd$.

By (1.1.7) we have here $\mathfrak{J}^p \subset \mathfrak{A}(\mathfrak{O}^p)$ and thus $\mathfrak{B}^p = \mathfrak{A}(\mathfrak{J}^p) \subset \mathfrak{A}(\mathfrak{O}^p)$. Every open set can be represented as the union of countably many open, bounded intervals (for example, with vertices having only rational coordinates). Every open, bounded interval $]a,b[$ is moreover the union of a sequence of intervals of $\mathfrak{J}^p$:

$$]a,b[\; = \bigcup_{n=1}^{\infty} [a_n,b[,$$

where

$$a_n \equiv \left(\min\left(\alpha_1 + \frac{1}{n}, \beta_1\right), \; . \; . \; . \; , \min\left(\alpha_p + \frac{1}{n}, \beta_p\right)\right)$$

and where $\alpha_1, \ldots, \alpha_p$ and $\beta_1, \ldots, \beta_p$ denote the coordinates of a and b, respectively. Every open set is thus the union of a sequence of intervals from $\mathfrak{J}^p$, and hence $\mathfrak{O}^p \subset \mathfrak{A}(\mathfrak{J}^p) = \mathfrak{B}^p$. Now it follows that $\mathfrak{A}(\mathfrak{O}^p) \subset \mathfrak{B}^p$, and hence, by the dual relation just proved, $\mathfrak{B}^p = \mathfrak{A}(\mathfrak{O}^p)$. ⌟

Several deeper properties of the L-B-measure are discussed in Section 1.8. In particular, there we shall obtain the existence of *non*-Borel subsets of $\mathbf{R}^p$, that is, we shall see that

$$\mathfrak{B}^p \neq \mathfrak{P}(\mathbf{R}^p).$$

For the present we shall content ourselves with the computation of $\lambda^p(B)$ for simple Borel sets $B \subset \mathbf{R}^p$.

Examples

1. Every hyperplane H orthogonal to one of the coordinate axes of $\mathbf{R}^p$ is an L-B-*nullset,* that is, $\lambda^p(H) = 0$. Suppose H is orthogonal to the ith coordinate axis, and hence

$$H = \{x = (\xi_1, \ldots, \xi_p) \in \mathbf{R}^p \colon \xi_i = \alpha\} \tag{1.6.4}$$

for a suitable $\alpha \in \mathbf{R}$ $(i = 1, \ldots, p)$. H is closed and hence Borel. For every $n \in \mathbf{N}$, let x_n and y_n denote the point in $\mathbf{R}^p$ whose various coordinates with the exception of the ith are equal to $-n$ and $+n$, respectively, and the ith coordinate is equal to α and $\alpha + 2^{-n}(2n)^{1-p}\epsilon$, respectively, for arbitrary $\epsilon > 0$. Then obviously

$$H \subset \bigcup_{n=1}^{\infty} [x_n,y_n[$$

and

$$\lambda^p([x_n,y_n[) = 2^{-n}\epsilon.$$

By (1.3.10) we then obtain

$$\lambda^p(H) \leqq \sum_{n=1}^{\infty} \lambda^p([x_n,y_n[) = \epsilon.$$

Since $\epsilon > 0$ was arbitrary, $\lambda^p(H) = 0$.

In particular it follows from the isotonicity of measures that $\lambda^p(B) = 0$ for all Borel subsets B of such hyperplanes H.

2. Every countable subset of $\mathbf{R}^p$ is an L-B-nullset.

Because of the σ-additivity of measures, it suffices to treat the case of a one-point subset $\{x\} \subset \mathbf{R}^p$. This is closed and therefore Borel; further, there are hyperplanes H of the form (1.6.4) with $\{x\} \subset H$.

3. For points $a, b \in \mathbf{R}^p$ with $a \leqq b$, consider the already defined intervals $[a,b[$ and $]a,b[$ as well as the compact interval

$$[a,b] \equiv \{x \in \mathbf{R}^p \colon a \leqq x \leqq b\} \tag{1.6.5}$$

and (in contrast to $[a,b[$) the left half-open interval

$$]a,b] \equiv \{x \in \mathbf{R}^p \colon a \lhd x \leqq b\}. \tag{1.6.6}$$

Then

$$\lambda^p([a,b[) = \lambda^p(]a,b[) = \lambda^p([a,b]) = \lambda^p(]a,b]). \tag{1.6.7}$$

By Theorem 1.6.4, the intervals $[a,b[$, $]a,b[$, and $[a,b]$ are Borel sets. As in the proof of Theorem 1.6.4, we show that $]a,b]$ is the intersection of a sequence of intervals $]a,b_n[$, that is, of open sets. But by Theorem 1.6.4, $]a,b]$ is then also Borel. Equation (1.6.7) now follows by means of the additivity of measures from Example 1. We only have to observe that from $[a,b]$ we can obtain all of the other three intervals, by deleting at most $2p$ Borel sets from the boundary of $[a,b]$, each of which is contained in a hyperplane with the equation $\xi_i = \alpha_i$ or $\xi_i = \beta_i$, respectively ($i = 1, \ldots, p$). Here we again denote the coordinates of a and b by α_i and β_i, respectively.

The choice of the right half-open intervals in the construction of λ^p is justified only by the fact that these intervals generate the ring $\mathfrak{F}^p$ of figures which is easy to describe.

PROBLEM

Prove: A set $A \in \mathfrak{B}^p$ is a L-B-null set if and only if for every $\epsilon > 0$ it can be covered by a sequence (I_n) of open (resp. compact, resp. half-open) intervals in $\mathbf{R}^p$ such that $\sum_{n=1}^{\infty} \lambda^p(I_n) \leqq \epsilon$. [*Hint:* Use formula (1.5.1).]

1.7 MEASURABLE MAPPINGS AND IMAGE MEASURES

The following considerations are simpler to formulate if we introduce some notation: If Ω is a set and $\mathfrak{A}$ a σ-algebra in Ω, then we call the pair

$$(\Omega,\mathfrak{A})$$

a *measurable space* and the sets of $\mathfrak{A}$ *measurable sets*. If in addition a measure μ is defined on $\mathfrak{A}$, then we call the triple

$$(\Omega,\mathfrak{A},\mu)$$

a *measure space*. Correspondingly, we call $(\mathbf{R}^p,\mathfrak{B}^p)$ the p-dimensional *Borel measurable space* and $(\mathbf{R}^p,\mathfrak{B}^p,\lambda^p)$ the p-dimensional *Lebesgue-Borel measure* space.

The notation of measurable space has a formal analogy to the concept of topological space. A topological space also is a pair consisting of a set and a system of subsets, namely that of open sets. In the sense of this analogy, the following concept of measurable mapping corresponds to the concept of continuous mapping in topology.

1.7.1. Definition. Let $(\Omega,\mathfrak{A})$ and $(\Omega',\mathfrak{A}')$ be measurable spaces, and let $T: \Omega \to \Omega'$ be a mapping of Ω into Ω'. We say that T is ($\mathfrak{A}$-$\mathfrak{A}'$-)*measurable* if

$$T^{-1}(A') \in \mathfrak{A} \qquad \text{for all } A' \in \mathfrak{A}'. \tag{1.7.1}$$

The $\mathfrak{A}$-$\mathfrak{A}'$-measurability of T is expressed symbolically by

$$T: (\Omega,\mathfrak{A}) \to (\Omega',\mathfrak{A}');$$

we then also speak of a measurable mapping of the first measurable space into the second. Using the notation introduced in (1.1.5), we write (1.7.1) in the form

$$T^{-1}(\mathfrak{A}') \subset \mathfrak{A}. \tag{1.7.1$'$}$$

Examples

1. Every *constant mapping* $T: \Omega \to \Omega'$ is $\mathfrak{A}$-$\mathfrak{A}'$-measurable.

2. Every *continuous* mapping $T: \mathbf{R}^p \to \mathbf{R}^q$ $(p, q = 1, 2, \ldots)$ is $\mathfrak{B}^p$-$\mathfrak{B}^q$-measurable or, briefly, *Borel-measurable*. By Theorem 1.6.4, the system $\mathfrak{O}^q$ of all open subsets of $\mathbf{R}^q$ is a generator of $\mathfrak{B}^q$. From the continuity of T, we have $T^{-1}(O) \in \mathfrak{O}^p \subset \mathfrak{B}^p$ for all $O \in \mathfrak{O}^q$. The measurability of T then follows from the next theorem.

(1.c)

1.7.2. Theorem. Let $(\Omega,\mathfrak{A})$ and $(\Omega',\mathfrak{A}')$ be measurable spaces; further let $\mathfrak{E}'$ be a generator of $\mathfrak{A}'$. A mapping $T: \Omega \to \Omega'$ is measurable if and only if

$$T^{-1}(A') \in \mathfrak{A} \qquad \text{for all } A' \in \mathfrak{E}'. \tag{1.7.2}$$

Proof. The system $\mathfrak{Q}'$ of all sets $Q' \in \mathfrak{P}(\Omega')$ with $T^{-1}(Q') \in \mathfrak{A}$ is a σ-algebra in Ω'. Consequently $\mathfrak{A}' \subset \mathfrak{Q}'$ if and only if $\mathfrak{E}' \subset \mathfrak{Q}'$. $\mathfrak{A}' \subset \mathfrak{Q}'$ is equivalent to the measurability of T; $\mathfrak{E}' \subset \mathfrak{Q}'$ is equivalent to (1.7.2). ┘

For the composition of measurable mappings we have:

1.7.3. Theorem. If $T_1: (\Omega_1,\mathfrak{A}_1) \to (\Omega_2,\mathfrak{A}_2)$ and $T_2: (\Omega_2,\mathfrak{A}_2) \to (\Omega_3,\mathfrak{A}_3)$ are measurable mappings, then the composed mapping $T_2 \circ T_1$ is $\mathfrak{A}_1$-$\mathfrak{A}_3$-measurable.

Proof. The assertion follows from the formula

$$(T_2 \circ T_1)^{-1}(A) = T_1^{-1}(T_2^{-1}(A)),$$

which holds for all $A \subset \Omega_3$ and in particular for $A \in \mathfrak{A}_3$. ┘

Now suppose we are given a family of measurable spaces $((\Omega_i,\mathfrak{A}_i))_{i\in I}$ and a family $(T_i)_{i\in I}$ of mappings $T_i: \Omega \to \Omega_i$ where T_i maps Ω into Ω_i for each $i \in I$. Then obviously the σ-algebra generated by $\bigcup_{i\in I} T_i^{-1}(\mathfrak{A}_i)$ in Ω is the smallest σ-algebra $\mathfrak{A}$ with respect to which every T_i is $\mathfrak{A}$-$\mathfrak{A}_i$-measurable. We denote this σ-algebra by $\mathfrak{A}(T_i; i \in I)$, that is,

$$\mathfrak{A}(T_i; i \in I) \equiv \mathfrak{A}\Big(\bigcup_{i\in I} T_i^{-1}(\mathfrak{A}_i)\Big). \tag{1.7.3}$$

$\mathfrak{A}(T_i; i \in I)$ is called the *σ-algebra generated by the mappings* T_i [and the measurable spaces $(\Omega_i,\mathfrak{A}_i)$]. If $I = \{1, \ldots ,n\}$ is finite, then we also denote it by $\mathfrak{A}(T_1, \ldots ,T_n)$. For $n = 1$, $\mathfrak{A}(T_1) = T_1^{-1}(\mathfrak{A}_1)$.

As a further application of Theorem 1.7.2, we show:

1.7.4. Theorem. Let $(T_i)_{i\in I}$ be a family of mappings $T_i: \Omega \to \Omega_i$ of a set Ω into measurable spaces $(\Omega_i,\mathfrak{A}_i)$. Further let $S: \Omega_0 \to \Omega$ be a mapping of a measurable space $(\Omega_0,\mathfrak{A}_0)$ into Ω. The mapping S is $\mathfrak{A}_0$-$\mathfrak{A}(T_i; i \in I)$-measurable if and only if each of the mappings $T_i \circ S$ is $\mathfrak{A}_0$-$\mathfrak{A}_i$-measurable $(i \in I)$.

Proof. The condition is necessary by Theorem 1.7.3. The following reasoning shows that it is also sufficient. By (1.7.3)

$$\mathfrak{E} \equiv \bigcup_{i\in I} T_i^{-1}(\mathfrak{A}_i)$$

is a generator of $\mathfrak{A}(T_i;\ i \in I)$. Every set $E \in \mathfrak{E}$ is of the form $E = T_i^{-1}(A_i)$ with $A_i \in \mathfrak{A}_i$ and $i \in I$. Thus we have

$$S^{-1}(E) = (T_i \circ S)^{-1}(A_i) \in \mathfrak{A}_0$$

because of the assumed measurability of $T_i \circ S$. Then by Theorem 1.7.2, S is $\mathfrak{A}_0$-$\mathfrak{A}(T_i;\ i \in I)$-measurable. ⌟

Finally, with the help of measurable mappings, we can also map measures:

1.7.5. Theorem. Let $T\colon (\Omega,\mathfrak{A}) \to (\Omega',\mathfrak{A}')$ be a measurable mapping. Then for every measure μ on $\mathfrak{A}$,

$$\mu'(A') = \mu(T^{-1}(A')) \tag{1.7.4}$$

defines a measure μ' on $\mathfrak{A}'$.

Proof. We need note only that with every sequence $(A'_n)_{n\in\mathbf{N}}$ of pairwise disjoint sets of $\mathfrak{A}'$, $(T^{-1}(A'_n))_{n\in\mathbf{N}}$ is also a sequence of pairwise disjoint sets (in $\mathfrak{A}$) and we have

$$T^{-1}\Big(\bigcup_{n=1}^{\infty} A'_n\Big) = \bigcup_{n=1}^{\infty} T^{-1}(A'_n). \quad ⌟$$

1.7.6. Definition. Suppose we have the situation described in Theorem 1.7.5. Then the measure μ' is called the *image* of μ under the mapping T; it is denoted by $T(\mu)$.

Thus by definition

$$T(\mu)(A') = \mu(T^{-1}(A')) \qquad (A' \in \mathfrak{A}'). \tag{1.7.5}$$

The formation of image measures is *transitive*, that is,

$$(T_2 \circ T_1)(\mu) = T_2(T_1(\mu)), \tag{1.7.6}$$

if we have the situation described in Theorem 1.7.3 and μ is a measure on $\mathfrak{A}_1$. For $T \equiv T_2 \circ T_1$ and every $A \in \mathfrak{A}_3$ we have the equality $T^{-1}(A) = T_1^{-1}(T_2^{-1}(A))$; here $T_2^{-1}(A)$ lies in $\mathfrak{A}_2$. Thus if we let $\mu' \equiv T_1(\mu)$ and $\mu'' \equiv T_2(\mu')$, then

$$T(\mu)(A) = \mu(T_1^{-1}(T_2^{-1}(A))) = \mu'(T_2^{-1}(A)) = \mu''(A)$$

for all $A \in \mathfrak{A}_3$ and hence $T(\mu) = \mu''$ and (1.7.6) follows.

Examples

3. Suppose in particular that $\Omega = \Omega' = \mathbf{R}^p$, $\mathfrak{A} = \mathfrak{A}' = \mathfrak{B}^p$, and that $\mu = \lambda^p$ is the p-dimensional L-B-measure. For every point $a \in \mathbf{R}^p$, the translation mapping $T_a: \mathbf{R}^p \to \mathbf{R}^p$ is defined by

$$T_a(x) \equiv a + x.$$

T_a is a continuous mapping and therefore measurable according to the second example of this section. We investigate the image measure $\lambda' \equiv T_a(\lambda^p)$.

The mapping T_a is bijective. Since $T_a^{-1} = T_{-a}$, for every interval $[c,d[\in \mathfrak{J}^p$ we have

$$T_a^{-1}([c,d[) = [c - a, d - a[$$

and hence $\lambda'([c,d[) = \lambda^p([c - a, d - a[) = \lambda^p([c,d[)$. Both measures λ^p and λ' associate with every interval of $\mathfrak{J}^p$ its p-dimensional elementary content. Then by Theorem 1.6.2 we have $\lambda^p = \lambda'$, and hence

$$T_a(\lambda^p) = \lambda^p \qquad (a \in \mathbf{R}^p). \tag{1.7.7}$$

This property is called the *translation-invariance* of λ^p.

If as usual we define

$$a + A = A + a \equiv T_a(A) = \{a + x : x \in A\} \tag{1.7.8}$$

for sets $A \in \mathfrak{P}(\mathbf{R}^p)$ and points $a \in \mathbf{R}^p$, then $T_a(\lambda^p)(A) = \lambda^p(-a + A)$ for arbitrary $A \in \mathfrak{B}^p$. Property (1.7.7) can now be written as follows:

$$\lambda^p(a + A) = \lambda^p(A), \qquad \text{for arbitrary } A \in \mathfrak{B}^p \text{ and } a \in \mathbf{R}^p. \tag{1.7.7'}$$

4. Given the situation of Example 3, we consider the *homothetic mapping* $H_r: \mathbf{R}^p \to \mathbf{R}^p$ associated with a real number $r \neq 0$ (continuous, hence Borel-measurable), as follows:

$$H_r(x) \equiv r \cdot x.$$

Then we have

$$H_r(\lambda^p) = \frac{1}{|r|^p} \lambda^p. \tag{1.7.9}$$

For every interval $[a,b[\in \mathfrak{J}^p$ we have

$$H_r^{-1}([a,b[) = \begin{cases} \left[\frac{1}{r} a, \frac{1}{r} b\right[, & r > 0 \\ \left]-\frac{1}{|r|} b, -\frac{1}{|r|} a\right], & r < 0 \end{cases}$$

and thus, by (1.6.7),

$$\lambda^p(H_r^{-1}([a,b[)) = |r|^{-p} \lambda^p([a,b[).$$

$H_r(\lambda^p)$ and $|r|^{-p}\lambda^p$ are therefore measures on $\mathfrak{B}^p$ which coincide on $\mathfrak{J}^p$. By Definition 1.6.1, $\mathfrak{J}^p$ is a generator of $\mathfrak{B}^p$ which obviously has all the properties of the generator $\mathfrak{E}$ of the Uniqueness Theorem 1.5.5 with respect to the two given measures. Thus (1.7.9) follows from Theorem 1.5.5.

For $r = -1$ we obtain $H_{-1}(\lambda^p) = \lambda^p$. Since H_{-1} is the reflection through zero, this property is called the *reflection-invariance* of λ^p.

PROBLEMS

1. Consider the measure space $(\Omega,\mathfrak{A},\mu)$ of Section 1.3, Example 2 for the set $\Omega \equiv \mathbf{R}$. Define $\Omega' \equiv \{0,1\}$, $\mathfrak{A}' \equiv \mathfrak{P}(\Omega')$, and $T\colon \Omega \to \Omega'$ by
$$T(\omega) = \begin{cases} 0, & \omega \text{ rational} \\ 1, & \omega \text{ irrational.} \end{cases}$$
Prove: T is $\mathfrak{A}$-$\mathfrak{A}'$-measurable. Determine $T(\mu)$.
2. Let Ω and Ω' be sets, $T\colon \Omega \to \Omega'$ a mapping, and let $\mathfrak{E}'$ be a subset of $\mathfrak{P}(\Omega')$. Prove:
$$T^{-1}(\mathfrak{A}(\mathfrak{E}')) = \mathfrak{A}(T^{-1}(\mathfrak{E}')).$$
3. Let K be a compact set in $\mathbf{R}^p$ such that the intersection $H_r(K) \cap H_{r'}(K)$ of two homothetic images of K has λ^p-measure zero whenever $0 < r < r' < 1$. (Observe that every sphere in $\mathbf{R}^p$ with center $\mathbf{0}$ has this property.) Prove: $\lambda^p(K) = 0$. [*Hint:* $H_r(K) \subset \tilde{K}$ for all $0 < r \leqq 1$, where $\tilde{K} \equiv \{tx\colon 0 \leqq t \leqq 1,\ x \in K\}$ is compact, and hence has finite λ^p-measure.]
4. Let $\mathbf{U}$ be the unit circle in $\mathbf{R}^2$, that is,
$$\mathbf{U} = \{(x,y) \in \mathbf{R}^2\colon x^2 + y^2 = 1\},$$
and denote by $\mathfrak{B}(\mathbf{U})$ the σ-algebra $\mathbf{U} \cap \mathfrak{B}^2$. Construct a finite measure $\nu \neq 0$ on $\mathfrak{B}(\mathbf{U})$ which is invariant with respect to all rotations of $\mathbf{U}$. [*Hint:* Construct ν as an image of λ^1.]

1.8 FURTHER PROPERTIES OF LEBESGUE-BOREL MEASURE

By Example 3 of the preceding section, the L-B-measure λ^p is translation-invariant on $\mathfrak{B}^p$. Most important is the statement that λ^p is uniquely determined by this property and the following normalization: For the p-dimensional *unit cube*
$$W \equiv [\mathbf{0},\mathbf{1}] \tag{1.8.1}$$
with the determining points $\mathbf{0} \equiv (0, \ldots ,0)$ and $\mathbf{1} \equiv (1, \ldots ,1)$ of $\mathbf{R}^p$, we have by (1.6.7)
$$\lambda^p(W) = 1. \tag{1.8.2}$$

As an application of Fubini's Theorem, in Section 3.2 we shall give the proof of the following theorem:

1.8.1. Theorem. The Lebesgue-Borel measure λ^p is the only measure μ on $\mathfrak{B}^p$ which is translation-invariant, that is, satisfies the condition

$$T_a(\mu) = \mu,$$

for every translation $x \to T_a(x) \equiv a + x$ of $\mathbf{R}^p$, and which satisfies the normalization condition

$$\mu(W) = 1. \tag{1.8.2'}$$

The following formulation is equivalent:

1.8.2. Corollary. If μ is a translation-invariant measure on $\mathfrak{B}^p$ with

$$\alpha \equiv \mu(W) < +\infty, \tag{1.8.3}$$

then

$$\mu = \alpha\lambda^p.^6 \tag{1.8.4}$$

Proof. Let $\alpha > 0$. Then $(1/a)\mu$ is a translation-invariant measure on $\mathfrak{B}^p$ which associates the measure 1 with W. By Theorem 1.8.1 we have $(1/\alpha)\mu = \lambda^p$ and hence (1.8.4). If $\alpha = 0$, we must show that $\mu(B) = 0$ for all $B \in \mathfrak{B}^p$. Now let $B \in \mathfrak{B}^p$ be bounded. The closure $\bar{B}$ is then bounded and closed, hence compact. Since $B \subset \bar{B}$, we have $\mu(B) = 0$ whenever $\mu(\bar{B}) = 0$. Thus it suffices to consider the case of a compact set B. The point $x_0 \equiv (\frac{1}{2}, \ldots, \frac{1}{2})$ of $\mathbf{R}^p$ whose coordinates are all equal to $\frac{1}{2}$ lies in the interior of W, namely, in the open interval $]\mathbf{0},\mathbf{1}[$. Therefore the compact interval

$$V \equiv [-x_0, \mathbf{1} - x_0] = -x_0 + W$$

is a neighborhood of $\mathbf{0}$ and for every $a \in \mathbf{R}^p$

$$a + V = [a - x_0, a - x_0 + \mathbf{1}]$$

is a neighborhood of a. We have

$$B \subset \bigcup_{a \in B} (a + V);$$

since B is compact, the *Heine-Borel covering theorem* yields the existence of finitely many points $a_1, \ldots, a_n \in B$ with

$$B \subset \bigcup_{i=1}^{n} (a_i + V).$$

[6] Conversely, of course, $\mu \equiv \beta\lambda^p$ for every $\beta \in \mathbf{R}_+$ is a translation-invariant measure on $\mathfrak{B}^p$ with $\mu(W) < \infty$.

By (1.3.10) we then have

$$0 \leqq \mu(B) \leqq \sum_{i=1}^{n} \mu(a_i + V). \tag{1.8.5}$$

Now $a + V = a - x_0 + W$ and μ is translation-invariant, that is,

$$\mu(a + V) = \mu(W) = 0 \qquad \text{for all } a \in \mathbf{R}^p. \tag{1.8.6}$$

Thus from the last inequality we obtain $\mu(B) = 0$. For an arbitrary set $B \in \mathfrak{B}^p$, we consider the sequence of compact spheres K_n of center $\mathbf{0}$ and radius $n \in \mathbf{N}$. Then $K_n \uparrow \mathbf{R}^p$ and hence $K_n \cap B \uparrow B$. Since μ is continuous from below, we have

$$\mu(B) = \lim_{n \to \infty} \mu(K_n \cap B).$$

Now each $K_n \cap B$ is a bounded Borel set and thus $\mu(K_n \cap B) = 0$. It follows that $\mu(B) = 0$. ⌟

This corollary states that λ^p is the Haar measure on the additive group $\mathbf{R}^p$ in the sense of the theory of locally compact groups. An analogous measure can be constructed on every locally compact group G; it is called the *Haar measure* of G. The reader interested in the theory of these measures is referred to, for example, Nachbin [7].

Remark. The proof of Corollary 1.8.2 shows, more generally, that a translation-invariant measure μ on $\mathfrak{B}^p$ with $\mu(W) < \infty$ associates with all bounded sets $B \in \mathfrak{B}^p$ a finite measure $\mu(B)$. The considerations in the case $\alpha = 0$ refer to the hypothesis $\alpha = 0$ first in (1.8.6) in the form $\mu(W) = 0$. For arbitrary $\alpha \in \mathbf{R}_+$, (1.8.5), as well as the first equality in (1.8.6) holds. Together with (1.8.5), this yields $\mu(B) < \infty$ for all compact and hence for all relatively compact, that is, here bounded sets $B \in \mathfrak{B}^p$. The proof of Theorem 1.8.1 will make use of this remark.]

We stated Theorem 1.8.1 at the beginning of the considerations of this section since in the following the translation invariance of λ^p is used in a crucial way.

Now we can derive a further invariance property of λ^p:

1.8.3. Theorem. The Lebesgue-Borel measure λ^p is *invariant* with respect to orthogonal transformations, that is, $T(\lambda^p) = \lambda^p$ for every orthogonal transformation[7] T of $\mathbf{R}^p$ ($p = 1, 2, \ldots$).

[7] By definition this means all mappings $T\colon \mathbf{R}^p \to \mathbf{R}^p$ which preserve the Euclidean distance of points.

Proof.[8] Let H be a not necessarily homogeneous hyperplane in $\mathbf{R}^p$, that is, not necessarily passing through the null vector $\mathbf{0}$, and let S_H be the reflection through H. Then, as is well known, every orthogonal transformation T of $\mathbf{R}^p$ can be represented by finitely many successive reflections $S_{H_1}, \ldots, S_{H_n}$:

$$T = S_{H_1} \circ \cdots \circ S_{H_n}.^9$$

Because of the transitivity (1.7.6) of image measures it thus suffices to consider the case $T = S_H$ of such a reflection. S_H is continuous, that is, Borel-measurable; hence, $\mu \equiv S_H(\lambda^p)$ is defined. We shall show that $\mu = \lambda^p$.

To this end, we consider an orthogonal transformation K of $\mathbf{R}^p$ which transforms the hyperplane

$$H_p \equiv \{(\xi_1, \ldots, \xi_p) \in \mathbf{R}^p: \xi_p = 0\} \tag{1.8.7}$$

into H:

$$K(H_p) = H.$$

K introduces a new orthogonal coordinate system, with H as coordinate hyperplane. K is a homeomorphism of $\mathbf{R}^p$ onto itself, and therefore K and K^{-1} are Borel-measurable. K is a bijective map of $\mathfrak{B}^p$ onto itself; therefore; $\mathfrak{B}^p$ is the smallest σ-algebra $\mathfrak{A}$ in $\mathbf{R}^p$ relative to which K^{-1} is $\mathfrak{A}$-$\mathfrak{B}^p$-measurable. By Theorem 1.7.2 for every generator $\mathfrak{E}$ of $\mathfrak{B}^p$, $K(\mathfrak{E})$ is also a generator of $\mathfrak{B}^p$. According to Section 1.6 in particular, the system of all open intervals $]a,b[$ is such a generator $\mathfrak{E}$. The elements of $\mathfrak{E}' \equiv K(\mathfrak{E})$ are all open parallelotopes whose sides are either parallel or orthogonal to H. Hence, for every $E \in \mathfrak{E}'$, $S_H(E)$ is obtained from E by a translation (orthogonal to H). Because of the translation invariance of λ^p and the fact that $S_H = S_H^{-1}$, we thus have

$$\lambda^p(E) = \lambda^p(S_H(E)) = \lambda^p(S_H^{-1}(E)) = \mu(E).$$

Hence λ^p and μ are two measures on $\mathfrak{B}^p$ which coincide on $\mathfrak{E}'$. The desired equality $\mu = \lambda^p$ now follows since $\mathfrak{E}'$ also has the remaining properties of the generator appearing in the Uniqueness Theorem 1.5.5; $\mathfrak{E}'$ is $\cap$-stable, contains sequences (E_n) with $E_n \uparrow \mathbf{R}^p$, and all sets $E \in \mathfrak{E}'$ are bounded, that is, of finite measure $\lambda^p(E)$. ┘

Since with every orthogonal transformation T of $\mathbf{R}^p$, T^{-1} is also an orthogonal transformation, the orthogonal transformation invariance can also be written in the following form: For every orthogonal transformation

[8] Following S. Guber [34].

[9] See S. Guber, *Lineare Algebra und Analytische Geometrie II*, Verlag R. Merkel, Univ.-Buchhandl. Erlangen (1968).

T of $\mathbf{R}^p$ and every Borel set $A \in \mathfrak{B}^p$ we have

$$\lambda^p(T(A)) = \lambda^p(A). \tag{1.8.8}$$

Examples

1. Every hyperplane $H \subset \mathbf{R}^p$ is a L-B-null set. This follows from Section 1.6, Example 1, since for H there exists an orthogonal transformation T which transforms, say, the hyperplane H_p of (1.8.7) into H.
2. Every closed or open rectangular parallelotope $Q \subset \mathbf{R}^p$ (= parallelotope with pairwise orthogonal edges) with edge lengths $l_1, \ldots, l_p$ has the Lebesgue measure

$$\lambda^p(Q) = l_1 \cdot \ldots \cdot l_p.$$

This follows analogously from Section 1.6, Example 3.

By a fundamental application of the axiom of choice of set theory, we finally prove the existence of non-Borel sets.

1.8.4. Theorem. We have $\mathfrak{B}^p \neq \mathfrak{P}(\mathbf{R}^p)$ for every $p = 1, 2, \ldots$.

Proof. Let $\mathbf{Q}^p$ be the set of all points $x = (\xi_1, \ldots, \xi_p) \in \mathbf{R}^p$ all of whose coordinates $\xi_1, \ldots, \xi_p$ are rational. Then $\mathbf{Q}^p$ is a subgroup of the additive group $\mathbf{R}^p$ and hence the congruence $x \sim y$ of points $x, y \in \mathbf{R}^p$ modulo $\mathbf{Q}^p$ is an equivalence relation.[10] Thus the space $\mathbf{R}^p$ is divided into the corresponding equivalence classes, each of which is a set $x + \mathbf{Q}^p$ with $x \in \mathbf{R}^p$. The congruence $x \sim y$ is equivalent to $x + \mathbf{Q}^p = y + \mathbf{Q}^p$. Since for every real number η there exists an integer n with $n \leqq \eta < n + 1$, that is, with $\eta - n \in [0,1[$, there is a point $x \in [\mathbf{0},\mathbf{1}[$ in every equivalence class. Then, by the axiom of choice there is a set

$$K \subset [\mathbf{0},\mathbf{1}[\tag{1.8.9}$$

such that K has exactly one element in common with every equivalence class; hence,

$$\mathbf{R}^p = \bigcup_{y \in \mathbf{Q}^p} (y + K) \tag{1.8.10}$$

and

$$y_1 \neq y_2 \Rightarrow (y_1 + K) \cap (y_2 + K) = \emptyset \qquad (y_1, y_2 \in \mathbf{Q}^p). \tag{1.8.11}$$

Then K cannot lie in $\mathfrak{B}^p$. Otherwise, since $\mathbf{Q}$ and thus also $\mathbf{Q}^p$ is countable, we would have from (1.8.10) and (1.8.11)

$$+\infty = \lambda^p(\mathbf{R}^p) = \sum_{y \in \mathbf{Q}^p} \lambda^p(y + K). \tag{1.8.12}$$

[10] Thus $x \sim y$ is equivalent to $x - y \in \mathbf{Q}^p$.

Since λ^p is translation-invariant,

$$\lambda^p(y + K) = \lambda^p(K), \qquad \text{for all } y \in \mathbf{Q}^p. \tag{1.8.13}$$

But now by (1.8.9),

$$\bigcup_{y \in [\mathbf{0},\mathbf{1}[\cap \mathbf{Q}^p} (y + K) \subset [\mathbf{0},\mathbf{2}[,$$

where **2** denotes the point whose p coordinates are equal to 2. Another application of the σ-additivity of λ^p now yields

$$\sum_{y \in [\mathbf{0},\mathbf{1}[\cap \mathbf{Q}^p} \lambda^p(y + K) \leqq \lambda^p([\mathbf{0},\mathbf{2}[) = \mathbf{2}^p < +\infty,$$

whence follows, by (1.8.13), $\lambda^p(K) = 0$. But this together with (1.8.13) contradicts (1.8.12). ┘

Remark. If μ is a measure on a σ-algebra $\mathfrak{A}$ in a set Ω and A is a μ-null set, that is, a set of $\mathfrak{A}$ with $\mu(A) = 0$, then every subset of $\mathfrak{A}$ belonging to A is a μ-null set by virtue of isotonicity, but not every subset of A necessarily belongs to $\mathfrak{A}$. As can be shown, this situation occurs for the Lebesgue-Borel measure.[11] If every subset of a μ-null set belongs to $\mathfrak{A}$, then we call μ a *complete* measure (compare Section 1.5, Problem 4). In Section 1.5, Problem 5 we discuss the possibility of completing a given measure μ [or measure space $(\Omega,\mathfrak{A},\mu)$] by the so-called *completion* μ_0 of μ [or $(\Omega,\mathfrak{A}_0,\mu_0)$ of $(\Omega,\mathfrak{A},\mu)$].

The completion of the Lebesgue-Borel measure in $\mathbf{R}^p$ is called the *Lebesgue measure* in $\mathbf{R}^p$; the sets of its domain are said to be *Lebesgue-measurable*. Obviously, in the transition from Borel sets to Lebesgue-measurable sets, the important property of Borel sets of being determined only by the topology of $\mathbf{R}^p$ is lost. Since we shall soon see that $\mathfrak{B}^p$ is also the domain of many other important measures, we omit an exact description of the transition from Lebesgue-Borel to Lebesgue measure. Only the Lebesgue-Borel measure is used in the following.

PROBLEMS

1. Let $T: (\Omega,\mathfrak{A}) \to (\Omega_0',\mathfrak{A}_0')$ be a measurable mapping, μ a measure on $\mathfrak{A}$, and $\mu' := T(\mu)$ its T-image. Let $(\Omega,\mathfrak{A}_0,\mu_0)$ [resp. $(\Omega',\mathfrak{A}_0',\mu_0')$] be the completion of $(\Omega,\mathfrak{A},\mu)$ [resp. $(\Omega',\mathfrak{A}',\mu')$]. Prove: T is $\mathfrak{A}_0$-$\mathfrak{A}_0'$-measurable, and $T(\mu_0) = \mu_0'$. Conclude that the Lebesgue measure is translation-invariant (and hence invariant with respect to all orthogonal transformations).
2. Prove that the set K constructed in the proof of Theorem 1.8.4 is not even Lebesgue-measurable.

[11] See Haupt-Aumann-Pauc [4], p. 66, No. 3.4.3.1.

2

INTEGRATION THEORY

Suppose we are given an arbitrary measure space $(\Omega,\mathfrak{A},\mu)$. We pose the problem of associating an integral, that is, a "mean value" over Ω formed relative to μ, with a set of numerical functions f on Ω, which is as extensive as possible. The solution follows step by step in Sections 2.2–2.4. Section 2.1 is of a preparatory nature.

2.1 MEASURABLE NUMERICAL FUNCTIONS

The σ-algebra $\mathfrak{B}^1$ of Borel sets is defined on the real line $\mathbf{R}$. If we compactify $\mathbf{R}$ in the usual way to $\bar{\mathbf{R}}$ by adjoining the "ideal" points $+\infty$ and $-\infty$, then we say that the sets $B \subset \bar{\mathbf{R}}$ for which $B \cap \mathbf{R} \in \mathfrak{B}^1$ are *Borel in* $\bar{\mathbf{R}}$. The sets that are Borel in $\bar{\mathbf{R}}$ are then precisely the sets B_0, $B_0 \cup \{+\infty\}$, $B_0 \cup \{-\infty\}$, $B_0 \cup \{+\infty\} \cup \{-\infty\}$ with $B_0 \in \mathfrak{B}^1$. The system $\bar{\mathfrak{B}}^1$ of these sets that are Borel in $\bar{\mathbf{R}}$ is obviously a σ-algebra in $\bar{\mathbf{R}}$ with

$$\mathbf{R} \cap \bar{\mathfrak{B}}^1 = \mathfrak{B}^1. \tag{2.1.1}$$

Now if $(\Omega,\mathfrak{A})$ is a measurable space, then $\mathfrak{A}$-$\bar{\mathfrak{B}}^1$-measurability of numerical functions $f: \Omega \to \bar{\mathbf{R}}$ is well defined. Such $\mathfrak{A}$-$\bar{\mathfrak{B}}^1$-measurable functions f will henceforth be called $(\mathfrak{A})$-*measurable numerical functions* on Ω. Every real function $f: \Omega \to \mathbf{R}$ is also a numerical function. Because of (2.1.1), $\mathfrak{A}$-$\bar{\mathfrak{B}}^1$-measurability has the same significance as $\mathfrak{A}$-$\mathfrak{B}^1$-measurability for every such function.

EXAMPLES

1. Let $(\Omega,\mathfrak{A})$ be a measurable space and A a subset of Ω. The function

$$1_A(\omega) \equiv \begin{cases} 1, & \omega \in A \\ 0, & \omega \notin A \end{cases} \qquad (\omega \in \Omega) \tag{2.1.2}$$

is called the *indicator function* (or characteristic function) of A. This real function 1_A defined on Ω is $\mathfrak{A}$-measurable if and only if A lies in $\mathfrak{A}$. Indeed, for every $B \subset \bar{\mathbf{R}}$, $1_A^{-1}(B)$ is one of the four sets: Ω, A, $\complement A$, $\emptyset$.

Sets A and their indicator functions 1_A are obviously in one-to-one correspondence. The following rules, valid for arbitrary sets $A, B \subset \Omega$ and families $(A_i)_{i\in I}$ of arbitrary subsets of Ω, will be used frequently and can be seen directly:

$$\begin{aligned} A \subset B &\Leftrightarrow 1_A \leqq 1_B; \\ 1_{\complement A} &= 1 - 1_A; \\ 1_{\bigcup_{i\in I} A_i} &= \sup_{i\in I} 1_{A_i}; \\ 1_{\bigcap_{i\in I} A_i} &= \inf_{i\in I} 1_{A_i}. \end{aligned}$$

2. Suppose $(\Omega,\mathfrak{A})$ is the p-dimensional Borel measurable space $(\mathbf{R}^p,\mathfrak{B}^p)$. The $\mathfrak{B}^p$-measurable numerical functions on $\mathbf{R}^p$ are said to be *Borel-measurable* or *Baire* functions on $\mathbf{R}^p$. The following theorem shows that all continuous numerical functions on $\mathbf{R}^p$ are Borel measurable (see also Section 1.7, Example 2).

Below we shall study measurable numerical functions on a given measure space $(\Omega,\mathfrak{A})$ in more detail.

2.1.1. Theorem. A numerical function f on Ω is $\mathfrak{A}$-measurable if and only if

$$\{\omega \in \Omega\colon f(\omega) \geqq \alpha\} \in \mathfrak{A} \qquad \text{for all } \alpha \in \mathbf{R}. \tag{2.1.3}$$

Proof. According to Theorem 1.7.2 we need to show only that the system $\mathfrak{E}$ of all intervals $[\alpha,+\infty]$ with $\alpha \in \mathbf{R}$ generates the σ-algebra $\bar{\mathfrak{B}}^1$ in $\bar{\mathbf{R}}$. Since $[\alpha,+\infty] \in \bar{\mathfrak{B}}^1$ for all $\alpha \in \mathbf{R}$ we obviously have $\bar{\mathfrak{D}} \subset \bar{\mathfrak{B}}^1$ for the σ-algebra $\bar{\mathfrak{D}}$ in $\bar{\mathbf{R}}$ generated by $\mathfrak{E}$. Since $[\alpha,\beta[\ = [\alpha,+\infty] \setminus [\beta,+\infty]$, all intervals $[\alpha,\beta[$ with $\alpha,\beta \in \mathbf{R}$, $\alpha \leqq \beta$, lie in $\mathbf{R} \cap \bar{\mathfrak{D}}$. It then follows from Definition 1.6.1 that $\mathfrak{B}^1 \subset \mathbf{R} \cap \bar{\mathfrak{D}}$. But the one-point sets

$$\{+\infty\} = \bigcap_{n=1}^{\infty} [n,+\infty]$$

and

$$\{-\infty\} = \bigcap_{n=1}^{\infty} [-\infty, -n[= \bigcap_{n=1}^{\infty} \complement[-n, +\infty]$$

lie in $\bar{\mathfrak{Q}}$. Consequently, when $Q \in \bar{\mathfrak{Q}}$, the set $\mathbf{R} \cap Q$ lies in $\bar{\mathfrak{Q}}$; in other words, $\mathbf{R} \cap \bar{\mathfrak{Q}} \subset \bar{\mathfrak{Q}}$ and thus $\mathfrak{B}^1 \subset \bar{\mathfrak{Q}}$. Since $\{+\infty\}, \{-\infty\} \in \bar{\mathfrak{Q}}$, this implies $\bar{\mathfrak{B}}^1 \subset \bar{\mathfrak{Q}}$ and thus $\bar{\mathfrak{Q}} = \bar{\mathfrak{B}}^1$. ┘

We now introduce the following abbreviations: If f, g are numerical functions on Ω, let

$$\{f \leqq g\} \equiv \{\omega \in \Omega: f(\omega) \leqq g(\omega)\}. \tag{2.1.4}$$

The sets $\{f < g\}$, $\{f = g\}$, $\{f \neq g\}$, and so on, are defined analogously. Condition (2.1.3) can now be written in the form: $\{f \geqq \alpha\} \in \mathfrak{A}$ for all $\alpha \in \mathbf{R}$. That we can equally well use $\{f > \alpha\}$, $\{f \leqq \alpha\}$, and so on, in the above criterion for measurability is shown by:

2.1.2. Theorem. Each of the following conditions is equivalent to the $\mathfrak{A}$-measurability of a numerical function f on Ω:

(a) $\{f \geqq \alpha\} \in \mathfrak{A}$ for all $\alpha \in \mathbf{R}$;
(b) $\{f > \alpha\} \in \mathfrak{A}$ for all $\alpha \in \mathbf{R}$;
(c) $\{f \leqq \alpha\} \in \mathfrak{A}$ for all $\alpha \in \mathbf{R}$;
(d) $\{f < \alpha\} \in \mathfrak{A}$ for all $\alpha \in \mathbf{R}$.[1]

Proof. We need to show only the equivalence of the four conditions. But this is a consequence of the following equalities, which hold for all $\alpha \in \mathbf{R}$:

$$\{f > \alpha\} = \bigcup_{n=1}^{\infty} \{f \geqq \alpha + n^{-1}\}; \qquad \{f \leqq \alpha\} = \complement\{f > \alpha\};$$

$$\{f < \alpha\} = \bigcup_{n=1}^{\infty} \{f \leqq \alpha - n^{-1}\}; \qquad \{f \geqq \alpha\} = \complement\{f < \alpha\}.$$ ┘

We now obtain a series of results on manipulation with measurable numerical functions.

2.1.3. Theorem. For any two $\mathfrak{A}$-measurable functions $f, g: \Omega \to \bar{\mathbf{R}}$, the sets $\{f < g\}$, $\{f \leqq g\}$, $\{f = g\}$, and $\{f \neq g\}$ lie in $\mathfrak{A}$.

[1] The corresponding conditions for all $\alpha \in \bar{\mathbf{R}}$ are of course equivalent to these.

Proof. Since the set $\mathbf{Q}$ of rational numbers is countable, the assertion follows (with the help of Theorem 2.1.2) from the equalities:

$$\{f < g\} = \bigcup_{\rho \in \mathbf{Q}} \{f < \rho\} \cap \{\rho < g\};$$
$$\{f \leqq g\} = \complement\{f > g\}; \qquad \{f = g\} = \{f \leqq g\} \cap \{g \leqq f\};$$
$$\{f \neq g\} = \complement\{f = g\}. \quad \lrcorner$$

2.1.4. Theorem. With $f, g: \Omega \to \bar{\mathbf{R}}$ the following functions are also $\mathfrak{A}$-measurable: $f \pm g$ (provided they are defined everywhere on Ω) and $f \cdot g$.

Proof. First, when g is measurable, $\sigma + \tau g$ is measurable for all $\sigma, \tau \in \mathbf{R}$. This follows from Theorem 2.1.2 since $\{\sigma + \tau g \geqq \alpha\} = \{g \geqq \tau^{-1}(\alpha - \sigma)\}$ or $= \{g \leqq \tau^{-1}(\alpha - \sigma)\}$ for $\tau > 0$ or $\tau < 0$, respectively. The case $\tau = 0$ is trivial. This observation, which again will follow from the final assertion, reduces the case $f - g$ to the case $f + g$. From

$$\{f + g \geqq \alpha\} = \{f \geqq \alpha - g\} \qquad (\alpha \in \mathbf{R}),$$

the observation above and Theorem 2.1.3 we have the measurability of $f + g$.

To study fg we first assume f and g to be real-valued. Then

$$fg = \tfrac{1}{4}(f + g)^2 - \tfrac{1}{4}(f - g)^2.$$

Referring to what has just been proved, we see that it suffices to treat the case $f = g$. But $\{f^2 \geqq \alpha\} = \Omega$ and $= \{f \geqq \sqrt{\alpha}\} \cup \{f \leqq -\sqrt{\alpha}\}$ for $\alpha \leqq 0$ and $\alpha > 0$, respectively. Hence f^2 is measurable.

If f and g are numerical functions, let $\Omega_1 \equiv \{fg = +\infty\}$, $\Omega_2 \equiv \{fg = -\infty\}$, $\Omega_3 \equiv \{fg = 0\}$, and $\Omega_4 \equiv \Omega \setminus \Omega_1 \cup \Omega_2 \cup \Omega_3$. Using theorem 2.1.3 we verify that these four pairwise disjoint sets lie in $\mathfrak{A}$. The restrictions f', g' of f, g to Ω_4 are $\Omega_4 \cap \mathfrak{A}$-measurable and real-valued. Thus $f'g'$ is $\Omega_4 \cap \mathfrak{A}$-measurable. But the $\mathfrak{A}$-measurability of fg now follows immediately. $\lrcorner$

2.1.5. Theorem. Let $(f_n)_{n \in \mathbf{N}}$ be a sequence of $\mathfrak{A}$-measurable numerical functions on Ω. Then each of the following functions is $\mathfrak{A}$-measurable:

$$\sup_{n \in \mathbf{N}} f_n, \ \inf_{n \in \mathbf{N}} f_n, \ \limsup_{n \to \infty} f_n, \ \liminf_{n \to \infty} f_n.$$

Proof. The function $s \equiv \sup f_n$ is measurable because

$$\{s \leqq \alpha\} = \bigcap_{n=1}^{\infty} \{f_n \leqq \alpha\} \qquad (\alpha \in \mathbf{R}).$$

Then $\inf f_n = -\sup(-f_n)$ is also measurable by Theorem 2.1.4. By definition we have

$$\limsup_{n\to\infty} f_n = \inf_{n\in\mathbf{N}} \sup_{m\geq n} f_m, \qquad \liminf_{n\to\infty} f_n = \sup_{n\in\mathbf{N}} \inf_{m\geq n} f_m.$$

From what has just been proved it follows that these functions are also measurable. ┘

2.1.6. Corollary 1. When finitely many numerical functions $f_1, \ldots, f_n$ on Ω are $\mathfrak{A}$-measurable, then their upper (lower) envelope $\sup(f_1, \ldots, f_n)$ ($\inf(f_1, \ldots, f_n)$) is also $\mathfrak{A}$-measurable.

Proof. It suffices to apply Theorem 2.1.5 to the sequence $f_1, \ldots, f_n, f_n, f_n, \ldots$. ┘

2.1.7. Corollary 2. If a sequence $(f_n)_{n\in\mathbf{N}}$ of $\mathfrak{A}$-measurable numerical functions on Ω is pointwise convergent, that is, $\lim_{n\to\infty} f_n(\omega)$ exists for all $\omega \in \Omega$, then $\lim_{n\to\infty} f_n$ is also $\mathfrak{A}$-measurable.

Proof. We have

$$\lim_{n\to\infty} f_n = \liminf_{n\to\infty} f_n = \limsup_{n\to\infty} f_n. \quad ┘$$

PROBLEMS

1. Let $(\Omega,\mathfrak{A})$ be a measurable space, and let D be a set of real numbers which is dense in $\mathbf{R}$. Prove that a numerical function f on Ω is $\mathfrak{A}$-measurable if and only if the conditions (a) to (d) of Theorem 2.1.2 only hold for all $\alpha \in D$.
2. Let $(f_n)_{n\in\mathbf{N}}$ be a sequence of $\mathfrak{A}$-measurable numerical functions on a measurable space $(\Omega,\mathfrak{A})$. Why is the set of all $\omega \in \Omega$ where the sequence $(f_n(\omega))$ is convergent (in $\bar{\mathbf{R}}$ or in $\mathbf{R}$) $\mathfrak{A}$-measurable?
3. Consider a real $\mathfrak{A}$-measurable function f on a measurable space $(\Omega,\mathfrak{A})$. Are the functions e^f and $\sin f$, that is, the functions $\omega \to e^{f(\omega)}$ and $\omega \to \sin f(\omega)$, $\mathfrak{A}$-measurable?
4. Consider the real-valued function $(x,y) \to \max(x,y)$ defined on $\mathbf{R}^2$. Use Theorem 2.1.1 in order to prove that this function is $\mathfrak{B}^2$-measurable. Deduce from this a new proof for Corollary 2.1.6.
5. Show that measurability of $|f|$ does not, in general, imply measurability of f.

2.2 ELEMENTARY FUNCTIONS AND THEIR INTEGRALS

2.2.1. Definition. A real function u on Ω is called an ($\mathfrak{A}$-) *elementary function*[2] if it is nonnegative and $\mathfrak{A}$-measurable and takes on only finitely many values.

Let $\mathcal{E} = \mathcal{E}(\mathfrak{A})$ denote *the set of all elementary functions.* If $\{\alpha_1, \ldots, \alpha_n\}$ is the set of values of a function $u \in \mathcal{E}$, then the sets $A_i \equiv u^{-1}(\alpha_i)$, $i = 1, \ldots, n$, are pairwise disjoint and are elements of $\mathfrak{A}$ as the inverse images of the Borel sets $\{\alpha_i\}$. Then, using the notation for indicator functions introduced in (2.1.2), we have

$$u = \sum_{i=1}^{n} \alpha_i 1_{A_i}. \tag{2.2.1}$$

Conversely, if we are given real numbers $\alpha_1 \geqq 0, \ldots, \alpha_n \geqq 0$ and sets $A_1, \ldots, A_n$ in $\mathfrak{A}$ and we define the function u by (2.2.1), then u is an elementary function since u is measurable by Theorem 2.1.4. Thus $\mathcal{E}$ is the set of all functions that have a representation (2.2.1), where the coefficients α_i lie in $\mathbf{R}_+$ and the sets A_i lie in $\mathfrak{A}$.

From Definition 2.2.1 and the results of Section 2.1, we immediately obtain the following further properties of $\mathcal{E}$:

$$u, v \in \mathcal{E} \Rightarrow \begin{cases} a\,u \in \mathcal{E}\ (a \in \mathbf{R}_+); \\ u + v \in \mathcal{E}; \\ u \cdot v \in \mathcal{E}; \\ \sup\,(u, v) \in \mathcal{E}; \\ \inf\,(u, v) \in \mathcal{E}. \end{cases} \tag{2.2.2}$$

The above proof of (2.2.1) also shows that every function $u \in \mathcal{E}$ has a representation (2.2.1), in which the sets $A_1, \ldots, A_n \in \mathfrak{A}$ are pairwise disjoint with Ω as union and thus form a *partition* of Ω. Representations (2.2.1) with this additional property will henceforth be called *normal representations* of u. It is easy to see that functions $u \in \mathcal{E}$ in general have different normal representations. However, we have:

2.2.2. Lemma. For any two normal representations

$$u = \sum_{i=1}^{m} \alpha_i 1_{A_i} = \sum_{j=1}^{n} \beta_i 1_{B_j}$$

[2] One also speaks of nonnegative *step-functions*.

of an elementary function $u \in \mathcal{E}$, we have

$$\sum_{i=1}^{m} \alpha_i \mu(A_i) = \sum_{j=1}^{n} \beta_j \mu(B_j).^3$$

Proof. Since $\Omega = A_1 \cup \cdots \cup A_m = B_1 \cup \cdots \cup B_n$, the additivity of μ yields

$$\mu(A_i) = \sum_{j=1}^{n} \mu(A_i \cap B_j)$$

and

$$\mu(B_j) = \sum_{i=1}^{m} \mu(A_i \cap B_j)$$

for all $i = 1, \ldots, m; j = 1, \ldots, n$. Hence we obtain

$$\sum \alpha_i \mu(A_i) = \sum_{i,j} \alpha_i \mu(A_i \cap B_j)$$

and

$$\sum \beta_j \mu(B_j) = \sum_{i,j} \beta_j \mu(A_i \cap B_j).$$

The assertion now follows if we observe the following: Since we started with two normal representations of u, $\alpha_i = \beta_j$ for every pair of indices i, j with $A_i \cap B_j \neq \emptyset$, that is, for every pair i, j with $\mu(A_i \cap B_j) \neq 0$. ┘

From this the following definition is justified:

2.2.3. Definition. Let u be an elementary function. Then the number

$$\int u \, d\mu \equiv \sum_{i=1}^{n} \alpha_i \mu(A_i), \tag{2.2.3}$$

independent of the chosen normal representation

$$u = \sum_{i=1}^{n} \alpha_i 1_{A_i},$$

is called the (μ-) *integral* of u (over Ω).

Thus $u \to \int u \, d\mu$ defines a mapping of $\mathcal{E}$ into $\bar{\mathbf{R}}_+$. We obviously have a mapping into $\mathbf{R}_+$ if and only if μ is finite. We collect the most important

[3] Here the conventions $0 \cdot \infty = 0$ and $\infty + \infty = \infty$ must be observed.

properties of this mapping:

$$\int 1_A \, d\mu = \mu(A) \qquad (A \in \mathfrak{A}); \tag{2.2.4}$$
$$\int (\alpha u) \, d\mu = \alpha \int u \, d\mu \qquad (u \in \mathcal{E}, \alpha \in \mathbf{R}_+); \tag{2.2.5}$$
$$\int (u + v) \, d\mu = \int u \, d\mu + \int v \, d\mu \qquad (u, v \in \mathcal{E}); \tag{2.2.6}$$
$$u \leqq v \Rightarrow \int u \, d\mu \leqq \int v \, d\mu \qquad (u, v \in \mathcal{E}). \tag{2.2.7}$$

Proof. The properties (2.2.4) and (2.2.5) follow directly from Definition 2.2.3. The next property is obtained as follows: Let

$$u = \sum_{i=1}^{m} \alpha_i 1_{A_i} \qquad \text{and} \qquad v = \sum_{j=1}^{n} \beta_j 1_{B_j}$$

be normal representations of $u, v \in \mathcal{E}$. Then

$$A_i = \bigcup_{j=1}^{n} (A_i \cap B_j) \qquad \text{and} \qquad B_j = \bigcup_{i=1}^{m} (A_i \cap B_j);$$

if we note that the sets $A_i \cap B_j$ are pairwise disjoint, we obtain

$$1_{A_i} = \sum_{j=1}^{n} 1_{A_i \cap B_j} \qquad \text{and} \qquad 1_{B_j} = \sum_{i=1}^{m} 1_{A_i \cap B_j}$$

for all $i = 1, \ldots, m$ and $j = 1, \ldots, n$, whence we obtain the new normal representations

$$u = \sum_{i,j} \alpha_i 1_{A_i \cap B_j}, \qquad v = \sum_{i,j} \beta_j 1_{A_i \cap B_j}$$

and

$$u + v = \sum_{i,j} (\alpha_i + \beta_j) 1_{A_i \cap B_j}.$$

Consequently,

$$\int (u + v) \, d\mu = \sum_{i,j} (\alpha_i + \beta_j) \mu(A_i \cap B_j),$$

$$\int u \, d\mu = \sum_{i,j} \alpha_i \mu(A_i \cap B_j),$$

and

$$\int v \, d\mu = \sum_{i,j} \beta_j \mu(A_i \cap B_j),$$

that is, (2.2.6) holds. The above reasoning shows that u, v always have normal representations

$$u = \sum_{i=1}^{k} \gamma_i 1_{C_i} \qquad \text{and} \qquad v = \sum_{i=1}^{k} \delta_i 1_{C_i}$$

with equal sets $C_1, \ldots, C_k$. Then $u \leqq v$ implies $\gamma_i \leqq \delta_i$ for every $i = 1, \ldots, k$ with $C_i \neq \emptyset$. Hence (2.2.7) follows. ⌋

Now let $u = \sum_{i=1}^n \alpha_i 1_{A_i}$ be an arbitrary representation of some $u \in \mathcal{E}$ by means of coefficients $\alpha_i \in \mathbf{R}_+$ and sets $A_i \in \mathfrak{A}$, not necessarily a normal representation. Then from (2.2.4) to (2.2.6) we have

$$\int u \, d\mu = \sum \alpha_i \mu(A_i).$$

For normal representations this equality served for the definition of $\int u \, d\mu$. Note that the introduction of normal representations was useful only for the purposes of proofs.

PROBLEMS

1. Let $(\Omega,\mathfrak{A},\mu)$ be a measure space and $(\Omega,\mathfrak{A}_0,\mu_0)$ its completion. Prove that for every $\mathfrak{A}_0$-elementary function u there exist $\mathfrak{A}$-elementary functions u_1 and u_2 satisfying $u_1 \leqq u \leqq u_2$ and $\mu(\{u_1 \neq u_2\}) = 0$. For all these functions u_1, u_2, one has $\int u_1 \, d\mu = \int u_2 \, d\mu = \int u \, d\mu_0$. (Compare this result with Section 1.5, Problem 5.)
2. Is the function

$$f(x) = \begin{cases} 1, & \text{if } x \text{ is rational} \\ 0, & \text{if } x \text{ is irrational} \end{cases}$$

(defined on $\mathbf{R}$) $\mathfrak{B}^1$-elementary?

2.3 THE INTEGRAL OF NONNEGATIVE MEASURABLE FUNCTIONS

The following result will enable us to proceed further.

2.3.1. Theorem. For every isotone sequence $(u_n)_{n \in \mathbf{N}}$ of functions in $\mathcal{E}$ and every $v \in \mathcal{E}$:

$$v \leqq \sup_{n \in \mathbf{N}} u_n \Rightarrow \int v \, d\mu \leqq \sup_{n \in \mathbf{N}} \int u_n \, d\mu.^4 \qquad (2.3.1)$$

Proof. For $v = 0$ we have $\int v \, d\mu = 0$ [for example, by (2.2.5)] and there is nothing to prove. Thus let $v \neq 0$ and Q be the set $\{v > 0\}$ which belongs to $\mathfrak{A}$ by Theorem 2.1.2. If we set

$$\alpha \equiv \inf v(Q) \qquad \text{and} \qquad \beta \equiv \sup v(Q),$$

[4] By (2.2.7), $\sup \int u_n \, d\mu = \lim \int u_n \, d\mu$.

then $0 < \alpha \leqq \beta < \infty$. Now let $\epsilon \in \mathbf{R}$ with $0 < \epsilon < \alpha$ be arbitrarily chosen. The set

$$A_n \equiv \{u_n \geqq v - \epsilon\} \cap Q$$

lies in $\mathfrak{A}$ by virtue of measurability. By the hypotheses on (u_n) we have: $A_n \uparrow Q$ which implies $\lim_n \mu(A_n) = \mu(Q)$ by Theorem 1.3.2. From the definition of A_n it follows that

$$u_n \geqq (v - \epsilon)1_{A_n} \geqq (\alpha - \epsilon)1_{A_n}$$

and hence

$$\int u_n \, d\mu \geqq (\alpha - \epsilon)\mu(A_n).$$

If $\mu(Q) = \infty$, then $\sup \int u_n \, d\mu = \infty$ and hence the assertion follows. Thus we can assume $\mu(Q) < \infty$. Then to both sides of the inequality $u_n \geqq (v - \epsilon)1_{A_n}$ we add the function $(v - \epsilon)1_{B_n}$ with $B_n \equiv Q \setminus A_n$ and obtain

$$u_n + (v - \epsilon)1_{B_n} \geqq (v - \epsilon)1_Q = v - \epsilon 1_Q$$

and since $v \leqq \beta$, finally,

$$u_n + (\beta - \epsilon)1_{B_n} + \epsilon 1_Q \geqq v.$$

From (2.2.4) to (2.2.7) we then obtain

$$\int u_n \, d\mu + (\beta - \epsilon)\mu(B_n) + \epsilon \, \mu(Q) \geqq \int v \, d\mu.$$

If we now note that $\mu(B_n) = \mu(Q) - \mu(A_n)$, then letting $n \to \infty$, it follows that

$$\sup \int u_n \, d\mu + \epsilon\mu(Q) \geqq \int v \, d\mu.$$

Since ϵ with $0 < \epsilon < \alpha$ is arbitrary, this yields the assertion. ⌟

2.3.2. Corollary. For any two isotone sequences $(u_n)_{n \in \mathbf{N}}$ and $(v_n)_{n \in \mathbf{N}}$ of functions in $\mathcal{E}$, we have

$$\sup_{n \in \mathbf{N}} u_n = \sup_{n \in \mathbf{N}} v_n \Rightarrow \sup_{n \in \mathbf{N}} \int u_n \, d\mu = \sup_{n \in \mathbf{N}} \int v_n \, d\mu. \tag{2.3.2}$$

Proof. For every $m = 1, 2, \ldots$ we have $v_m \leqq \sup u_n$ and $u_m \leqq \sup v_n$, whence, by Theorem 2.3.1, it follows that

$$\int v_m \, d\mu \leqq \sup \int u_n \, d\mu \qquad \text{and} \qquad \int u_m \, d\mu \leqq \sup \int v_n \, d\mu. \quad ⌟$$

Now let $\mathcal{E}^*$ denote *the set of all numerical functions $f \geqq 0$ on Ω for which there exists an isotone sequence (u_n) of elementary functions $u_n \in \mathcal{E}$* such that

$$f = \sup u_n. \tag{2.3.3}$$

Then by (2.3.2) the number

$$\sup \int u_n \, d\mu \in \bar{\mathbf{R}}_+$$

depends only on f and is independent of the particular sequence (u_n) in the representation (2.3.3) of f. Thus we have a situation similar to that of Definition 2.2.3. Therefore we give:

2.3.3. Definition. For an arbitrary function $f \in \mathcal{E}^*$ let $f = \sup u_n$ be a representation of f as the supremum of an isotone sequence $(u_n)_{n \in \mathbf{N}}$ of elementary functions. Then we call the number

$$\int f \, d\mu \equiv \sup_{n \in \mathbf{N}} \int u_n \, d\mu, \tag{2.3.4}$$

which is independent of the particular representation by means of (u_n) the (μ-) *integral* of f (over Ω).

Now obviously $\mathcal{E} \subset \mathcal{E}^*$ since $u = \sup u_n$ for every $u \in \mathcal{E}$ and the constant sequence $u_n = u$. Then 2.3.4 applied to $f = u$ shows that the new definition of integral is consistent with the old one, concerning only elementary functions. The mapping $f \to \int f \, d\mu$, defined first only on $\mathcal{E}$, has now been extended to a mapping of $\mathcal{E}^*$ into $\bar{\mathbf{R}}_+$. We shall now show that the previously derived properties of the integral are preserved.

We derive properties analogous to (2.2.2) and (2.2.5)–(2.2.7):

$$f,g \in \mathcal{E}^* \Rightarrow \begin{cases} \alpha f \in \mathcal{E}^* \quad (\alpha \in \mathbf{R}_+); \\ f + g \in \mathcal{E}^*; \\ f \cdot g \in \mathcal{E}^*; \\ \sup(f,g) \in \mathcal{E}^*; \\ \inf(f,g) \in \mathcal{E}^*; \end{cases} \tag{2.3.5}$$

$$\int (\alpha f) \, d\mu = \alpha \int f \, d\mu \qquad (f \in \mathcal{E}^*, \alpha \in \mathbf{R}_+); \tag{2.3.6}$$

$$\int (f + g) \, d\mu = \int f \, d\mu + \int g \, d\mu \qquad (f,g \in \mathcal{E}^*); \tag{2.3.7}$$

$$f \leqq g \Rightarrow \int f \, d\mu \leqq \int g \, d\mu \qquad (f,g \in \mathcal{E}^*). \tag{2.3.8}$$

Proof. (2.3.5) follows from the definition of $\mathcal{E}^*$ and (2.2.2). We need note only that $\sup u_n = \lim u_n$ for isotone sequences (u_n). The same conclusion leads to proofs of (2.3.6) and (2.3.7). We leave (2.3.6) to the reader and prove (2.3.7): Let $f = \sup u_n$ and $g = \sup v_n$ be representations of f, $g \in \mathcal{E}^*$ by means of isotone sequences (u_n) and (v_n) of elementary functions. Then, by definition,

$$\int (f + g) \, d\mu = \sup \int (u_n + v_n) \, d\mu,$$
$$\int f \, d\mu = \sup \int u_n \, d\mu, \quad \int g \, d\mu = \sup \int v_n \, d\mu.$$

Hence (2.3.7) follows from (2.2.6) since by virtue of isotonicity,

$$\sup \int (u_n + v_n) \, d\mu = \lim \left(\int u_n \, d\mu + \int v_n \, d\mu\right) = \int f \, d\mu + \int g \, d\mu.$$

If we assume in addition $f \leqq g$, then $u_m \leqq \sup v_n$ for every $m \in \mathbf{N}$. Now (2.3.8) follows from Theorem 2.3.1. ┘

(2.3.6)–(2.3.8) tell us that the integral is a *positive-homogeneous, additive,* and *isotone function* on $\mathcal{E}^*$.

Finally, it turns out that Theorem 2.3.1, so basic in our treatment, also holds in $\mathcal{E}^*$:

2.3.4. Theorem of Monotone Convergence (B. Levi). For every isotone sequence $(f_n)_{n \in \mathbf{N}}$ of functions of $\mathcal{E}^*$,

$$\sup_{n \in \mathbf{N}} f_n \in \mathcal{E}^* \qquad \text{and} \qquad \int \sup_{n \in \mathbf{N}} f_n \, d\mu = \sup_{n \in \mathbf{N}} \int f_n \, d\mu.$$

Proof. We set $f \equiv \sup f_n$. It suffices to verify the existence of an isotone sequence (v_n) of functions in $\mathcal{E}$ which satisfies the conditions:

$$\sup v_n = f \qquad \text{and} \qquad v_n \leqq f_n, \qquad \text{for all } n \in \mathbf{N}.$$

Indeed then f lies in $\mathcal{E}^*$ and by the definition of the integral we have $\int f \, d\mu = \sup \int v_n \, d\mu$ and $\int v_n \, d\mu \leqq \int f_n \, d\mu$. This implies $\int f \, d\mu \leqq \sup \int f_n \, d\mu$. Since $f_n \leqq f$, we also have $\sup \int f_n \, d\mu \leqq \int f \, d\mu$ by (2.3.8).

The existence of the sequence (v_n) is now obtained as follows: For every f_n there is by definition an isotone sequence $(u_{mn})_{m=1,2,\ldots}$ of functions in $\mathcal{E}$ with $\sup_{m \in \mathbf{N}} u_{mn} = f_n$. By (2.2.2) the function

$$v_m \equiv \sup (u_{m1}, \ldots, u_{mm})$$

lies in $\mathcal{E}$ ($m = 1, 2, \ldots$). The isotonicity of the sequences $(u_{mn})_{m=1,2,\ldots}$ implies the isotonicity of (v_m). From the isotonicity of (f_m) it follows that $v_m \leqq f_m$ for all m and hence $\sup v_m \leqq f$. For all $m \geqq n$ we have $u_{mn} \leqq v_m$ and thus $\sup u_{mn} = f_n \leqq \sup v_m$ ($n = 1, 2, \ldots$). Hence, $\sup v_m = f$ and (v_n) is a sequence with the desired properties. ⌋

2.3.5. Corollary. For every sequence $(f_n)_{n \in \mathbf{N}}$ of functions in $\mathcal{E}^*$,

$$\sum_{n=1}^{\infty} f_n \in \mathcal{E}^* \qquad \text{and} \qquad \int \Big(\sum_{n=1}^{\infty} f_n \Big) d\mu = \sum_{n=1}^{\infty} \int f_n \, d\mu.$$

Proof. Apply Theorem 2.3.4 to the sequence $(f_1 + \cdots + f_n)_{n \in \mathbf{N}}$ and note (2.3.7). ⌋

Examples

1. Let $(\Omega, \mathfrak{A})$ be an arbitrary measurable space and let ϵ_a be the measure defined on $\mathfrak{A}$ by the unit mass at the point $a \in \Omega$ (see Section 1.3, Example 1, and Section 1.5, Example 2). Then

$$\int f \, d\epsilon_a = f(a)$$

for every $f \in \mathcal{E}^*$. By Definition 2.3.3 we can assume $f \in \mathcal{E}$. But if $f = \sum \alpha_i 1_{A_i}$ is a normal representation of f, then a lies in exactly one A_i, say A_{i_0}. Consequently, $\int f d\epsilon_a = \sum \alpha_i \epsilon_a(A_i) = \alpha_{i_0} = f(a)$.

2. Let Ω be the set of natural numbers and $\mathfrak{A} = \mathfrak{P}(\Omega)$. By the σ-additivity, a measure μ on $\mathfrak{A}$ is uniquely determined by arbitrarily chosen numbers $\alpha_n = \mu(\{n\}) \in \bar{\mathbf{R}}_+$, $n = 1, 2, \ldots$. $\mathcal{E}^*$ consists of *all* numerical functions $f \geqq 0$ on Ω. Indeed, for every $n \in \mathbf{N}$, set

$$u_n(i) \equiv \begin{cases} \min(n, f(i)), & i = 1, \ldots, n; \\ 0, & i = n+1, n+2, \ldots. \end{cases}$$

Then (u_n) is an isotone sequence of functions in $\mathcal{E}$ with $f = \sup u_n$. We have $\int f\, d\mu = \sum_{n=1}^{\infty} f(n)\alpha_n$. In fact, if we set $f_n \equiv f(n)1_{\{n\}}$ for every $n \in \mathbf{N}$, then $f = \sum_{n=1}^{\infty} f_n$, and the assertion follows from Corollary 2.3.5.

3. Let $(\Omega, \mathfrak{A})$ be a measurable space, $(\mu_n)_{n \in \mathbf{N}}$ a sequence of measures on $\mathfrak{A}$, and μ the measure $\sum_{n=1}^{\infty} \mu_n$ (Section 1.3, Example 4). Then for every function $f \in \mathcal{E}^*$,

$$\int f\, d\mu = \sum_{n=1}^{\infty} \int f\, d\mu_n.$$

This is obviously true if f is an indicator function. Thus the assertion holds for all elementary functions. The transition to arbitrary $f \in \mathcal{E}^*$ is accomplished as follows: Let (u_n) be an isotone sequence in $\mathcal{E}$ with $\sup u_n = f$. Then the double sequence

$$\alpha_{mn} \equiv \sum_{i=1}^{n} \int u_m\, d\mu_i \qquad (m, n \in \mathbf{N})$$

is isotone in m and n, that is, $\sup_{m \in \mathbf{N}} (\sup_{n \in \mathbf{N}} \alpha_{mn}) = \sup_{n \in \mathbf{N}} (\sup_{m \in \mathbf{N}} \alpha_{mn}) (= \sup_{m,n \in \mathbf{N}} \alpha_{mn})$. But this is the asserted equality.

After verifying that the set of functions $\mathcal{E}^*$ is a natural generalization of $\mathcal{E}$, we seek a simple characterization of this set. Thus we finally show:

2.3.6. Theorem. $\mathcal{E}^*$ is the set of all $\mathfrak{A}$-measurable numerical functions $f \geqq 0$ on Ω.

Proof. Every elementary function is measurable and therefore, by Theorem 2.1.5, every function in $\mathcal{E}^*$ is measurable. Conversely, let f be a measurable numerical function $\geqq 0$. For every $n = 1, 2, \ldots$, the sets

$$A_{in} \equiv \begin{cases} \left\{f \geqq \dfrac{i}{2^n}\right\} \cap \left\{f < \dfrac{i+1}{2^n}\right\}, & i = 0, 1, \ldots, n2^n - 1 \\ \{f \geqq n\}, & i = n2^n \end{cases}$$

are pairwise disjoint and lie in $\mathfrak{A}$. Consequently,

$$u_n \equiv \sum_{i=0}^{n2^n} i2^{-n} 1_{A_{in}}$$

is a function in $\mathcal{E}$ (in normal representation). On A_{in}, u_{n+1} can take on only the values $2^{-n-1}(2i)$ and $2^{-n-1}(2i + 1)(i = 0, \ldots, n2^n - 1)$ and on $A_{n2^n,n}$ only values $\geqq n$. Hence the sequence (u_n) is isotone. We have $\sup u_n = f$, since for arbitrary $\omega \in \Omega$, either $f(\omega) = \infty$ and then $u_n(\omega) = n$ for all n or $f(\omega) < \infty$ and then $u_n(\omega) \leqq f(\omega) < u_n(\omega) + 2^{-n}$ for all $n > f(\omega)$. Hence f lies in $\mathcal{E}^*$. ┘

Example

4. Let Ω be an uncountable set and $\mathfrak{A}$ the σ-algebra (from Section 1.1, Example 2) of all sets $A \subset \Omega$ for which either A or $\complement A$ is countable. A numerical function f on Ω is $\mathfrak{A}$-measurable if and only if it is constant on the complement of a countable set. It is evident that this condition implies $\mathfrak{A}$-measurability. To prove the converse, we can assume $f \geqq 0$. The assertion then holds if we are dealing with an elementary function. This follows since in a finite family of pairwise disjoint sets of $\mathfrak{A}$ with Ω as union there is precisely one set with countable complement. For arbitrary $f \geqq 0$ let (u_n) be an isotone sequence of elementary functions with $f = \sup u_n$. Every function u_n is equal to a constant α_n on the complement of a countable set A_n. But then $f(\omega) = \sup \alpha_n$ for all $\omega \in \bigcap_{n=1}^{\infty} \complement A_n = \complement \bigcup_{n=1}^{\infty} A_n$. The set $\bigcup_{n=1}^{\infty} A_n$ is countable.

PROBLEM

Prove: Every bounded $\mathfrak{A}$-measurable function $f \geqq 0$ on a measurable space $(\Omega, \mathfrak{A})$ is the uniform limit of an increasing sequence of elementary functions.

2.4 INTEGRABILITY

The last theorem shows that until now the integral $\int f\,d\mu$ is defined for all $\mathfrak{A}$-measurable numerical functions $f \geqq 0$ on Ω. As a third and last step we now define $\int f\,d\mu$ also for certain measurable numerical functions f of arbitrary sign. To this end, for every function $f\colon \Omega \to \bar{\mathbf{R}}$, we set

$$f^+ \equiv \sup(f,0) \qquad \text{and} \qquad f^- \equiv (-f)^+ = -\inf(f,0). \tag{2.4.1}$$

Hence, $f^+(\omega) = f(\omega)$ if $f(\omega) \geqq 0$ and $f^+(\omega) = 0$ otherwise. Clearly

$$f = f^+ - f^- \qquad \text{and} \qquad |f| = f^+ + f^-. \tag{2.4.2}$$

We call f^+ the *positive* and f^- the *negative part* of f. Note, however, that $f^+ \geqq 0$ *and* $f^- \geqq 0$ hold.

By Theorems 2.1.4 and 2.1.5, f is measurable if and only if f^+ and f^- are measurable. This remark leads to the following definition.

2.4.1. Definition. A numerical function f on Ω is said to be (μ-) *integrable* if it is $\mathfrak{A}$-measurable and if $\int f^+\,d\mu$ and $\int f^-\,d\mu$ are finite. Then

$$\int f\,d\mu \equiv \int f^+\,d\mu - \int f^-\,d\mu \tag{2.4.3}$$

is called the (μ-) *integral* of f (over Ω).

If we want to emphasize the variable $\omega \in \Omega$, we may also write

$$\int f(\omega)\,\mu(d\omega) \qquad \text{or} \qquad \int f(\omega)\,d\mu(\omega)$$

in place of $\int f\,d\mu$.

Note: For functions $f \geqq 0$ we have $f^+ = f$ and $f^- = 0$. Thus (2.4.3) is consistent with the definition of integral thus far. Although $\int f\,d\mu$ was defined for all $f \in \mathcal{E}^*$, only those $f \in \mathcal{E}^*$ for which $\int f\,d\mu$ is finite are integrable.

For measurable f the right side of (2.4.3) is also meaningful if f^+ *or* f^- has a finite integral. Then we say that *the integral of f exists* and we again define $\int f\,d\mu$ by (2.4.3). We shall not dwell any further on this simple generalization.

Now we gather the most important properties of the new concepts:

2.4.2. Theorem. Each of the following four conditions is necessary and sufficient for integrability of a measurable numerical function f on Ω:

(a) f^+ and f^- are integrable.
(b) There are integrable functions $u \geqq 0$, $v \geqq 0$ with $f = u - v$.
(c) There is an integrable function g with $|f| \leqq g$.
(d) $|f|$ is integrable.

From (b) it follows that $\int f\,d\mu = \int u\,d\mu - \int v\,d\mu$.

Proof. We need to show only the equivalence of statements (a)–(d), since by definition f is integrable if and only if (a) holds.

(a) $\Rightarrow$ (b): $u \equiv f^+$ and $v \equiv f^-$ yield the desired result.

(b) $\Rightarrow$ (c): Since the integral is additive on $\mathcal{E}^*$, then whenever u and v are integrable, $u + v$ is also integrable. Since $f = u - v \leqq u \leqq u + v$ and $-f = v - u \leqq v \leqq u + v$, $g \equiv u + v$ yields the desired result.

(c) $\Rightarrow$ (d): This follows from the isotonicity of the integral on $\mathcal{E}^*$: $\int |f|\,d\mu \leqq \int g\,d\mu < \infty$.

(d) $\Rightarrow$ (a): Since $f^+ \leqq |f|$ and $f^- \leqq |f|$, this again follows from the isotonicity of the integral on $\mathcal{E}^*$.

We have $f = u - v = f^+ - f^-$ and thus $u + f^- = v + f^+$. By (2.3.7) it now follows that $\int u\,d\mu + \int f^-\,d\mu = \int v\,d\mu + \int f^+\,d\mu$ and hence we have the rest of the assertion. ┘

2.4.3. Theorem. Let f and g be integrable numerical functions on Ω. Then the functions αf (with $\alpha \in \mathbf{R}$) and $f + g$ (if defined everywhere on Ω) are integrable. We have

$$\int (\alpha f)\,d\mu = \alpha \int f\,d\mu \qquad \text{and} \qquad \int (f+g)\,d\mu = \int f\,d\mu + \int g\,d\mu. \tag{2.4.4}$$

Further, the functions

$$\sup(f,g) \qquad \text{and} \qquad \inf(f,g)$$

are also integrable.

Proof. For αf the assertion follows from (2.3.6), since

$$(\alpha f)^+ = \alpha f^+,\ (\alpha f)^- = \alpha f^-, \qquad \text{for } \alpha \geqq 0$$

and

$$(\alpha f)^+ = |\alpha| f^-,\ (\alpha f)^- = |\alpha| f^+, \qquad \text{for } \alpha \leqq 0.$$

For $f + g$ we draw the conclusion as follows: From $f = f^+ - f^-$ and $g = g^+ - g^-$ it follows that

$$f + g = f^+ + g^+ - (f^- + g^-).$$

By (2.3.7), $u \equiv f^+ + g^+$ and $v \equiv f^- + g^-$ are integrable along with f and g. But then, since $f + g = u - v$, the assertion for $f + g$ follows from

Theorem 2.4.2. Finally we have

$$|\sup(f,g)| \leqq |f| + |g| \qquad \text{and} \qquad |\inf(f,g)| \leqq |f| + |g|.$$

Hence the integrability of the measurable functions $\sup(f,g)$ and $\inf(f,g)$ follows from Theorem 2.4.2 (c) since $|f|$ and $|g|$ and therefore also $|f| + |g|$ are integrable functions. ⌟

2.4.4. Theorem. Let f and g be integrable numerical functions on Ω. Then

$$f \leqq g \Rightarrow \int f\,d\mu \leqq \int g\,d\mu; \tag{2.4.5}$$
$$\left|\int f\,d\mu\right| \leqq \int |f|\,d\mu. \tag{2.4.6}$$

Proof. From $f \leqq g$ it follows that $f^+ \leqq g^+$ and $f^- \geqq g^-$, and hence we have (2.4.5) from the isotonicity of the integral on $\mathcal{E}^*$. Since $f \leqq |f|$ and $-f \leqq |f|$, we obtain (2.4.6) from (2.4.5) by choosing $g = |f|$. ⌟

These relationships become especially clear if we consider only real-valued integrable functions. The following notation is in general use:

$$\mathcal{L}^1(\mu) \equiv \text{the set of all } \textit{real } \mu\text{-integrable functions on } \Omega. \tag{2.4.7}$$

Now we can express Theorem 2.4.3 and (2.4.5) as follows:

$\mathcal{L}^1(\mu)$ is a vector space[5] over the field $\mathbf{R}$ of real numbers.

$f \to \int f\,d\mu$ is an isotone[6] linear form on $\mathcal{L}^1(\mu)$.

Examples

1. Let $(\Omega,\mathfrak{A})$ be an arbitrary measurable space and ϵ_a the measure on $\mathfrak{A}$ defined by the unit mass at $a \in \Omega$. By Section 2.3, Example 1, precisely those $\mathfrak{A}$-measurable numerical functions f on Ω with $|f(a)| < \infty$ are ϵ_a-integrable. We have

$$\int f\,d\epsilon_a = f(a).$$

2. Let $(\Omega,\mathfrak{A},\mu)$ be the measure space defined in Section 2.3, Example 2, with $\mu(\{n\}) = \alpha_n$, $n = 1, 2, \ldots$. From what was shown there it follows that precisely those functions $f\colon \Omega \to \bar{\mathbf{R}}$ for which $\sum_{n=1}^{\infty} |f(n)|\alpha_n < \infty$ are μ-integrable. Further, we then have $\int f\,d\mu = \sum_{n=1}^{\infty} f(n)\alpha_n$.

[5] More precisely, we mean: With respect to the operations $(f + g)(\omega) = f(\omega) + g(\omega)$ and $(\alpha f)(\omega) = \alpha f(\omega)\ (\alpha \in \mathbf{R})$.

[6] Due to the linearity, the isotonicity is equivalent to the property: $f \in \mathcal{L}^1(\mu)$, $f \geqq 0 \Rightarrow \int f\,d\mu \geqq 0$. Therefore, one speaks of a *positive* linear form.

3. Let $(\Omega,\mathfrak{A},\mu)$ be the measure space defined in Section 1.3, Example 2 (see also Section 1.5, Example 1). A function $f\colon \Omega \to \bar{\mathbf{R}}$ is μ-integrable if and only if, on the complement of a countable subset of Ω, it is equal to a real constant α. Then we have $\int f\,d\mu = \alpha$. This follows from Section 2.3, Example 4.

4. Let $(\Omega,\mathfrak{A},\mu)$ be a measure space with *finite* measure μ, that is, with $\mu(\Omega) < \infty$. Then every constant real function, and hence, by Theorem 2.4.2 every *bounded*, measurable real function on Ω, is μ-integrable.

5. Let μ and ν be measures on a σ-algebra $\mathfrak{A}$ in Ω. A numerical function f on Ω is $(\mu + \nu)$-integrable if and only if it is both μ- and ν-integrable. Then we have

$$\int f\,d(\mu + \nu) = \int f\,d\mu + \int f\,d\nu.$$

In fact, for every $\mathfrak{A}$-measurable function $g \geqq 0$ on Ω, by Section 2.3, Example 3, we have $\int g\,d(\mu + \nu) = \int g\,d\mu + \int g\,d\nu$. For $g = |f|$ the above and Theorem 2.4.2 imply the first part, and for $g = f^+$ and $g = f^-$, the second part of the assertion is implied. In particular, we have

$$\mathfrak{L}^1(\mu + \nu) = \mathfrak{L}^1(\mu) \cap \mathfrak{L}^1(\nu).$$

PROBLEMS

1. Characterize those elementary functions which are integrable.
2. Let $(\Omega,\mathfrak{A},\mu)$ be a measure space. The indicator function 1_A of a set $A \in \mathfrak{A}$ is integrable if and only if $\mu(A) < +\infty$. These sets are called μ-integrable. Denote by $\mathfrak{R}$ the system of these sets. Prove: $\mathfrak{R}$ is a ring such that $R \cap A \in \mathfrak{R}$ whenever $R \in \mathfrak{R}$ and $A \in \mathfrak{A}$. Conversely, if μ is σ-finite, then a set $A \subset \Omega$ is in $\mathfrak{A}$ if $R \cap A \in \mathfrak{R}$ for all $R \in \mathfrak{R}$.
3. Is the function of Section 2.2, Problem 2 λ^1-integrable?
4. Let $(\Omega,\mathfrak{A})$ be the measurable space of Section 2.3, Example 4. Consider the measure μ on $\mathfrak{A}$ which is zero for all countable sets and $+\infty$ for all uncountable sets. Determine all integrable functions and their integrals.
5. Let $(\Omega,\mathfrak{A},\mu)$ be a measure space with finite measure μ. Prove that every real function f on Ω which is the uniform limit of a sequence (f_n) in $\mathfrak{L}^1(\mu)$ is itself in $\mathfrak{L}^1(\mu)$. Prove that this result fails already in the case where μ is σ-finite but not finite. [*Hint:* Construct a sequence (g_n) in $\mathfrak{L}^1(\mu)$ satisfying $0 \leqq g_n \leqq 1$ and $\int g_n\,d\mu \geqq n^2$ and take

$$f_n := \sum_{i=1}^{n} \frac{1}{i^2}\, g_i.]$$

2.5 ALMOST EVERYWHERE PROPERTIES

The concept of null set mentioned in Section 1.6 will play an important role in the further development of the theory. Here $N \subset \Omega$ is called a $(\mu$-)*null set* if $N \in \mathfrak{A}$ and $\mu(N) = 0$. The *union of every sequence* of null sets is again a null set by (1.3.10); similarly, every subset of a null set that belongs to $\mathfrak{A}$ is a null set by virtue of isotonicity.

2.5.1. Definition. Let E be a property such that every point of Ω either has property E or it does not. We shall say "$(\mu$-)*almost all* points $\omega \in \Omega$ have the property E" or "E holds $(\mu$-) *almost everywhere* on Ω" if there is a μ-null set N such that all points $\omega \in \complement N$ have the property E.

[*Note:* We do not require that the set N_E of all $\omega \in \Omega$ which do not have the property E be a μ-null set. Indeed, in general N_E will not belong to $\mathfrak{A}$. If for example A is a subset of Ω not belonging to $\mathfrak{A}$ and E is the property "ω is a point of A," then $N_E = \complement A$ is not a set of $\mathfrak{A}$.]

Examples of properties E that arise below are: equality of the values of two functions f and g at $\omega \in \Omega$, finiteness of the values of a function f at $\omega \in \Omega$, and so on. Accordingly we say in the following that f and g are $(\mu$-) almost everywhere equal on Ω; in symbols:

$$f = g(\mu\text{-}) \text{ almost everywhere,}$$

and that f is $(\mu$-) almost everywhere finite on Ω; in symbols:

$$|f| < +\infty \ (\mu\text{-}) \text{ almost everywhere,}$$

and so on.

The following theorems show the significance of these concepts for integration theory.

2.5.2. Theorem. For every function $f \in \mathcal{E}^*$,

$$\int f\, d\mu = 0 \Leftrightarrow f = 0 \ \mu\text{-almost everywhere.}$$

Proof. Because of the measurability of f we have $N \equiv \{f \neq 0\} \in \mathfrak{A}$. Thus we need to show

$$\int f\, d\mu = 0 \Leftrightarrow \mu(N) = 0.$$

Let $\int f\, d\mu = 0$. Each of the sets $A_n \equiv \{f \geqq n^{-1}\}$, $n = 1, 2, \ldots$, lies in $\mathfrak{A}$, and $A_n \uparrow N$. Since $\mu(N) = \lim \mu(A_n)$, we have to show that $\mu(A_n) = 0$ for all n. But now, obviously, $f \geqq n^{-1}1_{A_n}$ and hence

$$0 = \int f\, d\mu \geqq n^{-1}\mu(A_n) \geqq 0,$$

that is, $\mu(A_n) = 0$. Suppose conversely that $\mu(N) = 0$. Each of the functions $u_n \equiv n1_N$, $n = 1, 2, \ldots$, lies in $\mathcal{E}$ and we have $\int u_n\,d\mu = 0$. Thus if we set $g \equiv \sup u_n$, it follows that $g \in \mathcal{E}^*$ and $\int g\,d\mu = \sup \int u_n\,d\mu = 0$. Finally, $f \leqq g$ and hence $0 \leqq \int f\,d\mu \leqq \int g\,d\mu = 0$, that is, $\int f\,d\mu = 0$. ┘

2.5.3. Theorem. Let f and g be $\mathfrak{A}$-measurable numerical functions on Ω that are μ-almost everywhere equal on Ω. Then

(a) $$f \geqq 0,\ g \geqq 0 \Rightarrow \int f\,d\mu = \int g\,d\mu;$$
(b) $$f\text{-integrable} \Rightarrow g\text{-integrable},\ \int f\,d\mu = \int g\,d\mu.$$

Proof. PART (a): By hypothesis and by Theorem 2.1.3 we have $N \equiv \{f \neq g\}$ is a null set. If we set $M \equiv \complement N$, then the functions $f1_N$, $g1_N$, $f1_M$, and $g1_M$ lie in $\mathcal{E}^*$. Each of the first two functions is almost everywhere zero, that is, by Theorem 2.5.2,

$$\int f1_N\,d\mu = \int g1_N\,d\mu = 0;$$

on the other hand, $f1_M = g1_M$ and thus

$$\int f1_M\,d\mu = \int g1_M\,d\mu.$$

If we now observe that

$$f = f1_N + f1_M \qquad \text{and} \qquad g = g1_N + g1_M,$$

then (a) follows from the additivity of the integral on $\mathcal{E}^*$.

PART (b): From the hypothesis it follows that

$$f^+ = g^+ \text{ almost everywhere} \qquad \text{and} \qquad f^- = g^- \text{ almost everywhere.}$$

Then by (a) we have

$$\int f^+\,d\mu = \int g^+\,d\mu \qquad \text{and} \qquad \int f^-\,d\mu = \int g^-\,d\mu.$$

Due to the integrability of f both numbers above are real and nonnegative. The assertion now follows. ┘

Roughly speaking we have shown that integrability and integral of a function f are invariant with respect to measurable changes of f on null sets. From this we can obtain easy sharpenings of assertions proved earlier. For example, we show:

2.5.4. Corollary. Let f and g be $\mathfrak{A}$-measurable numerical functions on Ω with $|f| \leqq g$ μ-almost everywhere. Then f is μ-integrable if g is.

Proof. If we set $g^* \equiv \sup(g, |f|)$, then g^* is measurable, $g^* = g$ almost everywhere, and $|f| \leqq g^*$. Then by Theorem 2.5.3, g^* is integrable and consequently f is also, by Theorem 2.4.2. ⌋

Of importance also is:

2.5.5. Theorem. Every μ-integrable numerical function f on Ω is μ-almost everywhere finite on Ω.

Proof. Let N be the set $\{|f| = \infty\}$ lying in $\mathfrak{A}$. Then for every real $\alpha \geqq 0$, $\alpha 1_N \leqq |f|$ and hence $\alpha\mu(N) \leqq \int |f|\, d\mu < \infty$. But it then follows that $\mu(N) = 0$, that is, f is μ-almost everywhere finite. ⌋

Theorem 2.5.3 has the following consequence: Let N be a μ-null set and f^* a numerical function which is defined on $M \equiv \complement N$ and $M \cap \mathfrak{A}$-measurable. Such a function is said to be a (μ-)almost everywhere defined, ($\mathfrak{A}$-)measurable function. The restriction of every $\mathfrak{A}$-measurable function $f\colon \Omega \to \bar{\mathbf{R}}$ to M is such a function. Conversely, every such f^* can be extended in many ways to an $\mathfrak{A}$-measurable function f on Ω; for example,

$$f(\omega) \equiv \begin{cases} f^*(\omega), & \omega \in M \\ 0, & \omega \in N \end{cases}$$

is one such extension. Any two such extensions f_1 and f_2 are μ-almost everywhere equal. Then by Theorem 2.5.3 either none or all such extensions are integrable; in the latter case, all such extensions have the same μ-integral.

This observation justifies the following definition:

2.5.6. Definition. Let f^* be an $\mathfrak{A}$-measurable numerical function defined μ-almost everywhere on Ω. We say that f^* is (μ-)*integrable* (over Ω) if f^* can be extended to a μ-integrable function f defined on all of Ω. We then call $\int f\, d\mu$ the (μ-)*integral* of f^* and denote it by $\int f^*\, d\mu$.

This broadening of the concept of integral will be needed only occasionally. But its usefulness is shown by the following:

Remark. Let f and g be integrable numerical functions on Ω. By Theorem 2.5.5 both are almost everywhere finite. Since the union of two null sets is itself a null set, there is a null set N with $|f(\omega)| < \infty$ and $|g(\omega)| < \infty$ for all $\omega \notin N$. But then

$$\omega \to f(\omega) + g(\omega) \qquad (\omega \in \complement N)$$

is an almost everywhere defined measurable function. This result, together with the above, shows that the hypothesis of Theorem 2.4.3 that $f + g$ be everywhere defined is rather unimportant. For any two integrable numerical functions f and g on Ω, $f + g$ is almost everywhere defined and integrable in the sense of Definition 2.5.6. It is then always true that

$$\int (f + g)\, d\mu = \int f\, d\mu + \int g\, d\mu.$$

PROBLEMS

1. Let $(\Omega,\mathfrak{A},\mu)$ be a complete measure space, and let f and g be numerical functions on Ω such that $f = g$ μ-almost everywhere. Prove that measurability of f implies measurability of g.
2. Let $(\Omega,\mathfrak{A},\mu)$ be a measure space and $(\Omega,\mathfrak{A}_0,\mu_0)$ its completion. Prove: A numerical function f on Ω is $\mathfrak{A}_0$-measurable if and only if there exist $\mathfrak{A}$-measurable numerical functions f_1, f_2 on Ω such that $f_1 \leqq f \leqq f_2$ and $f_1 = f_2$ μ-almost everywhere. If in addition f is μ_0-integrable, then all functions f_1, f_2 with the above properties are μ-integrable and $\int f_1\, d\mu = \int f_2\, d\mu = \int f\, d\mu_0$. (This generalizes Section 1.5, Problem 5 and Section 2.2, Problem 1.)
3. Reconsider Problem 2 in the case where f is real-valued. Prove that f_1 and f_2 cannot be assumed to be real-valued, in general. Check this in the following case: Ω infinite set, $\mathfrak{A} \equiv \{\emptyset,\Omega\}$, $\mu = 0$.

2.6 THE SPACES $\mathfrak{L}^p(\mu)$

By Theorem 2.1.4 the product of any two measurable functions is again measurable. On the other hand, the product of two integrable functions is in general not again integrable.

Example

Let $(\Omega,\mathfrak{A},\mu)$ be the measure space described in Section 2.4, Example 2 and Section 2.3, Example 2; in particular, let $\alpha_n = n^{-p-1}$ $(n = 1, 2, \ldots)$; then the function f defined by $f(n) \equiv n$ is integrable but f^p is not $(1 < p < \infty)$.

This remark leads us to investigate those measurable numerical functions f on Ω for which $|f|^p$ is integrable.

In the following let p be a real number with $p \geqq 1$. For every $\mathfrak{A}$-measurable numerical function f on Ω, $|f|$ and also $|f|^p$ are measurable.[7] For if α is a real number, then

$$\{|f|^p \geqq \alpha\} = \begin{cases} \Omega, & \alpha \leqq 0 \\ \{|f| \geqq \alpha^{1/p}\}, & \alpha > 0. \end{cases}$$

[7] Here of course we let $\infty^p = \infty$.

Hence for every such f the following is defined:

$$N_p(f) \equiv (\int |f|^p \, d\mu)^{1/p}. \tag{2.6.1}$$

We have $0 \leqq N_p(f) \leqq \infty$. By definition it follows immediately that

$$N_p(\alpha f) = |\alpha| N_p(f) \qquad (\alpha \in \mathbf{R}). \tag{2.6.2}$$

We now show two deeper properties:

2.6.1. Theorem. Let p be a real number with $1 < p < \infty$ and let q be defined by

$$\frac{1}{p} + \frac{1}{q} = 1.$$

Then for any two measurable numerical functions f, g on Ω:

$$N_1(fg) \leqq N_p(f) N_q(g) \qquad \text{(Hölder's inequality).} \tag{2.6.3}$$

Proof. By (2.6.1), we may assume $f \geqq 0$ and $g \geqq 0$ without loss of generality. If we set

$$\sigma \equiv N_p(f) \qquad \text{and} \qquad \tau \equiv N_q(g),$$

then we can assume $\sigma > 0$ and $\tau > 0$. For if σ or τ is equal to 0, then by Theorem 2.5.2, f^p or g^p and thus f or g, as well as fg, is almost everywhere equal to zero. Again by Theorem 2.5.2 we have $N_1(fg) = 0$, that is, (2.6.3) holds. But if σ and τ are positive, then we can obviously further assume $\sigma < \infty$ and $\tau < \infty$.

If we apply the Mean Value Theorem of differential calculus to the function $\eta \to (1 + \eta)^{1/p}$, we immediately obtain the well-known Bernoulli inequality

$$(1 + \eta)^{1/p} \leqq \frac{\eta}{p} + 1, \qquad \text{for all } \eta \in \mathbf{R}_+$$

or

$$\xi^{1/p} \leqq \frac{\xi}{p} + \frac{1}{q}, \qquad \text{for all } \xi \geqq 1.$$

Now if $\alpha > 0$ and $\beta > 0$ are real numbers, then either $\alpha\beta^{-1}$ or $\alpha^{-1}\beta \geqq 1$ is such a number ξ. Substitution of this ξ in the last inequality (in the second case, after interchanging p and q) yields

$$\alpha^{1/p}\beta^{1/q} \leqq \frac{\alpha}{p} + \frac{\beta}{q}.$$

But this inequality then holds immediately for all $\alpha, \beta \in \mathbf{R}_+$. If we choose $\alpha \equiv (\sigma^{-1}f(\omega))^p$ and $\beta \equiv (\tau^{-1}g(\omega))^q$ with $\omega \in \Omega$, then it follows that

$$\frac{1}{\sigma\tau} fg \leqq \frac{1}{\sigma^p p} f^p + \frac{1}{\tau^q q} g^q$$

and hence by integration we have (2.6.3). ┘

2.6.2. Theorem. For any two measurable numerical functions f, g on Ω with sum $f + g$ defined everywhere on Ω and for every p with $1 \leqq p < \infty$ we have

$$N_p(f + g) \leqq N_p(f) + N_p(g) \qquad \text{(Minkowski's inequality)}. \qquad (2.6.4)$$

Proof. Because $|f + g| \leqq |f| + |g|$ we have $N_p(f + g) \leqq N_p(|f| + |g|)$; thus we can assume $f \geqq 0$ and $g \geqq 0$. If $p = 1$, equality holds in (2.6.4) [by (2.3.7)]. Therefore, we can further assume $1 < p < \infty$ and again define q by $p^{-1} + q^{-1} = 1$. Furthermore we can obviously assume that $N_p(f)$ and $N_p(g)$ are finite, that is, f^p and g^p are integrable. By virtue of Theorem 2.4.2, the inequalities

$$(f + g)^p \leqq [2 \sup (f,g)]^p = 2^p \sup (f^p,g^p) \leqq 2^p(f^p + g^p),$$

imply that $(f + g)^p$ is integrable and hence $N_p(f + g) < \infty$.

Now we can conclude as follows: We have

$$\int (f + g)^p \, d\mu = \int f(f + g)^{p-1} \, d\mu + \int g(f + g)^{p-1} \, d\mu.$$

The Hölder inequality applied to both summands yields

$$\int (f + g)^p \, d\mu \leqq N_p(f)N_q((f + g)^{p-1}) + N_p(g)N_q((f + g)^{p-1})$$

and hence

$$(N_p(f + g))^p \leqq [N_p(f) + N_p(g)](N_p(f + g))^{p-1}.$$

Since $N_p(f + g) < \infty$, (2.6.4) now follows. ┘

2.6.3. Definition. A numerical function f on Ω is said to be an L^p-*function* or to be (μ-)*integrable in the pth power* $(1 \leqq p < +\infty)$ if it is measurable and if $|f|^p$ is μ-integrable.

The L^1-functions coincide with the integrable functions by Theorem 2.4.2. In the case $p = 2$ we also speak of *square integrable functions*.

It follows directly from the definition that a *measurable* function f is an L^p-function if and only if $|f|$ is an L^p-function. The following condition is equivalent: There is an L^p-function $g \geqq 0$ with $|f| \leqq g$. A further property, well known for $p = 1$, is:

2.6.4. Theorem. If f and g are L^p-functions on Ω, then

$$\alpha f \qquad (\alpha \in \mathbf{R}) \qquad \text{and} \qquad f + g$$

(provided the latter function is defined everywhere on Ω) and

$$\sup(f,g) \qquad \text{and} \qquad \inf(f,g)$$

are also L^p-functions ($1 \leqq p < +\infty$).

Proof. Since a function f is an L^p-function if and only if it is measurable and $N_p(f)$ is finite, the first part of the assertion follows from (2.6.2) and (2.6.4). That $\sup(f,g)$ and $\inf(f,g)$ are L^p-functions follows, as in the case $p = 1$, from the inequalities

$$|\sup(f,g)| \leqq |f| + |g| \qquad \text{and} \qquad |\inf(f,g)| \leqq |f| + |g|. \quad \lrcorner$$

2.6.5. Corollary. A numerical function f on Ω is an L^p-function if and only if both its positive part f^+ and its negative part f^- are L^p-functions ($1 \leqq p < +\infty$).

Proof. Since $f = f^+ - f^-$, by Theorem 2.6.4 f is an L^p-function if f^+ and f^- are L^p-functions. The converse follows from Theorem 2.6.4 since

$$f^+ = \sup(f,0) \qquad \text{and} \qquad f^- = \sup(-f,0). \quad \lrcorner$$

Now for every p with $1 \leqq p < \infty$ we define

$$\mathcal{L}^p(\mu) \equiv \text{the set of all } \textit{real } L^p\text{-functions on } \Omega. \tag{2.6.5}$$

From Theorem 2.6.4 it follows that $\mathcal{L}^p(\mu)$ is a vector space over $\mathbf{R}$. We already know this for $p = 1$.

It follows directly from (2.6.3) that:

2.6.6. Theorem. Let $1 < p < \infty$ and $1/p + 1/q = 1$. Then the product of an L^p- and an L^q-function is integrable.

In particular, then, the product of two square integrable functions is an integrable function.

2.6.7. Corollary. Let the measure μ be finite. Then every L^p-function is integrable ($1 \leqq p < \infty$).

Proof. Since $\mu(\Omega) < \infty$, every constant real function on Ω is an L^q-function ($1 \leqq q < \infty$). For $1 < p < \infty$ the assertion then follows from Theorem 2.6.6; for $p = 1$ there is nothing to show. ┘

Remark. The assertion of Corollary 2.6.7 does not hold in general without the hypothesis $\mu(\Omega) < \infty$. For instance, in Section 2.4, Example 2, we could choose the measure μ such that $\alpha_n = n^{-1/2}$. Then the function $f(n) = \alpha_n$, $n = 1, 2, \ldots$ lies in $\mathfrak{L}^2(\mu)$ but not in $\mathfrak{L}^1(\mu)$.

More generally, for finite μ and $1 \leqq p' \leqq p < \infty$, every L^p-function is an $L^{p'}$-function (compare Problem 3).

In connection with Theorem 2.6.6 we now show:

2.6.8. Theorem. Let $f: \Omega \to \bar{\mathbf{R}}$ be integrable and let $g: \Omega \to \bar{\mathbf{R}}$ be a measurable, almost everywhere bounded function. Then fg is also integrable.

Proof. There is an $\alpha \in \mathbf{R}_+$ with $|g| \leqq \alpha$ almost everywhere. But then $|fg| \leqq \alpha|f|$ almost everywhere. Hence, by the integrability of $\alpha|f|$, Corollary 2.5.4 gives us the assertion. ┘

Theorems 2.6.6 and 2.6.8 lead us to the following definition:

$$\mathfrak{L}^\infty(\mu) \equiv \text{the set of all real } \mathfrak{A}\text{-measurable } \mu\text{-almost everywhere bounded functions on } \Omega.^8 \tag{2.6.6}$$

We see immediately that $\mathfrak{L}^\infty(\mu)$ is also a vector space over $\mathbf{R}$. Then Theorems 2.6.6 and 2.6.8 yield the assertion:

$$f \in \mathfrak{L}^p(\mu),\ g \in \mathfrak{L}^q(\mu) \Rightarrow fg \in \mathfrak{L}^1(\mu) \qquad (1 \leqq p \leqq \infty,\ p^{-1} + q^{-1} = 1). \tag{2.6.7}$$

Remark. If $\Omega = \{1, \ldots, n\}$, $\mathfrak{A} = \mathfrak{P}(\Omega)$ and $\mu(\{i\}) = 1$ for every $i \in \Omega$, then (2.6.3) [(2.6.4)] becomes the "classical" Hölder [Minkowski] inequality. When $p = 2$ we obtain the well-known inequalities of analytic geometry.

PROBLEMS

1. Let $(\Omega, \mathfrak{A}, \mu)$ be a measure space with finite measure μ. Prove that every real function f on Ω which is the uniform limit of a sequence (f_n) in $\mathfrak{L}^p(\mu)$ is itself in $\mathfrak{L}^p(\mu)$ ($1 \leqq p < +\infty$).

[8] Instead of μ-almost everywhere bounded, one says also μ-essentially bounded.

2. Prove for an arbitrary measure space $(\Omega,\mathfrak{A},\mu)$: A real function f on Ω is in $\mathcal{L}^p(\mu)$ if and only if $f\,|f|^{p-1}$ is in $\mathcal{L}^1(\mu)$ $(1 \leqq p < +\infty)$.
3. Let $(\Omega,\mathfrak{A},\mu)$ be a measure space with finite measure μ and f a measurable numerical function on Ω. Prove: If $1 \leqq p' \leqq p < +\infty$, then

$$N_{p'}(f) \leqq N_p(f)\ (\mu(\Omega))^{1/p'-1/p}$$

and $\mathcal{L}^p(\mu) \subset \mathcal{L}^{p'}(\mu)$.

2.7 CONVERGENCE THEOREMS

Let $1 \leqq p < \infty$. The function N_p is finite on the vector space $\mathcal{L}^p(\mu)$; we have a mapping $N_p\colon \mathcal{L}^p(\mu) \to \mathbf{R}_+$ with properties (2.6.2) and (2.6.4). From this last it follows that the function

$$d_p(f,g) \equiv N_p(f - g) \qquad (f,g \in \mathcal{L}^p(\mu)) \tag{2.7.1}$$

has all the properties of a metric[9] on $\mathcal{L}^p(\mu)$ with the following exception: By Theorems 2.5.2 and 2.5.3,

$$d_p(f,g) = 0$$

is equivalent only to

$$f = g \text{ almost everywhere}$$

but not to $f = g$. Because of this missing property, we speak of a *pseudometric*. We call d_p the pseudometric of *convergence in pth mean* (with respect to μ). In the case $p = 1$ $(p = 2)$ we speak of convergence *in mean* (convergence in *square-mean*). From now on let $\mathcal{L}^p(\mu)$ always be equipped with this pseudometric (respectively with the corresponding topology, which does not satisfy the Hausdorff separation axiom). Then (2.6.4) implies

$$|N_p(f) - N_p(g)| \leqq N_p(f + g) \qquad (f,g \in \mathcal{L}^p(\mu)). \tag{2.7.2}$$

In particular the convergence in pth mean of a sequence (f_n) from $\mathcal{L}^p(\mu)$ to an $f \in \mathcal{L}^p(\mu)$ is defined; this means $\lim d_p(f_n,f) = \lim N_p(f_n - f) = 0$. By the above, the limit f is only almost everywhere uniquely determined. Because of properties (2.6.2) and (2.6.4) of N_p, we have the usual rules for linear operations with convergent sequences in $\mathcal{L}^p(\mu)$.[10]

The following results are technical preliminaries to the main theorems.

[9] The Minkowski inequality (2.6.4) here becomes the *triangle inequality* for d_p.

[10] More precisely, the topology in $\mathcal{L}^p(\mu)$ defined by d_p is compatible with the vector space structure of $\mathcal{L}^p(\mu)$.

2.7.1. Lemma of Fatou. For every sequence $(f_n)_{n\in\mathbf{N}}$ of functions from $\mathcal{E}^*$:

$$\int \liminf_{n\to\infty} f_n \, d\mu \leqq \liminf_{n\to\infty} \int f_n \, d\mu.$$

Proof. By Theorems 2.1.5 and 2.3.6, $f \equiv \liminf f_n$ lies in $\mathcal{E}^*$. By the same theorems, for every $n = 1, 2, \ldots,$

$$g_n \equiv \inf_{m\geqq n} f_m$$

lies in $\mathcal{E}^*$. The sequence (g_n) is isotone and, by the definition of limit inferior, has f as supremum. Then by Theorem 2.3.4,

$$\int f \, d\mu = \sup_{n\in\mathbf{N}} \int g_n \, d\mu.$$

From this our result follows since, by virtue of isotonicity,

$$\int g_n \, d\mu \leqq \inf_{m\geqq n} \int f_m \, d\mu$$

for every $n \in \mathbf{N}$. ┘

If we choose in particular a sequence (1_{A_n}) of indicator functions of sets $A_n \subset \Omega$, then $\liminf_{n\to\infty} 1_{A_n}$ is the indicator function of the set

$$\liminf_{n\to\infty} A_n \equiv \bigcup_{n=1}^{\infty} \bigcap_{m=n}^{\infty} A_m. \tag{2.7.3}$$

Hence $\liminf A_n$ is the set of all elements $\omega \in \Omega$ which lie in every set A_n, for n *sufficiently large*. Dually to this, we define

$$\limsup_{n\to\infty} A_n \equiv \bigcap_{n=1}^{\infty} \bigcup_{m=n}^{\infty} A_m. \tag{2.7.4}$$

This is the set of all $\omega \in \Omega$ which lie in *infinitely many* of the sets A_n, that is, for which $\omega \in A_n$ holds for infinitely many $n \in \mathbf{N}$. We obviously have the equality

$$\complement\,(\limsup_{n\to\infty} A_n) = \liminf_{n\to\infty} \complement A_n. \tag{2.7.5}$$

From this we obtain the following corollary:

2.7.2. Corollary. For every sequence $(A_n)_{n\in\mathbf{N}}$ of sets in the σ-algebra $\mathfrak{A}$, we have

$$\mu(\liminf_{n\to\infty} A_n) \leqq \liminf_{n\to\infty} \mu(A_n). \tag{2.7.6}$$

If the measure μ is finite, then moreover

$$\mu(\limsup_{n\to\infty} A_n) \geqq \limsup_{n\to\infty} \mu(A_n). \tag{2.7.7}$$

Proof. (2.7.6) is a direct consequence of Theorem 2.7.1. If we apply (2.7.6) to the sequence $(\complement A_n)$, then by (2.7.5),

$$\begin{aligned}\mu(\Omega) - \mu(\limsup A_n) &= \mu(\complement \limsup A_n) = \mu(\liminf \complement A_n) \\ &\leqq \liminf \mu(\complement A_n) = \mu(\Omega) - \limsup \mu(A_n)\end{aligned}$$

and hence we have (2.7.7). ┘

2.7.3. Lemma. For every sequence $(f_n)_{n\in\mathbf{N}}$ of functions from $\mathcal{E}^*$,

$$N_p\left(\sum_{n=1}^{\infty} f_n\right) \leqq \sum_{n=1}^{\infty} N_p(f_n).$$

Proof. If we set $s_n \equiv f_1 + \cdots + f_n$ for every n, then by (2.6.4),

$$N_p(s_n) \leqq \sum_{i=1}^{n} N_p(f_i) \leqq \sum_{n=1}^{\infty} N_p(f_n).$$

The sequence (s_n) is isotone and has $\sum_{n=1}^{\infty} f_n$ as supremum; the same is then true for the pth powers. Hence by Theorem 2.3.4,

$$N_p\left(\sum_{n=1}^{\infty} f_n\right) = \sup N_p(s_n)$$

and thus we have the assertion. ┘

Now we come to the convergence theorems:

2.7.4. Theorem of Dominated Convergence (H. Lebesgue). Let $(f_n)_{n\in\mathbf{N}}$ be a sequence of functions from $\mathcal{L}^p(\mu)$ which converges almost everywhere on Ω. Suppose there exists an L^p-function $g \geqq 0$ on Ω with

$$|f_n| \leqq g \qquad (n \in \mathbf{N}).$$

Then there is a real, measurable function f on Ω to which (f_n) converges almost everywhere. Every such f lies in $\mathcal{L}^p(\mu)$, and the sequence (f_n) converges in pth mean to f $(1 \leqq p < \infty)$.

Proof. By hypothesis there is a null set M_1 such that $\lim f_n(\omega)$ exists for all $\omega \notin M_1$. Due to the integrability of g^p there is, by Theorem 2.5.5, another null set M_2 with $g(\omega) < \infty$ for all $\omega \notin M_2$. If we set

$$f(\omega) \equiv \begin{cases} \lim\limits_{n\to\infty} f_n(\omega), & \omega \notin M_1 \cup M_2, \\ 0, & \omega \in M_1 \cup M_2, \end{cases}$$

then f is real-valued and $\mathfrak{A}$-measurable. (f_n) converges almost everywhere to f.

Now let f be an arbitrary function with the above properties. Then $|f| \leqq g$ almost everywhere; by Corollary 2.5.4, when g^p is integrable so is $|f|^p$, that is, $f \in \mathcal{L}^p(\mu)$. If we set

$$g_n \equiv |f_n - f|^p,$$

then we must show that $\int g_n \, d\mu = 0$. We have

$$0 \leqq g_n \leqq (|f_n| + |f|)^p \leqq (|f| + g)^p.$$

Hence $h \equiv (|f| + g)^p$ and every g_n are also integrable (by Theorems 2.6.4 and 2.4.2). We can apply Lemma 2.7.1 to the sequence of functions $(h - g_n)$; then we obtain

$$\int \liminf_{n\to\infty} (h - g_n) \, d\mu \leqq \liminf_{n\to\infty} \int (h - g_n) \, d\mu$$
$$= \int h \, d\mu - \limsup_{n\to\infty} \int g_n \, d\mu.$$

Since (f_n) converges to f almost everywhere, $(h - g_n)$ converges almost everywhere to h. Consequently we have almost everywhere

$$\liminf_{n\to\infty} (h - g_n) = h,$$

that is,

$$\int \liminf (h - g_n) \, d\mu = \int h \, d\mu.$$

Hence it follows that $\limsup \int g_n \, d\mu \leqq 0$. Since $g_n \geqq 0$ this is equivalent to $\lim \int g_n \, d\mu = 0$. ┘

The concept of a Cauchy sequence is also meaningful in pseudometric spaces, and in particular in $\mathcal{L}^p(\mu)$. A sequence (f_n) of functions from $\mathcal{L}^p(\mu)$ is called a *Cauchy sequence* in $\mathcal{L}^p(\mu)$ if for every $\epsilon > 0$ there exists $n_0 \in \mathbf{N}$ such that

$$d_p(f_n, f_m) = N_p(f_n - f_m) \leqq \epsilon \qquad \text{for all } m, n \geqq n_0.$$

According to the Minkowski inequality, every convergent sequence in $\mathcal{L}^p(\mu)$ is a Cauchy sequence. The second convergence theorem asserts that the converse is also true. Thus the space $\mathcal{L}^p(\mu)$ is *complete.*

2.7.5. Theorem. Every Cauchy sequence $(f_n)_{n\in\mathbf{N}}$ in $\mathcal{L}^p(\mu)$ converges in pth mean to an $f \in \mathcal{L}^p(\mu)$. There exists a subsequence of (f_n) which converges almost everywhere to f $(1 \leqq p < \infty)$.

Proof. By the definition of Cauchy sequence there is a subsequence $(f_{n_k})_{k\in\mathbf{N}}$ with $N_p(f_{n_{k+1}} - f_{n_k}) \leqq 2^{-k}$ for all k. We set

$$g_k \equiv f_{n_{k+1}} - f_{n_k} \qquad \text{and} \qquad g \equiv \sum_{k=1}^{\infty} |g_k|.$$

Then by Lemma 2.7.3

$$N_p(g) \leqq \sum_{k=1}^{\infty} N_p(g_k) \leqq \sum_{k=1}^{\infty} 2^{-k} = 1.$$

Consequently, the function g lying in $\mathcal{E}^*$ is an L^p-function and, by Theorem 2.5.5 almost everywhere finite, that is, the series Σg_k converges almost everywhere absolutely. Thus, (f_{n_k}) converges almost everywhere on Ω. Further,

$$|f_{n_{k+1}}| = |g_1 + \cdot \cdot \cdot + g_k + f_{n_1}| \leqq g + |f_{n_1}|,$$

where $g + |f_{n_1}|$ is an L^p-function by Theorem 2.6.4. Thus, (f_{n_k}) satisfies all the hypotheses of Theorem 2.5.4. Accordingly, (f_{n_k}) converges in pth mean to an $f \in \mathfrak{L}^p(\mu)$, and we have

$$\lim_{k \to \infty} f_{n_k} = f \text{ almost everywhere.}$$

Since (f_n) is a Cauchy sequence, (f_n) also converges in pth mean to f.[11] ⌟

The following example shows that in Theorem 2.7.5, we cannot get along without the transition to a subsequence in order to obtain convergence almost everywhere.

Example

Let $\Omega \equiv [0,1]$, $\mathfrak{A} \equiv \Omega \cap \mathfrak{B}^1$, and $\mu \equiv \lambda_\Omega^1$. Set $f_n \equiv 1_{A_n}$, where $A_n = [k2^{-h}, (k+1)2^{-h}]$ and $n = 2^h + k$ with $0 \leqq k < 2^h$ (h, k integers). The sequence (f_n) converges in pth mean to 0 and is therefore a Cauchy sequence in $\mathfrak{L}^p(\mu)$ $(1 \leqq p < \infty)$. However the sequence $(f_n(\omega))$ is not convergent for any $\omega \in \Omega$.

The linear form $f \to \int f\,d\mu$ is defined on the space $\mathfrak{L}^1(\mu)$. If $(f_n)_{n \in \mathbf{N}}$ is a sequence of functions from $\mathfrak{L}^1(\mu)$ which converges in mean to $f \in \mathfrak{L}^1(\mu)$, then $\lim_{n \to \infty} \int f_n\,d\mu = \int f\,d\mu$. This is given by the following theorem.

2.7.6. Theorem. The linear form $f \to \int f\,d\mu$ is continuous on the space $\mathfrak{L}^1(\mu)$.

Proof. For any two functions $f, g \in \mathfrak{L}^1(\mu)$,

$$\left|\int f\,d\mu - \int g\,d\mu\right| \leqq \int |f - g|\,d\mu = N_1(f - g) = d_1(f,g).$$

Hence even uniform continuity of the given linear form follows. ⌟

[11] See, for example, Franz [32], Theorem 21.3, p. 84.

From (2.7.2) we obtain an analogous property:

The function $f \to N_p(f)$ is continuous on the space $\mathcal{L}^p(\mu)$ $(1 \leqq p < \infty)$. (2.7.8)

Analogous to Corollary 2.6.7, we have

2.7.7. Theorem. Let the measure μ be finite. Then every sequence $(f_n)_{n \in \mathbf{N}}$ in $\mathcal{L}^p(\mu)$ which converges in pth mean to an $f \in \mathcal{L}^p(\mu)$ also converges in mean to f $(1 \leqq p < \infty)$.

Proof. For $p = 1$ there is nothing to show. For $1 < p < \infty$ the assertion follows from Theorem 2.6.6 since the constant function 1 is in $\mathcal{L}^q(\mu)$ $(p^{-1} + q^{-1} = 1)$. ┘

As in the remark pertaining to Corollary 2.6.7 the reader should make it clear by an example that the converse of the above is not true.

Remarks. 1. The properties (2.6.2) and (2.6.4) of

$$N_p: \mathcal{L}^p(\mu) \to \mathbf{R}_+$$

state that N_p is a seminorm on the vector space $\mathcal{L}^p(\mu)$. The linear subspace

$$\mathfrak{N} \equiv N_p^{-1}(0)$$

is independent of p since $\mathfrak{N}$ consists of all measurable real functions f on Ω with $f = 0$ almost everywhere. If we go over to the quotient space

$$L^p(\mu) \equiv \mathcal{L}^p(\mu)/\mathfrak{N}$$

then, in a natural way $L^p(\mu)$ becomes a normed space: If we let $f \to \tilde{f}$ denote the canonical mapping of $\mathcal{L}^p(\mu)$ onto $L^p(\mu)$, then we set

$$\|\tilde{f}\|_p \equiv N_p(f)$$

for all $\tilde{f} \in L^p(\mu)$. We can verify immediately that $\tilde{f} \to \|\tilde{f}\|_p$ is a norm on $L^p(\mu)$.

Theorem 2.7.5 states that $L^p(\mu)$ is complete relative to this norm, that is, it is a *Banach space* $(1 \leqq p < \infty)$.

$L^2(\mu)$ even is a *Hilbert space*. By Theorem 2.6.6, the product-function fg is integrable for any two functions $f, g \in \mathcal{L}^2(\mu)$. The integral $\int fg\, d\mu$ depends only on the images $\tilde{f}$ and $\tilde{g}$ of f and g for the canonical mapping. Hence,

$$(\tilde{f},\tilde{g}) \to \int fg\, d\mu$$

is a scalar product in $L^2(\mu)$.

2. For every $f \in \mathcal{L}^\infty(\mu)$, the set W_f of all $\alpha \in \mathbf{R}_+$ for which $|f| \leqq \alpha$ holds almost everywhere is nonempty by definition. We set

$$N_\infty(f) \equiv \inf W_f$$

and see easily that $N_\infty\colon \mathcal{L}^\infty(\mu) \to \mathbf{R}_+$ is a seminorm on $\mathcal{L}^\infty(\mu)$. We can verify directly that $N_\infty^{-1}(0)$ is identical with the space $\mathfrak{N}$ described in Remark 1. By a transition to

$$L^\infty(\mu) \equiv \mathcal{L}^\infty(\mu)/\mathfrak{N},$$

we define a norm $\|\tilde{f}\|_\infty$ on $L^\infty(\mu)$ by means of N_∞ as above. We can easily verify that $L^\infty(\mu)$ is then also a *Banach space*.

PROBLEMS

1. Let (f_n) be a sequence of real measurable functions on a measure space $(\Omega,\mathfrak{A},\mu)$; assume the existence of functions $g \in \mathcal{L}^1(\mu)$ such that $|f_n| \leqq g$ for all $n \in \mathbf{N}$. Prove: $\liminf f_n$ and $\limsup f_n$ are μ-integrable functions which satisfy the following inequalities:

$$\int \liminf f_n \, d\mu \leqq \liminf \int f_n \, d\mu \leqq \limsup \int f_n \, d\mu \leqq \int \limsup f_n \, d\mu.$$

Prove that these inequalities do not hold in general if the function g above does not exist.

2. Let μ be a finite measure and $1 \leqq p' \leqq p < +\infty$. Prove that every sequence in $\mathcal{L}^p(\mu)$ which converges in pth mean to a function $f \in \mathcal{L}^p(\mu)$ also converges in p'th mean to f. [*Hint:* Use Section 2.6, Problem 3.]
3. Let M be a metric space, $(\Omega,\mathfrak{A},\mu)$ a measure space, and $f\colon M \times \Omega \to \mathbf{R}$ a real-valued function on $M \times \Omega$ having the following properties:
   (a) $x \to f(x,\omega)$ is continuous on M for all $\omega \in \Omega$;
   (b) $\omega \to f(x,\omega)$ is μ-integrable for all $x \in M$;
   (c) there exists $g \in \mathcal{L}^1(\mu)$ such that

$$|f(x,\omega)| \leqq g(\omega), \qquad \text{for all } (x,\omega) \in M \times \Omega.$$

Prove the continuity of the function $x \to \int f(x,\omega)\mu(d\omega)$ on M.
4. Prove the details of the preceding Remarks 1 and 2.

2.8 REMARKS ON LEBESGUE AND RIEMANN INTEGRAL

We can easily rid ourselves of the restriction that integrals are extended over all of Ω. Two possibilities for this present themselves: Either we can note by (2.3.5) that when any two functions are in $\mathcal{E}^*$ their product is in

$\mathcal{E}^*$. Thus, in particular, $f \in \mathcal{E}^*$ and $A \in \mathfrak{A}$ imply $f1_A \in \mathcal{E}^*$. We define

$$\int_A f\,d\mu \equiv \int f1_A\,d\mu \tag{2.8.1}$$

and call $\int_A f\,d\mu$ the *μ-integral of f over A*. In particular, then,

$$\int_\Omega f\,d\mu = \int f\,d\mu.$$

Alternately we may consider the restriction μ_A of the measure μ to the trace $A \cap \mathfrak{A}$ of the σ-algebra $\mathfrak{A}$ in $A \in \mathfrak{A}$ and integrate the restriction f_A of $f \in \mathcal{E}^*$ to A relative to this new measure. The following lemma shows that we arrive at the same result.

2.8.1. Lemma. For a function $f \in \mathcal{E}^*$ and a set $A \in \mathfrak{A}$, let

$$\mu_A \equiv \mathrm{rest}_A\,\mu \qquad \text{and} \qquad f_A \equiv \mathrm{rest}_A\,f. \tag{2.8.2}$$

Then f_A is measurable on A relative to the σ-algebra $A \cap \mathfrak{A}$ and

$$\int f_A\,d\mu_A = \int_A f\,d\mu. \tag{2.8.3}$$

Proof. Since $\mathcal{E}^*$ is the set of all $\mathfrak{A}$-measurable numerical functions $\geqq 0$ on Ω, the asserted measurability of f_A follows from the equality

$$f_A^{-1}(B) = A \cap f^{-1}(B).$$

which holds for arbitrary Borel sets B in $\bar{\mathbf{R}}$. By the definition of the functions of $\mathcal{E}^*$, there is an isotone sequence (u_n) of $\mathfrak{A}$-elementary functions with $f1_A = \sup u_n$. If we set

$$u_n^* \equiv \mathrm{rest}_A\,u_n \qquad (n \in \mathbf{N}),$$

then (u_n^*) is obviously an isotone sequence of $A \cap \mathfrak{A}$-elementary functions with $\sup u_n^* = f_A$. Hence,

$$\int_A f\,d\mu = \sup \int u_n\,d\mu \qquad \text{and} \qquad \int f_A\,d\mu_A = \sup \int u_n^*\,d\mu_A. \tag{2.8.4}$$

Since $0 \leqq u_n \leqq f1_A$ we have $u_n(\omega) = 0$ for all $\omega \in \Omega \setminus A$, that is, $u_n = u_n1_A$, whence u_n is of the form

$$u_n = \sum_{i=1}^{k_n} \alpha_i 1_{A_i}$$

with sets $A_i \in \mathfrak{A}$ (dependent on n) and each a subset of A (and coefficients $\alpha_i \geqq 0$ dependent on n). Consequently,

$$u_n^* = \sum_{i=1}^{k_n} \alpha_i 1_{A_i}^*,$$

where 1_Q^* denotes the indicator function of a set $Q \subset A$ relative to A. Hence we obtain

$$\int u_n \, d\mu = \int u_n^* \, d\mu_A, \qquad \text{for all } n \in \mathbf{N},$$

from (2.8.4) we obtain the rest of the assertion. ⌟

The p-dimensional Lebesgue-Borel measure λ_B^p on B (more precisely, on $B \cap \mathfrak{B}^p$) was defined in Section 1.6 for every p-dimensional Borel set $B \in \mathfrak{B}^p$. Then $\int f \, d\lambda_B^p$ is defined for suitable Borel-measurable numerical functions f on B. We call this integral the *Lebesgue integral* of f over B. We frequently also write

$$\int_B f(x) \, dx \equiv \int f \, d\lambda_B^p. \tag{2.8.5}$$

In the case $p = 1$, the notations $\int_\alpha^\beta f(x) \, dx$, $\int_{-\infty}^\alpha f(x) \, dx$, $\int_{-\infty}^{+\infty} f(x) \, dx$ for $B = [\alpha,\beta], =]-\infty,\alpha], = \mathbf{R}$, and so on, are also usual.

Since elementary texts often treat the Riemann integral exclusively, the following observations may be useful.

2.8.2. Theorem. Let f be a Borel-measurable, bounded real function defined on a compact interval $\Omega = [\alpha,\beta]$ in $\mathbf{R}$. Then if f is Riemann-integrable, f is also Lebesgue-integrable. The Riemann integral is equal to the Lebesgue integral of f over Ω.

Proof. To every finite subdivision of Ω

$$\mathfrak{z} \equiv (\alpha = \alpha_0 \leqq \alpha_1 \leqq \cdots \leqq \alpha_n = \beta)$$

correspond the lower and upper sums

$$L_\mathfrak{z} \equiv \sum_{i=1}^n \gamma_i(\alpha_i - \alpha_{i-1}) \qquad \text{and} \qquad U_\mathfrak{z} \equiv \sum_{i=1}^n \Gamma_i(\alpha_i - \alpha_{i-1}),$$

where

$$\gamma_i \equiv \inf f([\alpha_{i-1},\alpha_i]) \qquad \text{and} \qquad \Gamma_i \equiv \sup f([\alpha_{i-1},\alpha_i]).$$

If we set $A_i \equiv [\alpha_{i-1},\alpha_i]$, then

$$l_\mathfrak{z} \equiv \sum_{i=1}^n \gamma_i 1_{A_i} \qquad \text{and} \qquad u_\mathfrak{z} \equiv \sum_{i=1}^n \Gamma_i 1_{A_i}$$

are Borel measurable functions on Ω with

$$L_\mathfrak{z} = \int l_\mathfrak{z} \, d\mu \qquad \text{and} \qquad U_\mathfrak{z} = \int u_\mathfrak{z} \, d\mu,$$

where μ denotes the L-B-measure on Ω. Because of the Riemann integrability of f there is a sequence $(\mathfrak{z}_n)$ of subdivisions of Ω with the following properties: $\mathfrak{z}_{n+1}$ is a refinement of $\mathfrak{z}_n$; the sequences $(L_{\mathfrak{z}_n})$ and $(U_{\mathfrak{z}_n})$ converge to the same finite limit, the Riemann integral ρ of f over Ω. By the first property of $(\mathfrak{z}_n)$, $(l_{\mathfrak{z}_n})$ is an isotone sequence and $(u_{\mathfrak{z}_n})$ an antitone sequence, that is, there exists

$$q \equiv \lim (u_{\mathfrak{z}_n} - l_{\mathfrak{z}_n})$$

on Ω. Then if we apply Fatou's Lemma 2.7.1 to the sequence $(u_{\mathfrak{z}_n} - l_{\mathfrak{z}_n})$, which is possible because $u_{\mathfrak{z}_n} - l_{\mathfrak{z}_n} \geqq 0$, we have

$$0 \leqq \int q \, d\mu \leqq \lim (U_{\mathfrak{z}_n} - L_{\mathfrak{z}_n}) = 0.$$

Thus, $q = 0$ μ-almost everywhere on Ω. Since $l_{\mathfrak{z}_n} \leqq f \leqq u_{\mathfrak{z}_n}$, we have

$$\lim l_{\mathfrak{z}_n} = f \text{ } \mu\text{-almost everywhere.}$$

Finally if we note that, because f is bounded, the sequence $(|l_{\mathfrak{z}_n}|)$ is majorized by a constant and hence μ-integrable function, we obtain the μ-integrability of f from Theorem 2.7.4 as well as the convergence in mean of $(l_{\mathfrak{z}_n})$ to f. Thus from Definition 1.5.6 we have

$$\int f \, d\mu = \lim \int u_{\mathfrak{z}_r} \, d\mu = \lim L_{\mathfrak{z}_n} = \rho. \quad \lrcorner$$

Remarks. 1. If we do not require the Borel measurability of f, then the above proof shows that $\lim l_{\mathfrak{z}_n} = f$ μ-almost everywhere, that is, f is μ-almost everywhere equal to the Borel-measurable function $\lim l_{\mathfrak{z}_n}$. But f itself need not necessarily be Borel-measurable.[12]

2. Suppose f is the well-known "*Dirichlet jump function*" on the unit interval $[0,1]$: $f(x) = 1$ for every rational and $f(x) = 0$ for every irrational $x \in [0,1]$. The function f is obviously Borel-measurable and, with respect to the L-B measure $\lambda^1_{[0,1]}$, almost everywhere equal to zero. Consequently, f is Lebesgue-integrable and $\int_0^1 f(x) \, dx = 0$. But we know that f is *not* Riemann-integrable. Thus the converse of the assertion of Theorem 2.8.2 is not true.

2.8.3. Corollary. Suppose a Borel-measurable real function $f \geqq 0$ defined on $\mathbf{R}$ is Riemann-integrable on every compact interval. Then f is Lebesgue-integrable if and only if the improper Riemann integral ρ exists. In this case,

$$\rho = \int f \, d\lambda^1.$$

Proof. Let ρ_n be the Riemann integral of f over $A_n \equiv [-n, +n]$. By Lemma 2.8.1 and Theorem 2.8.2, $f1_{A_n}$ is λ^1-integrable and

$$\rho_n = \int f 1_{A_n} \, d\lambda^1.$$

[12] See Haupt-Aumann-Pauc [4], p. 66, No. 3.4.3.1.

The sequence $(f1_{A_n})$ is isotone and has f as supremum. Then by Theorem 2.3.4,

$$\sup \rho_n = \int f \, d\lambda^1.$$

By definition the improper Riemann integral exists if and only if $\sup \rho_n < \infty$. In that case, $\rho = \sup \rho_n$ and the assertion follows. ⌋

With the help of the decomposition $f = f^+ - f^-$ into positive and negative parts, it follows from Theorem 2.2.2 and Corollary 2.8.3 that every Borel-measurable real function f on $\mathbf{R}$ is Lebesgue-integrable if it has an *absolutely convergent* improper Riemann integral. In that case $\int f \, d\lambda^1$ is equal to the improper Riemann integral of f.

Of course we may replace $\mathbf{R}$ by any open or half-open interval $I \neq \emptyset$ in Corollary 2.8.3.

On the other hand, even for continuous f, the existence of the improper Riemann integral does *not* imply the Lebesgue-integrability of f. For example we know that the improper Riemann integral

$$\lim_{a \to +\infty} \int_0^a \frac{\sin x}{x} \, dx$$

exists. But the function $x \to \sin x/x$ is not Lebesgue-integrable over $\mathbf{R}_+$. By Theorem 2.2.2 its absolute value over $\mathbf{R}_+$ would then be Lebesgue-integrable. But since the inequality

$$\frac{|\sin x|}{x} \geqq \frac{1}{(k+1)\pi} |\sin x| \qquad (k \in \mathbf{N})$$

holds on every interval $[k\pi, (k+1)\pi[$, we have

$$\int_{\mathbf{R}_+} \frac{|\sin x|}{x} \, dx \geqq \sum_{k=1}^{n} \frac{1}{(k+1)\pi} \int_{k\pi}^{(k+1)\pi} |\sin x| \, dx$$

$$= \frac{1}{\pi} \int_0^{\pi} |\sin x| \, dx \cdot \sum_{k=1}^{n} \frac{1}{k+1}$$

for every $n \in \mathbf{N}$. But the harmonic series is divergent.

The above results show that integrals that are known to the reader perhaps only as Riemann integrals can immediately be interpreted as Lebesgue integrals under the given hypotheses. Well-known formulas such as

$$\int_{-\infty}^{+\infty} e^{-x^2/2} \, dx = \sqrt{2\pi} \qquad (2.8.6)$$

are thus now available to us also in the Lebesgue theory. We shall no longer make special reference to such considerations below. For a proof of (2.8.6) see also Section 3.2, Problem 6.

PROBLEMS

1. Check which of the following two functions f are integrable or square integrable with respect to the Lebesgue-Borel measure on Ω:

   (a) $\Omega \equiv [1,+\infty[,\ f(x) \equiv \dfrac{1}{x}$;

   (b) $\Omega \equiv\ ]0,1],\ f(x) \equiv \dfrac{1}{\sqrt{x}}$.
2. Prove that $x \to e^{-x^\alpha}$ is λ^1-integrable over $\mathbf{R}_+$ for every real number $\alpha > 0$. [*Hint:* Use suitable majorizations.]
3. Prove that $x \to e^{-\alpha x}\ (\sin x/x)^3$ is λ^1-integrable over $]0,+\infty[$ for all $\alpha > 0$, and that

$$\alpha \to \int_0^\infty e^{-\alpha x}\left(\frac{\sin x}{x}\right)^3 \lambda^1(dx)$$

   is continuous on $]0,+\infty[$. [*Hint:* Use Section 2.7, Problem 3.]
4. Do the Convergence Theorems 2.3.4 and 2.7.4 hold in the theory of the Riemann integral?

2.9 MEASURES WITH DENSITIES

Again let $(\Omega,\mathfrak{A},\mu)$ be an arbitrary measure space and $\mathcal{E}^*$ the set of all $\mathfrak{A}$-measurable numerical functions $f \geqq 0$ on Ω. In (2.8.1) we defined the integral of f over A for every function $f \in \mathcal{E}^*$ and every set $A \in \mathfrak{A}$. We are now interested in the behavior of the function $A \to \int_A f\,d\mu$ defined on $\mathfrak{A}$.

2.9.1. Theorem. For every function $f \in \mathcal{E}^*$,

$$A \to \int_A f\,d\mu \tag{2.9.1}$$

defines a measure ν on $\mathfrak{A}$.

Proof. We have $\nu(\emptyset) = 0$ and $\nu \geqq 0$. For every sequence $(A_n)_{n \in \mathbf{N}}$ of pairwise disjoint sets of $\mathfrak{A}$ with $A \equiv \bigcup_{n=1}^{\infty} A_n$,

$$f1_A = \sum_{n=1}^{\infty} f1_{A_n}$$

and hence by Corollary 2.3.5,

$$\nu(A) = \sum_{n=1}^{\infty} \nu(A_n).$$

Thus ν is a measure on $\mathfrak{A}$. ┘

2.9.2. Definition. If f is a nonnegative $\mathfrak{A}$-measurable numerical function on Ω, then the measure ν on $\mathfrak{A}$ defined by (2.9.1) is called the measure with *density* f relative to μ. It is denoted also by

$$\nu = f\mu. \tag{2.9.2}$$

As for the relationship between ν- and μ-integrals, we show:

2.9.3. Theorem. Let $\nu \equiv f\mu$ with $f \in \mathcal{E}^*$. Then for every function $\varphi \in \mathcal{E}^*$,

$$\int \varphi \, d\nu = \int \varphi f \, d\mu.^{13} \tag{2.9.3}$$

An $\mathfrak{A}$-measurable function $\varphi: \Omega \to \bar{\mathbf{R}}$ is ν-integrable if and only if φf is μ-integrable. In this case (2.9.3) again holds.

Proof. First let $\varphi = \sum_{i=1}^{n} \alpha_i 1_{A_i}$ be a function in $\mathcal{E}$. Then (2.9.3) holds, since

$$\int \varphi \, d\nu = \sum_{i=1}^{n} \alpha_i \nu(A_i) = \sum_{i=1}^{n} \alpha_i \int f1_{A_i} \, d\mu = \int \varphi f \, d\mu.$$

An arbitrary $\varphi \in \mathcal{E}^*$ is the supremum of an isotone sequence (u_n) of functions in $\mathcal{E}$. But then the sequence $(u_n f)$ is also isotone and $\sup (u_n f) = \varphi f$. Hence (2.9.3) is a consequence of Theorem 2.3.4.

Now let φ be an $\mathfrak{A}$-measurable numerical function on Ω. We already know that

$$\int \varphi^+ \, d\nu = \int \varphi^+ f \, d\mu \quad \text{and} \quad \int \varphi^- \, d\nu = \int \varphi^- f \, d\mu.$$

Hence and from the definition of integrability the second part of the assertion follows. ┘

[13] (2.9.3) is easier to remember in the notation $\int \varphi d(f\mu) = \int \varphi f \, d\mu$.

With regard to the uniqueness of the density f, we show:

2.9.4. Theorem. For any two functions $f, g \in \mathcal{E}^*$,

$$f = g \ \mu\text{-almost everywhere} \Rightarrow f\mu = g\mu. \qquad (2.9.4)$$

If f or g is μ-integrable, then the converse of the above implication also holds.

Proof. For every set $A \in \mathfrak{A}$ the functions $f1_A$ and $g1_A$ coincide with f and g μ-almost everywhere. Consequently,

$$\int f1_A \, d\mu = \int g1_A \, d\mu,$$

for all $A \in \mathfrak{A}$, that is, $f\mu = g\mu$.

Now let f be μ-integrable and $f\mu = g\mu$. Since $g \geqq 0$ and $\int g \, d\mu = \int f \, d\mu < +\infty$, g is also μ-integrable. We show that the set

$$N \equiv \{f > g\},$$

which is in $\mathfrak{A}$, is a μ-null set. $f(\omega) - g(\omega)$ is defined and >0 for all $\omega \in N$, therefore the definition

$$h \equiv f1_N - g1_N$$

is meaningful. Since $f1_N \leqq f$ and $g1_N \leqq g$, the functions $f1_N$ and $g1_N$ are also μ-integrable; since $f\mu = g\mu$, they have the same μ-integral. Hence we obtain

$$\int h \, d\mu = \int_N f \, d\mu - \int_N g \, d\mu = 0.$$

It now follows that $\mu(N) = 0$ by Theorem 2.5.2 since $N = \{h > 0\}$. By interchanging f and g we obtain $\mu(N') = 0$ for the set $N' \equiv \{g > f\}$ and hence, since $\{f \neq g\} = N \cup N'$, the assertion follows. ⌟

The converse of the implication (2.9.4) does not hold without appropriate additional hypotheses on the densities f and g. This is shown by the following:

Example

1. As in Section 1.3, Example 2 let Ω be an uncountable set and let $\mathfrak{A}$ be the σ-algebra of Section 1.1, Example 2. Let the measure μ be defined as follows: Let $\mu(A) = 0$ or $= +\infty$ depending on whether A or $\complement A$ is countable. Let f denote the constant function 1 and g the constant function 2 on Ω, then $\mu = f\mu = g\mu$; but $f(\omega) \neq g(\omega)$ for all $\omega \in \Omega$. By Theorem 2.9.4 neither f nor g is μ-integrable.

Now suppose two measures ν and μ are given on the σ-algebra $\mathfrak{A}$ of the measurable space $(\Omega, \mathfrak{A})$. We pose the problem of how to decide whether

ν possesses a density relative to μ, that is, whether there is an $\mathfrak{A}$-measurable numerical function $f \geqq 0$ on Ω with $\nu = f\mu$ so that

$$\nu(A) = \int_A f\,d\mu, \qquad \text{for all } A \in \mathfrak{A}. \tag{2.9.5}$$

The condition that every μ-null set $A \in \mathfrak{A}$ is also a ν-null set is necessary here. This follows immediately from (2.9.5), taking Theorem 2.5.3 into account.

2.9.5. Definition. A measure ν on $\mathfrak{A}$ is said to be absolutely continuous relative to a measure μ on $\mathfrak{A}$, abbreviated *μ-continuous*, if every μ-null set of $\mathfrak{A}$ is also a ν-null set.[14]

For the case of a finite measure ν, we immediately give a condition equivalent to μ-continuity:

2.9.6. Theorem. A finite measure ν on $\mathfrak{A}$ is μ-continuous if and only if for every number $\epsilon > 0$ there exists a number $\delta > 0$ such that for every $A \in \mathfrak{A}$,

$$\mu(A) \leqq \delta \Rightarrow \nu(A) \leqq \epsilon. \tag{2.9.6}$$

Proof. From the given condition it follows that $\nu(A) \leqq \epsilon$ for every μ-null set $A \in \mathfrak{A}$ and every $\epsilon > 0$. Hence $\nu(A) = 0$ and therefore ν (even without the hypothesis of finiteness) is a μ-continuous measure. Conversely, we show that ν is not μ-continuous if the given condition is not satisfied. Then there is an $\epsilon > 0$ and a sequence $(A_n)_{n\in\mathbf{N}}$ in $\mathfrak{A}$ with the properties

$$\mu(A_n) \leqq 2^{-n} \qquad \text{and} \qquad \nu(A_n) > \epsilon, \qquad \text{for all } n \in \mathbf{N}.$$

If we set

$$A \equiv \limsup A_n = \bigcap_{n=1}^{\infty} \bigcup_{m=n}^{\infty} A_m,$$

then for this set from $\mathfrak{A}$ we have on the one hand,

$$\mu(A) \leqq \mu\left(\bigcup_{m=n}^{\infty} A_m\right) \leqq \sum_{m=n}^{\infty} \mu(A_m) \leqq \sum_{m=n}^{\infty} 2^{-m} = 2^{-n+1}, \qquad \text{for all } n \in \mathbf{N},$$

that is, $\mu(A) = 0$, and on the other hand, because ν is finite,

$$\nu(A) \geqq \limsup \nu(A_n) \geqq \epsilon > 0$$

according to Corollary 2.7.2. Hence the measure ν is not μ-continuous. ┘

[14] This is often denoted by $\nu \ll \mu$.

Examples

2. Let Ω be an uncountable set and $\mathfrak{A}$ the σ-algebra of all sets $A \subset \Omega$ for which either A or $\complement A$ is countable. As in Example 1, we consider the measure ν on $\mathfrak{A}$ with $\nu(A) = 0$ or $= +\infty$ depending on whether A is countable or not. Let μ be the counting measure on $\mathfrak{A}$ for which $\mu(A)$ is equal to the number of elements of A for finite sets $A \in \mathfrak{A}$ and is otherwise equal to $+\infty$. Then $\emptyset$ is the only μ-null set and hence ν is trivially μ-continuous.

However, ν cannot possess a density relative to μ. From $\nu = f\mu$ with $f \in \mathcal{E}^*$ it would then follow for every $\omega \in \Omega$ that

$$0 = \nu(\{\omega\}) = \int_{\{\omega\}} f\,d\mu = f(\omega)\mu(\{\omega\}) = f(\omega).$$

But then we would have $\nu = 0$, which is false.

3. Let $(\mathbf{R},\mathfrak{B}^1,\mu)$ be the 1-dimensional Lebesgue-Borel measure space, that is, $\mu = \lambda^1$, and let $\mathfrak{N}$ be the system of all μ-null sets. Then $\mathfrak{N}$ is a so-called *σ-ideal* in $\mathfrak{B}^1$, that is, the union of each sequence of sets of $\mathfrak{N}$ lies in $\mathfrak{N}$ and along with every set of $\mathfrak{N}$, each of its Borel subsets also belongs to $\mathfrak{N}$. On the basis of these properties,

$$\nu(A) \equiv \begin{cases} 0, & \text{if } A \in \mathfrak{N} \\ +\infty, & \text{if } A \in \mathfrak{B}^1 \setminus \mathfrak{N} \end{cases}$$

defines a measure on $\mathfrak{B}^1$. This is μ-continuous by construction. However, the condition in Theorem 2.9.6 is not satisfied. For every $\delta > 0$,

$$\mu([0,\delta[) = \delta \qquad \text{and} \qquad \nu([0,\delta[) = +\infty.$$

Thus the hypothesis of finiteness of ν is not superfluous in Theorem 2.9.6. Example 2 tells us that the μ-continuity of ν is a necessary but not a sufficient condition for the existence of a density $f \in \mathcal{E}^*$ with $\nu = f\mu$. The following theorem of Radon-Nikodym is thus the more noteworthy; we precede its proof with a lemma.

2.9.7. Lemma. Let σ and τ be two finite measures on a σ-algebra $\mathfrak{A}$ in Ω with $\sigma(\Omega) < \tau(\Omega)$. Then there is a set $\Omega' \in \mathfrak{A}$ with the following properties:

(a) $\quad \sigma(\Omega') < \tau(\Omega')$;

(b) $\quad \sigma(A) \leqq \tau(A), \qquad$ for all $A \in \Omega' \cap \mathfrak{A}$.

Proof. The function $\delta \equiv \tau - \sigma$ is bounded on $\mathfrak{A}$ since $-\sigma(\Omega) \leqq \delta(A) \leqq \tau(\Omega)$ for all $A \in \mathfrak{A}$ and the measures σ and τ are finite. Therefore, using induction, we can define sequences $(A_n)_{n=0,1,\ldots}$ and $(\Omega_n)_{n=0,1,\ldots}$ in $\mathfrak{A}$ as follows: Let $A_0 \equiv \emptyset$ and $\Omega_0 \equiv \Omega = \Omega \setminus A_0$. If $A_0, \ldots, A_n$ and $\Omega_0, \ldots, \Omega_n$ have already been defined, we set

$$\alpha_n \equiv \inf_{A \in \Omega_n \cap \mathfrak{A}} \delta(A)$$

and proceed as follows: In the case $\alpha_n \geqq 0$ we set $A_{n+1} \equiv \emptyset$ and $\Omega_{n+1} \equiv \Omega_n = \Omega_n \setminus A_{n+1}$. If $\alpha_n < 0$, we choose A_{n+1} in $\Omega_n \cap A$ such that $\delta(A_{n+1}) \leqq \frac{1}{2}\alpha_n$, and again set $\Omega_{n+1} \equiv \Omega_n \setminus A_{n+1}$. The sequence (A_n) then consists of pairwise disjoint sets. Hence the series $\sum_{n=0}^{\infty} \delta(A_n)$ is convergent [to $\tau\left(\bigcup A_n\right) - \sigma\left(\bigcup A_n\right)$], whence $\lim_{n\to\infty} \delta(A_n) = 0$ and hence obviously $\lim_{n\to\infty} \alpha_n = 0$. If we now set

$$\Omega' \equiv \bigcup_{n=0}^{\infty} \Omega_n,$$

then from the antitonicity of (Ω_n) and the continuity properties of the measures σ and τ, it follows that

$$\delta(\Omega') = \lim_{n\to\infty} \delta(\Omega_n).$$

This set Ω' in $\mathfrak{A}$ yields the desired result. Indeed $\delta(A_{n+1}) \leqq 0$ for all n and hence

$$\delta(\Omega_{n+1}) = \delta(\Omega_n) - \delta(A_{n+1}) \geqq \delta(\Omega_n) \geqq \delta(\Omega_{n-1}) \geqq \cdots \geqq \delta(\Omega_0) = \delta(\Omega) > 0,$$

which implies

$$\delta(\Omega') = \lim \delta(\Omega_n) > 0.$$

But this is property (a).

To verify (b), let A in $\Omega' \cap \mathfrak{A}$ be chosen arbitrarily. Then $A \in \Omega_n \cap \mathfrak{A}$ and thus $\delta(A) \geqq \alpha_n$ for all n. Recalling the convergence of (α_n) to zero, it follows that $\delta(A) \geqq 0$. ┘

Now we are able to prove:

2.9.8. Theorem of Radon-Nikodym. Let μ and ν be measures on a σ-algebra $\mathfrak{A}$ in a set Ω. Then if μ is σ-finite, the following assertions are equivalent:

I. ν possesses a density relative to μ.
II. ν is μ-continuous.

Proof. Only the implication "II $\Rightarrow$ I" is still to be proved. We distinguish three cases.

FIRST CASE: The measures μ and ν are finite. Then we consider the set $\mathcal{G}$ of all $\mathfrak{A}$-measurable numerical functions $g \geqq 0$ on Ω which satisfy the condition

$$g\mu \leqq \nu$$

[that is, $\int_A g\,d\mu \leqq \nu(A)$ for all $A \in \mathfrak{A}$]. The constant function $g = 0$ obviously lies in $\mathcal{G}$, so that $\mathcal{G}$ is nonempty. Moreover, $\mathcal{G}$ is sup-stable, that is, for every two functions $g, h \in \mathcal{G}$, $\sup(g,h)$ also lies in $\mathcal{G}$. To see this, consider the sets $A_1 \equiv \{g \geqq h\}$ and $A_2 \equiv \complement A_1$. But then for every $A \in \mathfrak{A}$ we obtain the desired inequality

$$\int_A \sup(g,h)\,d\mu = \int_{A\cap A_1} g\,d\mu + \int_{A\cap A_2} h\,d\mu$$
$$\leqq \nu(A \cap A_1) + \nu(A \cap A_2) = \nu(A).$$

Now set

$$\gamma \equiv \sup_{g\in\mathcal{G}} \int g\,d\mu;$$

then $\gamma < +\infty$ since $\int g\,d\mu \leqq \nu(\Omega) < +\infty$ for all $g \in \mathcal{G}$; and there is a sequence (g_n^*) in $\mathcal{G}$ with $\lim\limits_{n\to\infty} \int g_n^*\,d\mu = \gamma$. Then by the above the sequence of functions $g_n \equiv \sup(g_1^*, \ldots, g_n^*)$ also lies in $\mathcal{G}$; since $g_n^* \leqq g_n$, we have the inequalities $\int g_n^*\,d\mu \leqq \int g_n\,d\mu$ $(n = 1, 2, \ldots)$ which imply that $\lim \int g_n\,d\mu = \gamma$. Since the sequence (g_n) is isotone, we apply the Monotone Convergence Theorem. Accordingly, $f \equiv \sup g_n$ is a function in $\mathcal{G}$ with $\int f\,d\mu = \gamma$. Thus, we have shown that the function $g \to \int g\,d\mu$ attains its maximum on $\mathcal{G}$ at f.

We show that $\nu = f\mu$. We have $f\mu \leqq \nu$. Hence,

$$\tau \equiv \nu - f\mu$$

is a finite measure on $\mathfrak{A}$ which is μ-continuous by hypothesis. We must show $\tau = 0$. Suppose that $\tau(\Omega) > 0$ holds. Because of the μ-continuity of τ we also have $\mu(\Omega) > 0$, that is,

$$0 < \beta \equiv \frac{1}{2} \cdot \frac{\tau(\Omega)}{\mu(\Omega)} < +\infty$$

and thus $\tau(\Omega) = 2\beta\mu(\Omega) > \beta\mu(\Omega)$. Now the above lemma (for τ and $\sigma = \beta\mu$) yields the existence of a set $\Omega' \in \mathfrak{A}$ with $\tau(\Omega') > \beta\mu(\Omega')$ and $\tau(A) \geqq \beta\mu(A)$ for all $A \in \Omega' \cap \mathfrak{A}$. For the $\mathfrak{A}$-measurable function $f_0 = f + \beta 1_{\Omega'}$ we then have

$$\int_A f_0\,d\mu = \int_A f\,d\mu + \beta\mu(A \cap \Omega') \leqq \int_A f\,d\mu + \tau(A) = \nu(A)$$

for arbitrary $A \in \mathfrak{A}$. Therefore, f_0 is a function in $\mathcal{G}$. For f_0 we have

$$\int f_0\,d\mu = \int f\,d\mu + \beta\mu(\Omega') = \gamma + \beta\mu(\Omega') > \gamma,$$

since $\mu(\Omega')$ must be positive because of $\tau(\Omega') > \beta\mu(\Omega')$ and the μ-continuity of τ. But the inequality $\int f_0\, d\mu > \gamma$ contradicts the definition of γ. Therefore, the assumption $\tau(\Omega) > 0$ leads to a contradiction.

SECOND CASE: Let $\mu(\Omega) < \nu(\Omega) = +\infty$. Then, as we shall show, there is a partition $\Omega = \bigcup_{n=0}^{\infty} \Omega_n$ of Ω into pairwise disjoint sets $\Omega_0, \Omega_1, \ldots \in \mathfrak{A}$ with the following properties:

(i) For every $A \in \Omega_0 \cap \mathfrak{A}$, either $\mu(A) = \nu(A) = 0$ or $\mu(A) > 0$ and $\nu(A) = \infty$.

(ii) $$\nu(\Omega_n) < \infty, \qquad \text{for all } n = 1, 2, \ldots .$$

To this end let $\mathfrak{Q}$ be the system of all $Q \in \mathfrak{A}$ with $\nu(Q) < \infty$, and $\alpha \equiv \sup_{Q \in \mathfrak{Q}} \mu(Q)$. Thus there is a sequence $(Q_m)_{m=1,2,\ldots}$ in $\mathfrak{Q}$ with $\alpha = \lim_{m\to\infty} \mu(Q_m)$. Since $\mathfrak{Q}$ is $\cup$-stable, we can assume that (Q_m) is isotone. Then $Q_0 \equiv \bigcup_{m=1}^{\infty} Q_m$ is a set in $\mathfrak{A}$ with $\mu(Q_0) = \alpha$, and we shall now show that $\Omega_0 \equiv \complement Q_0$ has property (i). Let A be a set in $\Omega_0 \cap \mathfrak{A}$ with $\nu(A) < \infty$. Then $Q_m \cup A \in \mathfrak{Q}$ and therefore $\mu(Q_m \cup A) \leqq \alpha$ for all m, and thus

$$\mu(Q_0 \cup A) = \lim \mu(Q_m \cup A) \leqq \alpha.$$

Now A is disjoint from Q_0, that is, $\mu(Q_0 \cup A) = \alpha + \mu(A) \leqq \alpha$, and it follows that $\mu(A) = 0$. This and the μ-continuity of ν yield the alternative (i). Finally, we set $\Omega_1 \equiv Q_1, \Omega_m \equiv Q_m \setminus Q_1 \cup \cdots \cup Q_{m-1}$ $(m = 2, 3, \ldots)$ in order to obtain a partition of Ω with the desired properties.

Now letting μ_n and ν_n denote the restrictions of μ and ν respectively to the σ-algebra $\Omega_n \cap \mathfrak{A}$ in Ω_n, we see that ν_n is a μ_n-continuous measure on $\Omega_n \cap \mathfrak{A}$ for all $n = 0, 1, \ldots$. μ_n and ν_n are finite for all $n \geqq 1$. According to the first case there is an $\Omega_n \cap \mathfrak{A}$-measurable function $f_n \geqq 0$ on Ω_n with $\nu_n = f_n\mu_n$. By (i) we have $\nu_0 = f_0\mu_0$, where f_0 denotes the constant function $+\infty$ on Ω_0. A simple "putting together of all the pieces" now yields the result: $\nu = f\mu$, where the $\mathfrak{A}$-measurable function $f \geqq 0$ on Ω is so defined that it coincides with f_n on each of the sets Ω_n.

THIRD CASE: Now we require only the σ-finiteness of μ. Then according to Lemma 1.5.8, there is a partition $\Omega = \bigcup_{n=1}^{\infty} \Omega_n$ of Ω into pairwise disjoint sets $\Omega_n \in \mathfrak{A}$ with $\mu(\Omega_n) < +\infty$. Thus, on each of the measurable spaces $(\Omega_n, \Omega_n \cap \mathfrak{A})$, the restrictions μ_n and ν_n of μ and ν respectively to $\Omega_n \cap \mathfrak{A}$ are covered by Cases 1 and 2, that is, $\nu_n = f_n\mu_n$ with a suitable $(\Omega_n \cap \mathfrak{A})$-measurable function $f_n \geqq 0$ on Ω_n. Combining all the pieces again finally yields the assertion. ┘

The question arises of whether, in the situation of Theorem 2.9.8, the density f of ν relative to μ is μ-almost everywhere uniquely determined. By Theorem 2.9.4 this is true if f is μ-integrable, that is, if ν is a finite measure. More generally we can prove:

2.9.9. Theorem. Let $\nu = f\mu$ be a measure with density f relative to a σ-finite measure μ on $\mathfrak{A}$. Then f is μ-almost everywhere uniquely determined. Moreover ν is σ-finite if and only if f is μ-almost everywhere finite.

Proof. To treat both problems, it suffices to deal with the case of a finite measure μ. In the general case we may partition Ω into a sequence of pairwise disjoint sets $\Omega_n \in \mathfrak{A}$ with $\mu(\Omega_n) < \infty$ and by restricting the measures μ and ν to $\Omega_n \cap \mathfrak{A}$, we reduce the general case back to this particular one. Thus let $\mu(\Omega) < \infty$.

According to Theorem 2.9.4, we can also assume $\nu(\Omega) = \infty$ for the proof of the first assertion. But then the proof of the Radon-Nikodym Theorem (second case) tells us that it suffices to assume that μ and ν satisfy the following alternatives: For every $A \in \mathfrak{A}$, either $\mu(A) = \nu(A) = 0$ or $\mu(A) > 0$ and $\nu(A) = \infty$. Then the constant function $+\infty$ is a density of ν relative to μ. Thus, we must show that $f = +\infty$ μ-almost everywhere. To this end it suffices to show that for every $n = 1, 2, \ldots$,

$$\mu(\{f \leqq n\}) = 0.$$

But we have

$$\nu(\{f \leqq n\}) = \int_{\{f \leqq n\}} f\,d\mu \leqq n\mu(\{f \leqq n\}),$$

and therefore from the alternative we must have $\mu(\{f \leqq n\}) = 0$.

If ν is finite, then f is μ-integrable, that is, μ-almost everywhere finite. The same also holds then in the σ-finite case; we again partition Ω into a sequence of pairwise disjoint sets in $\mathfrak{A}$ each of finite ν-measure. Conversely, if f is μ-almost everywhere finite, then ν is σ-finite. To see this let $\Omega = \bigcup_{n=0}^{\infty} \Omega_n$ be a partition of Ω into a sequence of pairwise disjoint sets of $\mathfrak{A}$ with $\mu(\Omega_n) < \infty$. Further, let $A_n \equiv \{n - 1 \leqq f < n\}$ for every $n = 1, 2, \ldots$ and $A_0 \equiv \{f = \infty\}$. Then A_0 has μ-measure zero by hypothesis. But then $\Omega = \bigcup_{i,j=0}^{\infty} \Omega_i \cap A_j$ is a partition of Ω into a sequence of pairwise disjoint sets in $\mathfrak{A}$ with $\nu(\Omega_i \cap A_j) < \infty$. Indeed, $\nu(\Omega_i \cap A_0) = 0$ and $\nu(\Omega_i \cap A_j) \leqq j\mu(\Omega_i) < \infty$ for all $i = 0, 1, \ldots$ and $j = 1, 2, \ldots$. ⌟

Theorems 2.9.8 and 2.9.9 now justify calling the density of a continuous measure ν relative to a σ-finite measure μ *the Radon-Nikodym integrand* or the Radon-Nikodym *density* of ν relative to μ.

PROBLEMS

1. Prove that the measure ϵ_x on $\mathfrak{B}^p$ for a point $x \in \mathbf{R}^p$ cannot have a density with respect to λ^p. (Physicists sometimes use the existence of such a density d_x and call d_x the "Dirac function" for x. The mathematical object they really have in mind is ϵ_x. Therefore ϵ_x is also called *Dirac measure* for x.)
2. Let ρ, σ, τ be measures on a σ-algebra $\mathfrak{A}$. Assume $\rho = f\sigma$ and $\sigma = g\tau$ with densities $f, g \in \mathcal{E}^*$. Prove that ρ has fg as density with respect to τ. (For σ-finite measures σ and τ, this can be written in the form $d\rho/d\tau = d\rho/d\sigma \cdot d\sigma/d\tau$, where $d\nu/d\mu$ denotes the Radon-Nikodym integrand of a measure ν with respect to a σ-finite measure μ.)
3. Prove that the relation $\ll$ is reflexive and transitive on the set of measures on a σ-algebra $\mathfrak{A}$. Define $\mu \sim \nu$ to mean $\mu \ll \nu$ and $\nu \ll \mu$. (That is, μ and ν have the same null sets.) Prove that $\sim$ is an equivalence relation. Now let μ and ν be σ-finite measures on $\mathfrak{A}$. Prove: $\mu \sim \nu$ if and only if $\mu = f\nu$, where the density f satisfies $0 < f(\omega) < \infty$ ν-almost everywhere (or even everywhere).
4. Let μ be a σ-finite measure on a σ-algebra $\mathfrak{A}$. Prove the existence of a finite measure ν on $\mathfrak{A}$ having the same null sets as μ.
5. Let $(\Omega,\mathfrak{A},\mu)$ be a measure space, and consider an $\mathfrak{A}$-$\mathfrak{A}$-measurable mapping $T: \Omega \to \Omega$. Prove: $T(\mu) \ll \mu$ if and only if $T^{-1}(A)$ is a μ-null set for every μ-null set A.

2.10 INTEGRATION WITH RESPECT TO IMAGE MEASURES

Along with the measure space $(\Omega,\mathfrak{A},\mu)$ suppose we are also given a measurable space $(\Omega',\mathfrak{A}')$ and an $\mathfrak{A}$-$\mathfrak{A}'$-measurable mapping

$$T: \Omega \to \Omega'.$$

Then the image measure

$$\mu' \equiv T(\mu)$$

is defined according to (1.7.5). To show the connection between μ- and μ'-integrals, we prove:

2.10.1. Theorem. For every $\mathfrak{A}'$-measurable numerical function $f' \geqq 0$ on Ω',

$$\int f' \, dT(\mu) = \int f' \circ T \, d\mu. \tag{2.10.1}$$

Proof. By Theorem 1.7.3, the function $f' \circ T$ is $\mathfrak{A}$-measurable and $\geqq 0$. The integral on the right side of (2.10.1) is thus defined. To prove the equality (2.10.1) we choose as f' an $\mathfrak{A}'$-elementary function

$$f' = \sum_{i=1}^{n} \alpha_i 1_{A'_i}$$

(with coefficients $\alpha_i \in \mathbf{R}_+$ and sets $A'_i \in \mathfrak{A}'$). Then obviously

$$f' \circ T = \sum_{i=1}^{n} \alpha_i 1_{A_i}$$

with the sets $A_i \equiv T^{-1}(A'_i)$ is an $\mathfrak{A}$-elementary function. Since, by definition,

$$T(\mu)(A'_i) = \mu(A_i) \qquad \text{for } i = 1, \ldots, n,$$

(2.10.1) is satisfied. For arbitrary $\mathfrak{A}'$-measurable $f' \geqq 0$ we choose an isotone sequence (u'_n) of $\mathfrak{A}'$-elementary functions with $\sup u'_n = f'$. Then $(u'_n \circ T)$ is an isotone sequence of $\mathfrak{A}$-elementary functions with

$$\sup u'_n \circ T = f' \circ T.$$

Then (2.10.1) also follows for this case from the definition of the integral. ⌟

2.10.2. Corollary. Let f' be an $\mathfrak{A}'$-measurable numerical function on Ω'. Then the $T(\mu)$-integrability of f' implies the μ-integrability of $f' \circ T$ and conversely. In this case we also have

$$\int f' \, dT(\mu) = \int f' \circ T \, d\mu.$$

Proof. By Theorem 2.10.1,

$$\int f'^+ \, dT(\mu) = \int f'^+ \circ T \, d\mu \qquad \text{and} \qquad \int f'^- \, dT(\mu) = \int f'^- \circ T \, d\mu;$$

further, it is obvious that

$$(f \circ T)^+ = f'^+ \circ T \qquad \text{and} \qquad (f' \circ T)^- = f'^- \circ T.$$

Hence the assertion follows directly from the definition of the integral. ⌟

The assertions of Theorem 2.10.1 and Corollary 2.10.2 are called the "*transformation theorem* for integrals."

Finally, we have the almost trivial but useful theorem:

2.10.3. Theorem. Let $(\Omega,\mathfrak{A},\mu)$ be a measure space and $T: \Omega \to \Omega$ a bijection of Ω onto itself which together with its inverse is $\mathfrak{A}$-measurable. Then for every $f \in \mathcal{E}^*$:

$$T(f\mu) = (f \circ T^{-1})T(\mu). \tag{2.10.2}$$

Proof. If we use the abbreviations

$$\nu \equiv T(f\mu) \qquad \text{and} \qquad \mu' \equiv T(\mu),$$

then

$$\begin{aligned}\nu(A) &= \int_{T^{-1}(A)} f\,d\mu = \int f(1_A \circ T)\,d\mu = \int (f \circ T^{-1} \circ T)(1_A \circ T)\,d\mu \\ &= \int [(f \circ T^{-1})1_A] \circ T\,d\mu = \int (f \circ T^{-1})1_A\,d\mu' = \int_A f \circ T^{-1}\,d\mu'\end{aligned}$$

for every $A \in \mathfrak{A}$, that is, $\nu = (f \circ T^{-1})\mu'$. ⌟

PROBLEMS

1. Let $T\colon \mathbf{R}^p \to \mathbf{R}^p$ be an orthogonal transformation of $\mathbf{R}^p$. Prove: A function f on $\mathbf{R}^p$ is λ^p-integrable if and only if $f \circ T$ is λ^p-integrable; furthermore

$$\int_A f\,d\lambda^p = \int_{T^{-1}(A)} f \circ T\,d\lambda^p, \qquad \text{for all } A \in \mathfrak{B}^p.$$

2. Let $(\Omega,\mathfrak{A},\mu)$ be a σ-finite measure space and $T\colon \Omega \to \Omega$ an $\mathfrak{A}$-$\mathfrak{A}$-measurable mapping such that $T^{-1}(A)$ is a μ-null set for each μ-null set A. Prove the existence of a measurable function $q \geqq 0$ such that

$$\int_A f\,d\mu = \int_{T^{-1}(A)} (f \circ T)q\,d\mu$$

holds for all $f \in \mathcal{L}^1(\mu)$ and all $A \in \mathfrak{A}$.

2.11 STOCHASTIC CONVERGENCE

We return to the study of L^p-functions begun in Section 2.6. Our goal is to replace the concept of convergence almost everywhere of a sequence of functions, so important for Theorem 2.7.4 of dominated convergence and for Theorem 2.7.5, by a weaker notion of convergence. This new concept of convergence will be suggested by a simple but very useful inequality.

The arguments again involve an arbitrary *measure space* $(\Omega,\mathfrak{A},\mu)$.

2.11.1. Lemma. For every measurable numerical function f on Ω and every pair of real numbers $p > 0$, $\alpha > 0$, we have the *Chebyshev-Markov inequality*

$$\mu(\{|f| \geqq \alpha\}) \leqq \frac{1}{\alpha^p} \int |f|^p\,d\mu. \tag{2.11.1}$$

Proof. If we set $A_\alpha \equiv \{|f| \geqq \alpha\}$, then A_α lies in $\mathfrak{A}$ and

$$\int |f|^p \, d\mu \geqq \int |f|^p 1_{A_\alpha} \, d\mu \geqq \alpha^p \int 1_{A_\alpha} \, d\mu = \alpha^p \mu(A_\alpha).$$

From this (2.11.1) follows. ┘

When the integral $\int |f|^p \, d\mu$ is finite, for $p \geqq 1$ this just says that f is an L^p-function, we have from (2.11.1)

$$\lim_{\alpha \to +\infty} \mu(\{|f| \geqq \alpha\}) = 0. \tag{2.11.2}$$

On the basis of this result, we give:

2.11.2. Definition. A sequence $(f_n)_{n \in \mathbf{N}}$ of measurable real functions on Ω is said to be (μ-)*stochastically convergent* (or convergent in measure) to a measurable real function f on Ω if for every real number $\alpha > 0$ and every set $A \in \mathfrak{A}$ of finite measure $\mu(A)$ we have

$$\lim_{n \to \infty} \mu(\{|f_n - f| \geqq \alpha\} \cap A) = 0. \tag{2.11.3}$$

Then we also write

$$\mu\text{-}\lim_{n \to \infty} f_n = f \tag{2.11.4}$$

and we call f the (μ-)*stochastic limit* of the sequence (f_n).

Remarks. 1. For a *finite* measure μ we may set $A = \Omega$ in (2.11.3). Then the stochastic convergence of (f_n) to f is equivalent to the requirement

$$\lim_{n \to \infty} \mu(\{|f_n - f| \geqq \alpha\}) = 0, \qquad \text{for all } \alpha > 0. \tag{2.11.5}$$

The more complicated requirement (2.11.3) is needed to include infinite, and in particular σ-finite, measures.

2. If f^* is a measurable real function on Ω and $f = f^*$ μ-almost everywhere *on every set* $A \in \mathfrak{A}$ *of finite measure*, then f^* is also a stochastic limit of the sequence (f_n) of Definition 2.11.2. The sets $\{|f_n - f| \geqq \alpha\} \cap A$ and $\{|f_n - f^*| \geqq \alpha\} \cap A$ then differ only by a null set independent of n. Conversely, if f and f^* are stochastic limits of the same sequence (f_n), then by the triangle inequality

$$\{|f - f^*| \geqq \alpha\} \subset \left\{|f_n - f| \geqq \frac{\alpha}{2}\right\} \cup \left\{|f_n - f^*| \geqq \frac{\alpha}{2}\right\},$$

that is,

$$\mu(\{|f - f^*| \geqq \alpha\} \cap A) \leqq \mu\left(\left\{|f_n - f| \geqq \frac{\alpha}{2}\right\} \cap A\right) + \mu\left(\left\{|f_n - f^*| \geqq \frac{\alpha}{2}\right\} \cap A\right).$$

By passing to the limit $n \to \infty$ we obtain

$$\mu(\{|f - f^*| \geqq \alpha\} \cap A) = 0,$$

for all $\alpha > 0$ and all sets $A \in \mathfrak{A}$ of finite measure. But then $f = f^*$ μ-almost everywhere on each such set A, since

$$\{f \neq f^*\} \cap A = \bigcup_{k=1}^{\infty} \left\{|f - f^*| \geqq \frac{1}{k}\right\} \cap A$$

is a μ-null set. Thus the stochastic limit is μ-almost everywhere uniquely determined for every *σ-finite measure* μ.

3. Two stochastic limits f and f^* of a sequence (f_n) are almost everywhere equal even for an arbitrary measure μ provided both functions f and f^* are in $\mathfrak{L}^p(\mu)$ for some $p \in [1, \infty[$. Indeed, by (2.11.1), $\{|f - f^*| \geqq \alpha\}$ is a set of finite measure for every $\alpha > 0$, that is, $f = f^*$ μ-almost everywhere on $\{|f - f^*| \geqq \alpha\}$ and hence on $\{|f - f^*| > 0\} = \bigcup_{n=1}^{\infty} \{|f - f^*| \geqq 1/n\}$. But then $f = f^*$ μ-almost everywhere on Ω.

Example

1. Let $\Omega = \{\omega_0, \omega_1, \ldots\}$ be a countably infinite set, $\mathfrak{A} = \mathfrak{P}(\Omega)$ the σ-algebra of all subsets of Ω, and μ the measure on $\mathfrak{A}$ uniquely determined by the requirement

$$\mu(\{\omega_0\}) \equiv +\infty, \ \mu(\{\omega_n\}) = 2^{-n} \qquad (n = 1, 2, \ldots).$$

Then μ is not σ-finite. We shall show that the stochastic limit of a sequence (f_n) of real (and thus $\mathfrak{A}$-measurable) functions on Ω is not μ-almost everywhere uniquely determined. It suffices to set

$$f_n(\omega_i) \equiv \begin{cases} n, & i = 0 \\ 0, & i = 1, \ldots, n \\ 1, & i = n+1, n+2, \ldots \end{cases} \qquad (n \in \mathbf{N}).$$

Then obviously all stochastic limits of this sequence are given by

$$f^\alpha(\omega_i) \equiv \begin{cases} \alpha, & i = 0 \\ 0, & i = 1, 2, \ldots, \end{cases}$$

with $\alpha \in \mathbf{R}$. But for $\alpha \neq \beta$ we never have $f^\alpha = f^\beta$ μ-almost everywhere.

The following considerations lead to an important class of stochastic convergent sequences:

2.11.3. Theorem. If a sequence $(f_n)_{n \in \mathbf{N}}$ of functions in $\mathfrak{L}^p(\mu)$ converges to a function $f \in \mathfrak{L}^p(\mu)$ in pth mean ($1 \leqq p < +\infty$), then the sequence (f_n) also converges μ-stochastically to f.

Proof. By the Chebyshev-Markov inequality, for every set $A \in \mathfrak{A}$, every $\alpha > 0$ and every natural number n,

$$\mu(\{|f_n - f| \geqq \alpha\} \cap A) \leqq \mu(\{|f_n - f| \geqq \alpha\}) \leqq \frac{1}{\alpha^p} \int |f_n - f|^p \, d\mu.$$

From this the result follows, since by definition of convergence in pth mean we have

$$\lim_{n \to \infty} \int |f_n - f|^p \, d\mu = 0. \quad \lrcorner$$

The following theorem shows that almost everywhere convergent sequences always converge stochastically.

2.11.4. Theorem. If a sequence $(f_n)_{n \in \mathbf{N}}$ of measurable real functions on Ω converges μ-almost everywhere to a measurable real function f on Ω, then the sequence (f_n) also converges μ-stochastically to f.

Proof. We have

$$\{|f_n - f| \geqq \alpha\} \subset \{\sup_{m \geqq n} |f_m - f| \geqq \alpha\}$$

and thus

$$\mu(\{|f_n - f| \geqq \alpha\} \cap A) \leqq \mu(\{\sup_{m \geqq n} |f_m - f| \geqq \alpha\} \cap A)$$

for all $\alpha > 0$ and $A \in \mathfrak{A}$. The theorem then follows from the next lemma, since $B \to \mu(B \cap A)$ is a finite and μ-continuous measure on $\mathfrak{A}$ when $\mu(A) < \infty$. $\lrcorner$[15]

2.11.5. Lemma. If the measure μ is finite, then a sequence $(f_n)_{n \in \mathbf{N}}$ of measurable real functions on Ω converges μ-almost everywhere to 0 if and only if for every real $\alpha > 0$,

$$\lim_{n \to \infty} \mu(\{\sup_{m \geqq n} |f_m| \geqq \alpha\}) = 0. \tag{2.11.6}$$

Proof. For every $\alpha > 0$ and every $n \in \mathbf{N}$, we set

$$A_n^\alpha \equiv \{\sup_{m \geqq n} |f_m| \geqq \alpha\}.$$

Then $n \to A_n^\alpha$ and $\alpha \to A_n^\alpha$ are obviously antitone mappings. Consequently, $k \to A_n^{1/k} (k \in \mathbf{N})$ is isotone. If we now set

$$A \equiv \{\omega \in \Omega : \lim_{n \to \infty} f_n(\omega) = 0\},$$

[15] This conclusion shows that it suffices to require the almost everywhere convergence of (f_n) to f only on every set $A \in \mathfrak{A}$ of finite measure.

then this set is in $\mathfrak{A}$ since

$$A = \bigcap_{k=1}^{\infty} \bigcup_{n=1}^{\infty} \complement A_n^{1/k} = \bigcap_{\alpha>0} \bigcup_{n=1}^{\infty} \complement A_n^{\alpha}.$$

By passing to complements we obtain

$$\complement A = \bigcup_{k=1}^{\infty} \bigcap_{n=1}^{\infty} A_n^{1/k} = \bigcup_{\alpha>0} \bigcap_{n=1}^{\infty} A_n^{\alpha}. \tag{2.11.7}$$

Hence it follows that $\bigcap_{n=1}^{\infty} A_n^{\alpha} \subset \complement A$ and because μ is finite we have

$$\lim_{n\to\infty} \mu(A_n^{\alpha}) = \mu\left(\bigcap_{n=1}^{\infty} A_n^{\alpha}\right) \leqq \mu(\complement A).$$

If (f_n) converges almost everywhere to 0, $\mu(\complement A) = 0$ and (2.11.6) is satisfied for every $\alpha > 0$. But by using the monotonicity properties of A_n^{α}, it follows from (2.11.7) that

$$\mu(\complement A) = \lim_{k\to\infty} \mu\left(\bigcap_{n=1}^{\infty} A_n^{1/k}\right) = \lim_{k\to\infty} \lim_{n\to\infty} \mu(A_n^{1/k}).$$

Hence, if conversely (2.11.6) is satisfied for all $\alpha > 0$, and in particular for $\alpha = 1/k$, $k = 1, 2, \ldots$, then it follows that $\mu(\complement A) = 0$, that is, we have convergence almost everywhere to 0. ⌟

The conditions for stochastic convergence in the two theorems above are sufficient, but not necessary. This is shown by the following examples.

Examples

2. Let $\Omega \equiv [0,1]$, $\mathfrak{A} \equiv \Omega \cap \mathfrak{B}^1$, and $\mu \equiv \lambda_\Omega^1$. Then μ is a finite measure. The sequence $(n \cdot 1_{A_n})_{n\in\mathbf{N}}$ with $A_n \equiv]0,1/n[$ converges pointwise to 0 on all of Ω. By Theorem 2.11.4 or because of

$$\mu(\{n \cdot 1_{A_n} \geqq \alpha\}) = \mu(A_n) = \frac{1}{n} \qquad (0 < \alpha \leqq n)$$

it also converges stochastically to 0. On the other hand, because

$$\int (n \cdot 1_{A_n})^p \, d\mu = n^p \mu(A_n) = n^{p-1},$$

the sequence does not converge in pth mean to 0 for any real $p \geqq 1$.

3. Let $(\Omega,\mathfrak{A},\mu)$ be the measure space of the preceding example. By the example of Section 2.7, the sequence $(f_n(\omega))_{n\in\mathbf{N}}$ with $f_n \equiv 1_{A_n}$ and

$$A_n \equiv [k2^{-h},(k+1)2^{-h}],$$

where $n = 2^h + k$, $0 \leqq k < 2^h$, h and k integers $\geqq 0$, converges for no $\omega \in \Omega$. But the sequence (f_n) converges stochastically to 0 since for every $\alpha > 0$ and all $n \in \mathbf{N}$,

$$\mu(\{|f_n| \geqq \alpha\}) \leqq 2^{-h} \leqq \frac{2}{n}.$$

The relationship between stochastic convergence and almost everywhere convergence is, however, closer than we might expect on the basis of this last example.

2.11.6. Theorem. If the measure μ is σ-finite, then a sequence $(f_n)_{n\in\mathbf{N}}$ of measurable real functions converges stochastically to a measurable real function f if and only if from every subsequence of (f_n) we can extract a further subsequence which converges almost everywhere to f.

Proof. The above condition is sufficient for the stochastic convergence of the sequence (f_n) to f. Every subsequence $(f_{n_k})_{k\in\mathbf{N}}$ of $(f_n)_{n\in\mathbf{N}}$ contains a further subsequence which converges almost everywhere to f, hence, according to Theorem 2.11.4 also converges stochastically. Thus, for every $\alpha > 0$ and every set $A \in \mathfrak{A}$ with $\mu(A) < +\infty$, a subsequence converging to zero is contained in the sequence of numbers

$$(\mu(\{|f_{n_k} - f| \geqq \alpha\} \cap A))_{k\in\mathbf{N}}.$$

Since this is true for every subsequence (f_{n_k}),

$$(\mu(\{|f_n - f| \geqq \alpha\} \cap A))_{n\in\mathbf{N}}$$

must be a null sequence, that is, (f_n) must converge stochastically to f.[16] Conversely, we now assume that (f_n) converges stochastically to f. Then every subsequence (f_{n_k}) obviously also converges stochastically to f. Thus we will have shown the necessity of the above condition if we can choose a subsequence converging almost everywhere to f from every sequence (f_n) of measurable real functions which converges stochastically to f. We proceed as follows:

We first assume that the measure is *finite*. By the triangle inequality, for arbitrary $\alpha > 0$,

$$\{|f_m - f_n| \geqq \alpha\} \subset \left\{|f_m - f| \geqq \frac{\alpha}{2}\right\} \cup \left\{|f_n - f| \geqq \frac{\alpha}{2}\right\}$$

for all $m, n \in \mathbf{N}$; hence, because of the assumed stochastic convergence $\mu(\{|f_m - f_n| \geqq \alpha\})$ is arbitrarily small for sufficiently large m and n. Now if $(\eta_k)_{k\in\mathbf{N}}$ is a sequence of positive real numbers with $\sum_{k=1}^{\infty} \eta_k < +\infty$,

[16] The σ-finiteness of μ has not been used in this part of the proof.

then for each $k \in \mathbf{N}$ there is a natural number n_k with

$$\mu(\{|f_m - f_{n_k}| \geqq \eta_k\}) \leqq \eta_k, \qquad \text{for all } m \geqq n_k \text{ and } k \in \mathbf{N}.$$

We can assume without loss of generality that $n_k < n_{k+1}$ for all $k \in \mathbf{N}$. If we also set

$$A_k \equiv \{|f_{n_{k+1}} - f_{n_k}| \geqq \eta_k\},$$

then

$$\sum_{k=1}^{\infty} \mu(A_k) \leqq \sum_{k=1}^{\infty} \eta_k < +\infty$$

and hence

$$\lim_{n \to \infty} \sum_{k=n}^{\infty} \mu(A_k) = 0.$$

Let $A'_n \equiv \bigcup_{k=n}^{\infty} A_k$. Then because μ is finite and (A'_n) is an antitone sequence we have for

$$A \equiv \bigcap_{n=1}^{\infty} A'_n = \limsup_{n \to \infty} A_n$$

that

$$\mu(A) = \lim_{n \to \infty} \mu(A'_n) \leqq \lim_{n \to \infty} \sum_{k=n}^{\infty} \mu(A_k),$$

and hence,

$$\mu(A) = 0.$$

By the definition of A, we have for every $\omega \in \complement A$ the inequality

$$|f_{n_{k+1}}(\omega) - f_{n_k}(\omega)| \geqq \eta_k$$

for at most finitely many $k \in \mathbf{N}$; therefore, the series

$$\sum_{k=1}^{\infty} (f_{n_{k+1}}(\omega) - f_{n_k}(\omega))$$

is absolutely convergent since $\Sigma\eta_k < \infty$, and the sequence $(f_{n_k}(\omega))$ is convergent. Thus, the sequence $(f_{n_k})_{k \in \mathbf{N}}$ converges almost everywhere to a measurable real function f^* on Ω. Since by Theorem 2.11.4 f^* is also the stochastic limit of (f_{n_k}), and since (f_{n_k}) as a subsequence of (f_n) converges stochastically to f, by Remark 2 at the end of Definition 2.11.2 we have

$$f = f^* \text{ almost everywhere.}$$

Thus (f_{n_k}) converges almost everywhere to f.

Now let μ be an arbitrary σ-finite measure; then there is a sequence $(A_n)_{n \in \mathbf{N}}$ in $\mathfrak{A}$ with $A_n \uparrow \Omega$ and $\mu(A_n) < +\infty$ for all $n \in \mathbf{N}$. Every measure

μ_n on $\mathfrak{A}$ defined by

$$\mu_n(A) \equiv \mu(A \cap A_n) \qquad (A \in \mathfrak{A}) \tag{2.11.8}$$

is finite; since μ is continuous from below it follows that

$$\lim \mu_n(A) = \mu(A) \qquad (A \in \mathfrak{A}). \tag{2.11.9}$$

The sequence (f_n) converges μ-stochastically and from the definition (2.11.8), it converges μ_k-stochastically to f for every $k \in \mathbf{N}$. The corresponding result then holds for every subsequence of (f_n). According to the first part of the proof there is a subsequence $(f_n^{(1)})$ of (f_n) which converges μ_1-almost everywhere to f. Likewise, there is a subsequence $(f_n^{(2)})$ of $(f_n^{(1)})$ which converges μ_2-almost everywhere to f. In general, for every $k \in \mathbf{N}$, there is a subsequence $(f_n^{(k)})$ of $(f_n^{(k-1)})$ which converges μ_k-almost everywhere to f. Here $(f_n^{(0)})$ denotes the sequence (f_n). Then the diagonal sequence $(f_n^{(n)})_{n \in \mathbf{N}}$ is, except for finitely many terms, a subsequence of each sequence $(f_n^{(k)})_{n \in \mathbf{N}}$ and of the original sequence (f_n), and thus converges μ_k-almost everywhere to f for every $k \in \mathbf{N}$. But then $(f_n^{(n)})$ converges μ-almost everywhere to f, and hence is a subsequence of the original sequence (f_n) with the desired property. In fact, the set

$$Q \equiv \{\limsup_{n\to\infty} f_n^{(n)} \neq f\} \cup \{\liminf_{n\to\infty} f_n^{(n)} \neq f\}$$

is in $\mathfrak{A}$ and has μ_k-measure zero because of the μ_k-almost everywhere convergence of $(f_n^{(n)})$ to f. Thus from (2.11.9) we obtain $\mu(Q) = \lim \mu_k(Q) = 0$, that is, $\lim f_n^{(n)} = f$ holds μ-almost everywhere. ⌟

Remarks. 4. By combining two stochastically convergent sequences with limits that do not coincide almost everywhere, we can see immediately that the existence of *one* almost everywhere convergent subsequence does not imply the stochastic convergence of the original sequence.

5. The second part of the proof of Theorem 2.11.6 shows that a *Cauchy* criterion holds for *finite measures* μ for the stochastic convergence of a sequence (f_n): A necessary and sufficient condition for the stochastic convergence of a sequence (f_n) of measurable real functions on Ω is the condition

$$\lim_{m,n\to\infty} \mu(\{|f_m - f_n| \geqq \alpha\}) = 0, \tag{2.11.10}$$

for all $\alpha > 0$.

6. Example 1 shows that the conclusion of Theorem 2.11.6 fails for measures that are *not* σ-finite.

PROBLEMS

1. Let (f_n) and (g_n) be sequences of measurable real functions on a measure space which are stochastically convergent to f and g, respectively. Prove: $(\alpha f_n + \beta g_n)$ converges stochastically to $\alpha f + \beta g$ $(\alpha, \beta \in \mathbf{R})$, $(|f_n|)$ converges stochastically to $|f|$.
2. Let $(\Omega,\mathfrak{A},\mu)$ be a measure space with finite measure μ, and let d be the pseudometric on $\mathfrak{A}$ as defined in Section 1.3, Problem 5. Prove: A sequence (A_n) in $\mathfrak{A}$ is d-convergent to a set $A \in \mathfrak{A}$ if and only if the sequence (1_{A_n}) converges stochastically to 1_A.
3. Let f and g be measurable real functions on a measure space $(\Omega,\mathfrak{A},\mu)$ with finite measure μ. Define
$$d_\mu(f,g) := \inf\{\epsilon > 0 : \mu(\{|f-g| \geqq \epsilon\}) \leqq \epsilon\}$$
and prove:
   (a) d_μ is a pseudometric on the set $\mathfrak{M}$ of all measurable real-valued functions.
   (b) a sequence (f_n) in $\mathfrak{M}$ converges stochastically to $f \in \mathfrak{M}$ if and only if $d_\mu(f_n,f) \to 0$.
   (c) $\mathfrak{M}$ is complete with respect to d_μ, that is, every d_μ-Cauchy sequence is convergent. How is d_μ related to the pseudometric d of Problem 2?
4. Let $(\Omega,\mathfrak{A},\mu)$ again be a finite measure space. Prove that also
$$d_0(f,g) := \int \frac{|f-g|}{1+|f-g|}\, d\mu$$
defines a pseudometric on $\mathfrak{M}$, and that d_0 has the properties (a) to (c) of d_μ in the preceding problem.

2.12 UNIFORM INTEGRABILITY

The sufficient condition for convergence in pth mean in the dominated convergence theorem will now be extended into a necessary and sufficient condition with the help of stochastic convergence. For this we shall need the concept of uniform integrability. Below, let $(\Omega,\mathfrak{A},\mu)$ be an arbitrary measure space; further let p be a real number with $1 \leqq p < +\infty$.

We start with the following simple observation: Let f be a measurable numerical function on Ω. Then f is an L^p-function if and only if for every number $\epsilon > 0$ there is an L^p-function $g \geqq 0$ such that

$$\int_{\{|f| \geqq g\}} |f|^p \, d\mu \leqq \epsilon \tag{2.12.1}$$

In fact, if f is an L^p-function, we may choose $g \equiv 2|f|$. Then $\{|f| \geqq g\} = \{f = 0\} \cup \{|f| = +\infty\}$ and hence, according to Theorem 2.5.2, the integral in (2.12.1) is zero. Conversely, from (2.12.1) it follows that

$$\int |f|^p \, d\mu = \int_{\{|f| \geqq g\}} |f|^p \, d\mu + \int_{\{|f| < g\}} |f|^p \, d\mu \leqq \epsilon + \int g^p \, d\mu < +\infty,$$

that is, f is an L^p-function.

Motivated by this observation, we give:

2.12.1. Definition. *A set* $\mathfrak{F}$ *of* $\mathfrak{A}$*-measurable numerical functions on* Ω *is said to be uniformly* (μ-)*integrable of order* p *if for every real number* $\epsilon > 0$ *there exists an* L^p*-function* $g \geqq 0$ *on* Ω *such that*

$$\int_{\{|f| \geqq g\}} |f|^p \, d\mu \leqq \epsilon, \qquad \text{for all } f \in \mathfrak{F}. \tag{2.12.2}$$

When $p = 1$ we say simply that $\mathfrak{F}$ is uniformly integrable.

Correspondingly, a *family* $(f_i)_{i \in I}$ of measurable numerical functions on Ω is said to be uniformly integrable of order p if the set

$$\{f_i \colon i \in I\}$$

is uniformly integrable of order p.

By the observation above, every function f of a uniformly integrable set of order p is an L^p-function.

The L^p-function $g \geqq 0$ with property (2.12.2), which exists for every $\epsilon > 0$ according to Definition 2.12.1, will henceforth be called an *ϵ-bound of order p* for $\mathfrak{F}$. Along with g, every L^p-function $g^* \geqq g$ is obviously also an ϵ-bound of order p for $\mathfrak{F}$.

Examples

1. When finitely many sets $\mathfrak{F}_1, \ldots, \mathfrak{F}_n$ of measurable functions on Ω are all uniformly μ-integrable of order p, then

$$\mathfrak{F}_1 \cup \cdots \cup \mathfrak{F}_n$$

is also uniformly μ-integrable of order p. If g_i is an ϵ-bound of order p for $\mathfrak{F}_i$, $i = 1, \ldots, n$, then $\sup(g_1, \ldots, g_n)$ is an ϵ-bound of order p for $\mathfrak{F}_1 \cup \cdots \cup \mathfrak{F}_n$.

2. Every finite set $\{f_1, \ldots, f_n\}$ of L^p-functions is uniformly μ-integrable of order p. This follows from Example 1 since, by (2.12.1), every set $\{f\}$ consisting of only one L^p-function f is uniformly integrable of order p.

3. Every set $\mathfrak{F}$ of measurable numerical functions on Ω for which there exists an L^p-function $g \geqq 0$ with

$$|f| \leqq g \ \mu\text{-almost everywhere}, \qquad \text{for all } f \in \mathfrak{F}$$

is uniformly μ-integrable of order p. Every ϵ-bound h of order p for the set $\{g\}$ is an ϵ-bound of order p for $\mathfrak{F}$ because

$$\int_{\{|f| \geqq h\}} |f|^p \, d\mu \leqq \int_{\{g \geqq h\}} g^p \, d\mu \qquad (f \in \mathfrak{F}).$$

4. Let $(\Omega, \mathfrak{A}, \mu)$ be the measure space of Section 2.11, Example 2, and let $(f_n)_{n \in \mathbf{N}}$ be the sequence of functions $f_n = n \cdot 1_{A_n}$ considered there with $A_n =]0,1/n[$. Then this sequence is *not* uniformly integrable of order p $(1 \leqq p < +\infty)$. For every L^p-function $g \geqq 0$ and every $n \in \mathbf{N}$,

$$\int_{\{|f_n| \geqq g\}} |f_n|^p \, d\mu = \int_{A_n \cap \{n \geqq g\}} n^p \, d\mu = \int_{A_n} n^p \, d\mu - \int_{A_n \cap \{n < g\}} n^p \, d\mu$$
$$\geqq n^{p-1} - \int_{A_n} g^p \, d\mu.$$

Since $A_n \downarrow \varnothing$ and the measure $g^p \mu$ is finite, it follows that

$$\liminf_{n \to \infty} \int_{\{|f_n| \geqq g\}} |f_n|^p \, d\mu \geqq 1.$$

Therefore, for $\epsilon \in]0,1[$ no ϵ-bound g of order p exists.

The following characterization of uniformly integrable sets is useful:

2.12.2. Theorem. A set $\mathfrak{F}$ of measurable numerical functions on Ω is uniformly μ-integrable of order p if and only if it satisfies the following two conditions:

$$\sup_{f \in \mathfrak{F}} \int |f|^p \, d\mu < \infty; \qquad (2.12.3)$$

for every $\epsilon > 0$ there is an L^p-function $h \geqq 0$ and a $\delta > 0$ such that for every $A \in \mathfrak{A}$,

$$\int_A h^p \, d\mu \leqq \delta \Rightarrow \int_A |f|^p \, d\mu \leqq \epsilon, \qquad \text{for all } f \in \mathfrak{F}. \qquad (2.12.4)$$

Proof. For every set $A \in \mathfrak{A}$, for every measurable numerical function f on Ω and for every L^p-function $g \geqq 0$, we have

$$\int_A |f|^p \, d\mu = \int_{A \cap \{|f| \geqq g\}} |f|^p \, d\mu + \int_{A \cap \{|f| < g\}} |f|^p \, d\mu \leqq \int_{\{|f| \geqq g\}} |f|^p \, d\mu + \int_A g^p \, d\mu,$$

and in particular, for $A = \Omega$,

$$\int |f|^p \, d\mu \leqq \int_{\{|f| \geqq \alpha h\}} |f|^p \, d\mu + \int g^p \, d\mu.$$

$< \varepsilon/2 + \varepsilon/2 = \varepsilon$

If we choose for g an $(\epsilon/2)$-bound of order p for the set $\mathfrak{F}$ assumed to be uniformly integrable of order p, and if we set $h \equiv g$ and $\delta \equiv \epsilon/2$, then the properties (2.12.3) and (2.12.4) follow. Conversely, if the conditions (2.12.3) and (2.12.4) are satisfied, then for $\epsilon > 0$ we choose the function h according to (2.12.4) and note the inequality

$$\int |f|^p\, d\mu \geqq \int_{\{|f| \geqq \alpha h\}} |f|^p\, d\mu \geqq \alpha^p \int_{\{|f| \geqq \alpha h\}} h^p\, d\mu$$

valid for all $f \in \mathfrak{F}$ and every real $\alpha > 0$, or equivalently,

$$\int_{\{|f| \geqq \alpha h\}} h^p\, d\mu \leqq \frac{1}{\alpha^p} \int |f|^p\, d\mu.$$

By (2.12.3) there is an $\alpha > 0$ such that

$$\int_{\{|f| \geqq \alpha h\}} h^p\, d\mu \leqq \delta, \qquad \text{for all } f \in \mathfrak{F}.$$

Then by (2.12.4), $g \equiv \alpha h$ is an ϵ-bound of order p for $\mathfrak{F}$. Hence $\mathfrak{F}$ is uniformly integrable of order p. ┘

2.12.3. Corollary. Along with every set $\mathfrak{F}$ of measurable real functions on Ω, the set

$$\mathfrak{F}^* \equiv \{\alpha f + \beta g \colon f,g \in \mathfrak{F};\ \alpha,\beta \in \mathbf{R},\ |\alpha| \leqq 1,\ |\beta| \leqq 1\}$$

is also uniformly μ-integrable of order p.

Proof. For every set $A \in \mathfrak{A}$ and every function $f \in \mathfrak{L}^p(\mu)$, $f1_A$ is also in $\mathfrak{L}^p(\mu)$ since $|f1_A| \leqq |f|$. According to the Minkowski inequality, for arbitrary f_1, $f_2 \in \mathfrak{L}^p(\mu)$ and $A \in \mathfrak{A}$,

$$N_p(f_1 1_A + f_2 1_A) \leqq N_p(f_1 1_A) + N_p(f_2 1_A),$$

that is,

$$\int_A |f_1 + f_2|^p\, d\mu \leqq \left[\left(\int_A |f_1|^p\, d\mu\right)^{1/p} + \left(\int_A |f_2|^p\, d\mu\right)^{1/p}\right]^p.$$

By applying the inequality to the functions $f_1 \equiv \alpha f$ and $f_2 \equiv \beta g$ with f, $g \in \mathfrak{F}$ and $|\alpha| \leqq 1$, $|\beta| \leqq 1$ and observing Theorem 2.12.2, we see that $\mathfrak{F}^*$ also satisfies the conditions (2.12.3) and (2.12.4). ┘

Now we obtain the generalization of the dominated convergence theorem announced in the introduction. Example 3 and Theorem 2.11.4 show that we indeed obtain a generalization.

2.12.4. Theorem. For every sequence $(f_n)_{n \in \mathbf{N}}$ of functions in $\mathfrak{L}^p(\mu)$, $1 \leqq p < +\infty$, the following two statements are equivalent:

I. The sequence (f_n) is convergent in pth mean.
II. The sequence (f_n) converges μ-stochastically and is uniformly μ-integrable of order p.

Proof. (I) $\Rightarrow$ (II): Suppose (f_n) converges in pth mean to $f \in \mathfrak{L}^p(\mu)$, that is, in the notation of Section 2.6,

$$\lim_{n \to \infty} N_p(f_n - f) = 0.$$

By Theorem 2.11.3 only uniform integrability of order p is still to be proved. By (2.7.8), the sequence $(N_p(f_n))_{n \in \mathbf{N}}$ converges to $N_p(f)$ and is therefore bounded; therefore, the sequence $(N_p(f_n)^p)_{n \in \mathbf{N}}$ is also bounded and thus the condition (2.12.3) is satisfied for the set $\mathfrak{F} \equiv \{f_n : n \in \mathbf{N}\}$.

For every set $A \in \mathfrak{A}$, the functions $f_n 1_A$ and $f 1_A$ also are in $\mathfrak{L}^p(\mu)$. From the Minkowski inequality for N_p we then obtain

$$N_p(f_n 1_A) \leqq N_p((f_n - f)1_A) + N^p(f 1_A) \leqq N_p(f_n - f) + N_p(f 1_A).$$

Since $(1_A)^p = 1_A$, this can be rewritten as

$$\left(\int_A |f_n|^p \, d\mu\right)^{1/p} \leqq N_p(f_n - f) + \left(\int_A |f|^p \, d\mu\right)^{1/p}, \qquad \text{for all } n \in \mathbf{N}.$$

For every $\epsilon > 0$ there is a natural number n_0 such that $N_p(f_n - f) < \epsilon^{1/p}/2$ for all $n > n_0$. If we set $\delta \equiv \epsilon/2^p$ and

$$h \equiv \sup(|f_1|, \ldots, |f_{n_0}|, |f|),$$

then condition (2.12.4) is also satisfied for $\mathfrak{F}$.

(II) $\Rightarrow$ (I): From the stochastic convergence of the sequence (f_n) and by the triangle inequality, it follows as in the proof of Theorem 2.11.6 that

$$\lim_{m,n \to \infty} \mu(\{|f_m - f_n| \geqq \alpha\} \cap A) = 0$$

for every set $A \in \mathfrak{A}$ with $\mu(A) < \infty$ and every real $\alpha > 0$. We shall show that (f_n) is a Cauchy sequence in $\mathfrak{L}^p(\mu)$, that is,

$$\lim_{m,n \to \infty} \int |f_{mn}|^p \, d\mu = 0$$

holds for the double sequence of functions $f_{mn} \equiv f_m - f_n$. Now by Corollary 2.12.3 this double sequence $(f_{mn})_{m,n \in \mathbf{N}}$ is uniformly integrable of order p. Thus, for every $\epsilon > 0$, there is an L^p-function $g \geqq 0$ with

$$\int_{\{|f_{mn}| \geqq g\}} |f_{mn}|^p \, d\mu \leqq \epsilon, \qquad \text{for all } m, n \in \mathbf{N}. \tag{2.12.5}$$

Since

$$\int |f_{mn}|^p \, d\mu = \int_{\{|f_{mn}| \geqq g\}} |f_{mn}|^p \, d\mu + \int_{\{|f_{mn}| < g\}} |f_{mn}|^p \, d\mu,$$

it suffices to show that

$$\int_{\{|f_{mn}| < g\}} |f_{mn}|^p \, d\mu \leqq 3\epsilon, \qquad \text{for } m, n \in \mathbf{N} \text{ sufficiently large.} \tag{2.12.6}$$

Now $g^p\mu$ is a finite measure on $\mathfrak{A}$ and thus is continuous from above. Since $\bigcap_{\eta>0} \{g < \eta\} = \{g = 0\}$, we may choose $\eta > 0$ so that

$$\int_{\{g<\eta\}} g^p \, d\mu \leqq \epsilon.$$

Consequently,

$$\int_{\{|f_{mn}|<g\}\cap\{g<\eta\}} |f_{mn}|^p \, d\mu \leqq \int_{\{g<\eta\}} g^p \, d\mu \leqq \epsilon \qquad (m,\, n \in \mathbf{N}). \quad (2.12.7)$$

By the Chebyshev-Markov inequality, the set $\{g \geqq \eta\}$ has finite μ-measure. Hence for the double sequence of sets

$$A_{mn} \equiv \{|f_{mn}| \geqq \alpha\} \cap \{g \geqq \eta\} \qquad (m,\, n \in \mathbf{N})$$

we have for fixed $\alpha > 0$

$$\lim_{m,n\to\infty} \mu(A_{mn}) = 0.$$

Choose the number α so that

$$\left(\frac{\alpha}{\eta}\right)^p \int g^p \, d\mu \leqq \epsilon.$$

By virtue of the μ-continuity of the finite measure $g^p\mu$, we can then determine an $n_0 \in \mathbf{N}$ according to Theorem 2.9.6 so that

$$\int_{A_{mn}} g^p \, d\mu \leqq \epsilon, \qquad \text{for all } m \geqq n_0,\ n \geqq n_0.$$

Hence,

$$\int_{\{|f_{mn}|<g\}\cap A_{mn}} |f_{mn}|^p \, d\mu \leqq \int_{A_{mn}} g^p \, d\mu \leqq \epsilon \qquad (m \geqq n_0,\ n \geqq n_0). \quad (2.12.8)$$

Finally, with the help of the Chebyshev-Markov inequality, we obtain the inequality

$$\int_{\{|f_{mn}|<g\}\cap\{|f_{mn}|<\alpha\}\cap\{g\geqq\eta\}} |f_{mn}|^p \, d\mu \leqq \alpha^p\mu(\{g \geqq \eta\})$$
$$\leqq \left(\frac{\alpha}{\eta}\right)^p \int g^p \, d\mu \leqq \epsilon \qquad (m,\, n \in \mathbf{N}). \quad (2.12.9)$$

By adding the inequalities (2.12.7)–(2.12.9) we obtain the inequality (2.12.6), which was to be shown. ┘

2.12.5. Corollary. Let $(f_n)_{n\in\mathbf{N}}$ be a sequence of functions $f_n \geqq 0$ in $\mathfrak{L}^1(\mu)$ which converges μ-stochastically to a measurable function $f \geqq 0$ in $\mathfrak{L}^1(\mu)$. The sequence (f_n) is uniformly μ-integrable if and only if

$$\lim_{n\to\infty} \int f_n \, d\mu = \int f \, d\mu. \quad (2.12.10)$$

Proof. The uniform integrability of the sequence (f_n) implies, by Theorem 2.12.4, the convergence in mean to a function $f^* \in \mathcal{L}^1(\mu)$, and thus in particular

$$\lim \int f_n \, d\mu = \int f^* \, d\mu.$$

But f^* is also the stochastic limit of (f_n) according to Theorem 2.11.8 and hence $f = f^*$ almost everywhere by Section 2.11, Remark 3; hence, $\int f^* \, d\mu = \int f \, d\mu$. Only in proving the converse we use the positivity of the functions f_n and f. According to Example 3, the sequence $(\inf (f_n,f))_{n \in \mathbf{N}}$ is uniformly integrable; since

$$\{f - \inf (f_n,f) \geqq \alpha\} \subset \{|f - f_n| \geqq \alpha\} \qquad (\alpha > 0)$$

it also converges stochastically to f. By Theorem 2.12.4 (again taking note of Remark 3 of Section 2.11) it now follows that

$$\lim_{n \to \infty} \int (f - \inf (f_n,f)) \, d\mu = 0, \tag{2.12.11}$$

that is, we have convergence in mean of the sequence $(\inf (f_n,f))$ to f. But now $f + f_n = \sup (f_n,f) + \inf (f_n,f)$ and hence

$$\int f \, d\mu + \int f_n \, d\mu = \int \sup (f_n,f) \, d\mu + \int \inf (f_n,f) \, d\mu$$

for all n. Then by (2.12.10) and (2.12.11),

$$\lim_{n \to \infty} \int (\sup (f_n,f) - f) \, d\mu = 0$$

and hence

$$\lim_{n \to \infty} \int |f_n - f| \, d\mu = 0$$

since $|f_n - f| = \sup (f_n,f) - \inf (f_n,f)$. Thus, the sequence (f_n) converges in mean to f; by Theorem 2.12.4 it is uniformly integrable. ⌟

For σ-finite measures μ, uniformly integrable sets of functions can be written in a form especially convenient for applications. Here σ-finiteness is used in the following form:

2.12.6. Lemma. Let $(\Omega,\mathfrak{A},\mu)$ be a measure space. The measure μ is σ-finite if and only if there is a function $h \in \mathcal{L}^1(\mu)$ with

$$h(\omega) > 0, \qquad \text{for all } \omega \in \Omega. \tag{2.12.12}$$

Proof. If μ is σ-finite, then there is a sequence $(A_n)_{n \in \mathbf{N}}$ in $\mathfrak{A}$ with $A_n \uparrow \Omega$ and $\mu(A_n) < +\infty$ for all $n \in \mathbf{N}$. If we choose a sequence (η_n) of real numbers >0 such that $\eta_n \mu(A_n) \leqq 2^{-n}$ for all $n \in \mathbf{N}$, then the function

$$h \equiv \sum_{n=1}^{\infty} \eta_n 1_{A_n}$$

gives us the desired result. If, conversely, h is a strictly positive function from $\mathcal{L}^1(\mu)$, then

$$\left\{h \geqq \frac{1}{n}\right\} \uparrow \Omega;$$

by the Chebyshev-Markov inequality, each of the sets $\{h \geqq 1/n\}$, $n \in \mathbf{N}$, has finite measure. Thus μ is σ-finite. ⌟

If h is a strictly positive function in $\mathcal{L}^1(\mu)$, then $q \equiv h^{1/p}$ is a strictly positive function in $\mathcal{L}^p(\mu)$; conversely, for every strictly positive $q \in \mathcal{L}^p(\mu)$, the pth power $h \equiv q^p$ is a strictly positive function in $\mathcal{L}^1(\mu)$ $(1 \leqq p < +\infty)$. Thus for σ-finite measures μ we can put Definition 2.12.1 and Theorem 2.12.2 into the following equivalent form:

2.12.7. Theorem. Let μ be a σ-finite measure and let h be a strictly positive function in $\mathcal{L}^p(\mu)$. For every set $\mathfrak{F}$ of $\mathfrak{A}$-measurable numerical functions on Ω, the following three statements are equivalent:

I. $\mathfrak{F}$ is uniformly integrable of order p.

II. For every real $\epsilon > 0$ there is a real number $\alpha > 0$ such that αh is an ϵ-bound of order p for $\mathfrak{F}$.

III. We have

$$\sup_{f \in \mathfrak{F}} \int |f|^p \, d\mu < +\infty.$$

Further, for every $\epsilon > 0$ there is a $\delta > 0$ such that for every $A \in \mathfrak{A}$,

$$\int_A h^p \, d\mu \leqq \delta \Rightarrow \int_A |f|^p \, d\mu \leqq \epsilon, \qquad \text{for all } f \in \mathfrak{F}.$$

Condition II says precisely that

$$\lim_{\alpha \to \infty} \int_{\{|f| \geqq \alpha h\}} |f|^p \, d\mu = 0 \tag{2.12.13}$$

holds *uniformly* for the functions $f \in \mathfrak{F}$. Because of Theorem 2.9.6 the second condition of III is also called the *equi-$(h^p\mu)$-continuity of the measures* $|f|^p\mu$ with $f \in \mathfrak{F}$.

Proof. (I) ⇒ (II): Let g be an arbitrary $\epsilon/2$-bound of order p for $\mathfrak{F}$. Then for all $f \in \mathfrak{F}$ and $\alpha > 0$ we have the inequalities

$$\begin{aligned}\int_{\{|f| \geqq \alpha h\}} |f|^p \, d\mu &= \int_{\{|f| \geqq \alpha h\} \cap \{|f| \geqq g\}} |f|^p \, d\mu + \int_{\{|f| \geqq \alpha h\} \cap \{|f| < g\}} |f|^p \, d\mu \\ &\leqq \int_{\{|f| \geqq g\}} |f|^p \, d\mu + \int_{\{g > \alpha h\}} g^p \, d\mu \leqq \frac{\epsilon}{2} + \int_{\{g > \alpha h\}} g^p \, d\mu.\end{aligned}$$

This proves the assertion, since $g^p\mu$ is a finite measure on $\mathfrak{A}$ and

$$\bigcap_{\alpha>0} \{g > \alpha h\} = \bigcap_{n=1}^{\infty} \{g > nh\} = \{g = +\infty\}$$

holds due to the strict positivity of h. By Theorem 2.5.5, g^p is μ-almost everywhere finite, that is, $\mu(\{g = +\infty\}) = 0$.

(II) $\Rightarrow$ (III): This follows from the inequality given at the beginning of the proof of Theorem 2.12.2:

$$\int_A |f|^p \, d\mu \leqq \int_{\{|f| \geqq \alpha h\}} |f|^p \, d\mu + \alpha^p \int_A h^p \, d\mu \qquad (f \in \mathfrak{F}).$$

(III) $\Rightarrow$ (I): This is simply Theorem 2.12.2. ┘

Theorem 2.12.7 is particularly important for *finite* measures μ. Then for h we may choose the constant function 1. This choice is usually made in probability theory, where we have $\mu(\Omega) = 1$.

PROBLEMS

1. Prove: A set $\mathfrak{F}$ of measurable numerical functions is uniformly integrable of order p if and only if the set $\mathfrak{F}^p = \{|f|^p : f \in \mathfrak{F}\}$ is uniformly integrable (of order 1).
2. Prove: A set $\mathfrak{F}$ of measurable numerical functions is uniformly integrable if and only if for every $\epsilon > 0$ there exists an integrable function $h \geqq 0$ such that $\int (|f| - h)^+ \, d\mu \leqq \epsilon$ for all $f \in \mathfrak{F}$. [*Hint:* $g = \eta h$ is a 2ϵ-bound (of order one) for $\mathfrak{F}$ if $\eta > 0$ is sufficiently large.]
3. Let μ be a finite measure and $\mathfrak{F} \subset \mathfrak{L}^1(\mu)$. For all $n = 0, 1, \ldots$ and all $f \in \mathfrak{F}$, we define $a_n(f) \equiv n\mu(\{n \leqq |f| < n + 1\})$. Prove that $\mathfrak{F}$ is uniformly integrable if and only if the series $\sum_{n=0}^{\infty} a_n(f)$ is uniformly convergent on $\mathfrak{F}$.
4. Let μ be a finite measure and $\mathfrak{F} \subset \mathfrak{L}^1(\mu)$. Prove: $\mathfrak{F}$ is uniformly integrable if there exists a Borel-measurable function $q \geqq 0$ on $\mathbf{R}_+$ having the following two properties:

   (a) $\displaystyle\lim_{t \to +\infty} \frac{q(t)}{t} = +\infty$;

   (b) $\displaystyle\sup_{f \in \mathfrak{F}} \int q \circ |f| \, d\mu < +\infty$.

3

PRODUCT MEASURES

3.1 PRODUCTS OF σ-ALGEBRAS AND UNIQUENESS OF THE PRODUCT MEASURE

Suppose we are given finitely many measurable spaces $(\Omega_i,\mathfrak{A}_i)$, $i = 1, \ldots, n$. We consider the product set

$$\Omega \equiv \prod_{i=1}^{n} \Omega_i = \Omega_1 \times \cdots \times \Omega_n$$

and, for every i, the *projection mapping*

$$p_i \colon \Omega \to \Omega_i,$$

which maps the point $(\omega_1, \ldots, \omega_n) \in \Omega$ onto its ith coordinate ω_i. We call the σ-algebra

$$\bigotimes_{i=1}^{n} \mathfrak{A}_i = \mathfrak{A}_1 \otimes \cdots \otimes \mathfrak{A}_n \equiv \mathfrak{A}(p_1, \ldots, p_n) \tag{3.1.1}$$

generated by the mappings $p_1, \ldots, p_n$ the *product of the σ-algebras* $\mathfrak{A}_1, \ldots, \mathfrak{A}_n$. According to (2.9.3) we are dealing with the smallest σ-algebra $\mathfrak{A}$ in Ω such that every p_i is $\mathfrak{A}$-$\mathfrak{A}_i$-measurable.[1] We then immediately have:

3.1.1. Theorem. For every $i = 1, \ldots, n$, let $\mathfrak{E}_i$ be a generator of the σ-algebra $\mathfrak{A}_i$ in Ω_i which contains a sequence $(E_{ik})_{k\in\mathbf{N}}$ of sets with

[1] The product of finitely many topological spaces is defined similarly.

$E_{ik} \uparrow \Omega_i$. Then the σ-algebra $\mathfrak{A}_1 \otimes \cdots \otimes \mathfrak{A}_n$ is generated by the system of all sets

$$E_1 \times \cdots \times E_n, \qquad \text{with } E_i \in \mathfrak{E}_i \qquad (i = 1, \ldots, n).$$

Proof. Let $\mathfrak{A}$ be an arbitrary σ-algebra in Ω. We have to show: Each of the mappings p_i is $\mathfrak{A}$-$\mathfrak{A}_i$-measurable if and only if $E_1 \times \cdots \times E_n \in \mathfrak{A}$ holds for arbitrary $E_i \in \mathfrak{E}_i$. By Theorem 1.7.2, p_i is $\mathfrak{A}$-$\mathfrak{A}_i$-measurable if and only if $E_i \in \mathfrak{E}_i$ implies $p_i^{-1}(E_i) \in \mathfrak{A}$. If this condition is satisfied for all $i = 1, \ldots, n$, then

$$E_1 \times \cdots \times E_n = p_1^{-1}(E_1) \cap \cdots \cap p_n^{-1}(E_n)$$

also lies in $\mathfrak{A}$. Conversely, if all sets $E_1 \times \cdots \times E_n$ with $E_i \in \mathfrak{E}_i$ lie in $\mathfrak{A}$, then for arbitrary $i = 1, \ldots, n$ and $E_i \in \mathfrak{E}_i$, the sets

$$F_k \equiv E_{1k} \times \cdots \times E_{i-1,k} \times E_i \times E_{i+1,k} \times \cdots \times E_{nk}$$

lie in $\mathfrak{A}$ for all $k \in \mathbf{N}$. Since

$$F_k \uparrow \Omega_1 \times \cdots \times \Omega_{i-1} \times E_i \times \Omega_{i+1} \times \cdots \times \Omega_n = p_i^{-1}(E_i),$$

we have that $p_i^{-1}(E_i)$ is in $\mathfrak{A}$. This proves the theorem. ┘

Remark. We cannot simply drop the restricting hypotheses on the generators $\mathfrak{E}_i$: Suppose we chose $n = 2$, $\mathfrak{A}_1 = \{\varnothing, \Omega_1\}$, $\mathfrak{E}_1 = \{\varnothing\}$ and $\mathfrak{E}_2 = \mathfrak{A}_2$ where the σ-algebra $\mathfrak{A}_2$ contains at least 4 elements.

In particular, $\mathfrak{A}_1 \otimes \cdots \otimes \mathfrak{A}_n$ is generated by all sets $A_1 \times \cdots \times A_n$ with $A_i \in \mathfrak{A}_i$. The direction of our investigation is now determined by the following example:

Example

Let $\Omega_1 = \cdots = \Omega_n = \mathbf{R}$, $\mathfrak{A}_1 = \cdots = \mathfrak{A}_n = \mathfrak{B}^1$, and $\mathfrak{E}_1 = \cdots = \mathfrak{E}_n = \mathfrak{J}^1$. The system of all sets $E_1 \times \cdots \times E_n$ with $E_i \in \mathfrak{J}^1$ obviously coincides with the system $\mathfrak{J}^n$ of all right half-open intervals in $\mathbf{R}^n$. By Definition 1.6.1, $\mathfrak{J}^n$ generates the σ-algebra $\mathfrak{B}^n$ of n-dimensional Borel sets in $\mathbf{R}^n$. Together with Theorem 3.1.1—the hypotheses are obviously satisfied—this yields the result

$$\mathfrak{B}^n = \mathfrak{B}^1 \otimes \cdots \otimes \mathfrak{B}^1 \qquad (n \text{ factors}). \tag{3.1.2}$$

By Theorem 1.6.2, λ^n is the only measure on $\mathfrak{B}^n$ with

$$\lambda^n(I_1 \times \cdots \times I_n) = \lambda^1(I_1) \cdot \ldots \cdot \lambda^1(I_n)$$

for all $I_1, \ldots, I_n \in \mathfrak{J}^1$. This remark and the above example now lead to the following general question.

Suppose we are given n measure spaces $(\Omega_i, \mathfrak{A}_i, \mu_i)$, $i = 1, \ldots, n$

($n \geqq 2$) and for each of the σ-algebras $\mathfrak{A}_i$ we are given a generator $\mathfrak{E}_i$. Under what hypotheses can we prove the existence of a measure π on $\mathfrak{A}_1 \otimes \cdots \otimes \mathfrak{A}_n$ which satisfies the condition

$$\pi(E_1 \times \cdots \times E_n) = \mu_1(E_1) \cdot \ldots \cdot \mu_n(E_n) \tag{3.1.3}$$

for arbitrary $E_i \in \mathfrak{E}_i$?

With regard to *uniqueness* we can immediately show:

3.1.2. Theorem. Suppose each of the generators $\mathfrak{E}_i$ is $\cap$-stable and contains a sequence $(E_{ik})_{k \in \mathbf{N}}$ of sets with $E_{ik} \uparrow \Omega_i$ and $\mu_i(E_{ik}) < \infty$ for every $k = 1, 2, \ldots$. Then there is at most one measure π on $\mathfrak{A}_1 \otimes \cdots \otimes \mathfrak{A}_n$ with property (3.1.3).

Proof. Let $\mathfrak{E}$ be the system of all sets $E_1 \times \cdots \times E_n$ with $E_i \in \mathfrak{E}_i$ which, by Theorem 3.1.1, generates the σ-algebra $\mathfrak{A}_1 \otimes \cdots \otimes \mathfrak{A}_n$. $\mathfrak{E}$ is $\cap$-stable, since

$$\left(\prod_{i=1}^{n} E_i\right) \cap \left(\prod_{i=1}^{n} F_i\right) = \prod_{i=1}^{n} (E_i \cap F_i)$$

and since every $\mathfrak{E}_i$ is $\cap$-stable. Further, $E_k \equiv E_{1k} \times \cdots \times E_{nk}$ ($k = 1, 2, \ldots$) is a sequence in $\mathfrak{E}$ with

$$E_k \uparrow \Omega_1 \times \cdots \times \Omega_n.$$

The assertion now follows from Theorem 1.5.5 if we note that every $\mu_i(E_{ik})$ is finite. ⌟

We can also prove the *existence* of the measure π under the hypotheses of Theorem 3.1.2 which imply the σ-finiteness of each μ_i. We carry out this proof for $n = 2$ in the next section and consider the general case later.

PROBLEM

Let $\mathfrak{A}_i$ be an algebra in a set Ω_i, $i = 1, \ldots, n$. Prove: The algebra in $\Omega_1 \times \cdots \times \Omega_n$ generated by all sets $A_1 \times \cdots \times A_n$ (where $A_i \in \mathfrak{A}_i$) consists of all finite unions of such sets.

3.2 EXISTENCE AND PROPERTIES OF THE PRODUCT OF TWO MEASURES

Suppose that we are now given two measure spaces $(\Omega_i, \mathfrak{A}_i, \mu_i)$, $i = 1, 2$. For every set $Q \subset \Omega_1 \times \Omega_2$ and every $\omega_i \in \Omega_i$, we define the *ω_i-section Q_{ω_i} of Q* as follows:

$$\begin{aligned} Q_{\omega_1} &\equiv \{\omega_2 \in \Omega_2 \colon (\omega_1, \omega_2) \in Q\}, \\ Q_{\omega_2} &\equiv \{\omega_1 \in \Omega_1 \colon (\omega_1, \omega_2) \in Q\}. \end{aligned} \tag{3.2.1}$$

3.2.1. Lemma. All ω_1- and ω_2-sections of each set $Q \in \mathfrak{A}_1 \otimes \mathfrak{A}_2$ lie in $\mathfrak{A}_2$ and $\mathfrak{A}_1$, respectively.

Proof. For arbitrary subsets $Q, Q_1, Q_2, \ldots$ of $\Omega \equiv \Omega_1 \times \Omega_2$ and $\omega_1 \in \Omega_1$,

$$(\Omega \setminus Q)_{\omega_1} = \Omega_2 \setminus Q_{\omega_1}$$

and

$$(\bigcup Q_n)_{\omega_1} = \bigcup (Q_n)_{\omega_1}.$$

Further, $\Omega_{\omega_1} = \Omega_2$ and, more generally,

$$(A_1 \times A_2)_{\omega_1} = A_2 \qquad \text{or } \emptyset,$$

depending on whether ω_1 lies in A_1 or not ($A_i \subset \Omega_i$, $i = 1, 2$). For every $\omega_1 \in \Omega_1$, the system of all sets $Q \subset \Omega$ with $Q_{\omega_1} \in \mathfrak{A}_2$ is a σ-algebra in Ω which contains all sets $A_1 \times A_2$ with $A_i \in \mathfrak{A}_i$ ($i = 1, 2$). By Theorem 3.1.1, $\mathfrak{A}_1 \otimes \mathfrak{A}_2$ is the smallest σ-algebra containing all such sets $A_1 \times A_2$. This proves the part of the lemma involving the ω_1-sections. The ω_2-sections are treated in the same way. ┘

Since $\mu_2(Q_{\omega_1})$ and $\mu_1(Q_{\omega_2})$ are now defined for all $Q \in \mathfrak{A}_1 \otimes \mathfrak{A}_2$, $\omega_1 \in \Omega_1$, and $\omega_2 \in \Omega_2$, we can show further:

3.2.2. Lemma. Let the measures μ_1 and μ_2 be σ-finite. Then for every $Q \in \mathfrak{A}_1 \otimes \mathfrak{A}_2$, the function

$$\omega_1 \to \mu_2(Q_{\omega_1})$$

is $\mathfrak{A}_1$-measurable on Ω_1;

$$\omega_2 \to \mu_1(Q_{\omega_2})$$

is $\mathfrak{A}_2$-measurable on Ω_2.

Proof. Let us denote the function $\omega_1 \to \mu_2(Q_{\omega_1})$ by s_Q. We shall show the $\mathfrak{A}_1$-measurability of s_Q for every $Q \in \mathfrak{A}_1 \otimes \mathfrak{A}_2$. The measurability of the second function is proved analogously.

First let $\mu_2(\Omega_2) < \infty$. Then the set $\mathfrak{D}$ of all $D \in \mathfrak{A}_1 \otimes \mathfrak{A}_2$ such that s_D is an $\mathfrak{A}_1$-measurable function is a Dynkin system in $\Omega \equiv \Omega_1 \times \Omega_2$. This is obtained from the following easily verified statements:

$$\begin{aligned} s_\Omega &= \mu_2(\Omega_2); \\ s_{E \setminus D} &= s_E - s_D, \qquad \text{if } D, E \in \mathfrak{D}, \text{ and } D \subset E; \\ s_{\cup D_n} &= \sum s_{D_n} \end{aligned}$$

for every sequence (D_n) of pairwise disjoint sets from $\mathfrak{D}$. Further, $\mathfrak{D}$ contains all sets $A_1 \times A_2$ with $A_i \in \mathfrak{A}_i$ ($i = 1, 2$); since $s_{A_1 \times A_2} = \mu_2(A_2) 1_{A_1}$. The system $\mathfrak{E}$ of these sets $A_1 \times A_2$ is $\cap$-stable and generates $\mathfrak{A}_1 \otimes \mathfrak{A}_2$ by Theorem 3.1.1; then by Theorem 1.2.3, $\mathfrak{A}_1 \otimes \mathfrak{A}_2$ is the Dynkin system generated by $\mathfrak{E}$. Thus, $\mathfrak{E} \subset \mathfrak{D} \subset \mathfrak{A}_1 \otimes \mathfrak{A}_2$ implies $\mathfrak{D} = \mathfrak{A}_1 \otimes \mathfrak{A}_2$ and we have the desired result.

If μ_2 is σ-finite, then there exists a sequence (B_n) of sets of $\mathfrak{A}_2$ with $B_n \uparrow \Omega_2$ and $\mu_2(B_n) < \infty$. Then for every n, $A_2 \to \mu_2(A_2 \cap B_n)$ is a finite measure $\mu_{2,n}$ on $\mathfrak{A}_2$.[2] From what was proved above, each of the functions $\omega_1 \to \mu_{2,n}(Q_{\omega_1})$ with $Q \in \mathfrak{A}_1 \otimes \mathfrak{A}_2$ is $\mathfrak{A}_1$-measurable. Now we have

$$\sup_n \mu_{2,n}(Q_{\omega_1}) = \mu_2(Q_{\omega_1}),$$

since μ_2 is continuous from below. Then $\omega_1 \to \mu_2(Q_{\omega_1})$ is also $\mathfrak{A}_1$-measurable by Theorem 2.1.5. ┘

Now the existence of the desired measure π is easy to obtain:

3.2.3. Theorem. Let $(\Omega_i,\mathfrak{A}_i,\mu_i)$, $i = 1, 2$, be two σ-finite measure spaces.[3] Then there is exactly one measure π on $\mathfrak{A}_1 \otimes \mathfrak{A}_2$ with

$$\pi(A_1 \times A_2) = \mu_1(A_1)\mu_2(A_2), \qquad \text{for all } A_i \in \mathfrak{A}_i. \tag{3.2.2}$$

For every set $Q \in \mathfrak{A}_1 \otimes \mathfrak{A}_2$,

$$\pi(Q) = \int \mu_2(Q_{\omega_1})\mu_1(d\omega_1) = \int \mu_1(Q_{\omega_2})\mu_2(d\omega_2). \tag{3.2.3}$$

Proof. For every $Q \in \mathfrak{A}_1 \otimes \mathfrak{A}_2$, let s_Q again denote the $\mathfrak{A}_1$-measurable function $\omega_1 \to \mu_2(Q_{\omega_1})$ on Ω_1. Now $s_Q \geqq 0$, and thus

$$Q \to \pi(Q) \equiv \int s_Q \, d\mu_1$$

is defined as a function on $\mathfrak{A}_1 \otimes \mathfrak{A}_2$. For every sequence (Q_n) of pairwise disjoint sets from $\mathfrak{A}_1 \otimes \mathfrak{A}_2$, $s_{\bigcup Q_n} = \Sigma s_{Q_n}$ and hence, by Corollary 2.3.5, $\pi(\bigcup Q_n) = \Sigma \pi(Q_n)$. Since $s_\varnothing = 0$ we also have $\pi(\varnothing) = 0$ and thus π is a measure on $\mathfrak{A}_1 \otimes \mathfrak{A}_2$. It has the property (3.2.2) since $s_{A_1 \times A_2} = \mu_2(A_2)1_{A_1}$ and hence $\pi(A_1 \times A_2) = \mu_1(A_1)\mu_2(A_2)$. Analogously, we obtain that $Q \to \int \mu_1(Q_{\omega_1})\mu_2(d\omega_2)$ is another measure π' on $\mathfrak{A}_1 \otimes \mathfrak{A}_2$ with this property. But by Theorem 3.1.2 applied to $\mathfrak{E}_1 = \mathfrak{A}_1$ and $\mathfrak{E}_2 = \mathfrak{A}_2$, there is at most one such measure. Hence $\pi = \pi'$ and the remainder of the assertion follows.

3.2.4. Definition. The measure π on $\mathfrak{A}_1 \otimes \mathfrak{A}_2$ uniquely determined by property (3.2.2) for any two σ-finite measure spaces $(\Omega_i,\mathfrak{A}_i,\mu_i)$, $i = 1, 2$, is called the *product of the measures* μ_1 and μ_2 and is denoted by $\mu_1 \otimes \mu_2$.

The question raised in Section 3.1 for two σ-finite measures μ_1 and μ_2 has been answered by Theorem 3.2.3. Indeed if $\mathfrak{E}_i$ is a generator of $\mathfrak{A}_i$

[2] Note that $\mu_{2,n} = 1_{B_n}\mu_2$.

[3] That is, more precisely, two measure spaces each with a σ-finite measure.

($i = 1, 2$) with the properties stated in Theorem 3.1.2, then by Theorems 3.1.2 and 3.2.3, $\mu_1 \otimes \mu_2$ is the only measure π satisfying the condition (3.1.3).

The example of Section 3.1 now tells us that $\lambda^2 = \lambda^1 \otimes \lambda^1$. Analogous reasoning yields the result

$$\lambda^{m+n} = \lambda^m \otimes \lambda^n \qquad (m, n = 1, 2, \ldots).$$

It also follows from property (3.2.2) that when μ_1 and μ_2 are σ-finite, $\mu_1 \otimes \mu_2$ is also a *σ-finite* measure.

We now proceed to investigate integrability with respect to $\mu_1 \otimes \mu_2$, and for this we shall use the following abbreviation: For a given mapping $f\colon \Omega \to \Omega'$ of $\Omega \equiv \Omega_1 \times \Omega_2$ into a set Ω', we use f_{ω_1} and f_{ω_2} ($\omega_1 \in \Omega_1$, $\omega_2 \in \Omega_2$) to denote the following mappings of Ω_2 into Ω' and of Ω_1 into Ω', respectively:

$$\begin{aligned} \omega_2 &\to f_{\omega_1}(\omega_2) \equiv f(\omega_1,\omega_2), \\ \omega_1 &\to f_{\omega_2}(\omega_1) \equiv f(\omega_1,\omega_2). \end{aligned} \tag{3.2.4}$$

In particular, if we choose $f = 1_Q$ with $Q \subset \Omega$, then obviously

$$(1_Q)_{\omega_1} = 1_{Q_{\omega_1}},\ (1_Q)_{\omega_2} = 1_{Q_{\omega_2}}.^4 \tag{3.2.5}$$

We call the mapping f_{ω_i} the *ω_i-section of f*.

3.2.5. Lemma. For every measurable space $(\Omega',\mathfrak{A}')$ and every measurable mapping

$$f\colon (\Omega_1 \times \Omega_2,\ \mathfrak{A}_1 \otimes \mathfrak{A}_2) \to (\Omega',\mathfrak{A}'),$$

each of the mappings f_{ω_1} and f_{ω_2} is $\mathfrak{A}_2$-$\mathfrak{A}'$- and $\mathfrak{A}_1$-$\mathfrak{A}'$-measurable, respectively.

Proof. For every set $A' \in \mathfrak{A}'$,

$$f_{\omega_1}^{-1}(A') = \{\omega_2\colon (\omega_1,\omega_2) \in f^{-1}(A')\} = \{f^{-1}(A')\}_{\omega_1}$$

and correspondingly

$$f_{\omega_2}^{-1}(A') = \{f^{-1}(A')\}_{\omega_2}.$$

The assertion then follows from Lemma 3.2.1. ┘

The following theorem, which extends formula (3.2.3) from indicator functions to arbitrary measurable functions ≥ 0, is crucial:

3.2.6. Theorem. Let $(\Omega_i,\mathfrak{A}_i,\mu_i)$, $i = 1, 2$, be σ-finite measure spaces and let

$$f\colon \Omega_1 \times \Omega_2 \to \bar{\mathbf{R}}_+$$

[4] Note that these indicator functions refer to different sets.

be an $\mathfrak{A}_1 \otimes \mathfrak{A}_2$-measurable function. Then the functions $\omega_2 \to \int f_{\omega_2}\, d\mu_1$ and $\omega_1 \to \int f_{\omega_1}\, d\mu_2$ are $\mathfrak{A}_2$- and $\mathfrak{A}_1$-measurable, respectively, and we have

$$\int f\, d(\mu_1 \otimes \mu_2) = \int \left(\int f_{\omega_2}\, d\mu_1\right)\mu_2(d\omega_2) = \int \left(\int f_{\omega_1}\, d\mu_2\right)\mu_1(d\omega_1). \quad (3.2.6)$$

Proof. Let $\Omega \equiv \Omega_1 \times \Omega_2$, $\mathfrak{A} \equiv \mathfrak{A}_1 \otimes \mathfrak{A}_2$, and $\pi \equiv \mu_1 \otimes \mu_2$. First choose an $\mathfrak{A}$-elementary function

$$f = \sum_{i=1}^{n} \alpha_i 1_{Q^i} \qquad (\alpha_i \geqq 0,\ Q^i \in \mathfrak{A}.)$$

Then $f_{\omega_2} = \Sigma \alpha_i 1_{Q^i_{\omega_2}}$ by (3.2.5) and thus

$$\int f_{\omega_2}\, d\mu_1 = \sum_{i=1}^{n} \alpha_i \mu_1(Q^i_{\omega_2}).$$

Consequently, $\omega_2 \to \int f_{\omega_2}\, d\mu_1$ is an $\mathfrak{A}_2$-measurable function on Ω_2 by Lemma 3.2.2. Taking (3.2.3) into account, integration with respect to μ_2 yields

$$\int \left(\int f_{\omega_2}\, d\mu_1\right) \mu_2(d\omega_2) = \sum_{i=1}^{n} \alpha_i \pi(Q^i) = \int f\, d\pi,$$

that is, the first equality in (3.2.6). For an arbitrary $\mathfrak{A}$-measurable $f \geqq 0$, let $(u^n)_{n \in \mathbf{N}}$ be an isotone sequence of $\mathfrak{A}$-elementary functions with $\sup u^n = f$. Then, by the first part of the proof, $(u^n_{\omega_2})$ is a sequence of $\mathfrak{A}_1$-elementary functions which is obviously isotone and has f_{ω_2} as upper envelope ($\omega_2 \in \Omega_2$). But then the sequence

$$\omega_2 \to \varphi^n(\omega_2) = \int u^n_{\omega_2}\, d\mu_1, \qquad n = 1, 2, \ldots,$$

which according to what was just proved consists of $\mathfrak{A}_2$-measurable functions φ^n, is isotone and has

$$\omega_2 \to \int f_{\omega_2}\, d\mu_1$$

as supremum. Thus, this function is also $\mathfrak{A}_2$-measurable, and by Theorem 2.3.4 on monotone convergence, we have

$$\int \left(\int f_{\omega_2}\, d\mu_1\right)\mu_2(d\omega_2) = \sup \int \varphi^n\, d\mu_2.$$

From the first part of the proof we have $\int \varphi^n\, d\mu_2 = \int u^n\, d\pi$. According to the choice of the sequence (u^n) we thus obtain $\sup \int \varphi^n\, d\mu_2 = \int f\, d\pi$. Hence,

$$\int \left(\int f_{\omega_2}\, d\mu_1\right)\mu_2(d\omega_2) = \int f\, d\pi,$$

which is the first equality in (3.2.6). An analogous treatment of f_{ω_1} completes the proof. ┘

3.2.7. Corollary. Let $(\Omega_i,\mathfrak{A}_i,\mu_i)$, $i = 1, 2$, be σ-finite measure spaces and let f be a $\mu_1 \otimes \mu_2$-integrable numerical function on $\Omega_1 \times \Omega_2$. Then the function f_{ω_1} is μ_2-integrable for μ_1-almost all ω_1 and f_{ω_2} is μ_1-integrable for μ_2-almost all ω_2. Further, the function

$$\omega_1 \to \int f_{\omega_1}\, d\mu_2 \qquad \text{or} \qquad \omega_2 \to \int f_{\omega_1}\, d\mu_1$$

defined μ_1-almost everywhere or μ_2-almost everywhere, is μ_1- or μ_2-integrable, respectively, and (3.2.6) holds.

Proof. First, we obviously have

$$|f|_{\omega_i} = |f_{\omega_i}|,\ (f^+)_{\omega_i} = (f_{\omega_i})^+ \text{ and } (f^-)_{\omega_i} = (f_{\omega_i})^-.$$

Therefore, we write $f^+_{\omega_i}$ for $(f^+)_{\omega_i}$, and so on. ($\omega_i \in \Omega_i$; $i = 1, 2$). For $\pi \equiv \mu_1 \otimes \mu_2$ it follows from (3.2.6) that

$$\int\left(\int |f_{\omega_1}|\, d\mu_2\right)\mu_1(d\omega_1) = \int\left(\int |f_{\omega_2}|\, d\mu_1\right)\mu_2(d\omega_2) = \int |f|\, d\pi < \infty.$$

Then according to Theorem 2.5.5, the μ_1-integrable function $\omega_1 \to \int |f_{\omega_1}|\, d\mu_2$ is μ_1-almost everywhere finite; in other words, f_{ω_1} is μ_1-almost everywhere μ_2-integrable. By Theorem 3.2.6, the function

$$\omega_1 \to \int f_{\omega_1}\, d\mu_2 = \int f^+_{\omega_1}\, d\mu_2 - \int f^-_{\omega_1}\, d\mu_2$$

is μ_1-almost everywhere defined and $\mathfrak{A}_1$-measurable. Applying Theorem 3.2.6 to f^+ and f^-, we obtain the μ_1-integrability of this function, and

$$\begin{aligned}\int\left(\int f_{\omega_1}\, d\mu_2\right)\mu_1(d\omega_1) &= \int\left(\int f^+_{\omega_1}\, d\mu_2\right)\mu_1(d\omega_1) - \int\left(\int f^-_{\omega_1}\, d\mu_2\right)\mu_1(d\omega_1)\\ &= \int f^+\, d\pi - \int f^-\, d\pi = \int f\, d\pi.\end{aligned}$$

Interchange of the roles of ω_1 and ω_2 gives the remainder of the proof. ⌟

The conclusions contained in Theorem 3.2.6 and Corollary 3.2.7 are known as *Fubini's Theorem*. Because of this theorem, we shall henceforth write (3.2.6) also in the form

$$\int f\, d(\mu_1 \otimes \mu_2) = \int\int f(\omega_1,\omega_2)\mu_1(d\omega_1)\mu_2(d\omega_2) = \int\int f(\omega_1,\omega_2)\mu_2(d\omega_2)\mu_1(d\omega_1).$$

This theorem then also says in particular that under the given conditions the *order of integration* does not matter.

Remark. For arbitrary, not necessarily σ-finite measures μ_1 and μ_2, one can use other methods to prove the existence but no longer the uniqueness of a product measure. If one of the measures μ_1 and μ_2 is not σ-finite, the second equality in (3.2.3) already fails to hold. See Problem 1, taken from Halmos [3], p. 145, and Hahn-Rosenthal [35], Chapter IV, Section 16.

As a simple application of Fubini's Theorem, we now present the previously delayed proof of Theorem 1.8.1, which states that λ^p is the only translation-invariant measure μ on $\mathfrak{B}^p$ which associates the measure 1 with the unit cube $W = [\mathbf{0},\mathbf{1}]$.

To do this, consider two $\mathfrak{B}^p$-measurable numerical functions $f \geqq 0$ and $g \geqq 0$ on $\mathbf{R}^p$. Since the mapping $(x,y) \to x + y$ of $\mathbf{R}^p \times \mathbf{R}^p$ into $\mathbf{R}^p$ is continuous and, by Section 1.7, Example 2, $\mathfrak{B}^p \otimes \mathfrak{B}^p$-$\mathfrak{B}^p$-measurable, the composed mapping $(x,y) \to f(x + y)$ is also $\mathfrak{B}^p \otimes \mathfrak{B}^p$-measurable. Since the projection $(x,y) \to y$ and thus $(x,y) \to g(y)$ is $\mathfrak{B}^p \otimes \mathfrak{B}^p$-measurable, we can apply (3.2.6) to the $\mathfrak{B}^p \otimes \mathfrak{B}^p$-measurable function $(x,y) \to g(y)f(x + y)$ and, for the same reasons, to $(x,y) \to g(y - x)f(y)$. Then

$$\int\int g(y)f(x + y)\mu(dx)\lambda^p(dy) = \int\int g(y)f(x + y)\lambda^p(dy)\mu(dx) \quad (3.2.7)$$

and

$$\int\int g(y - x)f(y)\lambda^p(dy)\mu(dx) = \int\int g(y - x)f(y)\mu(dx)\lambda^p(dy). \quad (3.2.8)$$

Because of the translation-invariance of μ, the integral on the left side of (3.2.7) is equal to

$$\int g(y)\left(\int f(x + y)\mu(dx)\right)\lambda^p(dy) = \left(\int f\,d\mu\right)\left(\int g\,d\lambda^p\right);$$

because of the translation-invariance of λ^p, the integral on the right side of (3.2.7) is equal to the one on the left side of (3.2.8). Hence, it follows that

$$\int f\,d\mu \cdot \int g\,d\lambda^p = \int\int g(y - x)f(y)\mu(dx)\lambda^p(dy)$$
$$= \int f(y)\left(\int g(y - x)\mu(dx)\right)\lambda^p(dy);$$

by another use of the translation-invariance of μ we have

$$\int f\,d\mu \cdot \int g\,d\lambda^p = \int f\,d\lambda^p \cdot \int g(-x)\mu(dx).$$

If in particular we choose $f = 1_B$ with $B \in \mathfrak{B}^p$ and $g = 1_W$, we obtain

$$\mu(B) = \alpha\lambda^p(B),$$

where $\alpha = \int g(-x)\mu(dx) = \mu(-W)$. By the remark after the proof of Corollary 1.8.2, α is a real number. When $B = W$, it follows that $\alpha = 1$ and hence we have the assertion. ⌟

PROBLEMS

1. Let $\Omega_1 = \Omega_2$ be the real line $\mathbf{R}$ and $\mathfrak{A}_1 = \mathfrak{A}_2$ be the σ-algebra $\mathfrak{B}^1$ of all Borel sets. Take for μ_1 the L-B-measure λ^1 and for μ_2 the counting measure of Section 1.3, Problem 2 (which is not σ-finite). Prove that Fubini's formula [second equality in (3.2.3)] fails for $Q = D$ where $D \equiv \{(\omega,\omega): \omega \in \Omega\}$ is the "diagonal" of $\Omega_1 \times \Omega_2$. Why is D $\mathfrak{A}_1 \otimes \mathfrak{A}_2$-measurable?

2. Use Fubini's Theorem in order to prove that the function

$$(x,y) \to e^{-xy} - 2e^{-2xy}$$

is not λ^2-integrable over $A \equiv [1,+\infty[\times [0,1]$.

3. Let $|x|$ denote the Euclidean norm $(x_1^2 + \cdots + x_p^2)^{1/2}$ of a vector $x = (x_1, \ldots, x_p) \in \mathbf{R}^p$. Prove that the function $x \to e^{-|x|^\alpha}$ is λ^p-integrable for every $\alpha > 0$. [*Hint:* Use Section 2.8, Problem 2.] Prove that for $\alpha = 2$ the λ^p-integral equals $(\int_{-\infty}^{+\infty} e^{-x^2}\,dx)^p$.

4. Denote by $B_r(x_0)$ the closed ball of center x_0 and radius r in $\mathbf{R}^p$. Let α_p be the Lebesgue measure of $B_1(0)$. Prove:
   (a) $\lambda^p(B_r(x_0)) = \alpha_p r^p$;
   (b) $\alpha_{2q} = \dfrac{1}{q!}\pi^q,\ \alpha_{2q-1} = \dfrac{2^q}{1 \cdot 3 \cdot \ldots \cdot (2q-1)}\pi^{q-1}\ (q = 1, 2, \ldots)$.
   [*Hint:* Use (1.7.9) and observe that every x_p-section of $B_1(0)$ is either empty or a $(p-1)$-dimensional closed ball. This allows application of Fubini's Theorem and gives a recursion formula for α_p.]

5. Let h be a continuous real function on a compact interval $[\alpha,\beta] \subset \mathbf{R}_+$. Prove:

$$\int_{R(\alpha,\beta)} h(|x|)\lambda^p(dx) = p\alpha_p \int_\alpha^\beta h(t)t^{p-1}\,dt,$$

where $R(\alpha,\beta) \equiv \{x \in \mathbf{R}^p\colon \alpha \leqq |x| \leqq \beta\}$, $|x|$ is again the Euclidean norm in $\mathbf{R}^p$, and α_p is the λ^p-measure of the unit ball in $\mathbf{R}^p$ as defined in Problem 4. [*Hint:* Prove that the function

$$H(t) := \int_{R(\alpha,t)} h(|x|)\lambda^p(dx) \qquad (\alpha \leqq t \leqq \beta)$$

is differentiable on $[\alpha,\beta]$ and has $t \to p\alpha_p h(t)t^{p-1}$ as derivative.]

6. Apply Problem 5 to $p = 2$ and $h(t) = e^{-t^2}$ in order to prove

$$\int_{-\infty}^{+\infty} e^{-x^2/2}\,dx = \sqrt{2\pi}$$

via Problem 3.

3.3 EXTENSION TO THE CASE OF FINITELY MANY FACTORS

Now we return to the general case of Section 3.1. Suppose we are given finitely many *σ-finite* measure spaces $(\Omega_i,\mathfrak{A}_i,\mu_i)$, $i = 1, \ldots, n$ $(n \geqq 2)$.

We identify the two product sets $(\Omega_1 \times \cdots \times \Omega_{n-1}) \times \Omega_n$ and $\Omega_1 \times \cdots \times \Omega_n$ as is customary by means of the bijection

$$((\omega_1, \ldots, \omega_{n-1}), \omega_n) \to (\omega_1, \ldots, \omega_n).$$

The equality of these sets thus established then immediately implies the equality of the corresponding products of σ-algebras:

$$(\mathfrak{A}_1 \otimes \cdots \otimes \mathfrak{A}_{n-1}) \otimes \mathfrak{A}_n = \mathfrak{A}_1 \otimes \cdots \otimes \mathfrak{A}_n. \tag{3.3.1}$$

In fact, the sets $A_1 \times \cdots \times A_{n-1}$ with $A_i \in \mathfrak{A}_i$ generate

$$\mathfrak{A}_1 \otimes \cdots \otimes \mathfrak{A}_{n-1}$$

according to Theorem 3.1.1; by the same theorem, the sets

$$(A_1 \times \cdots \times A_{n-1}) \times A_n = A_1 \times \cdots \times A_n$$

with $A_i \in \mathfrak{A}_i$ $(i = 1, \ldots, n)$ generate both $(\mathfrak{A}_1 \otimes \cdots \otimes \mathfrak{A}_{n-1}) \otimes \mathfrak{A}_n$ and $\mathfrak{A}_1 \otimes \cdots \otimes \mathfrak{A}_n$.

Quite analogously, one proves the *associativity* of the product $\otimes$ in general:

$$\left(\bigotimes_{i=1}^{m} \mathfrak{A}_i\right) \otimes \left(\bigotimes_{i=m+1}^{n} \mathfrak{A}_i\right) = \bigotimes_{i=1}^{n} \mathfrak{A}_i \qquad (1 \leqq m < n). \tag{3.3.2}$$

The result (3.3.1) now makes it possible for us to verify the existence of the product measure for arbitrary $n \geqq 2$ by induction.

3.3.1. Theorem. There is exactly one measure π on $\mathfrak{A}_1 \otimes \cdots \otimes \mathfrak{A}_n$ with

$$\pi(A_1 \times \cdots \times A_n) = \mu_1(A_1) \cdot \ldots \cdot \mu_n(A_n) \tag{3.3.3}$$

for all $A_i \in \mathfrak{A}_i$ $(i = 1, \ldots, n)$.

Corresponding to Definition 3.2.4, we call π the *product* of the σ-finite measures $\mu_1, \ldots, \mu_n$ and denote it by

$$\bigotimes_{i=1}^{n} \mu_i = \mu_1 \otimes \cdots \otimes \mu_n.$$

Theorems 3.2.2 and 3.3.1 now answer the question raised in Section 3.1.

Proof. If in Theorem 3.1.2 we choose the σ-algebra $\mathfrak{A}_i$ itself as the generator $\mathfrak{E}_i$, then the uniqueness of π follows. For $n = 2$ the existence was shown in Theorem 3.2.3. We assume that $\pi' \equiv \mu_1 \otimes \cdots \otimes \mu_{n-1}$ exists for some $n > 2$ and derive the existence of $\mu_1 \otimes \cdots \otimes \mu_n$ as follows. When $\mu_1, \ldots, \mu_{n-1}$ are σ-finite, π' is also σ-finite and hence $\pi \equiv \pi' \otimes \mu_n$ is defined. By Theorem 3.2.3, π is a measure on the σ-algebra $(\mathfrak{A}_1 \otimes \cdots \otimes \mathfrak{A}_{n-1}) \otimes \mathfrak{A}_n$ with

$$\pi(Q' \times A_n) = \pi'(Q')\mu_n(A_n)$$

for all $Q' \in \mathfrak{A}_1 \otimes \cdots \otimes \mathfrak{A}_{n-1}$ and all $A_n \in \mathfrak{A}_n$. By (3.3.1), π is thus the desired measure. ┘

The proof also yields the equality

$$(\mu_1 \otimes \cdots \otimes \mu_{n-1}) \otimes \mu_n = \mu_1 \otimes \cdots \otimes \mu_n. \tag{3.3.4}$$

Analogous reasoning shows the associativity of the product in general:

$$(\bigotimes_{i=1}^{m} \mu_i) \otimes (\bigotimes_{i=m+1}^{n} \mu_i) = \bigotimes_{i=1}^{n} \mu_i \qquad (1 \leqq m < n). \tag{3.3.5}$$

In particular,

$$\lambda^n = \lambda^1 \otimes \cdots \otimes \lambda^1 \qquad (n \text{ factors}).$$

Using (3.3.5) and induction, we can now extend Fubini's Theorem. We formulate only the analog of Theorem 3.2.6:

Let $f \geqq 0$ be an $\mathfrak{A}_1 \otimes \cdots \otimes \mathfrak{A}_n$-measurable numerical function on the product $\Omega_1 \times \cdots \times \Omega_n$. Then for every permutation $i_1, \ldots, i_n$ of $1, \ldots, n$:

$$\int f \, d(\mu_1 \otimes \cdots \otimes \mu_n)$$
$$= \int(\cdots (\int(\int f(\omega_1, \ldots, \omega_n)\mu_{i_1}(d\omega_{i_1}))\mu_{i_2}(d\omega_{i_2})) \cdots)\mu_{i_n}(d\omega_{i_n}). \tag{3.3.6}$$

All the integrands involved here are measurable relative to the product of suitable σ-algebras $\mathfrak{A}_i$. Instead of the right side of (3.3.6) we use the abbreviation

$$\int \cdots \int f(\omega_1, \ldots, \omega_n)\mu_{i_1}(d\omega_{i_1}) \cdot \ldots \cdot \mu_{i_n}(d\omega_{i_n}).$$

The proof of this theorem (by induction) and the formulation and proof of the analog of Corollary 3.2.7 are left to the reader. Finally we give:

3.3.2. Definition. For any finite collection of σ-finite measure spaces $(\Omega_i, \mathfrak{A}_i, \mu_i)$, $i = 1, \ldots, n$, $\left(\prod_{i=1}^{n} \Omega_i, \bigotimes_{i=1}^{n} \mathfrak{A}_i, \bigotimes_{i=1}^{n} \mu_i\right)$ is called the *product* of there measure spaces and is denoted by

$$\bigotimes_{i=1}^{n} (\Omega_i, \mathfrak{A}_i, \mu_i).$$

3.4 CONVOLUTION OF FINITE BOREL MEASURES

We consider the p-dimensional Borel-measurable space $(\mathbf{R}^p, \mathfrak{B}^p)$. Every finite measure μ on $\mathfrak{B}^p$ will be called a *finite Borel measure;* we denote the set of these measures by $\mathfrak{M}^e(\mathbf{R}^p)$. For every $\mu \in \mathfrak{M}^e(\mathbf{R}^p)$ we call the number

$$\|\mu\| \equiv \mu(\mathbf{R}^p) \tag{3.4.1}$$

the *total mass of* μ.

By essential use of the group structure of $\mathbf{R}^p$ [5] with respect to the addition of points (vectors), we can associate with finitely many measures $\mu_1, \ldots, \mu_n \in \mathfrak{M}^e(\mathbf{R}^p)$ the so-called convolution product which, in contrast to the product measure, is again a measure on $\mathfrak{B}^p$ and indeed is an element of $\mathfrak{M}^e(\mathbf{R}^p)$.

Let $\mu_1 \otimes \cdots \otimes \mu_n$ be the product measure defined in Section 3.3; since $\mathfrak{B}^{np} = \mathfrak{B}^p \otimes \cdots \otimes \mathfrak{B}^p$, we have $\mu_1 \otimes \cdots \otimes \mu_n \in \mathfrak{M}^e(\mathbf{R}^{np})$. Now

$$(x_1, \ldots, x_n) \to A_n(x_1, \ldots, x_n) \equiv x_1 + \cdots + x_n$$

is a continuous and hence $\mathfrak{B}^{np}$-$\mathfrak{B}^p$-measurable mapping A_n of $\mathbf{R}^{np}$ onto $\mathbf{R}^p$; therefore, the following definition is meaningful:

3.4.1. Definition. For finitely many measures $\mu_1, \ldots, \mu_n \in \mathfrak{M}^e(\mathbf{R}^p)$, the image of $\mu_1 \otimes \cdots \otimes \mu_n$ under the mapping A_n is called the *convolution* (*product*) of these measures; in symbols:

$$\mu_1 * \cdots * \mu_n \equiv A_n(\mu_1 \otimes \cdots \otimes \mu_n). \tag{3.4.2}$$

By combining the theorems on product and image measures, we obtain the most important properties of the *convolution operation* $*$. First, $\mu_1 * \cdots * \mu_n$ is again an element of $\mathfrak{M}^e(\mathbf{R}^p)$; we have

$$\mu_1 * \cdots * \mu_n(\mathbf{R}^p) = \mu_1 \otimes \cdots \otimes \mu_n(\mathbf{R}^{np}) = \|\mu_1\| \cdot \ldots \cdot \|\mu_n\|$$

and hence

$$\|\mu_1 * \cdots * \mu_n\| = \|\mu_1\| \cdot \ldots \cdot \|\mu_n\|. \tag{3.4.3}$$

In studying the convolution product, it suffices to restrict consideration to the case $n = 2$, since

$$\mu_1 * \cdots * \mu_{n+1} = (\mu_1 * \cdots * \mu_n) * \mu_{n+1} \tag{3.4.4}$$

for any $n + 1$ measures in $\mathfrak{M}^e(\mathbf{R}^p)$. Indeed we have $A_{n+1} = A_2 \circ B_{n+1}$ where B_{n+1} denotes the continuous mapping

$$(x_1, \ldots, x_{n+1}) \to (x_1 + \cdots + x_n, x_{n+1})$$

of $\mathbf{R}^{(n+1)p}$ onto $\mathbf{R}^{2p}$. If we note that

$$B_{n+1}(\mu_1 \otimes \cdots \otimes \mu_{n+1}) = A_n(\mu_1 \otimes \cdots \otimes \mu_n) \otimes \mu_{n+1},$$

[5] The following considerations hold in general for arbitrary *Abelian locally compact groups*. See W. Rudin [42] and Hewitt-Ross [36].

then by the transitivity of image measures, it follows that

$$\mu_1 * \cdots * \mu_{n+1} = A_2(B_{n+1}(\mu_1 \otimes \cdots \otimes \mu_{n+1})) \\ = A_2((\mu_1 * \cdots * \mu_n) \otimes \mu_{n+1})$$

and hence we have (3.4.4). Henceforth, then, let $n = 2$.

For any two measures $\mu, \nu \in \mathfrak{M}^e(\mathbf{R}^p)$ and every $\mathfrak{B}^p$-measurable numerical function $f \geqq 0$ on $\mathbf{R}^p$, it then follows from Fubini's Theorem and Theorem 2.10.1 that

$$\int f\, d(\mu * \nu) = \int f \circ A_2\, d(\mu \otimes \nu) \\ = \int\int f(x + y)\mu(dx)\nu(dy) = \int\int f(x + y)\nu(dy)\mu(dx). \quad (3.4.5)$$

In particular, (3.4.5) yields for the indicator function $f = 1_B$ of any set $B \in \mathfrak{B}^p$:

$$\mu * \nu(B) = \int \mu(B - y)\nu(dy) = \int \nu(B - x)\mu(dx).^6 \quad (3.4.6)$$

Consequently, $*$ is a *commutative* and by (3.4.4) also an *associative* operation in $\mathfrak{M}^e(\mathbf{R}^p)$.

By Corollaries 2.10.2 and 3.2.7 it follows that (3.4.5) also holds for every $\mu * \nu$-integrable numerical function f on $\mathbf{R}^p$. (3.4.6) is often used for the definition of $\mu * \nu$. When $\mathfrak{M}^e(\mathbf{R}^p)$ contains any two measures ν_1 and ν_2, it obviously contains their sum $\nu_1 + \nu_2$; when it contains any measure ν, it contains the product $\alpha\nu$ for every number $\alpha \in \mathbf{R}_+$. Then from (3.4.6) we have immediately

$$\mu * (\nu_1 + \nu_2) = \mu * \nu_1 + \mu * \nu_2, \quad (3.4.7)$$

$$(\mu, \nu_1, \nu_2, \nu \in \mathfrak{M}^e(\mathbf{R}^p);\ \alpha \in \mathbf{R}_+).$$

$$\mu * (\alpha\nu) = (\alpha\mu) * \nu = \alpha(\mu * \nu) \quad (3.4.8)$$

The *distributive law* (3.4.7) also holds in the following generality: For every sequence (ν_n) of measures in $\mathfrak{M}^e(\mathbf{R}^p)$ with $\Sigma\|\nu_n\| < \infty$, $\Sigma\nu_n$ is also a measure in $\mathfrak{M}^e(\mathbf{R}^p)$ (see Section 1.3, Example 4). Then, in view of Corollary 2.3.5, (3.4.6) implies

$$\mu * \left(\sum_{n=1}^{\infty} \nu_n\right) = \sum_{n=1}^{\infty} \mu * \nu_n \quad (3.4.9)$$

for every $\mu \in \mathfrak{M}^e(\mathbf{R}^p)$.

Now we compute $\mu * \nu$ in several particular cases.

1. Let T_a again denote the translation mapping $x \to x + a$ of $\mathbf{R}^p$ onto itself ($a \in \mathbf{R}^p$). Further, let ϵ_a be the measure on $\mathfrak{B}^p$ defined by the unit mass at a; we have $\epsilon_a \in \mathfrak{M}^e(\mathbf{R}^p)$ and $\|\epsilon_a\| = 1$. Then it follows from

[6] Here $B - x = -x + B$ by (1.7.8).

(3.4.6)[7] that $\epsilon_a * \mu(B) = \mu(B - a) = \mu(T_a^{-1}(B))$ and hence

$$\epsilon_a * \mu = T_a(\mu) \qquad (\mu \in \mathfrak{M}^e(\mathbf{R}^p)). \tag{3.4.10}$$

Now T_0 is the identity mapping, so that ϵ_0 is a *unit* and obviously the only one with respect to convolution.

In particular, for $\mu = \epsilon_b$ we obtain

$$\epsilon_a * \epsilon_b = \epsilon_{a+b} \qquad (a,b \in \mathbf{R}^p). \tag{3.4.10'}$$

2. Let $f \geqq 0$ be a λ^p-integrable numerical function on $\mathbf{R}^p$ and $\mu = f\lambda^p$. Since $\|\mu\| = \int f\,d\lambda^p < \infty$, μ then lies in $\mathfrak{M}^e(\mathbf{R}^p)$. We compute $\mu * \nu$ for an arbitrary $\nu \in \mathfrak{M}^e(\mathbf{R}^p)$. By Theorem 2.9.3 using the translation-invariance of λ^p and Theorem 2.10.1 for every $B \in \mathfrak{B}^p$, we obtain

$$\begin{aligned} \mu * \nu(B) &= \int\int 1_B(x + y)f(x)\lambda^p(dx)\nu(dy) \\ &= \int\int 1_B(x + y)f(x)T_{-y}(\lambda^p)(dx)\nu(dy) \\ &\qquad = \int\int 1_B(x)f(x - y)\lambda^p(dx)\nu(dy). \end{aligned}$$

With the help of Fubini's Theorem 3.2.6 it now follows that

$$\mu * \nu(B) = \int 1_B(x)g(x)\lambda^p(dx) = \int_B g\,d\lambda^p,$$

where $x \to g(x) \equiv \int f(x - y)\,\nu(dy)$ is a $\mathfrak{B}^p$-measurable function $\geqq 0$. Since $\int g\,d\lambda^p = \|\mu * \nu\| < \infty$, it is λ^p-integrable. Along with μ, $\mu * \nu$ thus also has a density relative to λ^p. We set $f * \nu \equiv g$, that is, we define

$$f * \nu(x) \equiv \int f(x - y)\nu(dy) \qquad (x \in \mathbf{R}^p) \tag{3.4.11}$$

and we can thus write the result in the form

$$(f\lambda^p) * \nu = (f * \nu)\lambda^p, \tag{3.4.12}$$

which is easy to remember. We call $f * \nu$ the *convolution* of f and ν.

3. Now suppose not only $\mu = f\lambda^p$ but also $\nu = g\lambda^p$ has a λ^p-integrable density $g \geqq 0$ relative to λ^p. Then, by Theorem 2.9.3

$$f * (g\lambda^p)(x) = \int f(x - y)g(y)\,\lambda^p(dy) \qquad (x \in \mathbf{R}^p)$$

is the density of $\mu * \nu$ relative to λ^p. We denote it by $f * g$, that is, we set

$$f * g(x) \equiv \int f(x - y)g(y)\lambda^p(dy) \qquad (x \in \mathbf{R}^p) \tag{3.4.13}$$

and obtain

$$(f\lambda^p) * (g\lambda^p) = (f * g)\lambda^p. \tag{3.4.14}$$

We call $f * g$ the *convolution* of f and g. For every two λ^p-integrable functions $\geqq 0$, it is defined and is itself such a function. It follows from (3.4.13),

[7] See Example 1 in Sections 2.3 and 2.4.

with the help of the reflection- and translation-invariance of λ^p for every $x \in \mathbf{R}^p$, that

$$f * g(x) = \int f(x - y)g(y)\lambda^p(dy) = \int f(x + y)g(-y)\lambda^p(dy)$$
$$= \int f(y)g(x - y)\lambda^p(dy) = g * f(x),$$

that is, this operation is *commutative:*

$$f * g = g * f. \tag{3.4.15}$$

Analogous reasoning yields the *associativity*, that is,

$$(f * g) * h = f * (g * h) \tag{3.4.16}$$

for any three λ^p-integrable functions $f \geqq 0$, $g \geqq 0$, $h \geqq 0$. The *distributive law*

$$f * (g + h) = f * g + f * h \tag{3.4.17}$$

and

$$f * (\alpha g) = (\alpha f) * g = \alpha(f * g) \qquad (\alpha \in \mathbf{R}_+) \tag{3.4.18}$$

follow directly from (3.4.13) for such functions.

Remark. For functions $f \geqq 0$, $g \geqq 0$ in $\mathcal{L}^1(\lambda^p)$, $f * g$ is not necessarily finite. It suffices to choose for f a function $\geqq 0$ in $\mathcal{L}^1(\lambda^p)$ symmetric with respect to $x = 0$ which does not lie in $\mathcal{L}^2(\lambda^p)$. Then $f * f(0) = \infty$. For $p = 1$, we may choose, for example, the function

$$f(x) = \begin{cases} 0, & \text{if } |x| > 1 \text{ or } x = 0 \\ \dfrac{1}{\sqrt{|x|}}, & \text{if } 0 < |x| \leqq 1. \end{cases}$$

4. For arbitrary functions $f, g \in \mathcal{L}^1(\lambda^p)$, by using the decomposition into positive and negative parts, we obtain from 3 above that

$$x \to \int f(x - y)g(y)\lambda^p(dy)$$

is only λ^p-almost everywhere defined, but is λ^p-integrable as before.

Therefore we set

$$f * g(x) \equiv \int f(x - y)g(y)\lambda^p(dy)$$

now only for all $x \in \mathbf{R}^p$ for which the right side exists. We again call $f * g$ the convolution of f and g.

Remark. The difficulties in defining $f * g$ disappear in the transition to $L^1(\lambda^p)$. Indeed, let $f \to \tilde{f}$ be the canonical mapping of $\mathcal{L}^1(\lambda^p)$ onto $L^1(\lambda^p)$. We define $\tilde{f} * \tilde{g}$ for arbitrary $\tilde{f}, \tilde{g} \in L^1(\lambda^p)$ as the image $\tilde{h}$ of any function $h \in \mathcal{L}^1(\lambda^p)$ which coincides λ^p-almost everywhere with $f * g$. This definition is independent of the particular choice of f, g, and h ($f,g \in \mathfrak{L}^1(\lambda^p)$). With the additional operation $*$, the vector space $L^1(\lambda^p)$ becomes an *algebra* over **R**.

PROBLEMS

1. Prove that

$$N_1(f * g) \leqq N_1(f)N_1(g) \qquad \text{for all } f,\, g \in \mathfrak{L}^1(\lambda^p)$$

   and that equality holds for $f \geqq 0$ and $g \geqq 0$, where [as introduced in (2.6.1)] $N_1(f) = \int |f| \, d\lambda^p$.

2. Prove the details of the preceding remark and show that $\|f * g\|_1 \leqq \|f\|_1 \|g\|_1$ holds for all elements f, g of the Banach space $L^1(\lambda^p)$.

part II

Probability Theory

4

FUNDAMENTAL CONCEPTS OF THE THEORY

4.1 PROBABILITY SPACES

Probability theory will provide us with mathematical models for describing experiments with random outcomes. In such an experiment we are interested most of all in (random) "events" and their "probabilities." Accordingly, one of the first goals of the theory will be to develop a mathematical model for these two concepts.

1. To this end we consider one of the simplest experiments with random outcome, namely, tossing a fair die. The events to be studied are then, first of all, those described colloquially by "a k is thrown" ($k = 1, 2, \ldots, 6$). These particular events, which will later be called elementary events, are in one-to-one correspondence with the natural numbers $1, \ldots, 6$. But more complicated events are also of interest; for example, the result of a toss is even, odd, or not a one. These events can obviously be identified with sets of numbers $1, 2, \ldots, 6$, namely, $\{2,4,6\}$, $\{1,3,5\}$, or $\{2,3,4,5,6\}$. Thus, these events turn out to be subsets of the set $\Omega \equiv \{1,2, \ldots ,6\}$. In fact, every event in this experiment can be identified by means of the "number of dots" in a natural way with a subset of Ω and conversely, every subset of Ω can be interpreted as an event. For example, $\varnothing$ and Ω are such subsets; they represent the impossible event (say, throwing a 7) and the sure event (namely throwing 1 or 2 or . . . or 6). Thus, the set of all events is identified with the power set $\mathfrak{P}(\Omega)$ of Ω.

This identification has a noteworthy property. If A and B are subsets of Ω, that is, events, then the meanings of the events "A or B occurs,"

"A and B occur simultaneously," and "A does not occur" are intuitively clear. In the above identification, these events are represented by the sets $A \cup B$, $A \cap B$ and $\complement A$. Therefore, we can also speak of the algebra and obviously even the *σ-algebra* $\mathfrak{P}(\Omega)$ of the events.

In this die-tossing experiment we further speak of the probability of an event E or the probability of the occurrence of the event E. This probability is a real number $P(E)$ associated with E; thus, we have a real function $P: \mathfrak{P}(\Omega) \to \mathbf{R}$. If in a series of trials we now toss n times in a row, then in k of these n trials the event E will occur, and in the remaining $n - k$ trials the event "not E" $= \complement E$ will occur. It is an experimentally established fact that the quotient k/n, which measures the relative frequency of the occurrence of E, exhibits a deviation from $\frac{1}{6}|E|$ which approaches zero for large n. Here $|E|$ denotes the number of elements of the set $E \subset \Omega$. This fact leads us to speak of these quotients as the probability of the event E, and thus to define

$$P(E) \equiv \frac{|E|}{|\Omega|} \qquad (E \in \mathfrak{P}(\Omega)). \tag{4.1.1}$$

From (4.4.1) we immediately obtain the following properties of this function P: $P \geqq 0$, $P(\varnothing) = 0$, $P(\Omega) = 1$, and $P(E \cup F) = P(E) + P(F)$ for any two "inconsistent" events E, F—events with $E \cap F = \varnothing$, that is, whose simultaneous occurrence is impossible. Thus P is a content and, since Ω is finite, a *measure* on $\mathfrak{P}(\Omega)$ with the *normalization property* $P(\Omega) = 1$. Moreover, P is the only such content for which all "elementary events" $\{k\}$, $k = 1, \ldots, 6$, are "equally probable."

2. If we discuss another experiment in place of this one, say tossing a coin, or throwing a die three times, we would *formally* arrive at the same result: in the first case, Ω would be the set $\{H,T\}$ of the two (different) symbols H = "head" and T = "tail", and in the second case it would be the set of all ordered triples (k_1,k_2,k_3) of numbers $1, 2, \ldots, 6$. In both cases $\mathfrak{P}(\Omega)$ would be identified with the set of all events. According to (4.1.1) the probability would turn out to be a function $P: \mathfrak{P}(\Omega) \to \mathbf{R}$ and would be a measure on $\mathfrak{P}(\Omega)$ normalized to one.

In the examples so far Ω was always a finite set and all subsets of Ω were interpreted as events. The following example shows that we can have a different situation for infinite Ω.

3. Suppose a gun is shot at a target. Notwithstanding all previous practice, we can view the successful hit of the target as a random event; it does, after all, depend on many nonobvious influences such as unsureness of eye and hand of the marksman, mechanical failure of the gun, movement of air along the firing range, and so on. The events in which we are

interested are then of the following sort: The successful hit lies in a given subset E of the target Ω (for example, in one of the 12 rings). This formulation already makes it clear that such events can be identified in a natural fashion with the subsets E of Ω. As an approximation[1] we can say that the probability $P(E)$ of such an event E is proportional to the 2-dimensional Lebesgue measure $\lambda^2(E)$ of E, that is,

$$P(E) \equiv \frac{\lambda^2(E)}{\lambda^2(\Omega)}. \tag{4.1.2}$$

Of course E must then be a Borel (or at least a Lebesgue-measurable) subset of Ω. But (4.1.2) nevertheless is of a structure analogous to (4.1.1).

Although we have so far been inclined to interpret *all* subsets of Ω as events, the restricted domain of definition of the Lebesgue-Borel measure forces us to allow only *special*, namely Borel, subsets of Ω (that is, hits in such subsets) as events. This turns out to be completely adequate for practical purposes. It is then crucial that we formally have the same situation despite the difference from the preceding examples: The set $\mathfrak{A}$ of these (Borel) events is a σ-algebra in Ω as well as $\mathfrak{P}(\Omega)$; the function P defined on $\mathfrak{A}$ according to (4.1.2) is again a measure normalized to one on this σ-algebra of events. It is this formal analogy which led to the foundation of probility theory which is now commonly accepted.

We know from the axiomatic foundations of geometry, algebra, topology, and other mathematical disciplines that concepts such as point and line, number, neighborhood, and so on, are not defined explicitly. Similarly, it has turned out that an explicit definition of concepts such as "event" and "probability" is not necessary for a development of probability theory, and indeed is even not worth attempting, because of the logical difficulties involved, in order to obtain a comprehensive and easily applicable theory. As in the other mathematical disciplines mentioned above, in probability theory we are concerned only with the formal properties of these concepts. The Russian mathematician A. N. Kolmogorov [38] is credited with being the first to pursue this argumentation consistently and successfully.

In probability theory we consider three mathematical objects, as motivated by the heuristic preliminary considerations:

1. an abstract set Ω,
2. a σ-algebra $\mathfrak{A}$ in Ω, and
3. a measure P on $\mathfrak{A}$ with $P(\Omega) = 1$.

Henceforth every such triple $(\Omega,\mathfrak{A},P)$ will be called a *probability space.*

[1] We can obtain a better approximation with the help of the concept of distribution to be introduced in Section 4.3.

The elements ω of Ω are called *elementary events*. These are intuitively the possible random outcomes of an experiment or an observation. The elements E of $\mathfrak{A}$ are called *events; $P(E)$* is called the *probability of E*. We call P a *probability measure.*[2] Every such measure attains its values in the interval [0,1], as follows immediately from the isotonicity of measures.

We should note the following: In many concrete cases, for every elementary event $\omega \in \Omega$, the one-point set $\{\omega\}$ will belong to $\mathfrak{A}$, that is, it will be an event. However, this is not required in the definition of a probability space and also does not follow from the definition, which can easily be seen in examples.

In view of the intuitive interpretation of the concept "event," the following manner of speaking, which suggest a certain intuition, are also common (besides the pure set-theoretic ones): $\emptyset$ is called the *impossible event* and Ω the *sure event.* If E and F are events in $\mathfrak{A}$, then we say E *implies* F if $E \subset F$; E and F are said to be *inconsistent* if $E \cap F = \emptyset$; $E \setminus F$ is designated as the event "*E occurs but not F.*" If $E_1, E_2, \ldots$ are events, then $\bigcup E_n$ is said to be the event "*at least one E_n occurs,*" and $\bigcap E_n$ the event "*all E_n occur* (simultaneously)." Events $E \in \mathfrak{A}$ with $P(E) = 1$, $P(E) = 0$ are called *almost sure, almost impossible* events, respectively. Instead of (P-) almost everywhere we say (P-) *almost surely* [abbreviated as (P-) a.s.] or *with probability one.*

In the introductory examples, we gave for intuitive experiments with random outcome the probability spaces needed for their mathematical description. In general, the mathematical treatment of a concrete problem in probability theory is based on the explicit assignment of a suitable probability space. We now discuss an additional example which will be fundamental for the following.

4. We consider a finite number n of experiments with random outcomes; for the mathematical description of the ith experiment $\mathcal{E}_i$ we use the probability space $(\Omega_i, \mathfrak{A}_i, P_i)$, $i = 1, \ldots, n$. We are interested in a new experiment $\mathcal{E}$ with random outcome which consists of performing the experiments $\mathcal{E}_1, \ldots, \mathcal{E}_n$ "without mutual influence" one after another or simultaneously. (For example, we might have $\mathcal{E}_1 = \cdots = \mathcal{E}_n$ and each $\mathcal{E}_i$ might describe tossing a fair die. Then the experiment $\mathcal{E}$ consists of tossing n times in succession or tossing n dice simultaneously, which amounts to the same thing. We already considered the case $n = 3$ in 2.) The random outcomes of the experiment $\mathcal{E}$ can be represented by the points of the product set $\Omega_1 \times \cdots \times \Omega_n$, that is, by the n-tuple $(\omega_1, \ldots, \omega_n)$ of

[2] The notation P for probability measures has been widely adopted in probability theory.

elements $\omega_i \in \Omega_i$. Now if $A_i \in \mathfrak{A}_i$ for every $i = 1, \ldots, n$ is an event in the experiment $\mathcal{E}_i$, then in the experiment $\mathcal{E}$ we are interested in whether the events $A_1, A_2, \ldots, A_n$ occur in succession. We are thus interested in the random event of the experiment $\mathcal{E}$ which is given by the subsets $A_1 \times \cdots \times A_n$ of $\Omega_1 \times \cdots \times \Omega_n$. Since there should be no "mutual influence," we should assign the product $P_1(A_1) \cdot \ldots \cdot P_n(A_n)$ as the probability of the event $A_1 \times \cdots \times A_n$. But these sets

$$A_1 \times \cdots \times A_n$$

generate precisely the σ-algebra $\mathfrak{A}_1 \otimes \cdots \otimes \mathfrak{A}_n$, and according to Theorem 3.3.1 the product measure $P_1 \otimes \cdots \otimes P_n$ is the only measure P on it satisfying

$$P(A_1 \times \cdots \times A_n) = P_1(A_1) \cdot \ldots \cdot P_n(A_n)$$

for arbitrary $A_i \in \mathfrak{A}_i$. We see immediately that P is a probability measure. Therefore, we consider

$$\bigotimes_{i=1}^{n} (\Omega_i, \mathfrak{A}_i, P_i)$$

as the probability space adequate for describing the experiment $\mathcal{E}$. We can see that for finite sets Ω_i and $\mathfrak{A}_i = \mathfrak{P}(\Omega_i)$ we obtain: $\mathfrak{A}_1 \otimes \cdots \otimes \mathfrak{A}_n = \mathfrak{P}(\Omega_1 \times \cdots \times \Omega_n)$. In the case of n tosses of a die we thus obtain the probability space described above in Example 2 for $n = 3$.

4.2 TREATMENT OF SEVERAL ELEMENTARY PROBLEMS

What we said in Section 4.1 will now be illustrated in several examples which are typical of problems posed in elementary probability theory and which are also encountered in many problems arising in practice. Here we shall always be dealing, directly or indirectly, with so-called *Laplace experiments*. These are experiments with finitely many possible random outcomes which, moreover, are considered to be equally probable. The corresponding mathematical model is a probability space $(\Omega, \mathfrak{A}, P)$ in which Ω is a nonempty finite set of N elements, $\mathfrak{A} = \mathfrak{P}(\Omega)$ and P is the only probability measure with $P(\{\omega\}) = N^{-1}$ for every $\omega \in \Omega$. This probability space is called *Laplace space of order N*. Computation of the probability of an event E is then based upon combinatorial considerations. The starting point for this is formula (4.1.1), where $P(E)$ is equal to "*number of cases favorable to E*" divided by "*number N of possible cases*."

Examples

1. In an urn there are b black balls and w white balls ($b + w = n$) which are well mixed. We choose at random $m \leqq n$ balls and want to know the

probability of obtaining exactly $k \leqq m$ black balls. There are two important cases:

(a) *Drawing without replacement.* The m balls are drawn successively and left outside the urn. If we index the balls with the natural numbers 1 through n, then Ω consists of all sequences $(a_1, \ldots, a_m)$ of m distinct numbers from 1 through n, that is, it has

$$N \equiv n(n-1) \cdot \ldots \cdot (n-m+1)$$

elements. Under the hypothesis that we are dealing with a Laplace experiment, that is, $(\Omega, \mathfrak{A}, P)$ is the Laplace space of order N, we obtain the solution as follows: The event E being studied consists of all sequences $(a_1, \ldots, a_m)$ of the kind described in which the black balls are indexed by exactly k of the numbers $a_1, \ldots, a_m$. We can associate k distinct numbers in $\binom{m}{k}$ ways with the available indices $1, \ldots, m$. k black balls can be drawn in exactly $b(b-1) \cdot \ldots \cdot (b-k+1)$ different ways in succession. The remaining $m-k$ indices can be associated in $w(w-1) \cdot \ldots \cdot (w-m+k+1)$ further ways with numbers on white balls. The desired probability is therefore

$$\binom{m}{k} \frac{b(b-1) \cdot \ldots \cdot (b-k+1) w(w-1) \cdot \ldots \cdot (w-m+k+1)}{n(n-1) \cdot \ldots \cdot (n-m+1)}$$

$$= \frac{\binom{b}{k}\binom{w}{m-k}}{\binom{n}{m}}.$$

But we can also interpret the problem as follows: The m balls are not drawn successively from the urn, but are taken *in one draw*. Then there are obviously $\binom{n}{m}$ possible and $\binom{b}{k}\binom{w}{m-k}$ favorable cases. We then arrive at the same probability.

Case 1(a) is also encountered, in a different form, in the following practical problem. Instead of an urn with balls, suppose we are discussing a day's production of a mass-produced article. We translate "black" as "defective" and "white" as "nondefective." Then we have just computed the probability that in a random sample of size m from the day's production we find exactly k defectives. Of course this involves the number b of all defectives which is in general unknown. If we want to eliminate the unknown b, we are led to problems posed in mathematical statistics.

(b) *Drawing with replacement.* Every ball drawn is immediately replaced in the urn; after another mixing of the contents of the urn, the next ball is drawn at random. Here we obviously have the situation described in Section 4.1, (4): The Laplace experiment "draw one ball" is repeated m times without mutual influence. The single experiment is described by the Laplace space $(\Omega_0,\mathfrak{A}_0,P_0)$ of order n, the combined experiment by the product $(\Omega,\mathfrak{A},P)$ of m copies of $(\Omega_0,\mathfrak{A}_0,P_0)$. Then $(\Omega,\mathfrak{A},P)$ is the Laplace space of order n^m. If A_i denotes the event of drawing a black ball on the ith draw, then $p \equiv P_0(A_i) = b/n$ and $P_0(\complement A_i) = 1 - p$. The event $A_{i_1 \ldots i_k} \in \mathfrak{A}$, of drawing a black ball on the i_1st, . . . , i_kth draws $(1 \leqq i_1 < \cdots < i_k \leqq m)$ and a white one on the remaining $m - k$ draws is the product $B_1 \times \cdots \times B_n$, where $B_{i_\varkappa} = A_{i_\varkappa}$ $(\varkappa = 1, \ldots, k)$ and $B_j = \complement A_j$ for the remaining indices j. Consequently, $P(A_{i_1 \ldots i_k}) = p^k(1 - p)^{m-k}$, and by the additivity of P,

$$\binom{m}{k} p^k(1 - p)^{m-k} \qquad \text{with } p = \frac{b}{n}$$

is the desired probability.

2. Each of m persons chooses a natural number from the set $\{1, \ldots ,n\}$ at random and without knowledge of what anyone else does (with the same given n for all people). After all of them have made their choices, the events are reported. What is the probability that m different numbers were chosen?

The situation can obviously (always under the hypothesis that this is a Laplace experiment) also be described as follows: There are n balls in an urn. m balls are chosen at random with replacement. What is the probability that m different balls are drawn? Since there are obviously $n(n - 1) \cdot \ldots \cdot (n - m + 1)$ favorable cases, the answer is

$$\frac{n(n - 1) \cdot \ldots \cdot (n - m + 1)}{n^m}.$$

3. Two cards are chosen successively, without replacement, from a well-mixed deck of 52 cards. What is the probability that (a) the second card drawn is an ace, (b) the second card is an ace if the first draw already turned up an ace?

To answer (a) we naturally use the Laplace space $(\Omega,\mathfrak{A},P)$ of order $51 \cdot 52$. The number of cases favorable to the described event is $4 \cdot 51$. Thus the desired probability is 1/13. In case (b) we can use the Laplace space of order $4 \cdot 51$ and compute the desired probability as $4 \cdot 3/4 \cdot 51 = 1/17$. Although the same experiment was performed, we are using different probability spaces to answer the two questions. This seems impractical and leads us to the following method of solution using the same probability space: Let $A \in \mathfrak{A}$ be the event whose probability is desired in (a) and let $B \in \mathfrak{A}$ be the event that the first draw yields an ace. The answer to question (b) then goes as follows: $P(B)$ is the number of possible cases and $P(A \cap B)$ is the number of cases favorable to the desired event; hence, $P(A \cap B)/P(B)$ is the desired probability.

More generally, we have the following situation: Suppose we are given an arbitrary probability space $(\Omega,\mathfrak{A},P)$ and an event $B \in \mathfrak{A}$ with $P(B) > 0$. Then obviously

$$A \rightarrow \frac{P(A \cap B)}{P(B)} \qquad (A \in \mathfrak{A})$$

is again a probability measure P_B on $\mathfrak{A}$. We have $P_B(B) = 1$ even though $P(B)$ is not equal to 1 in general. Thus, by this simple procedure, B has become an event with probability 1. Accordingly, for every $A \in \mathfrak{A}$ we call $P_B(A)$ the *conditional probability of A given B* or under the *hypothesis* B and write $P(A|B)$ for $P_B(A)$; that is, we define

$$P(A \mid B) \equiv \frac{P(A \cap B)}{P(B)}. \tag{4.2.1}$$

Thus in Example 3(b), we were dealing with the computation of a conditional probability.

Equation (4.2.1) can be generalized immediately. Suppose we are given a (possibly finite) sequence $(B_n)_{n \in I}$ (where $I = \{1, \ldots ,n\}$ and $n \in \mathbf{N}$ or $I = \mathbf{N}$) of pairwise disjoint events $B_n \in \mathfrak{A}$ with $P(B_n) > 0$ for all n and $\Omega = \bigcup B_n$. Since $A = \bigcup (A \cap B_n)$, it then follows from the σ-additivity of P that $P(A) = \Sigma P(A \cap B_n)$ and

$$P(A) = \sum_{n \in I} P(B_n)P(A \mid B_n) \qquad (A \in \mathfrak{A}). \tag{4.2.2}$$

This is the *formula of total probability*. For $P(A) > 0$, we have

$$P(B_n \mid A) = \frac{P(B_n)P(A \mid B_n)}{\sum_{i \in I} P(B_i)P(A \mid B_i)} \qquad (n = 1, 2, \ldots), \tag{4.2.3}$$

since $P(B_n \mid A) = P(A \cap B_n)(P(A))^{-1}$. This is the *Bayes formula*. Both formulas are illustrated by the following example:

Example

4. Suppose we are given $N + 1$ urns $U_0, \ldots , U_N$. In each urn U_n there are N equally shaped balls, well mixed; n of them are black and $N - n$ are white ($n = 0, \ldots , N$). Suppose we choose an urn at random and draw one ball from it. What is the probability that we draw a black ball?

Let B_n be the event that the ball was drawn from the nth urn and A the event that a black ball was chosen. Under the assumption that all urns are equally probable, we have

$$P(B_n) = (N+1)^{-1} \qquad \text{and} \qquad P(A \mid B_n) = \frac{n}{N}.$$

Then by (4.2.2),

$$P(A) = \frac{1 + \cdots + N}{N(N+1)} = \frac{1}{2}.$$

We ask further: Assume that a black ball was drawn. What is the probability that it was drawn from the nth urn? The answer is given by (4.2.3):

$$P(B_n \mid A) = \frac{2n}{N(N+1)}.$$

As expected, this probability increases proportionally with n.

Since we are obviously dealing with a Laplace experiment, the usual combinatorial considerations would also have led to the same result. But the two formulas make our work easier. If we now imagine a random mechanism by which we choose the urn U_n with probability $\alpha_n > 0$ ($\alpha_0 + \cdots + \alpha_N = 1$), then we have a Laplace experiment in this example only for equal α_n. Nevertheless, (4.2.2) and (4.2.3) lead to the result: We now have $P(B_n) = \alpha_n$ and again $P(A \mid B_n) = n/N$. Thus, if we set $M \equiv \alpha_1 + 2\alpha_2 + \cdots + N\alpha_N$, we obtain

$$P(A) = \frac{M}{N} \qquad \text{and} \qquad P(B_n \mid A) = \frac{n\alpha_n}{M}.$$

We have a Laplace experiment here only after having chosen an urn. We therefore speak of a *relay-experiment.*

The detailed specification of the probability space used is also very instructive. It is based on the following general considerations: Let Ω be an arbitrary nonempty set and $(\Omega_i)_{i=1,\ldots,m}$ a partition of Ω into pairwise disjoint sets $\Omega_i \neq \varnothing$. Suppose each of these sets is the carrier of a probability space $(\Omega_i,\mathfrak{A}_i,P_i)$. The system $\mathfrak{A}$ of all sets $A \subset \Omega$ with $A \cap \Omega_i \in \mathfrak{A}_i$ for every $i = 1, \ldots, m$ is a σ-algebra in Ω consisting of all sets $A_1 \cup \cdots \cup A_m$ with $A_i \in \mathfrak{A}_i$.[3] Further, if $\alpha_1, \ldots, \alpha_m$ are positive real numbers with $\alpha_1 + \cdots + \alpha_m = 1$, then by (4.2.2),

$$A \to \sum_{i=1}^{m} \alpha_i P_i(A \cap \Omega_i)$$

[3] $(\Omega,\mathfrak{A})$ is the so-called *direct sum* of the measurable spaces $(\Omega_i,\mathfrak{A}_i)$. The construction can be extended immediately to the case of countably many $(\Omega_i, \mathfrak{A}_i)$.

is the only probability measure P on $\mathfrak{A}$ with $P(\Omega_i) = \alpha_i$ and $P(A \mid \Omega_i) = P_i(A \cap \Omega_i)$ for every $i = 1, \ldots, m$.

In our example, $m = N + 1$, every $(\Omega_i,\mathfrak{A}_i,P_i)$ is a Laplace space of order N with pairwise disjoint sets Ω_i, and $\Omega = \bigcup \Omega_i$. The probability space $(\Omega,\mathfrak{A},P)$ just constructed is then the mathematical model of the last-described relay-experiment.

PROBLEMS

1. A die is tossed n times. What is the probability that exactly at the nth toss a 4 is thrown for the kth time $(1 \leqq k \leqq n)$?
2. Suppose we are given three urns U_1, U_2, U_3 which contain, well mixed, equal balls of black, white, and red color. Assume that

   U_1 contains 2 black, 3 white, 5 red balls,
   U_2 contains 4 black, 2 white, 4 red balls,
   U_3 contains 2 black, 5 white, 3 red balls,

   (a) What is the probability to draw from U_1 without replacement first a black, then a black, and then a red ball?
   (b) What is the probability P_b (or P_w, or P_r) to draw, after choosing at random one of the urns, a black (or white, or red) ball? Prove: $P_b + P_w + P_r = 1$.
   (c) What is the probability to draw, after choosing at random one of the urns, 4 black balls successively without replacement? What is the probability that these 4 black balls come from the urn U_2?
3. What is the probability that 10 persons chosen at random have their birthdays in different months?
4. In an urn there are, well mixed, balls of r different colors, namely $k_i > 0$ balls of color $F_i (i = 1, \ldots, r)$. Suppose that n balls are taken in one draw $(1 \leqq n \leqq k_1 + \cdots + k_r)$. What is the probability of obtaining exactly n_i balls of color $F_i (n_i \geqq 0,\ n_1 + \cdots + n_r = n)$?
5. In an urn there are $b \geqq 1$ black and $w \geqq 1$ white balls, well mixed. We draw successively balls without replacement. What is the probability that x draws are necessary in order to obtain $k \leqq b$ black balls?
6. (*Polya's urn model.*) In an urn there are $b \geqq 1$ black and $w \geqq 1$ white balls, well mixed. A ball is drawn at random. It is replaced and, moreover, t balls of the color drawn are added. Then the next ball is drawn at random, and the above procedure is repeated. Define $N \equiv b + w$, and let n be a natural number. Prove: The probability that in n draws

k black and $n - k$ white balls appear ($k = 0, 1, \ldots, n$) equals

$$\binom{n}{k} \frac{b(b+t) \cdot \ldots \cdot (b+[k-1]t)w(w+t) \cdot \ldots \cdot (w+[n-k-1]t)}{N(N+t) \cdot \ldots \cdot (N+[n-1]t)}$$

Is Example 1(a) of Section 4.2 a special case of this result?

4.3 RANDOM VARIABLES, DISTRIBUTIONS, AND MOMENTS

In an experiment with random outcomes, we are often interested not only in the random outcome itself, but also in numbers and more general mathematical quantities determined by the random outcome of the experiment. Such quantities are called random variables. We might think, say, of the sum of tosses on three throws of a die or the distance of a hit from the center of a target in shooting (Section 4.1, Examples 2 and 3). If $(\Omega,\mathfrak{A},P)$ is the probability space constructed in Section 4.1 for the mathematical description of those experiments, then with every elementary event $\omega \in \Omega$ we associate a real number $X(\omega)$; in this case the sum of spots on the die or the distance of the hit from the bullseye. In both examples it is evident that we are dealing with an $\mathfrak{A}$-measurable mapping $X\colon \Omega \to \mathbf{R}$. Because of this measurability, the inverse image $X^{-1}(B)$ is an event for every Borel set $B \subset \mathbf{R}$. Thus $P(X^{-1}(B))$ is defined as the probability that X takes on a value in B. Motivated by these and similar examples, we define:

4.3.1. Definition. Let $(\Omega,\mathfrak{A},P)$ be a probability space and let $(\Omega',\mathfrak{A}')$ be a measurable space. Then every $\mathfrak{A}$-$\mathfrak{A}'$-measurable mapping $X\colon \Omega \to \Omega'$ is called a *random variable* (with values in Ω') or $(\Omega',\mathfrak{A}')$-random variable.[4]

In the cases $\Omega' = \mathbf{R}$, $\mathfrak{A}^1 = \mathfrak{B}^1$, or $\Omega' = \bar{\mathbf{R}}$, $\mathfrak{A}^1 = \bar{\mathfrak{B}}^1$ we also speak of *real* or *numerical* random variables, respectively. For every event $A \in \mathfrak{A}$, the indicator function 1_A is a real random variable. It is also called the *indicator variable of* A. Instead of elementary functions, in probability theory we speak of *elementary* or simple *random variables.*

In what follows, when we deal with events or random variables, and unless we explicitly state otherwise, we will always mean events or random variables from or on the same probability space. This probability space will then always be denoted by $(\Omega,\mathfrak{A},P)$.

We return to the general case of an $(\Omega',\mathfrak{A}')$-random variable X. In analogy with the notation of Section 2.1 and with a view to the intuitive

[4] Random variables are usually denoted by capital letters, frequently by X, Y, Z.

interpretation given in the introduction, we set

$$\{X \in A'\} \equiv X^{-1}(A') \tag{4.3.1}$$

$$P\{X \in A'\} \equiv P(X^{-1}(A')) \qquad (A' \in \mathfrak{A}'). \tag{4.3.2}$$

We call $\{X \in A'\}$ the event "X lies in A'," and $P\{X \in A'\}$ the probability of this event.

We have already encountered the mapping $A' \to P\{X \in A'\}$ in Section 1.7; it is just the image measure $X(P)$ of P. Since $P\{X \in \Omega'\} = P(\Omega) = 1$, we are dealing with a probability measure on $\mathfrak{A}'$.

4.3.2. Definition. Let X be an $(\Omega',\mathfrak{A}')$-random variable on a probability space $(\Omega,\mathfrak{A},P)$. Then the image measure

$$P_X \equiv X(P) \tag{4.3.3}$$

is called the *distribution* of X (with respect to P) or the probability law of X.

Thus we have

$$P_X(A') = P\{X \in A'\} \qquad (A' \in \mathfrak{A}'). \tag{4.3.4}$$

The distribution of a random variable X is of primary interest, for it makes it possible to compute the probabilities $P\{X \in A'\}$. Accordingly, we are particularly interested in concepts for random variables that can be formulated in terms of the corresponding distributions. Therefore, we sometimes designate them as *probability-theoretic concepts* and consider their study as a principal goal of probability theory.

The expected value to be defined now is such a concept. Thus, let X be a numerical random variable. Intuitively, to every elementary event $\omega \in \Omega$ corresponds the random value $X(\omega)$ of X. It is therefore natural to be interested in the "mean" or "expected" value of X. This is the integral $\int X\,dP$, provided it exists.

4.3.3. Definition. Let X be a numerical random variable on a probability space $(\Omega,\mathfrak{A},P)$. Then if $X \geqq 0$ or if X is P-integrable, we call

$$E(X) = E_P(X) \equiv \int X\,dP \tag{4.3.5}$$

the *expected value* of X.[5]

The properties of the expected value are, thus, those of the integral. In particular, by Theorem 2.4.2, the condition $E(|X|) < \infty$ is equivalent to the integrability of X.

[5] If $\int X\,dP$ exists in the sense mentioned in Definition 2.4.1, then we also say $E(X)$ *exists* and define $E(X)$ again by (4.3.5).

For the sake of simplicity, we restrict ourselves in further discussion to a *real* random variable X. Its distribution is then a probability measure on $\mathfrak{B}^1$. By the transformation theorem of Section 2.10 we have

$$E(f \circ X) = \int f \, dP_X, \tag{4.3.6}$$

which may also be written as

$$E_P(f \circ X) = E_{P_X}(f) \tag{4.3.6$'$}$$

for every Borel measurable real function f on $\mathbf{R}$ which is either $\geqq 0$ or P_X-integrable.

Thus if $X \geqq 0$ or if X is integrable and we choose $f(x) = x$ ($x \in \mathbf{R}$), then

$$E(X) = \int x P_X(dx). \tag{4.3.7}$$

In the case $X \geqq 0$ we should note that $f \geqq 0$ P_X-almost surely.

Equation (4.3.7) shows that we have a probability-theoretic concept in $E(X)$. The same is true of the variance to be defined now:

4.3.4. Definition. For every integrable real random variable X, we call

$$V(X) \equiv E([X - E(X)]^2) \tag{4.3.8}$$

the *variance* of X. The number

$$\sigma(X) \equiv \sqrt{V(X)} \tag{4.3.9}$$

is called the *standard deviation* of X.

We often write $\sigma^2(X)$ for $V(X)$.

4.3.5. Theorem. A real random variable X on a probability space $(\Omega,\mathfrak{A},P)$ is square integrable if and only if X is integrable and $V(X) < +\infty$. Then we have

$$V(X) = E(X^2) - E(X)^2 = \int x^2 P_X(dx) - \left(\int x P_X(dx)\right)^2. \tag{4.3.10}$$

Proof. For P we have, according to Corollary 2.6.7, $\mathfrak{L}^2(P) \subset \mathfrak{L}^1(P)$; $\mathfrak{L}^2(P)$ further contains all constant real functions on Ω. Thus $X \in \mathfrak{L}^2(P)$ implies the integrability of X as well as $V(X) < +\infty$. Conversely, if X is integrable and $V(X) < \infty$, then $X - E(X)$ lies in $\mathfrak{L}^2(P)$, that is, $X = X - E(X) + E(X)$ lies in $\mathfrak{L}^2(P)$. The rest of the assertion follows when we make use of the linearity of E and (4.3.6). ⌟

Remarks. 1. We call $E(X^p)$, $p = 1, 2, \ldots$, the pth *moment* of X (see Section 8.3). The statements "$X \in \mathfrak{L}^2(P)$" and "the second moment of X is finite" are thus equivalent.

2. It follows from (4.3.8) that X and $X - E(X)$ have the same variance; indeed,

$$E(X - E(X)) = E(X) - E(E(X)) = E(X) - E(X) = 0.$$

Real random variables Y with $E(Y) = 0$ are also said to be *centered*. In other words, by making the transition from X to $X - E(X)$ we thus obtain a centered random variable with the same variance. The transition $X \to X - E(X)$ is called *centering on the expected value.*

PROBLEMS

1. Let X be a $(\Omega',\mathfrak{A})$-random variable on $(\Omega,\mathfrak{A},P)$ which attains only a countable number of different values ω_i', $i \in I$ (I countable). Prove: $P_X = \Sigma_{i\in I} P\{X = \omega_i'\}\epsilon_{\omega_i'}$.
2. Consider the Laplace space $(\Omega,\mathfrak{A},P)$ discussed in Section 4.2, Example 1(b), where Ω is the product $\Omega_0 \times \cdots \times \Omega_0$ of m copies of a set Ω_0 of n elements (for example, of balls). Ω_0 is decomposed into two disjoint subsets Ω_0^b and Ω_0^w (for example, of black and white balls). For every $\omega = (\omega_1, \ldots, \omega_n) \in \Omega$ denote by $X(\omega)$ the number of indices $i = 1, \ldots, m$ for which $\omega_i \in \Omega_0^b$. Determine the distribution of the random variable X.

4.4 SOME SPECIAL DISTRIBUTIONS

For a given measurable space $(\Omega',\mathfrak{A}')$ all probability measures P' on $\mathfrak{A}$ appear as distributions of $(\Omega',\mathfrak{A}')$-random variables when we allow arbitrary probability spaces $(\Omega,\mathfrak{A},P)$. We need only choose $\Omega = \Omega'$, $\mathfrak{A} = \mathfrak{A}'$, $P = P'$, and the identity mapping of Ω onto itself for X in order to obtain $P_X = P'$. This is why one sometimes call probability measures on a σ-algebra *distributions.*

Below we shall discuss several important types of distributions for $(\mathbf{R}^k,\mathfrak{B}^k)$-random variables ($k = 1, 2, \ldots$).

1. For every $x \in \mathbf{R}^k$ let ϵ_x be the probability measure on $\mathfrak{B}^k$ defined by the unit mass at x. We say that ϵ_x is a *singular distribution.* Every random variable X with such a distribution is said to be singularly distributed or degenerate. X is singularly distributed if and only if X is almost surely constant.

2. Let $(x_n)_{n\in\mathbf{N}}$ be a point sequence in $\mathbf{R}^k$ and $(\alpha_n)_{n\in\mathbf{N}}$ a sequence of real numbers $\geqq 0$ satisfying $\Sigma_{n=1}^{\infty} \alpha_n = 1$. Then $\mu = \Sigma_{n=1}^{\infty} \alpha_n\epsilon_{x_n}$ is a probability

measure on $\mathfrak{B}^k$. Every such distribution is said to be *discrete;* every random variable with this distribution is said to be discretely distributed. In particular, every singular distribution is obviously discrete.

3. Every λ^k-continuous probability measure μ on $\mathfrak{B}^k$ is said to be *Lebesgue continuous.* By the Radon-Nikodym Theorem, all measures $\mu = f\lambda^k$ with Borel measurable densities $f \geqq 0$ and $\int f\, d\lambda^k = 1$ are Lebesgue continuous probability measures. We then call f the *probability density* of μ on $\mathbf{R}^k$.

Remark. There is no probability measure μ on $\mathfrak{B}^k$ that is simultaneously discrete and Lebesgue continuous. From the λ^k-continuity of such a measure μ we would have $\mu(A) = 0$ for all $A \in \mathfrak{B}^k$ with $\lambda^k(A) = 0$, that is, in particular, $\mu(\{a\}) = 0$ for all $a \in \mathbf{R}^k$. But for every discrete probability measure μ there is always an $a \in \mathbf{R}^k$ with $\mu(\{a\}) > 0$.

For the real line ($k = 1$) we shall now discuss special discrete and Lebesgue continuous distributions.

4. Let p be a real number with $0 \leqq p \leqq 1$ and $q = 1 - p$.

By the Binomial Theorem we have

$$\sum_{k=0}^{n} \binom{n}{k} p^k q^{n-k} = (p + q)^n = 1$$

and hence

$$\beta_n^p \equiv \sum_{k=0}^{n} \binom{n}{k} p^k q^{n-k} \epsilon_k \tag{4.4.1}$$

is a discrete probability measure on $\mathfrak{B}^1$. When $0 < p < 1$, β_n^p is called the *binomial* or *Bernoulli distribution* and is denoted by $B_n(p)$ in mathematical statistics.

With the help of (4.3.6) and several simple computations, we obtain for every real random variable with a β_n^p distribution

$$E(X) = \sum_{k=1}^{n} k \binom{n}{k} p^k q^{n-k} = np(p + q)^{n-1} = np, \tag{4.4.2}$$

$$\begin{aligned} E(X^2) &= \sum_{k=1}^{n} k^2 \binom{n}{k} p^k q^{n-k} = np \sum_{k=1}^{n} k \binom{n-1}{k-1} p^{k-1} q^{n-k} \\ &= np[(n-1)p + 1] = np(np + q), \end{aligned} \tag{4.4.3}$$

and hence by (4.3.10),

$$V(X) = npq. \tag{4.4.4}$$

We already encountered the distribution β_m^p in Section 4.2, Example 1(b).

5. Since

$$e^\alpha = \sum_{k=0}^{\infty} \frac{\alpha^k}{k!},$$

then for every $\alpha \geqq 0$

$$\pi_\alpha \equiv \sum_{k=0}^{\infty} e^{-\alpha} \frac{\alpha^k}{k!} \epsilon_k \tag{4.4.5}$$

is a discrete probability measure on $\mathfrak{B}^1$ which is equal to ϵ_0 for $\alpha = 0$. When $\alpha > 0$, we call π_α the *Poisson distribution* with *parameter* α. Again by (4.3.6) it follows that for a real random variable X with distribution π_α

$$E(X) = \sum_{k=0}^{\infty} e^{-\alpha} \frac{\alpha^k}{k!} k = \alpha; \tag{4.4.6}$$

$$\begin{aligned} E(X^2) &= \sum_{k=0}^{\infty} e^{-\alpha} \frac{\alpha^k}{k!} k^2 \\ &= \sum_{k=1}^{\infty} e^{-\alpha} \frac{\alpha^k}{(k-1)!} (k - 1 + 1) = \alpha^2 + \alpha; \end{aligned} \tag{4.4.7}$$

$$V(X) = \alpha. \tag{4.4.8}$$

In the further examples we shall be dealing with Lebesgue continuous distributions.

6. By (2.8.6),

$$x \to g_{\alpha,\sigma^2}(x) \equiv (2\pi\sigma^2)^{-1/2} e^{-(x-\alpha^2)/2\sigma^2} \tag{4.4.9}$$

is a probability density g_{α,σ^2} on $\mathbf{R}$ and hence

$$\nu_{\alpha,\sigma^2} \equiv g_{\alpha,\sigma^2}\lambda^1 \tag{4.4.10}$$

is a probability measure on $\mathfrak{B}^1$ ($\alpha \in \mathbf{R}$, $\sigma > 0$). This is called the *normal* or *Gaussian distribution* with parameters α, σ^2 and is denoted in mathematical statistics by $N(\alpha,\sigma^2)$. We call $\nu_{0,1}$ the *standard* normal distribution.

For a real random variable X with ν_{α,σ^2} distribution we obtain from (4.3.6)

$$E(X) = \int x g_{\alpha,\sigma^2}(x)\, dx = \alpha; \tag{4.4.11}$$

$$E(X^2) = \int x^2 g_{\alpha,\sigma^2}(x)\, dx = \sigma^2 + \alpha^2; \tag{4.4.12}$$

$$V(X) = \sigma^2. \tag{4.4.13}$$

This illuminates the probabilistic meaning of the two parameters α and σ^2.

[Derivation of (4.4.11) and (4.4.12): By the linear transformation $x \to \sigma^{-1}(x - \alpha)$ we reduce the integrals (4.4.11) and (4.4.12) to the corresponding integrals for the standard normal distribution. The integrals $\int x g_{0,1}(x)\, dx$ and $\int x^2 g_{0,1}(x)\, dx$ exist even as absolutely convergent improper Riemann integrals. The first of these vanishes since the integrand is odd. The second is reduced by partial integration to $\int g_{0,1}(x)\, dx$. Here we should note that $g_{0,1}$ is a primitive of $x \to -x g_{0,1}(x)$.]

The graph of the function g_{α,σ^2} is known as the *Gaussian bell-shaped curve.* A simple study of this curve shows that the maximal value $(2\pi\sigma^2)^{-1/2}$ is taken on only at $x = \alpha$ and an inflection point lies at $x = \alpha \pm \sigma$. The variance σ^2 is therefore a measure for the "width" of the bell curve.

7. Also the function

$$x \to \frac{\alpha}{\pi} (\alpha^2 + x^2)^{-1}$$

is a probability density c_α on $\mathbf{R}$ for every $\alpha > 0$; we have indeed

$$\int_{-\infty}^{+\infty} (1 + x^2)^{-1}\, dx = [\text{arc tg } x]_{-\infty}^{+\infty} = \pi.$$

The probability measure

$$\gamma_\alpha = c_\alpha \lambda^1 \tag{4.4.14}$$

is called the (standard) *Cauchy distribution* with parameter $\alpha > 0$. We can see immediately that a real random variable with the distribution γ_α is *not* integrable, that is, its expected value is not defined. It suffices to remark that

$$\int_0^a \frac{x}{1 + x^2}\, dx = \frac{1}{2} \log (1 + a^2) \qquad \text{for } a \in \mathbf{R}_+,$$

and hence

$$\int_{-\infty}^{+\infty} \frac{|x|}{1 + x^2}\, dx = +\infty.$$

PROBLEMS

1. Let X be a singularly distributed real random variable. Calculate $E(X)$ and $V(X)$.
2. Let n, b, m be natural numbers such that $1 \leqq b < n$ and $m \leqq n$. Prove:

(a)
$$\eta_{n,b,m} \equiv \frac{1}{\binom{n}{m}} \sum_{k=0}^{m} \binom{b}{k} \binom{n - s}{m - k} \epsilon_k$$

is a probability measure on $\mathfrak{B}^1$, called *hypergeometric distribution* (with parameters n, b, m).

(b) $\eta_{n,b,m}$ is the distribution of a random variable defined on the probability space of Section 4.2, Example 1(a). (Compare this result with the probabilistic interpretation of β_n^p given above.)

3. Let X be a real random variable with $\eta_{n,b,m}$ (of Problem 2) as distribution. Prove:

$$E(X) = mp \qquad \text{and} \qquad V(X) = \frac{n-m}{n-1}\,mpq,$$

where $p \equiv m/n$ and $q \equiv 1 - p$.

4. Let $p_1, \ldots, p_k$ be nonnegative real numbers satisfying $\Sigma_{i=1}^k p_i = 1$. Prove that for $n \in \mathbf{N}$

$$\sum_{\substack{n_1,\ldots,n_k \geq 0 \\ n_1+\cdots+n_k=n}}^{k} \frac{n!}{n_1! \ldots n_k!}\, p_1^{n_1} \ldots p_k^{n_k}\, \epsilon_{(n_1,\ldots,n_k)}$$

is a discrete distribution on $\mathfrak{B}^k$ (called *multinominal distribution*).

4.5 DISTRIBUTION FUNCTIONS

We can describe all probability measures on $\mathfrak{B}^1$ with the help of real functions defined on $\mathbf{R}$. Indeed, let μ be a probability measure on $\mathfrak{B}^1$. Since $]-\infty,x[\, \in \mathfrak{B}^1$,

$$F_\mu(x) \equiv \mu(]-\infty,x[) \tag{4.5.1}$$

defines a real function F_μ on $\mathbf{R}$. This is called the *distribution function* of μ. If μ is the distribution of a real random variable X (defined on a probability space), then we call F_μ also the distribution function of X. The following theorem characterizes distribution functions by intrinsic properties:

4.5.1. Theorem. A real function F on $\mathbf{R}$ is the distribution function of a probability measure μ on $\mathfrak{B}^1$ if and only if it has the following properties:

$$F \text{ is isotone;} \tag{4.5.2}$$
$$F \text{ is left continuous;} \tag{4.5.3}$$
$$\lim_{x\to-\infty} F(x) = 0, \qquad \lim_{x\to+\infty} F(x) = 1. \tag{4.5.4}$$

The probability measure μ is uniquely determined by F.

Proof. Every distribution function $F = F_\mu$ has the given three properties: (4.5.2) follows from the isotonicity of μ. For every isotone sequence (x_n) of real numbers with $x = \lim x_n$,

$$]-\infty, x_n[\uparrow]-\infty, x[$$

and

$$\lim F(x_n) = \mu(]-\infty, x[),$$

since μ is continuous from below. For $x \in \mathbf{R}$ this yields the left continuity of F in x and for $x = +\infty$ it yields the second part of (4.5.4). The first part follows analogously from the $\varnothing$-continuity of μ.

If μ and ν are probability measures on $\mathfrak{B}^1$ with $F_\mu = F_\nu$, then

$$\mu(I) = F_\mu(b) - F_\mu(a) = F_\nu(b) - F_\nu(a) = \nu(I)$$

for every half-open interval $I = [a,b[\in \mathfrak{J}^1$ by (1.3.7). By the Uniqueness Theorem 1.5.5 we thus have $\mu = \nu$.

Finally, let F be a real function on $\mathbf{R}$ with the properties (4.5.2)–(4.5.4). Then a repetition of the derivations carried out in the proof of Theorem 1.4.3 shows that exactly one content μ exists on the ring $\mathfrak{F}^1$ of 1-dimensional figures with $\mu([a,b[) = F(b) - F(a)$ for all $[a,b[\in \mathfrak{J}^1$. Here only (4.5.2) is used. It follows from (4.5.3) that for every $I = [a,b[\in \mathfrak{J}^1$ and every $\epsilon > 0$ there is a $J = [a,c[\in \mathfrak{J}^1$ satisfying $J \subset I$ and

$$\mu(I) - \mu(J) = \mu([c,b[) = F(b) - F(c) \leqq \epsilon.$$

But the method of proof of Theorem 1.4.4 then shows that μ is a finite and σ-finite premeasure on $\mathfrak{F}^1$. By Theorem 1.5.7 this can be extended to a measure $\tilde{\mu}$ on $\mathfrak{B}^1$ in exactly one way. Since $[-n,n[\uparrow \mathbf{R}$ and because of (4.5.4), $\tilde{\mu}$ is a probability measure. By construction, $F = F_{\tilde{\mu}}$. ┘

Example

The indicator function of $]0,+\infty[$ relative to $\mathbf{R}$ satisfies the conditions (4.5.2)–(4.5.4), that is, it is a distribution function F. Clearly $F = F_{\epsilon_0}$.

Theorem 4.5.1 tells us that the study of probability measures on $\mathfrak{B}^1$ can be replaced by the study of functions with the properties (4.5.2)–(4.5.4). Therefore, the concept of distribution function prevails in some presentations of probability theory. Since the corresponding concept in $\mathbf{R}^p$ for $p \geqq 2$ is difficult to handle and since we shall develop probability theory in spaces where no distribution functions exist (namely in those without order structure), distribution functions do not play a fundamental role in the development of probability theory from measure-theoretic foundations. The distributions themselves are the basic tool.

PROBLEMS

1. Let μ be a probability measure on $\mathfrak{B}^1$ and F_μ its distribution function. Prove: Continuity of F_μ at a point $x \in \mathbf{R}$ is equivalent to the statement $\mu(\{x\}) = 0$.
2. Prove: A real random variable X is singularly distributed if and only if for every $\alpha \in \mathbf{R}$, the probability $P\{X \leqq \alpha\}$ is either 0 or 1.

5

INDEPENDENCE

5.1 INDEPENDENT EVENTS AND σ-ALGEBRAS

In the following, let $(\Omega,\mathfrak{A},P)$ be an arbitrary probability space. In Section 4.2 we defined the conditional probability $P(A \mid B)$ of an event $A \in \mathfrak{A}$ under the hypothesis of an event $B \in \mathfrak{A}$ with $P(B) > 0$ by

$$P(A \mid B) = \frac{P(A \cap B)}{P(B)}.$$

Intuitively, $P(A \mid B)$ is the probability of A when it is already known that the event B has occurred. Now it can happen that this additional information about the event A due to the occurrence of B has no influence on the probability of the occurrence of A, that is, $P(A \mid B) = P(A)$ or equivalently, $P(A \cap B) = P(A)P(B)$. In this case we speak of *independent* events A and B.

If, for example, we consider two throws of a die and the corresponding Laplace space $(\Omega,\mathfrak{A},P)$, where Ω is the set of all ordered pairs of numbers $1, \ldots, 6$, then the events A = "a number <4 turned up on the first throw" and B = "a 6 turned up on the second throw" are independent. We have

$$P(A) = \frac{3 \cdot 6}{6 \cdot 6} = \frac{1}{2}, \qquad P(B) = \frac{6 \cdot 1}{6 \cdot 6} = \frac{1}{6}$$

and

$$P(A \cap B) = \frac{3 \cdot 1}{6 \cdot 6} = P(A)P(B).$$

Motivated by these considerations, we define:

5.1.1. Definition. A family $(A_i)_{i \in I}$ of events from $\mathfrak{A}$ is said to be *independent*[1] (relative to P) if for every nonempty finite subset $\{i_1, \ldots, i_n\}$ of I,

$$P(A_{i_1} \cap \cdots \cap A_{i_n}) = P(A_{i_1}) \cdot \ldots \cdot P(A_{i_n}). \tag{5.1.1}$$

Often we have $I = \{1, \ldots, n\}$ or $\{1,2, \ldots\}$. In this case we say that the events $A_1, \ldots, A_n$ or $A_1, A_2, \ldots$ are independent.

Examples

1. Let us again consider two throws of a die and the corresponding Laplace space just mentioned above. We investigate the following three events A_i ($i = 1, 2, 3$): Let A_1 and A_2 be the events that an odd number turned up on the first and second throws, respectively; let A_3 be the event that the sum of the throws is odd. We see immediately that any two of these events are independent, but the family of the three events is not independent. We have in fact

$$A_1 \cap A_2 \cap A_3 = \emptyset,$$

but $P(A_i) > 0$ for all $i = 1, 2, 3$. Therefore, we *cannot* derive independence from the *pairwise* independence.

2. Let $(\Omega_i, \mathfrak{A}_i, P_i)$, $i = 1, \ldots, n$, be finitely many probability spaces and let $(\Omega, \mathfrak{A}, P)$ be their product. For every $i = 1, \ldots, n$, let A_1' be an event in $\mathfrak{A}_i$. Then the events $A_1 \equiv A_1' \times \Omega_2 \times \cdots \times \Omega_n$, $A_2 \equiv \Omega_1 \times A_2' \times \Omega_3 \times \cdots \times \Omega_n, \ldots, \quad A_n \equiv \Omega_1 \times \cdots \times \Omega_{n-1} \times A_n'$ of $\mathfrak{A}$ are independent.

We shall soon see that this example forms the nucleus of the concept of independence.

3. Let $\Omega \equiv [0,1[$, $\mathfrak{A} \equiv \Omega \cap \mathfrak{B}^1$, and $P \equiv \lambda_\Omega^1$ be the L-B-measure on Ω. For every $i = 1, 2, \ldots$ let

$$A_n \equiv \left[0, \frac{1}{2^n}\right[\cup \left[\frac{2}{2^n}, \frac{3}{2^n}\right[\cup \cdots \cup \left[\frac{2^n - 2}{2^n}, \frac{2^n - 1}{2^n}\right[.$$

Then the sequence (A_n) of these events is independent. We have

$$P(A_n) = \tfrac{1}{2}$$

for all n and

$$P(A_{i_1} \cap \cdots \cap A_{i_n}) = \tfrac{1}{2} P(A_{i_1} \cap \cdots \cap A_{i_{n-1}});$$

[1] In contrast with other concepts of independence in mathematics, we also speak of *stochastic* independence.

hence

$$P(A_{i_1} \cap \cdots \cap A_{i_n}) = P(A_{i_1} \cap \cdots \cap A_{i_{n-1}}) \cdot P(A_{i_n}) \\ = \cdots = P(A_{i_1}) \cdot \ldots \cdot P(A_{i_n})$$

for every finite set $\{i_1, \ldots, i_n\}$ of natural numbers.

The following generalization of Definition 5.1.1 is useful mainly for technical reasons:

5.1.2. Definition. Let $(\mathfrak{E}_i)_{i \in I}$ be a family of sets $\mathfrak{E}_i \subset \mathfrak{A}$ of events. The family is said to be *independent* if for every nonempty finite subset $\{i_1, \ldots, i_n\}$ of I and every possible choice of events

$$A_{i_\nu} \in \mathfrak{E}_{i_\nu} \qquad (\nu = 1, \ldots, n)$$

equality (5.1.1) holds.

If all the sets $\mathfrak{E}_i$ are one-element sets, then Definitions 5.1.2 and 5.1.1 coincide.

From the definition we immediately obtain two *consequences:*

1. A family $(\mathfrak{E}_i)_{i \in I}$ is independent if and only if every *finite subfamily* is independent.

2. The independence remains valid if each of the sets $\mathfrak{E}_i$ is *reduced,* that is, if we consider subsets of each $\mathfrak{E}_i$.

It is now important to know how the independence of a family $(\mathfrak{E}_i)_{i \in I}$ persists if we *enlarge* every set $\mathfrak{E}_i$ to a set $\tilde{\mathfrak{E}}_i$.

5.1.3. Theorem. Let $(\mathfrak{E}_i)_{i \in I}$ be an independent family of sets $\mathfrak{E}_i \subset \mathfrak{A}$. For each $i \in I$ let $\mathfrak{D}(\mathfrak{E}_i)$ denote the Dynkin system generated by $\mathfrak{E}_i$ in Ω. Then the family $(\mathfrak{D}(\mathfrak{E}_i))_{i \in I}$ is also independent.

Proof. It follows from the first of the two remarks above that we can assume without loss of generality that the index set I is finite. Then for arbitrary $i_0 \in I$ let $\mathfrak{D}_{i_0}$ be the set of all events $E \in \mathfrak{A}$ with the following property: If we replace the set $\mathfrak{E}_{i_0}$ by $\{E\}$ in $(\mathfrak{E}_i)_{i \in I}$, then the new family is independent. We show that $\mathfrak{D}_{i_0}$ is then a Dynkin system in Ω. First, Ω lies in $\mathfrak{D}_{i_0}$, since for every nonempty set $\{i_1, \ldots, i_n\} \subset I \setminus \{i_0\}$,

$$P(A_{i_1} \cap \cdots A_{i_n} \cap \Omega) = P(A_{i_1} \cap \cdots \cap A_{i_n}) \\ = P(A_{i_1}) \cdot \ldots \cdot P(A_{i_n}) P(\Omega)$$

for arbitrarily chosen A_{i_ν} in $\mathfrak{E}_{i_\nu}$. Then, for the same choice of A_{i_ν}, $E, F \in \mathfrak{D}_{i_0}$ and $F \subset E$ imply

$$\begin{aligned} P(A_{i_1} \cap \cdots \cap A_{i_n} \cap (E \setminus F)) &= P(A_{i_1} \cap \cdots \cap A_{i_n} \cap E) - P(A_{i_1} \cap \cdots \cap A_{i_n} \cap F) \\ &= P(A_{i_1}) \cdot \ldots \cdot P(A_{i_n})P(E) - P(A_{i_1}) \cdot \ldots \cdot P(A_{i_n})P(F) \\ &= P(A_{i_1}) \cdot \ldots \cdot P(A_{i_n})P(E \setminus F) \end{aligned}$$

and hence $E \setminus F \in \mathfrak{D}_{i_0}$. Analogous reasoning yields the third property (1.2.3) of a Dynkin system. Each time we use the independence of the original family $(\mathfrak{E}_i)_{i \in I}$ and the fact that $P(\Omega) = 1$ together with subtractivity or σ-additivity of P.

According to the construction of $\mathfrak{D}_{i_0}$, every family obtained by replacing the set $\mathfrak{E}_{i_0}$ in $(\mathfrak{E}_i)_{i \in I}$ by $\mathfrak{D}_{i_0}$ is independent. Since $\mathfrak{D}(\mathfrak{E}_{i_0}) \subset \mathfrak{D}_{i_0}$, we may replace $\mathfrak{E}_{i_0}$ by $\mathfrak{D}(\mathfrak{E}_{i_0})$ in $\mathfrak{E}_i$ and still obtain an independent family. We reapply this result to the new family and to an index $i \neq i_0$ of I. The asserted independence of $(\mathfrak{D}(\mathfrak{E}_i))_{i \in I}$ now follows after finitely many steps. ┘

5.1.4. Corollary 1. Let $(\mathfrak{E}_i)_{i \in I}$ be an independent family of $\cap$-stable sets $\mathfrak{E}_i \subset \mathfrak{A}$. Then the family $(\mathfrak{A}(\mathfrak{E}_i))_{i \in I}$ of σ-algebras generated by the $\mathfrak{E}_i$ is independent.

The proof follows directly from Theorems 5.1.3 and 1.2.3. We generalize this result further:

5.1.5. Corollary 2. Let $(\mathfrak{E}_i)_{i \in I}$ be an independent family of $\cap$-stable sets $\mathfrak{E}_i \subset \mathfrak{A}$ and let $(I_j)_{j \in J}$ be a partition of I into pairwise disjoint sets I_j. Then if $\mathfrak{A}_j \equiv \mathfrak{A}(\bigcup_{i \in I_j} \mathfrak{E}_i)$ denotes the σ-algebra[2] generated by the sets $\mathfrak{E}_i$ with $i \in I_j$, the family $(\mathfrak{A}_j)_{j \in J}$ is also independent.

Proof. For every $j \in J$, let $\mathfrak{E}'_j$ denote the system of all sets

$$E_{i_1} \cap \cdots \cap E_{i_n},$$

where $\{i_1, \ldots, i_n\}$ is an arbitrary nonempty finite subset of I_j and E_{i_ν} is arbitrarily chosen in $\mathfrak{E}_{i_\nu} (\nu = 1, \ldots, n)$. Then $\mathfrak{E}'_j$ is $\cap$-stable. The family $(\mathfrak{E}'_j)_{j \in J}$ is independent, as follows immediately from the independence of $(\mathfrak{E}_i)_{i \in I}$. Further, we obviously have $\mathfrak{A}_j = \mathfrak{A}(\mathfrak{E}'_j)$. The assertion then follows from Corollary 5.1.4. ┘

[2] If I_j is empty, then $\mathfrak{A}_j$ is the smallest σ-algebra over the empty system, that is, it equals $\{\varnothing, \Omega\}$.

Remark. If we choose $J = I$ and $I_j = \{j\}$ for every $j \in J$, then Corollary 5.1.4 follows from Corollary 5.1.5.

Because of Corollary 5.1.4, assuming $\mathfrak{E}_i$ to be a σ-algebra $\subset \mathfrak{A}$ is no restriction. Therefore Corollary 5.1.5 is often formulated only for this case and it is called the *theorem on combining independent σ-algebras.* We immediately give an application which will be useful in later work.

5.1.6. Definition. Let $(\mathfrak{A}_n)_{n \in \mathbf{N}}$ be a sequence of σ-algebras of events in $\mathfrak{A}$ and

$$\mathfrak{T}_n \equiv \mathfrak{A}\left(\bigcup_{m=n}^{\infty} \mathfrak{A}_m\right)$$

the σ-algebra generated by $\mathfrak{A}_n, \mathfrak{A}_{n+1}, \ldots$. Then

$$\mathfrak{T}_\infty : \equiv \bigcap_{n=1}^{\infty} \mathfrak{T}_n \tag{5.1.2}$$

is called the σ-algebra of the *terminal* (or tail) events of the sequence $(\mathfrak{A}_n)$.

By Theorem 1.1.2, $\mathfrak{T}_\infty$ is in fact a σ-algebra.

5.1.7. Theorem (Zero-One Law of Kolmogorov). Let $(\mathfrak{A}_n)_{n \in \mathbf{N}}$ be an independent sequence of σ-algebras $\mathfrak{A}_n \subset \mathfrak{A}$. Then $P(A) = 0$ or $P(A) = 1$ for every terminal event A of the sequence.

Proof. Let $A \in \mathfrak{T}_\infty$ be arbitrary and $\mathfrak{D}$ be the associated system of all events independent of A, that is, all $D \in A$ satisfying

$$P(A \cap D) = P(A)P(D).$$

It suffices to show that A lies in $\mathfrak{D}$. In that case we may choose $D = A$, and the assertion follows from

$$P(A) = P(A)^2.$$

Analogous (in fact, simpler) reasoning to that in the proof of Theorem 5.1.3 shows that $\mathfrak{D}$ is a Dynkin system. By Corollary 5.1.5, the σ-algebra $\mathfrak{T}_{n+1}$ defined in Definition 5.1.6 is independent of the σ-algebra

$$\mathfrak{U}_n \equiv \mathfrak{A}(\mathfrak{A}_1 \cup \cdots \cup \mathfrak{A}_n).$$

Since $A \in \mathfrak{T}_{n+1}$, $\mathfrak{U}_n \subset \mathfrak{D}$ for every $n = 1, 2, \ldots$; we thus have

$$\mathfrak{U}_0 \equiv \bigcup_{n=1}^{\infty} \mathfrak{U}_n \subset \mathfrak{D}.$$

The sequence $(\mathfrak{U}_n)$ is obviously isotone. Any two events $E, F \in \mathfrak{U}_0$ therefore in one $\mathfrak{U}_n$; hence, $E \cap F \in \mathfrak{U}_0$ whenever $E \in \mathfrak{U}_0$ and $F \in \mathfrak{U}_0$.[3] By Theorem 1.2.3, $\mathfrak{A}(\mathfrak{U}_0) = \mathfrak{D}(\mathfrak{U}_0) \subset \mathfrak{D}$ then follows from $\mathfrak{U}_0 \subset \mathfrak{D}$. Further, every $\mathfrak{A}_n$ is in $\mathfrak{U}_0$, that is, every $\mathfrak{T}_n$ and thus finally $\mathfrak{T}_\infty$ is contained in $\mathfrak{A}(\mathfrak{U}_0)$. Hence

$$\mathfrak{T}_\infty \subset \mathfrak{A}(\mathfrak{U}_0) \subset \mathfrak{D},$$

that is, $A \in \mathfrak{D}$, which is what we wanted to show. ⌋

Example

4. Let $(A_n)_{n\in\mathbf{N}}$ be an independent sequence of events in $\mathfrak{A}$ and let

$$\limsup_{n\to\infty} A_n \equiv \bigcap_{n=1}^{\infty} \bigcup_{m=n}^{\infty} A_m \tag{5.1.3}$$

be the event "infinitely many of the events A_n occur." We show

$$P(\limsup_{n\to\infty} A_n) = 0 \qquad \text{or} \qquad = 1. \tag{5.1.4}$$

For the proof, let $\mathfrak{A}_n = \{\varnothing, A_n, \complement A_n, \Omega\}$ be the σ-algebra generated by A_n. By Corollary 5.1.4 the sequence $(\mathfrak{A}_n)$ is independent. In the notation of Definition 5.1.6, $Q_n \equiv \bigcup_{m=n}^{\infty} A_m$ lies in $\mathfrak{T}_n$ for all n and hence $Q_n, Q_{n+1}, \ldots$ lies in $\mathfrak{T}_n$ for all n since $(\mathfrak{T}_n)$ is antitone. Therefore,

$$\limsup_{n\to\infty} A_i = \bigcap_{k=n}^{\infty} Q_k \in \mathfrak{T}_n$$

for all n, that is, lim sup A_n lies in $\mathfrak{T}_\infty$. Now (5.1.4) follows from the 0-1 Law of Kolmogorov.

We shall return to this example in Section 6.2.

PROBLEMS

1. Prove:
   (a) Two events A and B are independent if and only if A and $\complement B$ are independent.
   (b) If the events A, B, C are independent, then $A \cup B$ and C are independent.
2. Let $\mathfrak{E}_1$ and $\mathfrak{E}_2$ be sets of events such that $\mathfrak{E}_1 \subset \mathfrak{E}_2$. Prove: $\mathfrak{E}_1$, $\mathfrak{E}_2$ are independent if and only if $P(A) = 0$ or $P(A) = 1$ for each $A \in \mathfrak{E}_1$.
3. Give an example for an independent family $(\mathfrak{E}_i)$ of sets $\mathfrak{E}_i \subset \mathfrak{A}$ for which the family $(\mathfrak{A}(\mathfrak{E}_i))$ of generated σ-algebras is not independent.

[3] Moreover, analogous reasoning shows that $\mathfrak{U}_0$ is an algebra.

5.2 INDEPENDENT RANDOM VARIABLES

Suppose we are again given a probability space $(\Omega,\mathfrak{A},P)$. By Corollary 5.1.4, a family $(A_i)_{i\in I}$ of events is independent if and only if the family $(\mathfrak{A}_i)_{i\in I}$ of σ-algebras $\mathfrak{A}_i = \{\varnothing, A_i, \complement A_i, \Omega\}$ generated by every A_i is independent. But $\mathfrak{A}_i$ is the σ-algebra generated by the indicator variables 1_{A_i}, and thus the independence of $(A_i)_{i\in I}$ is equivalent to that of $(\mathfrak{A}(1_{A_i}))_{i\in I}$. This result justifies the following definition:

5.2.1. Definition. A family $(X_i)_{i\in I}$ of random variables (whose ranges may depend on i) is said to be *independent* if the family $(\mathfrak{A}(X_i))_{i\in I}$ generated by the random variables X_i is independent.

Thus also the independence of an arbitrary family of random variables is reduced to the independence of its finite subfamilies.

5.2.2. Theorem. Let $(X_i)_{i=1,\ldots,n}$ be a finite family of $(\Omega_i,\mathfrak{A}_i)$-random variables and let $\mathfrak{Q}_i$ be a $\cap$-stable generator of $\mathfrak{A}_i$ with $\Omega_i \in \mathfrak{Q}_i$ $(i = 1, \ldots, n)$. The family $(X_i)_{i=1,\ldots,n}$ is independent if and only if

$$P\left(\bigcap_{i=1}^{n} X_i^{-1}(Q_i)\right) = \prod_{i=1}^{n} P(X_i^{-1}(Q_i)) \tag{5.2.1}$$

for every choice of sets $Q_i \in \mathfrak{Q}_i$ $(i = 1, \ldots, n)$.

Proof. If we set

$$\mathfrak{E}_i \equiv \{X_i^{-1}(Q_i) \colon Q_i \in \mathfrak{Q}_i\},$$

then by Theorem 1.7.2 $\mathfrak{E}_i$ is a generator of $\mathfrak{A}(X_i)$ which by virtue of the properties of $\mathfrak{Q}_i$, is $\cap$-stable and contains Ω as an element. Thus, according to Corollary 5.1.4, we have to show that the independence condition of Corollary 5.1.4 for $(\mathfrak{E}_i)_{i=1,\ldots,n}$ is equivalent to the validity of

$$P(E_1 \cap \cdots \cap E_n) = P(E_1) \cdot \ldots \cdot P(E_n)$$

for all choices of $E_i \in \mathfrak{E}_i$. Since we can set appropriate sets E_i equal to Ω, this is obvious. ┘

The following theorem is simple to prove, but most useful:

5.2.3. Theorem. Let $(X_i)_{i\in I}$ be a family of $(\Omega_i,\mathfrak{A}_i)$-random variables and let

$$Y_i \colon (\Omega_i,\mathfrak{A}_i) \to (\Omega_i',\mathfrak{A}_i')$$

be a measurable mapping into a measurable space $(\Omega_i',\mathfrak{A}_i')$ for every $i \in I$. Then the independence of $(X_i)_{i\in I}$ implies the independence of $(Y_i \circ X_i)_{i\in I}$.

Proof. For every set $A' \in \mathfrak{A}'_i$ we have $(Y_i \circ X_i)^{-1}(A') = X_i^{-1}(Y_i^{-1}(A'))$ and hence $\mathfrak{A}(Y_i \circ X_i) \subset \mathfrak{A}(X_i)$ $(i \in I)$. But then, whenever $(\mathfrak{A}(X_i))_{i \in I}$, is independent, $(\mathfrak{A}(Y_i \circ X_i))_{i \in I}$ is also independent by Section 5.1, Consequence 2. ⌋

An essential generalization of this result connected with Corollary 5.1.5 will be presented in Section 5.4.

Finally, we show that the independence of finitely many random variables is a probability-theoretic property, that is, a property of their distributions. To this end, let $X_1, \ldots, X_n$ be random variables; more precisely, let X_i be an $(\Omega_i,\mathfrak{A}_i)$-random variable $(i = 1, \ldots, n)$. We define the product mapping

$$X_1 \otimes \cdots \otimes X_n: \Omega \to \Omega_1 \times \cdots \times \Omega_n,$$

which we denote by Y, by

$$Y(\omega) = (X_1(\omega), \ldots, X_n(\omega)) \qquad (\omega \in \Omega). \tag{5.2.2}$$

Then for every set $A_1 \times \cdots \times A_n$ with $A_i \in \mathfrak{A}_i$ $(i = 1, \ldots, n)$,

$$Y^{-1}(A_1 \times \cdots \times A_n) = X_1^{-1}(A_1) \cap \cdots \cap X_n^{-1}(A_n), \tag{5.2.3}$$

and thus by Theorem 1.7.2 Y is a $\left(\prod_{i=1}^{n} \Omega_i, \bigotimes_{i=1}^{n} \mathfrak{A}_i\right)$-random variable on the probability space $(\Omega,\mathfrak{A},P)$. In particular, the distributions P_{X_i} $(i = 1, \ldots, n)$ and P_Y are now defined. We call $P_Y = P_{X_1 \otimes \cdots \otimes X_n}$ the *joint distribution* of $X_1, \ldots, X_n$; it is a probability measure on $\bigotimes_{i=1}^{n} \mathfrak{A}_i$.

In this notation we now have:

5.2.4. Theorem. Finitely many random variables $X_1, \ldots, X_n$ are independent if and only if their joint distribution is the product of their single distributions, that is, if and only if

$$P_{X_1 \otimes \cdots \otimes X_n} = P_{X_i} \otimes \cdots \otimes P_{X_n}. \tag{5.2.4}$$

Proof. For every $i = 1, \ldots, n$, let A_i be an event from $\mathfrak{A}_i$. It then follows from the definition of distribution, in particular that of $Y \equiv X_1 \otimes \cdots \otimes X_n$, and (5.2.3), that

$$P_Y\left(\prod_{i=1}^{n} A_i\right) = P\left(Y^{-1}\left(\prod_{i=1}^{n} A_i\right)\right) = P\left(\bigcap_{i=1}^{n} X_i^{-1}(A_i)\right)$$

and

$$P_{X_i}(A_i) = P(X_i^{-1}(A_i)) \qquad (i = 1, \ldots, n).$$

Now P_Y is precisely the product of the measures P_{X_i} if

$$P_Y(A_1 \times \cdots \times A_n) = P_{X_1}(A_1) \cdot \ldots \cdot P_{X_n}(A_n).$$

Hence the statement

$$P\Big(\bigcap_{i=1}^{n} X_i^{-1}(A_i)\Big) = \prod_{i=1}^{n} P(X_i^{-1}(A_i))$$

is equivalent to the above for all $A_i \in \mathfrak{A}_i$ $(i = 1, \ldots, n)$. According to Theorem 5.2.2 this is equivalent to the independence of $X_1, \ldots, X_n$. ⌟

Remarks. 1. Theorem 5.2.4 tells us in particular, for n given probability spaces $(\Omega_i,\mathfrak{A}_i,P_i)$, how to construct a probability space $(\Omega,\mathfrak{A},P)$ and $(\Omega_i,\mathfrak{A}_i)$-random variables X_i defined on it which are independent and have the given distributions P_i $(i = 1, \ldots, n)$. For $(\Omega,\mathfrak{A},P)$ we choose the product of the n probability spaces and for $X_1, \ldots, X_n$ the projections in the coordinate axes.

2. The theorem further justifies the model developed in Section 4.1, (4). Indeed, if each of the probability spaces $(\Omega_i,\mathfrak{A}_i,P_i)$ describes an experiment $\mathfrak{E}_i$ with a random outcome, then the product of these probability spaces should describe that experiment $\mathfrak{E}$ which consists of performing $\mathfrak{E}_1, \ldots, \mathfrak{E}_n$ one after the other or simultaneously "*without mutual influence.*" The terminology "without mutual influence" can and should henceforth be replaced by "*independently.*" The random variable X_i just defined describes the outcome of the experiment $\mathfrak{E}_i$ in the joint experiment $\mathfrak{E}$.

PROBLEMS

1. Let $(X_i)_{i\in I}$ be a family of real random variables on a probability space. Prove: The family is independent if and only if the family $(\{X_i \leqq \alpha_i\})_{i\in I}$ of events is independent for all families $(\alpha_i)_{i\in I}$ of real numbers α_i.
2. Let X, Y, Z be independent real random variables. Prove the independence of $X + Y$ and Z, and of XY and Z.
3. Let X be a real random variable and f a real Borel-measurable function on $\mathbf{R}$. Prove: X and $f \circ X$ are independent random variables if and only if $f \circ X$ is almost surely constant. [*Hint:* Use Problems 2 of Sections 4.5 and 5.1.]
4. Let $X_1, \ldots, X_n$ be independent real random variables. Prove: $X_1 + \cdots + X_n$ is almost surely constant if and only if each X_i is almost surely constant.

5.3 PRODUCTS AND SUMS OF INDEPENDENT RANDOM VARIABLES

The last theorem has a number of useful consequences for real- or vector-valued random variables for which we have available the additional algebraic structure of the range.

First we present an analogy to (5.1.1) for real random variables:

5.3.1. Theorem (Multiplication Theorem). Let $X_1, \ldots, X_n$ be independent real random variables. Then

$$E\left(\prod_{i=1}^{n} X_i\right) = \prod_{i=1}^{n} E(X_i), \tag{5.3.1}$$

if either all $X_i \geqq 0$ or all X_i are integrable. In the second case, the product $\prod_{i=1}^{n} X_i$ is also integrable.

Conversely, if $\prod_{i=1}^{n} X_i$ is integrable and each $X_i \neq 0$ almost surely, then every X_i is also integrable.

Proof. By Theorem 5.2.4 $Q \equiv \bigotimes_{i=1}^{n} P_{X_i}$ is the joint distribution of $X_1, \ldots, X_n$. Then by the transformation theorem for integrals and Fubini's Theorem,

$$\begin{aligned} E\left(\left|\prod_{i=1}^{n} X_i\right|\right) &= \int |x_1 \cdot \ldots \cdot x_n| Q(dx_1, \ldots, dx_n) \\ &= \int \cdots \int |x_1| \cdot \ldots \cdot |x_n| P_{X_1}(dx_1) \cdots P_{X_n}(dx_n) \\ &= \prod_{i=1}^{n} \int |x_i| P_{X_i}(dx_i) = \prod_{i=1}^{n} E(|X_i|). \end{aligned}$$

Hence the assertion follows for the case $X_i \geqq 0$ $(i = 1, \ldots, n)$, as well as the integrability of ΠX_i in the case in which each X_i is integrable. By Fubini's Theorem the computations above remain valid if we remove all the absolute value signs. Finally, if $E(|\Pi X_i|) < +\infty$ and $E(|X_i|) > 0$ for each i, then

$$\Pi E(|X_i|) = E(|\Pi X_i|) < +\infty$$

and no factor is zero and thus $E(|X_i|) < +\infty$, that is, every X_i is integrable. ┘

Remark 1. By definition, two indicator variables X and Y are independent if and only if

$$E(XY) = E(X)E(Y). \tag{5.3.2}$$

Thus Theorem 5.3.1 leads to the question of whether the independence of two real, integrable random variables X and Y is equivalent to (5.3.2) in general. The following example shows that this is not the case: Let $(\Omega,\mathfrak{A},P)$ be the Laplace space of order 3 with $\Omega = \{1,2,3\}$. Let X and Y be defined on Ω as follows: $X(1) = 1$, $X(2) = 0$, $X(3) = -1$, and $Y(1) = Y(3) = 0$, $Y(2) = 1$. Equation (5.3.2) is obviously satisfied since $E(X) = E(XY) = 0$. On the other hand,

$$P(X^{-1}(1) \cap Y^{-1}(1)) = 0 \qquad \text{and} \qquad P(X^{-1}(1)) = P(Y^{-1}(1)) = \tfrac{1}{3},$$

and therefore X and Y are not independent.

Consequently we define:

5.3.2. Definition. Two integrable real random variables X and Y are said to be *uncorrelated* if

$$E(XY) = E(X)E(Y). \tag{5.3.2}$$

Any two independent, integrable real random variables are thus uncorrelated by Theorem 5.3.1. The converse, however, does not hold.

We now obtain the following property of the variance:

5.3.3. Theorem (Bienaymé's Equality). For finitely many pairwise uncorrelated, integrable, real random variables $X_1, \ldots, X_n$,

$$V(X_1 + \cdots + X_n) = V(X_1) + \cdots + V(X_n). \tag{5.3.3}$$

Proof. If we replace each X_i by $X_i - E(X_i)$, then simple computations show that the new random variables are still pairwise uncorrelated. Thus X_i can, in addition, be assumed to be centered (Section 4.3, Remark 2). We then have $E(X_iX_j) = E(X_i)E(X_j) = 0$ for every pair of indices $i \neq j$. We now assume that

$$S \equiv X_1 + \cdots + X_n$$

is square integrable. Then

$$S^2 - \sum_{\substack{i,j=1 \\ i\neq j}}^{n} X_iX_j = \sum_{j=1}^{n} X_j^2$$

is integrable and

$$\sum_{i=1}^{n} V(X_i) = \sum_{i=1}^{n} E(X_i^2) = E\left(\sum_{i=1}^{n} X_i^2\right) = E(S^2) = V\left(\sum_{i=1}^{n} X_i\right).$$

If S does not lie in $\mathfrak{L}^2(P)$, we have $V(S) = E(S^2) = +\infty$, and by (2.6.4)

$$\sqrt{V(S)} = N_2(S) \leqq N_2(X_1) + \cdots + N_2(X_n).$$

From $V(S) = +\infty$ we obtain $N_2(X_i) = \sqrt{V(X_i)} = +\infty$ for at least one $i = 1, \ldots, n$. Hence (5.3.3) holds also in this case. ┘

Remark 2. For finitely many *square integrable* real random variables $X_1, \ldots, X_n$, we compute the variance of their sum as

$$V\left(\sum_{i=1}^{n} X_i\right) = \sum_{i=1}^{n} V(X_i) + \sum_{\substack{i,j=1\\ i\neq j}}^{n} \operatorname{Cov}(X_i,X_j),$$

where

$$\operatorname{Cov}(X_i,X_j) \equiv E((X_i - E(X_i))(X_j - E(X_j)))$$

is the so-called *covariance* of X_i and X_j.

Finally, we can also determine the distribution of $X_1 + \cdots + X_n$ in the case of independence:

5.3.4. Theorem. Let $X_1, \ldots, X_n$ be independent $(\mathbf{R}^p,\mathfrak{B}^p)$-random variables ($p = 1, 2, \ldots$). Then the distribution of the random variable $S \equiv X_1 + \cdots + X_n$ equals

$$P_S = P_{X_1} * \cdots * P_{X_n}. \tag{5.3.4}$$

Proof. Let Y again denote the mapping $X_1 \otimes \cdots \otimes X_n$ of Ω into $\mathbf{R}^p \times \cdots \times \mathbf{R}^p$ (n factors); further, let A_n: $\mathbf{R}^p \times \cdots \times \mathbf{R}^p \to \mathbf{R}^p$ again be the mapping $(x_1, \ldots ,x_n) \to x_1 + \cdots + x_n$. Then

$$S = A_n \circ Y,$$

and thus S is a random variable and

$$P_S = A_n \circ Y(P) = A_n(P_Y)$$

is the distribution of S. Now Theorem 5.2.4 tells us that $P_Y = \bigotimes_{i=1}^{n} P_{X_i}$. But then, by the definition of convolution, we have $A_n(P_Y) = P_{X_1} * \cdots * P_{X_n}$. ┘

Now we can demonstrate frequently used properties of the *binomial, Poisson,* and *normal distributions:*

Examples

1. Suppose that two independent real random variables X and Y are distributed according to β_m^p and β_n^p, respectively. Then $X + Y$ is distributed according to β_{m+n}^p ($0 \leqq p \leqq 1$; $m, n = 1, 2, \ldots$). By using the

distributivity of convolution and taking account of the equality $\epsilon_x * \epsilon_y = \epsilon_{x+y}$ [see (3.4.10′)], we obtain for the distribution of $X + Y$:

$$\beta_m^p * \beta_n^p = \sum_{i=0}^{m} \sum_{j=0}^{n} \binom{m}{i} \binom{n}{j} p^{i+j} q^{m+n-i-j} \epsilon_i * \epsilon_j = \sum_{k=0}^{m+n} \alpha_k p^k q^{m+n-k} \epsilon_k$$

with $q = 1 - p$ and

$$\alpha_k = \sum_{i=0}^{k} \binom{m}{i} \binom{n}{k-i} = \binom{m+n}{k} \qquad (k = 0, \ldots, m + n).$$

In fact, we thus have $\beta_m^p * \beta_n^p = \beta_{m+n}^p$.

2. If X and Y are independent and distributed according to π_α and π_β, respectively, then $X + Y$ is distributed according to $\pi_{\alpha+\beta}$ ($\alpha, \beta \in \mathbf{R}_+$). Indeed, by (3.4.9) we obtain

$$\pi_\alpha * \pi_\beta = \sum_{i=0}^{\infty} \sum_{j=0}^{\infty} e^{-(\alpha+\beta)} \frac{\alpha^i}{i!} \frac{\beta^j}{j!} \epsilon_i * \epsilon_j = e^{-(\alpha+\beta)} \sum_{k=0}^{\infty} \rho_k \epsilon_k,$$

where

$$\rho_k = \sum_{i=0}^{k} \frac{\alpha^i}{i!} \frac{\beta^{k-i}}{(k-i)!} = \frac{(\alpha+\beta)^k}{k!} \qquad (k = 0, 1, \ldots).$$

3. If X and Y are independent and distributed according to ν_{α,σ^2} and ν_{β,τ^2}, respectively, then $X + Y$ is distributed according to $\nu_{\alpha+\beta,\sigma^2+\tau^2}$ ($\alpha, \beta \in \mathbf{R}$; $\sigma, \tau > 0$). To simplify the computations, we can assume that X and Y are centered and thus $\alpha = \beta = 0$. Then the two distributions have the probability densities g_{0,σ^2} and g_{0,τ^2} defined in (4.4.9). By (3.4.14), $g_{0,\sigma^2} * g_{0,\tau^2}$ is the density g of $\nu_{0,\sigma^2} * \nu_{0,\tau^2}$. We have

$$g(x) = (2\pi\sigma\tau)^{-1} \int_{-\infty}^{+\infty} \exp\left[-\frac{(x-y)^2}{2\sigma^2} - \frac{y^2}{2\tau^2}\right] dy = (2\pi\sigma\tau)^{-1} \exp\left(-\frac{x^2}{2(\sigma^2+\tau^2)}\right) \Psi(x),$$

where

$$\Psi(x) = \int_{-\infty}^{+\infty} \exp\left[-\frac{\sigma^2+\tau^2}{2\sigma^2\tau^2}\left(y - \frac{\tau^2 x}{\sigma^2+\tau^2}\right)^2\right] dy.$$

According to (2.8.6), the remaining integral $\Psi(x)$ equals

$$\left(2\pi \frac{\sigma^2\tau^2}{\sigma^2 + \tau^2}\right)^{1/2},$$

whence follows $g = g_{0,\sigma^2+\tau^2}$.

Remark 3. With somewhat more computation, which we are able to avoid in Section 8.2 by using other methods, it can be shown that if X and Y are independent and have the *Cauchy distributions* γ_α and γ_β, respectively, then $X + Y$ has the distribution $\gamma_{\alpha+\beta}$ ($\alpha > 0$, $\beta > 0$).

4. A probability-measure μ on $\mathfrak{B}^1$ is said to be *infinitely divisible* if for every natural number n a probability measure μ_n exists on $\mathfrak{B}^1$ such that $\mu = \mu_n * \cdots * \mu_n$ (n factors). Examples 2 and 3 show that π_α ($\alpha \geqq 0$) and ν_{α,σ^2} ($\alpha \in \mathbf{R}$, $\sigma > 0$) are such probability measures μ. It suffices to choose $\mu_n = \pi_{\alpha/n}$ or $= \nu_{\alpha/n,\sigma^2/n}$. Remark 3 shows that $\gamma_\alpha(\alpha > 0)$ is also infinitely divisible.

PROBLEMS

1. Let X and Y be real nonnegative random variables on a probability space. Denote by $\mathcal{B}_+$ the set of all real Borel-measurable functions $\geqq 0$ defined on $\mathbf{R}_+$. Does the condition

$$E(f \circ X \cdot g \circ Y) = E(f \circ X) \cdot E(g \circ Y), \qquad \text{for all } f, g \in \mathcal{B}_+$$

characterize the independence of X and Y?

2. Let $X_1, \ldots, X_n$ be real, integrable random variables, and let X be a real random variable with finite variance such that the variables $X_1, \ldots, X_n$, $X - \sum_{i=1}^n X_i$ are pairwise uncorrolated. Prove: $V(X_i) < \infty$ for all $i \in \{1, \ldots, n\}$.

5.4 INFINITE PRODUCTS OF PROBABILITY SPACES

The concluding remark of Section 5.2 reminds us that we are still lacking a probability-theoretic model for the following experiment $\mathfrak{E}$. Suppose we are given a sequence $(\mathfrak{E}_n)_{n\in\mathbf{N}}$ of experiments with random outcome; $\mathfrak{E}$ consists of performing the experiments $\mathfrak{E}_1, \mathfrak{E}_2, \ldots$ "independently" one after the other or simultaneously. For example, all $\mathfrak{E}_n$ could stand for the same experiment. Then we are concerned with repeating this experiment, for example tossing a coin or a die, infinitely often and independently of previously obtained results. Since we are fully familiar with the case of finitely many such repetitions by virtue of the theory of the product of finitely many probability spaces, it seems now promising to undertake an extension of this theory to include infinitely many factors. We motivate proceeding in this direction as follows:

Suppose every experiment $\mathfrak{E}_n$ is described by a probability space $(\Omega_n,\mathfrak{A}_n,P_n)$ $(n = 1, 2, \ldots)$. We seek a suitable probability space $(\Omega,\mathfrak{A},P)$ for describing $\mathfrak{E}$. We expect it to have the following properties: Every elementary event $\omega \in \Omega$ should be a sequence (ω_n) of elementary events $\omega_n \in \Omega_n$, that is, $\Omega = \prod_{n=1}^{\infty} \Omega_n$. Then if $A_1 \in \mathfrak{A}_1, \ldots, A_n \in \mathfrak{A}_n$ are possible events in the first n experiments,

$$A \equiv A_1 \times A_2 \times \cdots \times A_n \times \Omega_{n+1} \times \Omega_{n+2} \times \cdots$$

can be considered as that event for the experiment $\mathfrak{E}$ which consists of the occurrence of the events $A_1, \ldots, A_n$ in the first n trials of the series of trials $(\mathfrak{E}_i)$. We shall want the above A to be always in $\mathfrak{A}$ and to have

$$P(A) = P_1(A_1) \cdot \ldots \cdot P_n(A_n)$$

as its probability. The following more precise investigation will show that the probability measure P is already uniquely determined by these requirements and an additional minimal property.

Suppose that we are given an index set $I \neq \emptyset$ and a family $((\Omega_i,\mathfrak{A}_i,P_i))_{i\in I}$ of probability spaces.[4] Then for each subset $K \subset I$ we define

$$\Omega_K \equiv \prod_{i\in K} \Omega_i \tag{5.4.1}$$

and in particular we define

$$\Omega \equiv \Omega_I. \tag{5.4.1'}$$

Here Ω_K is the set of all mappings $\omega_K\colon K \to \bigcup_{i\in K} \Omega_i$ with $\omega_K(i) \in \Omega_i$ for all $i \in K$. By associating with every such mapping its restriction to a subset $J \subset K$, we obtain the *projection mapping*

$$p_J^K\colon \Omega_K \to \Omega_J \tag{5.4.2}$$

of Ω_K onto Ω_J. For $K = I$ we set $p_J \equiv p_J^I$; for $J = \{i\}$ let $p_i^K \equiv p_{\{i\}}^K$, and in particular $p_i \equiv p_i^I$. We obviously have

$$p_J^L = p_J^K \circ p_K^L \tag{5.4.3}$$

for any three sets $J \subset K \subset L \subset I$; in particular, then,

$$p_J = p_J^K \circ p_K \tag{5.4.3'}$$

for $J \subset K \subset I$.

[4] In most applications I will be countable.

Let $\mathfrak{H} = \mathfrak{H}(I)$ denote the system of all *nonempty, finite* subsets of I. For every $J \in \mathfrak{H}$,

$$\mathfrak{A}_J \equiv \bigotimes_{i \in J} \mathfrak{A}_i \qquad \text{and} \qquad P_J \equiv \bigotimes_{i \in J} P_i \tag{5.4.4}$$

are defined.

In accord with the definition for finite I given in Section 3.1, and in analogy with the definition of the product topology, we now define:

5.4.1. Definition. The *product* $\bigotimes_{i \in I} \mathfrak{A}_i$ of the σ-algebras $\mathfrak{A}_i$, $i \in I$, is the smallest σ-algebra $\mathfrak{A}_0$ in Ω with respect to which each of the projection mappings p_i is $\mathfrak{A}_0$-$\mathfrak{A}_i$-measurable. Thus,

$$\mathfrak{A}_0 = \bigotimes_{i \in I} \mathfrak{A}_i \equiv \mathfrak{A}(p_i;\, i \in I). \tag{5.4.5}$$

Then by Theorem 1.7.4, p_J is also $\mathfrak{A}_0$-$\mathfrak{A}_J$-measurable for every $J \in \mathfrak{H}$, since by (5.4.3′), $p_i = p_i^J \circ p_J$ for every $i \in J$. Thus we complete (5.4.5) by the equality

$$\mathfrak{A}(p_i;\, i \in I) = \mathfrak{A}(p_J;\, J \in \mathfrak{H}). \tag{5.4.5′}$$

According to the introductory remark, we are seeking a probability measure P on $\mathfrak{A}_0$ satisfying

$$P\left(p_J^{-1}\left(\prod_{i \in J} A_i\right)\right) = \prod_{i \in J} P_i(A_i)$$

for every $J \in \mathfrak{H}$ and arbitrary events $A_i \in \mathfrak{A}_i$ $(i \in J)$. By the definition of image measures this means that $p_J(P)$ should associate with every such product set $\prod_{i \in J} A_i$ the value $\prod_{i \in J} P_i(A_i)$. But by Theorem 3.3.1 the measure P_J defined in (5.4.4) is the only such measure $p_J(P)$. Therefore, our question finally is: Is there a probability measure P on $\mathfrak{A}_0$ such that its image under the projection p_J is equal to P_J for all $J \in \mathfrak{H}$? The answer is:

5.4.2. Theorem. On the σ-algebra $\mathfrak{A}_0 = \bigotimes_{i \in I} \mathfrak{A}_i$ there exists exactly one measure P such that for every set $J \in \mathfrak{H}(I)$

$$p_J(P) = P_J. \tag{5.4.6}$$

P is a probability measure.

Proof. Preliminary remarks. We can assume that I is an infinite set. For finite I the result follows from Theorem 3.3.1. For any two sets $J, K \in \mathfrak{H}$ with $J \subset K$ the mapping

$$p_J^K \colon \Omega_K \to \Omega_J$$

is $\mathfrak{A}_K$-$\mathfrak{A}_J$-measurable. The sets $\prod_{i \in J} A_i$ with $A_i \in \mathfrak{A}_i$ $(i \in J)$ indeed generate $\mathfrak{A}_J$, and

$$(p_J^K)^{-1}\left(\prod_{i \in J} A_i\right) = \prod_{i \in K} A_i' \tag{5.4.7}$$

with $A_i' \equiv A_i$ for $i \in J$ and $A_i' \equiv \Omega_i$ for $i \in K \setminus J$. Since we are concerned only with probability measures,

$$\prod_{i \in K} P_i(A_i') = \prod_{i \in J} P_i(A_i),$$

and hence, according to Theorem 3.3.1,

$$p_J^K(P_K) = P_J \qquad (J \subset K;\, J,\, K \in \mathfrak{H}). \tag{5.4.8}$$

Now we set

$$\mathfrak{Z}_J \equiv p_J^{-1}(\mathfrak{A}_J) \qquad (J \in \mathfrak{H}) \tag{5.4.9}$$

and call $\mathfrak{Z}_J$ the σ-algebra of the *J-cylinders*. The measurability of p_J^K just established implies that $(p_J^K)^{-1}(\mathfrak{A}_J) \subset \mathfrak{A}_K$. Together with (5.4.3′), this yields

$$\mathfrak{Z}_J \subset \mathfrak{Z}_K \qquad (J \subset K;\, J,\, K \in \mathfrak{H}). \tag{5.4.10}$$

Finally we call

$$\mathfrak{Z} \equiv \bigcup_{J \in \mathfrak{H}} \mathfrak{Z}_J \tag{5.4.11}$$

simply the set of all *cylinders*. Any two cylinders $Z_1, Z_2 \in \mathfrak{Z}$ lie in the same σ-algebra $\mathfrak{Z}_J$ for a suitable $J \in \mathfrak{H}$; if Z_i lies in $\mathfrak{Z}_{J_i}$ $(i = 1, 2)$, then, for example, $J_1 \cup J_2$ is such a J by (5.4.10). Hence it follows that $\mathfrak{Z}$ is an *algebra* (but not generally a σ-algebra) in Ω. By (5.4.5) and (5.4.5′) we have

$$\mathfrak{A}_0 = \mathfrak{A}(\mathfrak{Z}). \tag{5.4.12}$$

We separate the following *main part* of the proof into four steps:

1. If the problem has a solution, then according to (5.4.6) the desired measure P must associate with each J-cylinder $Z = p_J^{-1}(A)$ the value $P_J(A)(J \in \mathfrak{H},\ A \in \mathfrak{A}_J)$. Hence, this value must depend only on Z but not on the particular representation $Z = p_J^{-1}(A)$. The following arguments show that this is the case. Let

$$Z = p_J^{-1}(A) = p_K^{-1}(B)$$

be two representations of Z by means of J, $K \in \mathfrak{H}$, $A \in \mathfrak{A}_J$, and $B \in \mathfrak{A}_K$. When $J \subset K$, then by (5.4.3′),

$$p_J^{-1}(A) = p_K^{-1}((p_J^K)^{-1}(A))$$

and hence

$$p_K^{-1}(B) = p_K^{-1}(B') \qquad \text{with} \qquad B' = (p_J^K)^{-1}(A).$$

As a projection, p_K maps the set Ω onto Ω_K; thus, it follows that

$$B = B' = (p_J^K)^{-1}(A)$$

and, by (5.4.8),

$$P_K(B) = P_J(A).$$

For arbitrary J and K, we set $L \equiv J \cup K$; since $J \subset L$ and $K \subset L$, by (5.4.10) there is a $C \in \mathfrak{A}_L$ such that $p_L^{-1}(C) = p_J^{-1}(A) = p_K^{-1}(B)$. According to what we have just proved, we now have

$$P_L(C) = P_J(A) \qquad \text{and} \qquad P_L(C) = P_K(B),$$

and again

$$P_J(A) = P_K(B).$$

Thus we have shown that

$$P_0(p_J^{-1}(A)) \equiv P_J(A) \qquad (J \in \mathfrak{H}; A \in \mathfrak{A}_J) \tag{5.4.13}$$

defines a function P_0 on $\mathfrak{Z}$.

2. This function P_0: $\mathfrak{Z} \to \mathbf{R}$ is a content on $\mathfrak{Z}$. Indeed, $P_0 \geqq 0$ and $P_0(\varnothing) = 0$. For any two disjoint cylinders $Y, Z \in \mathfrak{Z}$, by the preliminary remark, there exists a set $J \in \mathfrak{H}$ satisfying $Y = p_J^{-1}(A)$ and $Z = p_J^{-1}(B)$ for suitable $A, B \in \mathfrak{A}_J$. Whenever Y and Z are disjoint, A and B are also disjoint, and therefore

$$Y \cup Z = p_J^{-1}(A \cup B)$$

yields the relationship

$$P_0(Y \cup Z) = P_J(A \cup B) = P_J(A) + P_J(B) = P_0(Y) + P_0(Z),$$

that is, the finite additivity of P_0.

The proof will be complete when P_0 is shown to be σ-additive. In fact, existence and uniqueness of P then follow from Theorem 1.5.7. P is the only extension of P_0 to a measure on $\mathfrak{A}(\mathfrak{Z}) = \mathfrak{A}_0$. P is a probability measure since $\Omega = p_J^{-1}(\Omega_J)$ is a J-cylinder for every $J \in \mathfrak{H}$ and thus $P(\Omega) = P_0(\Omega) = P_J(\Omega_J) = 1$. To verify the σ-additivity of P_0, we now show:

3. Let $Z \in \mathfrak{Z}$ and $J \in \mathfrak{H}$. Then for every $\omega_J \in \Omega_J$,

$$Z^{\omega_J} \equiv \{\omega \in \Omega: (\omega_J, p_{I \setminus J}(\omega)) \in Z\}\ ^5 \tag{5.4.14}$$

[5] As usual, $\Omega_J \times \Omega_{I \setminus J}$ has to be identified with Ω. Z^{ω_J} consists of all $\omega \in \Omega$ with the following property: If in ω we replace all the coordinates (function values) belonging to indices $i \in J$ by the corresponding coordinates of ω_J, then we obtain a point in Z. When $J = I \in \mathfrak{H}$ we have, accordingly, $Z^{\omega_J} = Z$ or $Z^{\omega_J} = \varnothing$.

is also a cylinder, and we have

$$P_0(Z) = \int P_0(Z^{\omega_J})P_J(d\omega_J). \tag{5.4.15}$$

In fact, for $Z \in \mathfrak{Z}$ there exist $K \in \mathfrak{H}$ and $A \in \mathfrak{A}_K$ such that $Z = p_K^{-1}(A)$, that is, $P_0(Z) = P_K(A)$. We can assume that $J \subset K$ by (5.4.10) and $J \neq K$, since I is infinite. Then obviously $Z^{\omega_J} = p_{K\setminus J}^{-1}(A_{\omega_J})$ holds for the ω_J-section A_{ω_J} of A in Ω_K, that is, for the set of all $\omega' \in \Omega_{K\setminus J}$ satisfying $(\omega_J,\omega') \in A$. By Lemma 3.2.1 applied to $\mathfrak{A}_1 \equiv \mathfrak{A}_J$ and $\mathfrak{A}_2 \equiv \mathfrak{A}_{K\setminus J}$, we have $A_{\omega_J} \in \mathfrak{A}_{K\setminus J}$, that is, $Z^{\omega_J} = p_{K\setminus J}^{-1}(A_{\omega_J})$, is a $(K \setminus J)$-cylinder. Then by Fubini's Theorem, and more precisely by the special case presented in (3.2.3), it follows from $P_K = P_J \otimes P_{K\setminus J}$ that

$$P_0(Z) = P_K(A) = \int P_{K\setminus J}(A_{\omega_J})P_J(d\omega_J).$$

But this is formula (5.4.15), since $P_0(Z^{\omega_J}) = P_{K\setminus J}(A_{\omega_J})$ follows from $Z^{\omega_J} = p_{K\setminus J}^{-1}(A_{\omega_J})$.

4. Finally we show that P_0 is $\emptyset$-continuous, that is, by Theorem 1.3.2, it is σ-additive. Thus, let (Z_n) be an antitone sequence of cylinders with $\alpha \equiv \inf P_0(Z_n) > 0$. We show that $\bigcap_{n=1}^{\infty} Z_n$ is nonempty. Every Z_n is of the form $Z_n = p_{J_n}^{-1}(A_n)$, where $J_n \in \mathfrak{H}$ and $A_n \in \mathfrak{A}_{J_n}$. According to (5.4.10) we can assume that $J_1 \subset J_2 \subset \cdots$. We apply the result proved in 3 to $J = J_1$ and $Z = Z_n$. Due to the $\mathfrak{A}_{J_1}$-measurability of $\omega_{J_1} \to P_0(Z_n^{\omega_{J_1}})$, the set

$$Q_n \equiv \left\{\omega_{J_1} \in \Omega_{J_1}: P_0(Z_n^{\omega_{J_1}}) \geqq \frac{\alpha}{2}\right\}$$

lies in $\mathfrak{A}_{J_1}$. From (5.4.15) and the fact that we are dealing only with probability measures, we obtain

$$\alpha \leqq P_0(Z_n) \leqq P_{J_1}(Q_n) + \frac{\alpha}{2},$$

that is, $P_{J_1}(Q_n) \geqq \alpha/2 > 0$ for all $n = 1, 2, \ldots$. Since whenever (Z_n) is antitone the sequence (Q_n) is also antitone, and as a probability measure, P_{J_1} is $\emptyset$-continuous, then $\bigcap_{n=1}^{\infty} Q_n$ cannot be empty. In other words, there is an $\omega_{J_1} \in \bigcap_{n=1}^{\infty} Q_n$ such that

$$P_0(Z_n^{\omega_{J_1}}) \geqq \frac{\alpha}{2} > 0 \qquad (n = 1, 2, \ldots).$$

Now if $J_2 \neq J_1$, then another application of 3 yields the existence of an

$\omega_{J_2\setminus J_1}$ such that $P_0((Z_n^{\omega_{J_1}})^{\omega_{J_2}/J_1}) \geqq \alpha 2^{-2}$ holds for all n. If we set $\omega_{J_2} \equiv (\omega_{J_1}, \omega_{J_2\setminus J_1})$, then ω_{J_2} lies in Ω_{J_2}; we have $(Z_n^{\omega_{J_1}})^{\omega_{J_2}/J_1} = Z_n^{\omega_{J_2}}$ and hence

$$P_0(Z_n^{\omega_{J_2}}) \geqq \frac{\alpha}{2^2} > 0 \qquad \text{and} \qquad p_{J_1}^{J_2}(\omega_{J_2}) = \omega_{J_1} \qquad (n = 1, 2, \ldots).$$

When $J_1 = J_2$ we may choose $\omega_{J_2} = \omega_{J_1}$. Induction then yields: For every $k = 1, 2, \ldots$, there exists an $\omega_{J_k} \in \Omega_{J_k}$ with

$$P_0(Z_n^{\omega_{J_k}}) \geqq \frac{\alpha}{2^k} > 0 \qquad \text{and} \qquad p_{J_k}^{J_{k+1}}(\omega_{J_{k+1}}) = \omega_{J_k} \qquad (k, n = 1, 2, \ldots).$$

By the latter of these two properties, there exists an $\omega_0 \in \Omega$ with $p_{J_k}(\omega_0) = \omega_{J_k}$ for $k = 1, 2, \ldots$. By the first property we have $Z_n^{\omega_{J_n}} \neq \emptyset$, so that an $\tilde{\omega}_n \in \Omega$ exists satisfying $(\omega_{J_n}, p_{I\setminus J_n}(\tilde{\omega}_n)) \in Z_n$. Since Z_n is a J_n-cylinder, $(\omega_{J_n}, p_{I\setminus J_n}(\omega_0)) = \omega_0$ then also lies in Z_n. Thus ω_0 lies in the intersection of all the Z_n, that is, we have $\bigcap_{n=1}^{\infty} Z_n \neq \emptyset$. ┘

For a finite set I the measure P is equal to P_I by (5.4.8). In accordance with Definition 3.3.2 we thus define:

5.4.3. Definition. The probability measure P uniquely determined according to Theorem 5.4.2 is called the *product* of the probability measures $(P_i)_{i\in I}$ and is denoted by $\bigotimes_{i\in I} P_i$. The probability space

$$\left(\prod_{i\in I} \Omega_i, \bigotimes_{i\in I} \mathfrak{A}_i, \bigotimes_{i\in I} P_i\right)$$

is called the product of the probability spaces $((\Omega_i, \mathfrak{A}_i, P_i))_{i\in I}$; we denote it by

$$\bigotimes_{i\in I} (\Omega_i, \mathfrak{A}_i, P_i)$$

As an application, we can now extend Theorem 5.2.4 to arbitrary families $(X_i)_{i\in I}$ of random variables on a probability space $(\Omega, \mathfrak{A}, P)$. Here let every X_i be an $(\Omega_i, \mathfrak{A}_i)$-random variable and P_{X_i} their distribution. At the same time we define the product mapping

$$\bigotimes_{i\in I} X_i \colon \Omega \to \prod_{i\in I} \Omega_i$$

by associating with every $\omega \in \Omega$ the mapping $i \to X_i(\omega)$ of I into $\bigcup_{i\in I} \Omega_i$. Then $Y \equiv \bigotimes_{i\in I} X_i$ is a $\left(\prod_{i\in I} \Omega_i, \bigotimes_{i\in I} \mathfrak{A}_i\right)$-random variable by Theorem 1.7.4 since $p_i \circ Y = X_i$ is $\mathfrak{A}$-$\mathfrak{A}_i$-measurable. Therefore, the distribution P_Y is defined. As an extension of what we said in Section 5.2, P_Y is called the *joint distribution* of the family $(X_i)_{i \in I}$. The generalization of Theorem 5.2.4 then goes as follows:

5.4.4. Theorem. A family $(X_i)_{i \in I}$ of random variables is independent if and only if its joint distribution is the product of its single distributions, that is, if and only if

$$P_{\underset{i \in I}{\otimes} X_i} = \bigotimes_{i \in I} P_{X_i}. \tag{5.4.16}$$

Proof. For every nonempty, finite set $J \subset I$ we again denote by

$$p_J\colon \prod_{i \in I} \Omega_i \to \prod_{i \in J} \Omega_i$$

the projection mapping and by Y and $Y_J\colon \Omega \to \prod_{i \in J} \Omega_i$ the mappings $\bigotimes_{i \in I} X_i$ and $\bigotimes_{i \in J} X_i$, respectively. Then obviously $Y_J = p_J \circ Y$, and therefore

$$P_{Y_J} = p_J(P_Y)$$

follows from the transitivity property of image measures noted in (1.7.6). The independence of $(X_i)_{i \in I}$ is now equivalent to the independence of $(X_i)_{i \in J}$, that is, by Theorem 5.2.4 equivalent to the validity of

$$P_{Y_J} = \bigotimes_{i \in J} P_{X_i}$$

for all finite sets $J \subset I$. By Theorem 5.4.2, (5.4.16) is equivalent to the validity of $p_J(P_Y) = \bigotimes_{i \in J} P_{X_i}$ for all such sets J. These three statements together yield the assertion. ⌟

Theorems 5.4.2 and 5.4.4 in particular allow us to construct independent families of random variables with given distributions. The following corollary extends Remark 1 of Section 5.2 to arbitrary index sets.

5.4.5. Corollary. For every family $((\Omega_i,\mathfrak{A}_i,P_i))_{i \in I}$ of probability spaces there exists an independent family $(X_i)_{i \in I}$ of $(\Omega_i,\mathfrak{A}_i)$-random variables X_i on a suitable probability space $(\Omega,\mathfrak{A},P)$ such that for every $i \in I$ the distribution of X_i is given by

$$P_{X_i} = P_i. \tag{5.4.17}$$

Proof. It suffices to choose the product

$$\bigotimes_{i \in I} (\Omega_i,\mathfrak{A}_i,P_i)$$

for $(\Omega,\mathfrak{A},P)$, and the projection mapping

$$p_i\colon \prod_{i\in I} \Omega_i \to \Omega_i$$

for X_i. Then P_i is the distribution of X_i for every $i \in I$ by the definition of product measures. The independence of the family $(X_i)_{i\in I}$ follows from Theorem 5.4.4 since $\underset{i\in I}{\otimes} X_i$ is obviously the identity mapping of $\Omega = \prod_{i\in I} \Omega_i$ onto itself and hence

$$P_{\underset{i\in I}{\otimes} X_i} = P = \underset{i\in I}{\otimes} P_i = \underset{i\in I}{\otimes} P_{X_i}. \quad \lrcorner$$

Remark. Now Remark 2 of Section 5.2 carries over: If a family $(\mathfrak{E}_i)_{i\in I}$ of experiments with random outcome is described by the family $(\Omega_i,\mathfrak{A}_i,P_i)_{i\in I}$ of probability spaces, then the product space

$$\underset{i\in I}{\otimes} (\Omega_i,\mathfrak{A}_i,P_i)$$

describes that experiment which consists of independent performance of each experiment $\mathfrak{E}_i$.

As announced earlier, we can finally present an essential generalization of Theorem 5.2.3.

5.4.6. Theorem. Let $(X_i)_{i\in I}$ be an independent family of $(\Omega_i,\mathfrak{A}_i)$-random variables and let $(I_j)_{j\in J}$ be a partition of I into pairwise disjoint sets. Suppose further that for every $j \in J$,

$$Y_j\colon \prod_{i\in I_j} \Omega_i \to \Omega'_j$$

is a measurable mapping of the measurable space $\left(\prod_{i\in I_j} \Omega_i, \underset{i\in I_j}{\otimes} \mathfrak{A}_i\right)$ into a measurable space $(\Omega'_j,\mathfrak{A}'_j)$. Then the family of $(\Omega'_j,\mathfrak{A}'_j)$-random variables

$$Z_j \equiv Y_j \circ \underset{i\in I_j}{\otimes} X_i \qquad (j \in J)$$

is independent.

Proof. By Theorem 5.2.3 it suffices to prove the independence of the family of random variables

$$Z_j \equiv \underset{i\in I_j}{\otimes} X_i \qquad (j \in J).$$

In this particular case, every Y_j is the identity mapping of $\prod_{i\in I_j} \Omega_i$ onto itself. The σ-algebra $\mathfrak{A}(Z_j)$ generated by Z_j is, however, also generated by the family of random variables

$$p_i \circ Z_j = X_i \qquad \text{with } i \in I_j$$

by Theorem 1.7.4 and Definition 5.4.1. Here p_i again denotes the projection mapping of $\prod_{i \in I_j} \Omega_i$ onto Ω_i. By hypothesis, $(\mathfrak{A}(X_i))_{i \in I}$ is independent. The assertion therefore follows from Corollary 5.1.5. ⌟

PROBLEM

Prove the existence of an independent sequence $(X_n)_{n \in \mathbf{N}}$ of integrable real random variables on an appropriate probability space such that (a) or (b) is satisfied:

(a) $P\{X_n = 1\} = P\{X_n = -1\} = \frac{1}{2}(1 - 2^{-n}),$
$P\{X_n = 2^n\} = P\{X_n = -2^n\} = 2^{-n-1},$ for all $n \in \mathbf{N}$;

(b) $P\{X_n = n^\lambda\} = P\{X_n = -n^\lambda\} = \frac{1}{2},$ for all $n \in \mathbf{N}$,

where $\lambda \in \mathbf{R}$ is a given real number.

Calculate $V(X_n)$ in both cases.

6

THE LAW OF LARGE NUMBERS

6.1 STATEMENT OF THE PROBLEM

In experiments with random outcomes we are often interested only in whether a certain event A does or does not occur. Suppose the probability of the occurrence of A is p ($0 \leqq p \leqq 1$). Using the indicator variable of A and its distribution, we see immediately that the simplest probability space $(\Omega_0, \mathfrak{A}_0, P_0)$ for the mathematical description of this experiment is the following: $\Omega_0 = \{0,1\}$, $\mathfrak{A}_0 = \mathfrak{P}(\Omega_0)$, $A = \{1\}$, $P_0(A) = p$, $P_0(\complement A) = P_0(\{0\}) = q \equiv 1 - p$.

Suppose the experiment is now repeated countably often and independently. The joint experiment thus constructed is often called a *sequence of Bernoulli trials or observations*. It could, for example, consist of successively throwing a die infinitely often where we are interested only in the event "a 6 comes up" or "a 6 does not come up" on each throw. According to the remark following Corollary 5.4.5, the probability space $(\Omega, \mathfrak{A}, P)$ with

$$\Omega \equiv \prod_{n=1}^{\infty} \Omega_n, \qquad \mathfrak{A} \equiv \bigotimes_{n=1}^{\infty} \mathfrak{A}_n, \qquad P \equiv \bigotimes_{n=1}^{\infty} P_n,$$

where every $\Omega_n = \Omega_0$, $\mathfrak{A}_n = \mathfrak{A}_0$, $P_n = P_0$ ($n = 1, 2, \ldots$) is an appropriate mathematical model. Thus, Ω is the set of all sequences $\omega \equiv (\omega_n)_{n \in \mathbf{N}}$ of numbers 0 and 1. We consider the random variable X_n, which is defined by the projection of Ω onto Ω_n

$$X_n(\omega) \equiv \omega_n \qquad [\omega = (\omega_n) \in \Omega]. \tag{6.1.1}$$

It assumes only the values 1 or 0 and specifies whether "success" or "failure" was the result of the nth performance of the original experiment.

Then

$$S_n \equiv \sum_{i=1}^{n} X_i \tag{6.1.2}$$

is also a random variable. It assumes only the values 0, 1, . . . , n and specifies the *number of successes* in the first n performances of the original experiment. As mentioned in Section 4.1, intuition and experience tell us that the relative frequency of success, that is, the sequence of random variables $(1/n)S_n$, approaches p "with high probability." This feeling is confirmed in practice, for example, by the records of any gambling casino from the observed results at the roulette tables. Now the question arises of whether and how our mathematical theory reflects this limit behavior of $((1/n)S_n)_{n\in\mathbf{N}}$. Here the discovery of the right convergence concept is crucial. Obviously $((1/n)S_n(\omega))_{n\in\mathbf{N}}$ does not converge to p for every $\omega \in \Omega$; indeed for the constant sequence $\omega = (0,0, \ldots)$ or $\omega = (1,1, \ldots)$ we obtain $(1/n)S_n(\omega)$ converging to 0 or 1 respectively for all n. If we translate the colloquial expression "with high probability" to "with probability one" (or, equivalently, to "almost surely"), then the question arises of whether

$$\lim_{n\to\infty} \frac{1}{n} S_n = p, \qquad P\text{-almost surely} \tag{6.1.3}$$

holds. A theorem of E. Borel answers this question in the positive. It will be obtained as a special case of the essentially more general Theorem 6.4.2.

Each of the random variables X_n has distribution β_1^p (see Section 4.4) and thus has expected value $E(X_n) = p$. Therefore, statement (6.1.3) can also be put into the equivalent form

$$\lim_{n\to\infty} \frac{1}{n} \sum_{i=1}^{n} (X_i - E(X_i)) = 0, \qquad P\text{-almost surely.} \tag{6.1.4}$$

The fact that we can prove the statement (6.1.4) for many sequences (X_n) of real random variables now belongs among the most noteworthy results of probability theory and simultaneously gives one of the best justifications of this theory. Here we use only very general properties of X_n and not their particular form, as in the above case of the sequence of Bernoulli trials. Methods of mathematical statistics in which probabilities are estimated from relative frequencies are mathematically based on and justified by statements of the form (6.1.4). Therefore we define:

6.1.1. Definition. A sequence $(X_n)_{n\in\mathbf{N}}$ of integrable real random variables is said to satisfy the *strong law of large numbers* if (6.1.4) holds.

In Section 6.4 we present sufficient conditions for the strong law of large numbers to hold. We arrive at the weak law of large numbers in Section 6.5 by weakening the almost sure convergence.

In the case of a sequence of Bernoulli trials, we need a few more properties of the sequences (X_n) and (S_n) considered there. We have already noted: (a) Each of the variables X_n has distribution β_1^p. By Theorem 5.4.4 we have: (b) The sequence (X_n) is independent. We then have, by Section 5.3, Example 1: (c) S_n has distribution β_n^p.

PROBLEM

Assume that the sequence $(X_n)_{n \in \mathbf{N}}$ satisfies the strong law of large numbers, and that $E(X_n) = E(X_1)$ for every $n \in \mathbf{N}$. Let $f\colon \mathbf{R} \to \mathbf{R}$ be continuous and bounded.

Prove:

$$\lim_{n \to \infty} E\left(f\left(\frac{1}{n} \sum_{i=1}^{n} X_i\right)\right) = f(E(X_1)).$$

6.2 ZERO-ONE LAWS

We return momentarily to Example 4 at the conclusion of Section 5.1. There we found that for every independent sequence $(A_n)_{n \in \mathbf{N}}$ of events of a probability space $(\Omega, \mathfrak{A}, P)$, the probability of occurrence of infinitely many of the A_n can only be equal to either 0 or 1. In symbols,

$$P\,(\limsup_{n \to \infty} A_n) = 0 \qquad \text{or} \qquad = 1. \tag{6.2.1}$$

We now present a necessary and sufficient condition for the occurrence of each of these two cases.

6.2.1. Lemma of Borel-Cantelli. Let $(A_n)_{n \in \mathbf{N}}$ be a sequence of events and $A \equiv \limsup_{n \to \infty} A_n$. Then

$$\sum_{n=1}^{\infty} P(A_n) < +\infty \Rightarrow P(A) = 0. \tag{6.2.2}$$

If the sequence $(A_n)_{n \in \mathbf{N}}$ is independent, then the converse also holds:

$$\sum_{n=1}^{\infty} P(A_n) = +\infty \Rightarrow P(A) = 1. \tag{6.2.3}$$

Proof. From the definition of A it follows that $A \subset \bigcup_{m=n}^{\infty} A_m$ for every $n = 1, 2, \ldots$. Then by (1.3.10), $P(A) \leqq \Sigma_{m=n}^{\infty} P(A_m)$ for all n, whence (6.2.2) follows.

Preliminary to proving the converse we note the following: For every sequence (α_n) of real numbers satisfying $0 \leqq \alpha_n \leqq 1$ $(n \in \mathbf{N})$,

$$\sum_{n=1}^{\infty} \alpha_n = +\infty \Rightarrow \lim_{n\to\infty} \prod_{\nu=1}^{n} (1 - \alpha_\nu) = 0. \tag{6.2.4}$$

We may assume $0 \leqq \alpha_n < 1$ for all n; otherwise the assertion is trivial. Now $\log(1 - \alpha_n) \leqq -\alpha_n$,[1] and therefore,

$$\prod_{\nu=1}^{n} \log(1 - \alpha_\nu) = \log \sum_{\nu=1}^{n} (1 - \alpha_\nu) \leqq -\sum_{\nu=1}^{n} \alpha_\nu.$$

Consequently,

$$\prod_{\nu=1}^{n} (1 - \alpha_\nu) \leqq \exp\left(-\sum_{\nu=1}^{n} \alpha_\nu\right),$$

that is, (6.2.4) follows.

Now suppose the sequence (A_n) and thus also, by Corollary 5.1.4, the sequence $(\complement A_n)$ is independent. Then

$$P\left(\bigcap_{m=n}^{N} \complement A_m\right) = \prod_{m=n}^{N} P(\complement A_m) = \prod_{m=n}^{N} (1 - P(A_m)),$$

for any two natural numbers $n \leqq N$. It follows from the definition of A that

$$\complement A = \bigcup_{n=1}^{\infty} \bigcap_{m=n}^{\infty} \complement A_m;$$

and we obtain

$$\begin{aligned} 1 - P(A) = P(\complement A) &= \lim_{n\to\infty} P\left(\bigcap_{m=n}^{\infty} \complement A_m\right) \\ &= \lim_{n\to\infty} \lim_{N\to\infty} P\left(\bigcap_{m=n}^{N} \complement A_m\right) \end{aligned}$$

due to the continuity properties of a probability measure. Thus we obtain

$$1 - P(A) = \lim_{n\to\infty} \lim_{N\to\infty} \prod_{m=n}^{N} (1 - P(A_m)).$$

[1] This follows immediately from the power series expansion of $\log(1 - x)$ or from the Mean Value Theorem of differential calculus.

If we now apply the preliminary remark to the sequence $\alpha_i \equiv P(A_{i+n-1})$, $i = 1, 2, \ldots$, we obtain

$$\lim_{N\to\infty} \prod_{m=n}^{N} (1 - P(A_m)) = 0, \qquad \text{for each } n = 1, 2, \ldots$$

and thus the rest of the assertion. ┘

In the case of an *independent* sequence (A_n), the statement (6.2.1) together with the result obtained in Lemma 6.2.1 on the occurrence of both possible cases is called the *Zero-One Law* of E. Borel.

6.2.2. Corollary. If a sequence $(A_n)_{n\in\mathbf{N}}$ of events has an independent subsequence $(A_{n_k})_{k\in\mathbf{N}}$ such that $\sum_{k=1}^{\infty} P(A_{n_k}) = +\infty$, then

$$P\,(\limsup_{n\to\infty} A_n) = 1.$$

Proof. The assertion follows from Lemma 6.2.1 because $\limsup A_{n_k} \subset \limsup A_n$. ┘

Example

A coin is tossed successively infinitely often. We wish to find the probability that two successive heads are thrown infinitely often.

Solution. Let $(\Omega,\mathfrak{A},P)$ be the probability space defined at the beginning of Section 6.1 with $p = q = \frac{1}{2}$. Let A_n denote the event that a head turns up on both the nth and $(n + 1)$st toss. Obviously $P(A_n) = 4^{-1}$ and thus $\sum_{n=1}^{\infty} P(A_{2n}) = +\infty$.

$A \equiv \limsup A_n$ is the event of interest to us. By Corollary 6.2.2 we have $P(A) = 1$ since the sequence (A_{2n}) [but not (A_n)] is independent.

In Theorem 5.1.7 we presented the more general *Kolmogorov* 0-1-*Law*. In particular, for σ-algebras $\mathfrak{A}_n \equiv \mathfrak{A}(X_n)$ generated by random variables X_n, it reads:

6.2.3. Theorem. Let $(X_n)_{n\in\mathbf{N}}$ be an independent sequence of random variables (with values in arbitrary measurable spaces). Then for every terminal event, that is, for every event

$$A \in \bigcap_{n=1}^{\infty} \mathfrak{A}(X_m;\, m \geqq n)$$

we have $P(A) = 0$ or $P(A) = 1$.

We wish to consider how this 0–1-Law gives us new insight into the question of the validity of the strong law of large numbers. Thus, suppose in particular that $(X_n)_{n\in\mathbf{N}}$ is an independent sequence of *real* random variables on $(\Omega,\mathfrak{A},P)$. Further, let $(\tau_n)_{n\in\mathbf{N}}$ be a sequence of real numbers satisfying $\lim\limits_{n\to\infty} \tau_n = 0$. We define

$$Y_n \equiv \tau_n \sum_{i=1}^{n} X_i \qquad (n = 1, 2, \ldots) \tag{6.2.5}$$

and consider the set A of all elementary events $\omega \in \Omega$ such that $\lim Y_n(\omega) = 0$. Since

$$A = \{\limsup_{n\to\infty} Y_n = 0\} \cap \{\liminf_{n\to\infty} Y_n = 0\},$$

A is an event by Theorems 2.1.3 and 2.1.5. Since $\lim \tau_n = 0$ and

$$Y_n = \tau_n \sum_{i=m}^{n} X_i + \tau_n \sum_{i=1}^{m-1} X_i \qquad (2 \leqq m \leqq n),$$

it follows that the set A remains the same when we change finitely many random variables (for example, set some equal to zero) in the sequence (X_n). Hence for every $m = 1, 2, \ldots,$

$$A = \bigcap_{k=1}^{\infty} \bigcup_{N=m}^{\infty} \left\{ \sup_{n\geq N} \left| \tau_n \sum_{i=m}^{n} X_i \right| \leqq \frac{1}{k} \right\}.$$

Thus A is a terminal event of the sequence (X_n) and therefore $P(A)$ can only be 0 or 1.

If every X_n is also integrable, then the sequence $(X_n - E(X_n))$ is again independent (by Theorem 5.2.3). The above result applied to this sequence and $\tau_n = n^{-1}$ yields that

$$P\left\{ \lim_{n\to\infty} \frac{1}{n} \sum_{i=1}^{n} (X_i - E(X_i)) = 0 \right\}$$

is either 0 or 1. Thus the strong law of large numbers holds by definition when this probability is 1.

Analogously the probability of convergence of the sequence (6.2.5) can likewise be only 0 or 1.

PROBLEMS

1. A coin is tossed infinitely often. Prove: With probability 1 every prescribed finite sequence of "head" and "tail" appears infinitely often.

2. Let $(X_n)_{n\in\mathbf{N}}$ be an independent sequence of real random variables, and let $(\tau_n)_{n\in\mathbf{N}}$ be a sequence of real numbers converging to zero. Prove: With probability 1, the sequence (X_n) [or $(\tau_n \sum_{i=1}^n (X_i)$] converges everywhere or nowhere.
3. Let p be a real number in [0,1]. Construct a probability space $(\Omega,\mathfrak{A},P)$ and a sequence (A_n) of events in A such that

$$\sum_{n=1}^{\infty} P(A_n) = +\infty \qquad \text{and} \qquad P\,(\lim_{n\to\infty}\ \sup A_n) = p.$$

[*Hint:* Try $\Omega = [0,1]$ and $P = \lambda_\Omega^1$.]
4. Let $(X_n)_{n\in\mathbf{N}}$ be an independent sequence of centered integrable real random variables on $(\Omega,\mathfrak{A},P)$. Prove: If (X_n) satisfies the strong law of large numbers, then

$$\sum_{n=1}^{\infty} P\left\{\frac{1}{n}\,|X_n| \geqq \epsilon\right\} < +\infty,$$

for all $\epsilon > 0$.

6.3 THE HÁJEK-RÉNYI INEQUALITY

All further considerations concerning the strong law of large numbers are based on the following inequality due to J. Hájek and H. Rényi:

6.3.1. Theorem. Suppose we are given n independent integrable real random variables $X_1, \ldots, X_n$ and n real numbers $\gamma_1 \geqq \cdots \geqq \gamma_n > 0$. We define

$$S_i \equiv X_1 - E(X_1) + \cdots + X_i - E(X_i) \qquad (i = 1, \ldots, n). \quad (6.3.1)$$

Then for every $m = 1, \ldots, n$ and every real $\epsilon > 0$,

$$P\{\sup_{m\leq i\leq n} \gamma_i\,|\,S_i\,| \geqq \epsilon\} \leqq \frac{1}{\epsilon^2}\left(\gamma_m^2 \sum_{j=1}^{m} V(X_j) + \sum_{j=m+1}^{n} \gamma_j^2 V(X_j)\right). \quad (6.3.2)^2$$

Proof. By Section 4.3, Remark 2 and Theorem 5.2.3 we may assume all $X_1, \ldots, X_n$ to be centered. Then, in particular, $S_i = X_1 + \cdots + X_i$. Further, all the variances $V(X_i)$ can be assumed finite since otherwise there is nothing to prove.

[2] Of course, $\sum_{j=n+1}^{n} \alpha_j = 0$ is always assumed.

Denote by A the event on the left side of (6.3.2). If we set

$$A_m \equiv \{\gamma_m|S_m| \geqq \epsilon\},$$
$$A_i \equiv \{\gamma_i|S_i| \geqq \epsilon\} \cap \{\gamma_{i-1}|S_{i-1}| < \epsilon\} \cap \cdots \cap \{\gamma_m|S_m| < \epsilon\} \quad (i = m+1, \ldots, n), \quad (6.3.3)$$

then the events $A_m, \ldots, A_n$ are pairwise disjoint and $A = \bigcup_{i=m}^{n} A_i$, whence

$$P(A) = \sum_{i=m}^{n} P(A_i) \quad (6.3.4)$$

follows. We also set $\gamma_{n+1} \equiv 0$ and

$$Z \equiv \sum_{i=m}^{n} (\gamma_i^2 - \gamma_{i+1}^2)S_i^2. \quad (6.3.5)$$

Bienaymé's equality yields

$$E(S_i^2) = V(S_i) = V(X_1) + \cdots + V(X_i)$$

and consequently

$$E(Z) = \gamma_m^2 \sum_{j=1}^{m} V(X_j) + \sum_{j=m+1}^{n} \gamma_j^2 V(X_j); \quad (6.3.6)$$

thus $\epsilon^{-2}E(Z)$ is the right side of (6.3.2). If we further set

$$Y_i \equiv 1_{A_i} \quad (i = m, \ldots, n), \quad (6.3.7)$$

then $Z \geqq 0$ and $\Sigma_{i=m}^{n} Y_i = 1_A \leqq 1$ imply the inequality

$$Z \geqq \sum_{i=m}^{n} Y_i Z;$$

thus, due to the antitonicity of the sequence $(\gamma_i)_{i=1,\ldots,n+1}$, we obtain

$$E(Z) \geqq \sum_{i=m}^{n} E(Y_i Z) = \sum_{i=m}^{n} \sum_{j=m}^{n} (\gamma_j^2 - \gamma_{j+1}^2)E(Y_i S_j^2)$$
$$\geqq \sum_{i=m}^{n} \sum_{j=i}^{n} (\gamma_j^2 - \gamma_{j+1}^2)E(Y_i S_j^2). \quad (6.3.8)$$

Now if

$$E(Y_i S_j^2) \geqq \frac{\epsilon^2}{\gamma_i^2} P(A_i) \quad (i = m, \ldots, n; j = i, \ldots, n) \quad (6.3.9)$$

holds, then (6.3.8) yields the further bound

$$E(Z) \geqq \epsilon^2 \sum_{i=m}^{n} P(A_i),$$

that is, yields the desired inequality because of (6.3.4) and (6.3.6). We now show (6.3.9).

For every pair of indices i, j considered there we set

$$U_{ij} \equiv S_j - S_i = \sum_{k=i+1}^{j} X_k$$

and obtain

$$E(Y_i S_j^2) = E(Y_i S_i^2) + E(Y_i U_{ij}^2) + 2E(Y_i S_i U_{ij}). \qquad (6.3.10)$$

By definition S_i, Y_i and thus $S_i Y_i$ are $\mathfrak{A}(X_1, \ldots, X_i)$-measurable and U_{ij} is $\mathfrak{A}(X_{i+1}, \ldots, X_n)$-measurable (the latter for $i = 1, \ldots, n-1$). Due to the independence of the (X_i), both σ-algebras are independent by Corollary 5.1.5; thus, $Y_i S_i$ and U_{ij} are independent ($i = 1, \ldots, n-1$). Since $U_{nn} = 0$, this also holds for $i = n$. But then, by Theorem 5.3.1,

$$E(Y_i S_i U_{ij}) = E(Y_i S_i) E(U_{ij}) = 0.$$

If we substitute this result into (6.3.10), then from (6.3.3) we obtain

$$E(Y_i S_j^2) \geqq E(Y_i S_i^2) = \int_{A_i} S_i^2 \, dP \geqq \frac{\epsilon^2}{\gamma_i^2} P(A_i).$$

But this is the inequality (6.3.9). ┘

We emphasize two *special cases* of inequality (6.3.2):

(a) Suppose we choose $m = n = 1$ and $\gamma_1 = 1$. Then for every integrable real random variable X we have the *Chebyshev inequality:*

$$P\{|X - E(X)| \geqq \epsilon\} \leqq \frac{1}{\epsilon^2} V(X) \qquad (\epsilon > 0). \qquad (6.3.11)$$

This obviously also follows from the Chebyshev-Markov inequality derived in Lemma 2.11.1 by choosing $p = 2$. If X has finite variance, then (6.3.11) tells us that large deviations of the random variable X from the expected value have small probability.

(b) Suppose we choose $m = 1$ and $\gamma_1 = \cdots = \gamma_n = 1$. Then for any n independent real integrable random variables $X_1, \ldots, X_n$ and every $\epsilon > 0$ we obtain the *Kolmogorov inequality:*

$$P\left\{\sup_{1 \leq j \leq n} \left| \sum_{i=1}^{j} (X_i - E(X_i)) \right| \geqq \epsilon \right\} \leqq \frac{1}{\epsilon^2} \sum_{i=1}^{n} V(X_i). \qquad (6.3.12)$$

But the most important application of (6.3.2) for our purposes is contained in the proof of Theorem 6.4.1.

6.4 THE KOLMOGOROV THEOREMS

We now give a first answer to the question raised in Section 6.1:

6.4.1. Theorem (Kolmogorov[3]). Let $(X_n)_{n\in\mathbf{N}}$ be an independent sequence of integrable real random variables. If

$$\sum_{n=1}^{\infty} \frac{V(X_n)}{n^2} < \infty, \qquad (6.4.1)$$

then the sequence (X_n) satisfies the strong law of large numbers.

Proof. For $n = 1, 2, \ldots,$ we set

$$Y_n \equiv \frac{1}{n}\sum_{i=1}^{n} (X_i - E(X_i)) \qquad \text{and} \qquad V_n \equiv V(X_n).$$

If we choose $\gamma_i = i^{-1}$ $(i = 1, \ldots, n)$ in Theorem 6.3.1, then all the hypotheses there are obviously satisfied; it thus follows from inequality (6.3.2) that for every real $\epsilon > 0$ and every pair of natural numbers $m \leqq n$,

$$P\{\sup_{m\le i\le n} |Y_i| > \epsilon\} \leqq \frac{1}{\epsilon^2}\left(\frac{1}{m^2}\sum_{j=1}^{m} V_j + \sum_{j=m+1}^{n} \frac{V_j}{j^2}\right).$$

Letting $n \to \infty$ we have

$$\{\sup_{m\le i\le n} |Y_i| > \epsilon\}\uparrow\{\sup_{i\ge m} |Y_i| > \epsilon\},$$

which, together with the continuity property of P, yields

$$P\{\sup_{i\ge m} |Y_i| > \epsilon\} \leqq \frac{1}{\epsilon^2}\left(\frac{1}{m^2}\sum_{j=1}^{m} V_j + \sum_{j=m+1}^{\infty} \frac{V_j}{j^2}\right).$$

For every natural number $M \leqq m$ we can further majorize the right side:

$$P\{\sup_{i\ge m} |Y_i| > \epsilon\} \leqq \frac{1}{\epsilon^2}\left(\frac{1}{m^2}\sum_{j=1}^{M} V_j + \sum_{j=M+1}^{\infty} \frac{V_j}{j^2}\right).$$

Now $\lim_{m\to\infty} m^{-2} \Sigma_{j=1}^{M} V_j = 0$ for every M and

$$\lim_{M\to\infty} \sum_{j=M+1}^{\infty} \frac{V_j}{j^2} = 0$$

[3] Theorem 6.4.1 is often called the Kolmogorov criterion.

by (6.4.1). Thus

$$\lim_{m\to\infty} P\{\sup_{i\geq m} |Y_i| > \epsilon\} = 0$$

for every $\epsilon > 0$. Then by Lemma 2.11.5, (Y_n) approaches 0 almost surely, that is, the sequence (X_n) satisfies the strong law of large numbers. ┘

If the random variables X_n all have the same distribution μ and if the variance

$$V(X_n) = \int x^2\mu(dx) - (\int x\mu(dx))^2,$$

which is independent of n by (4.3.10), is finite, then (4.3.1) is satisfied. This holds for the sequence (X_n) of Section 6.1 describing the sequence of Bernoulli trials. The theorem of E. Borel given there thus follows from Theorem 6.1.1. The following theorem now shows that the finiteness of $V(X_n)$ is no longer needed when all X_n have the same distribution, that is, in our previous terminology when they are *identically distributed.*

6.4.2. Theorem (Kolmogorov). Every independent sequence $(X_n)_{n\in\mathbf{N}}$ of integrable real and identically distributed random variables satisfies the strong law of large numbers.

Proof. Let μ be the distribution of the X_n, that is,

$$P_{X_n} = \mu \qquad (n \in \mathbf{N}).$$

In part 1 of the proof the variables X_n will be "truncated" in such a way that Theorem 6.4.1 can be applied. We repeatedly use the transformation formula (4.3.6) without explicit mention.

1. For every natural number n, let

$$I_n \equiv]-n,+n[, \quad H_n \equiv \{x \in \mathbf{R}: n - 1 \leqq |x| < n\}$$

and

$$f_n(x) \equiv x1_{I_n}(x) \qquad (x \in \mathbf{R}).^4$$

By Theorem 5.2.3

$$Y_n \equiv f_n \circ X_n \qquad (n \in \mathbf{N})$$

is then an independent sequence of real random variables. The common distribution is generally lost in the transition from (X_n) to (Y_n), but the hypotheses of Theorem 6.4.1 are satisfied for the sequence (Y_n): Y_n is square integrable due to the boundedness of Y_n. By (4.3.10) we obtain for the variance of Y_n

$$V(Y_n) \leqq E(Y_n^2) = E(f_n^2 \circ X_n) = \int f_n^2\, d\mu = \int_{I_n} x^2\mu(dx) = \sum_{j=1}^{n} \int_{H_j} x^2\mu(dx)$$

[4] 1_{I_n} is of course the indicator function of I_n with respect to $\mathbf{R}$.

and, by substituting and rearranging,

$$\sum_{n=1}^{\infty} \frac{V(Y_n)}{n^2} \leqq \sum_{n=1}^{\infty} \frac{1}{n^2} \sum_{j=1}^{n} \int_{H_j} x^2\mu(dx) = \sum_{j=1}^{\infty} \left(\sum_{n=j}^{\infty} \frac{1}{n^2}\right)\left(\int_{H_j} x^2\mu(dx)\right).$$

If we take note of the inequality

$$\sum_{n=j}^{\infty} \frac{1}{n^2} < \frac{1}{j^2} + \frac{1}{j(j+1)} + \frac{1}{(j+1)(j+2)} + \cdots = \frac{1}{j^2} + \frac{1}{j} \leqq \frac{2}{j},$$

we finally obtain the condition (6.4.1) for (Y_n):

$$\begin{aligned} \sum_{n=1}^{\infty} \frac{V(Y_n)}{n^2} &\leqq 2 \sum_{j=1}^{\infty} \int_{H_j} |x| \left|\frac{x}{j}\right| \mu(dx) \\ &\leqq 2 \sum_{j=1}^{\infty} \int_{H_j} |x|\mu(dx) = 2 \int |x|\mu(dx) = 2E(|X_n|) < +\infty. \end{aligned}$$

By Theorem 6.4.1 the sequence (Y_n) then satisfies the strong law of large numbers, and thus there exists an event A such that $P(A) = 1$ and

$$\lim_{n\to\infty} \frac{1}{n} \sum_{i=1}^{n} (Y_i(\omega) - E(Y_i)) = 0, \qquad \text{for all } \omega \in A.$$

2. Now let B be the set of all elementary events $\omega \in \Omega$ satisfying $X_n(\omega) \neq Y_n(\omega)$ for at most finitely many $n = 1, 2, \ldots$. Then

$$\lim_{n\to\infty} \frac{1}{n} \sum_{i=1}^{n} (X_i(\omega) - E(Y_i)) = 0, \qquad \text{for all } \omega \in A \cap B,$$

which is shown by the same reasoning used in the lines following (6.2.5). Therefore, we also have

$$\lim_{n\to\infty} \frac{1}{n} \sum_{i=1}^{n} (X_i(\omega) - E(X_i)) = 0, \qquad \text{for all } \omega \in A \cap B,$$

provided

$$\lim_{n\to\infty} \frac{1}{n} \sum_{i=1}^{n} (E(Y_i) - E(X_i)) = 0.$$

But this can be seen as follows: All X_n are identically distributed and thus have the same expected value α. We have

$$\alpha = \int x\mu(dx) = \lim_{n\to\infty} \int f_n \, d\mu = \lim_{n\to\infty} E(f_n \circ X_n) = \lim_{n\to\infty} E(Y_n),$$

since

$$|f_n(x)| \leqq |x| \qquad \text{and} \qquad \int |x|\mu(dx) = E(|X_n|) < +\infty$$

allow us to use Theorem 2.7.4 on dominated convergence. But when $(E(Y_n) - \alpha)$ is a null sequence, the sequence of arithmetic means, $n^{-1} \sum_{i=1}^{n} (E(Y_i) - \alpha)$, is then also a null sequence. Thus, the theorem is proved if B is an almost sure event. Then it would follow from $P(A) = P(B) = 1$ that $P(A \cap B) = 1$, by (1.3.5).

3. Let

$$C_n \equiv \{X_n \neq Y_n\} \qquad \text{and} \qquad C \equiv \limsup_{n\to\infty} C_n.$$

Then $B = \Omega \setminus C$ is obviously an event and we want to show that $P(C) = 0$. By Lemma 6.2.1 it suffices to verify the convergence of the series $\sum_{i=1}^{\infty} P(C_n)$. We have $C_n = \{|X_n| \geqq n\}$ and thus

$$P(C_n) = \mu(\complement I_n) = \sum_{j=n+1}^{\infty} \mu(H_j).$$

Hence we obtain

$$\begin{aligned}\sum_{n=1}^{\infty} P(C_n) &= \sum_{n=1}^{\infty} \sum_{j=n+1}^{\infty} \mu(H_j) = \sum_{j=2}^{\infty} (j-1)\mu(H_j) \\ &= \sum_{j=2}^{\infty} (j-1) \int_{H_j} d\mu \leqq \sum_{j=2}^{\infty} \int_{H_j} |x|\mu(dx) \leqq \int |x|\mu(dx).\end{aligned} \tag{6.4.2}$$

Since $\int |x|\mu(dx) = E(|X_n|) < +\infty$, $n = 1, 2, \ldots$, this series is indeed convergent. ⌟

Under the hypotheses of Theorem 6.4.2, the strong law of large numbers tells us that $(n^{-1} \sum_{i=1}^{n} X_i)$ converges almost surely to a constant α, indeed, $\alpha = E(X_n)$. In this formulation the expected value is no longer used explicitly and we might try to eliminate the hypothesis of integrability of X_n. But this is not possible, as is shown by the next result:

6.4.3. Theorem. Suppose for an independent sequence $(X_n)_{n\in\mathbf{N}}$ of real identically distributed random variables, that the sequence $((1/n) \sum_{i=1}^{n} X_i)$ converges almost surely to a real random variable Y as

$n \to \infty$. Then every X_n is integrable and Y is almost surely constant, that is,

$$Y = E(X_n) \text{ almost surely} \qquad (n = 1, 2, \ldots). \tag{6.4.3}$$

Proof. The assumed almost sure convergence of the sequence

$$Y_n \equiv \frac{1}{n} \sum_{i=1}^{n} X_i \qquad (n \in \mathbf{N})$$

implies the almost sure convergence of the sequence

$$\frac{1}{n} X_n = Y_n - \frac{n-1}{n} Y_{n-1} \qquad (n \in \mathbf{N})$$

to 0. Thus $|X_n| \geqq n$ for infinitely many n only occurs with probability 0. If C_n again denotes the event $\{|X_n| \geqq n\}$, then $P(\limsup_{n\to\infty} C_n) = 0$. Since when (X_n) is an independent sequence, the sequence (C_n) is also independent, the series $\sum_{n=1}^{\infty} P(C_n)$ is convergent by the Borel-Cantelli Lemma. Now if $H_j, j = 1, 2, \ldots$, is again defined as in the beginning of the proof of Theorem 6.4.2, then by (6.4.2) we have for the common distribution μ of the X_n:

$$\sum_{j=2}^{\infty} (j-1)\mu(H_j) = \sum_{n=1}^{\infty} P(C_n).$$

The convergence of this series and

$$\begin{aligned} E(|X_n|) &= \int |x|\mu(dx) = \sum_{j=1}^{\infty} \int_{H_j} |x|\mu(dx) \\ &\leqq \sum_{j=1}^{\infty} j\mu(H_j) = 1 + \sum_{j=2}^{\infty} (j-1)\mu(H_j) \end{aligned}$$

imply the integrability of each of the X_n. But then Theorem 6.4.2 can be applied, and thus Y_n converges almost surely to the common expected value α of all the X_n. Hence $Y = \alpha$ almost surely. ⌟

Example

Let $(\Omega,\mathfrak{A},P)$ be the probability space defined in Section 5.1, Example 3 and let (A_n) be the independent sequence of events given there. We consider the independent sequence (X_n) of the associated indicator variables $X_n \equiv 1_{A_n}$. Since $P(A_n) = \frac{1}{2}$, every X_n has distribution β_1^p with $p = \frac{1}{2}$, and thus, by Theorem 6.4.2, (X_n) satisfies the strong law of large numbers. Therefore, P-almost surely,

$$\lim_{n\to\infty} \frac{1}{n} \sum_{i=1}^{n} X_i = \frac{1}{2}.$$

This result can be interpreted as follows:

For every natural number $g \geqq 2$, every number $x \in \Omega = [0,1[$ has exactly one g-adic expansion

$$x = \sum_{n=1}^{\infty} \xi_n g^{-n},$$

in which ξ_n can take on the values $0, 1, \ldots, g-1$ and, for n sufficiently large, not all ξ_n are equal to $g-1$. For every $\epsilon \in \{0, 1, \ldots, g-1\}$, let $S_n^{\epsilon,g}(x)$ denote the number of integers among $i = 1, \ldots, n$ such that $\xi_i = \epsilon$ in the g-adic expansion of x. We say that $x \in \Omega$ is *g-normal* if

$$\lim_{n\to\infty} \frac{1}{n} S_n^{\epsilon,g}(x) = \frac{1}{g}, \qquad \text{for each } \epsilon = 0, 1, \ldots, g-1.$$

The number x is said to be *absolutely normal* if x is g-normal for all $g \geqq 2$. For $g = 2$ we obviously have

$$S_n^{0,2}(x) = \sum_{i=1}^{n} X_i(x) \qquad \text{and} \qquad S_n^{1,2}(x) = n - S_n^{0,2}(x).$$

The strong law of large numbers says precisely that P-almost all numbers $x \in \Omega$ are 2-normal. It can be shown analogously that P-almost all numbers $x \in \Omega$ are g-normal for arbitrary $g \geqq 2$. Since the union of countably many null sets is again a null set, it finally follows that P-almost all $x \in \Omega$ are absolutely normal.

PROBLEMS

1. Let λ be a real number, and let $(X_n)_{n\in\mathbf{N}}$ be a corresponding independent sequence of real random variables with the properties stated in part (b) of the problem of Section 5.4. Prove:
   (a) $\displaystyle\sum_{n=1}^{\infty} \frac{V(X_n)}{n^2}$ converges if and only if $\lambda < \frac{1}{2}$.
   (b) The sequence satisfies the strong law of large numbers for $\lambda < \frac{1}{2}$.
   (c) For $\lambda \geqq 1$ the strong law fails.
   [*Hint:* Use Section 6.2, Problem 4.] (For the case $\frac{1}{2} \leqq \lambda < 1$ see Section 9.2, Problem 6.)

2. Let $(X_n)_{n \in \mathbf{N}}$ and $(Y_n)_{n \in \mathbf{N}}$ be sequences of real, integrable random variables such that $E(X_n) = E(Y_n)$ for all n. Prove: If (X_n) satisfies the strong law of large numbers and if

$$\sum_{n=1}^{\infty} P\{X_n \neq Y_n\} < +\infty,$$

then also (Y_n) satisfies the strong law.
3. Let $(X_n)_{n \in \mathbf{N}}$ be an independent sequence of real random variables with the properties mentioned in part (a) of the Problem of Section 5.4. Prove: The sequence satisfies the strong law of large numbers but $\Sigma_{n=1}^{\infty}[V(X_n)/n^2]$ diverges.
[*Hint:* Introduce $Y_n \equiv \sup(-2, \inf(X_n, 2))$ and use Problem 2.] [This proves that Kolmogorov's criterion (6.4.1) is sufficient but not necessary for the strong law.]
4. Let $(\sigma_n)_{n \in \mathbf{N}}$ be a sequence of positive real numbers such that $\Sigma_{n=1}^{\infty} (\sigma_n^2/n^2)$ diverges. Prove the existence of an independent sequence (X_n) of integrable random variables with variances $V(X_n) = \sigma_n^2$ $(n \in \mathbf{N})$ for which the strong law fails. [*Hint:* Introduce $\alpha_n \equiv \max(\sigma_n, n)$ and $\beta_n \equiv \min(\sigma_n, n)$; define X_n in such a way that

$$P\{X_n = \alpha_n\} = P\{X_n = -\alpha_n\} = \frac{1}{2}\left(\frac{\beta_n}{n}\right)^2, P\{X_n = 0\} = 1 - \left(\frac{\beta_n}{n}\right)^2,$$

and apply Section 6.2, Problem 4.]

6.5 WEAK LAW OF LARGE NUMBERS

We now turn back for a moment to the sequence of Bernoulli trials considered in Section 6.1 and the corresponding sequence (X_n) of independent random variables describing them. Since $S_n = \Sigma_{i=1}^n X_i$ has distribution β_n^p, by (4.4.2) and (4.4.4) we have

$$E(n^{-1}S_n) = p,$$

and

$$V(n^{-1}S_n) = n^{-2}V(S_n) = n^{-1}pq.$$

For every real $\epsilon > 0$ the Chebyshev inequality (6.3.11) yields

$$P\left\{\left|\frac{1}{n} S_n - p\right| \geqq \epsilon\right\} \leqq \frac{pq}{n\epsilon^2},$$

and therefore $P\{|(1/n)S_n - p| \geqq \epsilon\} \to 0$ as $n \to \infty$. This fact, already known to Jacob Bernoulli, again shows that $((1/n)S_n)$ approaches p "with high probability." What is new here is that the almost sure convergence is replaced by the stochastic convergence (with respect to P) introduced in Definition 2.11.2. We now define:

6.5.1. Definition. A sequence $(X_n)_{n \in \mathbf{N}}$ of integrable real random variables on a probability space $(\Omega, \mathfrak{A}, P)$ is said to satisfy the *weak law* of large numbers if

$$P\text{-}\lim \frac{1}{n} \sum_{i=1}^{n} (X_i - E(X_i)) = 0.^{5} \tag{6.5.1}$$

The designation "weak law" is justified by Theorem 2.11.4, according to which stochastic convergence can always be concluded from the almost sure convergence of a sequence of random variables. The validity of the strong law of large numbers thus always implies the validity of the weak law of large numbers. Hence, in particular, the above result of J. Bernoulli is a corollary to the theorem of E. Borel of Section 6.1.

Because of these relationships, it should not be surprising that relatively weak hypotheses are sufficient for the validity of the weak law of large numbers. As an example, we have the following result:

6.5.2. Theorem. If

$$\lim_{n \to \infty} \frac{1}{n^2} \sum_{i=1}^{n} V(X_i) = 0, \tag{6.5.2}$$

for a sequence $(X_n)_{n \in \mathbf{N}}$ of real, integrable, pairwise uncorrelated random variables, then the sequence satisfies the weak law of large numbers.

Proof. If we define $S_n \equiv X_1 - E(X_1) + \cdots + X_n - E(X_n)$, then by Theorem 5.3.3

$$V(S_n) = \sum_{i=1}^{n} V(X_i),$$

and thus

$$V\left(\frac{1}{n} S_n\right) = \frac{1}{n^2} \sum_{i=1}^{n} V(X_i)$$

for all $n \in \mathbf{N}$. The assertion then follows from the Chebyshev inequality. ⌟

Condition (6.5.2) is satisfied, for example, if the sequence $(V(X_n))$ of variances is bounded. This is true, for instance, if the random variables

[5] We thus have stochastic convergence to the constant random variable 0.

X_n are all identically distributed and square integrable. By reasoning similar to that in the proof of Theorem 6.4.2 we can eliminate the finiteness of the variances and thus obtain a theorem of A. J. Khinchin (see, for example, Rényi [19]).

PROBLEMS

1. Let $(X_n)_{n\in\mathbf{N}}$ be an independent sequence of centered integrable real random variables on $(\Omega,\mathfrak{A},P)$. Prove: If (X_n) satisfies the weak law of large numbers, then the sequence $((1/n)X_n)$ converges stochastically to 0, that is,

$$\lim P\left\{\frac{1}{n}|X_n| \geqq \epsilon\right\} = 0$$

for all $\epsilon > 0$. Compare this result with Section 6.2, Problem 4.

2. Apply Problem 1 to situation (c) of Section 6.4, Problem 1 and prove that even the weak law of large numbers fails.

part III

Continuation of Measure and Integration Theory

7

MEASURES ON TOPOLOGICAL SPACES

To develop probability theory further, we need new properties of finite Borel measures on $\mathbf{R}^p$. Here we shall use only such properties of $\mathbf{R}^p$ which follow, on the one hand, from the existence of a countable base and, on the other hand, from the local compactness and completeness of $\mathbf{R}^p$ with respect to the Euclidean metric. Therefore, we carry out our investigations both for locally compact spaces with countable base and the more general class of Polish spaces. At the same time we shall get new insight into the meaning of the concept of integral. Finally a concept of convergence particularly suitable to Borel measures will lead us back to probabilistic considerations.

7.1 THE DANIELL-STONE THEOREM

Suppose we are given a set Ω and a set $\mathfrak{F}$ of real functions on Ω. We assume that $\mathfrak{F}$ is a *vector space* (over $\mathbf{R}$), that is, along with $u, v \in \mathfrak{F}$ and $\alpha, \beta \in \mathbf{R}$, the function $\alpha u + \beta v$ defined on Ω by

$$\omega \to \alpha u(\omega) + \beta v(\omega)$$

also lies in $\mathfrak{F}$. Suppose further that when $\mathfrak{F}$ contains u, it also contains the function $|u|$ defined on Ω by

$$\omega \to |u(\omega)|.$$

Then with any two functions $u, v \in \mathfrak{F}$, their upper envelope sup (u,v) and lower envelope inf (u,v) also lie in $\mathfrak{F}$; indeed,

$$\begin{aligned} \sup (u,v) &= \tfrac{1}{2}(u + v + |u - v|), \\ \inf (u,v) &= -\sup (-u,-v) = \tfrac{1}{2}(u + v - |u - v|). \end{aligned}$$

Then, in particular, every function $u \in \mathfrak{F}$ is the difference of nonnegative functions in $\mathfrak{F}$, for example, the positive part

$$u^+ = \sup (u,0)$$

and the negative part

$$u^- = (-u)^+$$

of u. In the sense of lattice theory, $\mathfrak{F}$ is thus a *vector lattice* (or *Riesz space*). Let $\mathfrak{F}_+$ denote the set of functions $u \in \mathfrak{F}$ satisfying $u \geqq 0$. If $\mathfrak{F}$ also satisfies the following condition (7.1.1) of M. H. Stone [43], then we call $\mathfrak{F}$ a *Stone vector lattice* (*of real functions* on Ω). The condition is

$$\inf (u,1) \in \mathfrak{F}, \qquad \text{for all } u \in \mathfrak{F}. \tag{7.1.1}$$

In particular, it is satisfied when the constant function 1 (and hence every constant real function on Ω), lies in $\mathfrak{F}$.

If we note the equality

$$\inf (u,\alpha) = \alpha \inf \left(\frac{1}{\alpha} u,1\right),$$

which holds for all $\alpha > 0$, then it follows that

$$\inf (u,\alpha) \in \mathfrak{F}, \qquad \text{for all } u \in \mathfrak{F} \text{ and all } \alpha \in \mathbf{R}_+ \tag{7.1.2}$$

in every Stone vector lattice.

Examples

1. Let Ω be a set and $\mathfrak{R}$ a ring of sets in Ω. Then the set $\mathfrak{F} = \mathfrak{F}(\mathfrak{R})$ of all linear combinations

$$u = \sum_{i=1}^{n} \alpha_i 1_{A_i}$$

of indicator functions 1_{A_i} of sets $A_i \in \mathfrak{R}$ ($n \in \mathbf{N}$, $\alpha_1, \ldots, \alpha_n \in \mathbf{R}$) is a Stone vector lattice. To see this, we need only take the sets $A_1, \ldots, A_n$ as pairwise disjoint. That this is always possible is shown by the following reasoning: Each set A_i is the union of the $k = 2^{n-1}$ pairwise disjoint sets $B_1, \ldots, B_k \in \mathfrak{R}$, where each B_j is of the form $C_1 \cap \cdots \cap C_n$ with $C_\nu = A_\nu$ or $= \complement A_\nu$ ($\nu \neq i$; $\nu = 1, \ldots, n$) and with $C_i = A_i$. We then have

$$1_{A_i} = 1_{B_1} + \cdots + 1_{B_k}.$$

We note that the statements $1 \in \mathfrak{F}$ and $\Omega \in \mathfrak{R}$ are equivalent.

2. Let $\Omega = [0,1]$ and let $\mathfrak{F}$ be the set of all continuous real functions u on Ω which are differentiable (on the right) at the point 0 and satisfy $u(0) = 0$. $\mathfrak{F}$ is a Stone vector lattice.

3. For every measure space $(\Omega,\mathfrak{A},\mu)$ and every number $p \in [1,+\infty]$, $\mathfrak{L}^p(\mu)$ is a Stone vector lattice. It follows from Definition 2.6.3 and (2.6.6) that when the vector space $\mathfrak{L}^p(\mu)$ contains a function u, it also contains $|u|$. Since $|\inf (u,1)| \leqq |u|$, we have that $\inf (u,1)$ is an L^p-function $(1 \leqq p < +\infty)$ whenever u is, and μ-almost everywhere bounded $(p = +\infty)$ whenever u is. The constant function 1 lies in $\mathfrak{L}^p(\mu)$ only for a finite measure μ, provided $1 \leqq p < +\infty$.

Of particular interest to us will be the Stone vector lattice $\mathfrak{L}^1(\mu)$. According to Section 2.4, $f \to \int f\, d\mu$ is a positive linear form on $\mathfrak{L}^1(\mu)$ which, according to the theorem on monotone convergence, is an abstract integral in the sense of the following definition.

7.1.1. Definition. An *abstract integral* is a linear form I (that is, linear mapping $I: \mathfrak{F} \to \mathbf{R}$) defined on a Stone vector lattice $\mathfrak{F}$ of real functions with the following two properties:

$$I \text{ is positive, that is, } I(u) \geqq 0, \text{ for all } u \in \mathfrak{F}_+;\text{[1]} \tag{7.1.3}$$

for every isotone sequence $(u_n)_{n\in\mathbf{N}}$ in $\mathfrak{F}_+$ with upper envelope $\sup u_n$ in $\mathfrak{F}$,

$$I\,(\sup_{n\in\mathbf{N}} u_n) = \sup_{n\in\mathbf{N}} I\,(u_n). \tag{7.1.4}$$

Taken with the other properties of I, the "*continuity property*" (7.1.4) is equivalent to

For every antitone sequence $(u_n)_{n\in\mathbf{N}}$ in $\mathfrak{F}$ with $\inf\limits_{n\in\mathbf{N}} u_n = 0$,

$$\lim_{n\to\infty} I\,(u_n) = 0. \tag{7.1.4'}$$

In fact, if (u_n) is an isotone sequence in $\mathfrak{F}_+$ with $u = \sup u_n \in \mathfrak{F}$, then the sequence $(u - u_n)$ is antitone and $\inf (u - u_n) = 0$. Conversely, for every antitone sequence (v_n) in $\mathfrak{F}$ with $\inf v_n = 0$, the sequence $(v_1 - v_n)$ is isotone, $v_1 - v_n \in \mathfrak{F}_+$ for all $n \in \mathbf{N}$ and $v_1 = \sup (v_1 - v_n)$.

Justification of the designation "abstract integral" is given by the Daniell-Stone Theorem below, which shows in a very precise sense that every abstract integral is a "concrete" one.

The following concept serves as preparation for the proof and several consequences of this theorem:

7.1.2. Definition. Let $\mathfrak{F}$ be a Stone vector lattice of real functions on a set Ω. A set $G \subset \Omega$ is said to be $\mathfrak{F}$-*open* if there exists an isotone

[1] The *isotonicity* of I is equivalent to this, as follows immediately from $I(v) - I(u) = I(v-u)(u, v \in \mathfrak{F})$.

sequence $(u_n)_{n \in \mathbf{N}}$ in $\mathfrak{F}_+$ such that

$$1_G = \sup u_n.$$

Let $\mathfrak{G} = \mathfrak{G}(\mathfrak{F})$ denote the system of all $\mathfrak{F}$-open sets.

If (similarly to Section 2.3) we let $\mathfrak{F}_+^*$ denote the set of all numerical functions $f \geqq 0$ on Ω which are upper envelopes of isotone sequences (u_n) in $\mathfrak{F}_+$, then we can also formulate Definition 7.1.5 as

$$\mathfrak{G} = \{G \in \mathfrak{P}(\Omega) : 1_G \in \mathfrak{F}_+^*\}. \tag{7.1.5}$$

7.1.3. Lemma. The system $\mathfrak{G}$ of $\mathfrak{F}$-open sets has the following properties:

(1) $\{f > \alpha\} \in \mathfrak{G}$ for every $f \in \mathfrak{F}_+^*$ and every real number $\alpha \geqq 0$.

(2) $\mathfrak{G}$ is an $\cap$-stable generator of the smallest σ-algebra $\mathfrak{A}(\mathfrak{F})$ in Ω with respect to which all functions $u \in \mathfrak{F}$ are measurable.

(3) Whenever $(G_n)_{n \in \mathbf{N}}$ is a sequence of $\mathfrak{F}$-open sets, $\bigcup\limits_{n=1}^{\infty} G_n$ is also $\mathfrak{F}$-open.

Proof. (1) For $f \in \mathfrak{F}_+^*$ there exists an isotone sequence (u_n) in $\mathfrak{F}_+$ such that $f = \sup u_n$. As a consequence of the Stone condition, each of the functions

$$v_n \equiv n(u_n - \inf(u_n, \alpha)) = n(u_n - \alpha)^+ \qquad (n = 1, 2, \ldots)$$

lies in $\mathfrak{F}_+$. Obviously $\lim\limits_{n \to \infty} v_n(\omega) = 0$ for all $\omega \in \{f \leqq \alpha\}$ and $\lim\limits_{n \to \infty} v_n(\omega) = +\infty$ for all $\omega \in \{f > \alpha\}$; further, the sequence (v_n) is isotone But then $w_n \equiv \inf(v_n, 1)$, $n = 1, 2, \ldots$, is an isotone sequence in $\mathfrak{F}_+$ satisfying $\sup w_n = 1_{\{f > \alpha\}}$, that is, $\{f > \alpha\} \in \mathfrak{G}$.

(2) Every function $f \in \mathfrak{F}_+^*$ is a pointwise limit of a sequence in $\mathfrak{F}_+$ and thus is itself $\mathfrak{A}(\mathfrak{F})$-measurable. By (7.1.5) it now follows that $\mathfrak{G} \subset \mathfrak{A}(\mathfrak{F})$, that is, $\mathfrak{A}(\mathfrak{G}) \subset \mathfrak{A}(\mathfrak{F})$. By (1), every function from $\mathfrak{F}_+^*$, and in particular from $\mathfrak{F}_+$, is $\mathfrak{A}(\mathfrak{G})$-measurable. Since every $u \in \mathfrak{F}$ is the difference of functions from $\mathfrak{F}_+$, the $\mathfrak{A}(\mathfrak{G})$-measurability of all $u \in \mathfrak{F}$ now follows, that is, $\mathfrak{A}(\mathfrak{F}) \subset \mathfrak{A}(\mathfrak{G})$ and hence $\mathfrak{A}(\mathfrak{G}) = \mathfrak{A}(\mathfrak{F})$. Finally, $\mathfrak{G}$ is $\cap$-stable since $1_{G \cap H} = \inf(1_G, 1_H)$ holds for arbitrary sets G, $H \subset \Omega$ and $\mathfrak{F}_+^*$ obviously contains the infinum of any two functions f, $g \in \mathfrak{F}_+^*$.

(3) Whenever the sequence (f_n) lies in $\mathfrak{F}_+^*$, $\sup f_n$ also lies in $\mathfrak{F}_+^*$. This is proved similarly to the analogous assertion in Theorem 2.3.4. We note only that $\mathfrak{F}_+$ and hence also $\mathfrak{F}_+^*$ contains $\sup(f,g)$ as an element whenever it contains f and g. Hence we may assume that the sequence (f_n) is isotone. Therefore, whenever $\mathfrak{G}$ contains a sequence (G_n), it also contains $\bigcup G_n$, since $1_{\cup G_n} = \sup 1_{G_n}$.

Remark. In general Ω is not $\mathfrak{F}$-open. This is shown by Example 1 for the ring $\mathfrak{R} = \{\emptyset\}$. We then have $\mathfrak{G} = \{\emptyset\}$.

Finally we come to:

7.1.4. Theorem (P. Daniell-M. H. Stone). Let Ω be a set, $\mathfrak{F}$ a Stone vector lattice of real functions on Ω, and I an abstract integral on $\mathfrak{F}$. Then there exists exactly one measure μ on the σ-algebra $\mathfrak{A}(\mathfrak{F})$ with the following properties:

$$\mathfrak{F} \subset \mathfrak{L}^1(\mu), \tag{7.1.6}$$

$$I(u) = \int u\, d\mu, \qquad \text{for all } u \in \mathfrak{F}; \tag{7.1.7}$$

$$\mu(A) = \begin{cases} \inf\limits_{\substack{G \in \mathfrak{G} \\ A \subset G}} \mu(G), & \text{if } A \subset G \text{ for some } G \in \mathfrak{G}, \\ +\infty, & \text{if } A \subset G \text{ for no } G \in \mathfrak{G}, \end{cases} \tag{7.1.8}$$
$$\text{for all } A \in \mathfrak{A}(\mathfrak{F})$$

Proof. We proceed in several steps:

1. For every isotone sequence $(u_n) \in \mathfrak{F}_+$ and every $v \in \mathfrak{F}_+$,

$$v \leqq \sup u_n \Rightarrow I(v) \leqq \sup I(u_n). \tag{7.1.9}$$ [2]

If we set $v_n \equiv \inf(v,u_n)$, then (v_n) is an isotone sequence in $\mathfrak{F}_+$ such that $\sup v_n = \inf(v, \sup u_n) = v \in \mathfrak{F}_+$. By (7.1.4) it then follows that $I(v) = \sup I(v_n)$. Taking into account the inequality $v_n \leqq u_n$ which holds for all n, we have $I(v_n) \leqq I(u_n)$ and hence $I(v) \leqq \sup I(u_n)$.

2. For any two isotone sequences (u_n) and (v_n) in $\mathfrak{F}_+$,

$$\sup u_n = \sup v_n \Rightarrow \sup I(u_n) = \sup I(v_n). \tag{7.1.10}$$

This is obtained from (7.1.9) in the same way as Corollary 2.3.2 was obtained from Theorem 2.3.1.

Thus we can define a numerical function I^* on $\mathfrak{F}_+^*$ as follows: For $f \in \mathfrak{F}_+^*$ let (u_n) be an isotone sequence in $\mathfrak{F}_+$ with $f = \sup u_n$. By (7.1.10),

[2] The reader should note the analogy with Theorem 2.3.1.

$\sup I(u_n)$ is independent of the particular choice of (u_n). Thus, we can set

$$I^*(f) \equiv \sup I(u_n). \tag{7.1.11}$$

Then, obviously,

$$I^*(u) = I(u), \qquad \text{for all } u \in \mathfrak{F}_+. \tag{7.1.12}$$

We obtain the most important properties of I^* similarly to Section 2.3: The isotonicity of I^* follows from (7.1.9) and the additivity and positive-homogeneity from (7.1.11). As in the proof of Theorem 2.3.4, we show that

$$I^*(\sup f_n) = \sup I^*(f_n) \tag{7.1.13}$$

for every isotone sequence (f_n) in $\mathfrak{F}_+^*$.

3. Now we define an outer measure μ^* on Ω. Let

$$\mu^*(G) = I^*(1_G), \qquad \text{for } G \in \mathfrak{G}, \tag{7.1.14}$$

$$\mu^*(Q) \equiv \begin{cases} \inf\limits_{\substack{G \in \mathfrak{G} \\ Q \subset G}} \mu^*(G), & \text{if } Q \subset G \text{ for some } G \in \mathfrak{G}, \\ +\infty, & \text{if } Q \subset G \text{ for no } G \in \mathfrak{G}, \end{cases} \tag{7.1.15}$$
$$\text{for } Q \in \mathfrak{P}(\Omega).$$

The isotonicity of I^* shows the compatibility of Definitions (7.1.14) and (7.1.15) as well as the isotonicity of μ^*. Moreover, $0 = 1_\varnothing$ and $\varnothing \in \mathfrak{G}$, so that $\mu^*(\varnothing) = 0$, implying $\mu^* \geqq 0$. So we still have to show that

$$\mu^*\left(\bigcup_{n=1}^{\infty} Q_n\right) \leqq \sum_{n=1}^{\infty} \mu^*(Q_n)$$

for an arbitrary sequence (Q_n) in $\mathfrak{P}(\Omega)$. Since the union of every sequence of $\mathfrak{F}$-open sets is $\mathfrak{F}$-open, it obviously suffices to deal with the case in which each Q_n is $\mathfrak{F}$-open. But then, using the properties of I^* already established in 2, we conclude $1_{\cup Q_n} \leqq \Sigma\, 1_{Q_n}$ and thus

$$\begin{aligned} \mu^*\left(\bigcup_{n=1}^{\infty} Q_n\right) &= I^*(1_{\cup Q_n}) \leqq I^*\left(\sum_{n=1}^{\infty} 1_{Q_n}\right) = I^*\left(\sup_{n\in\mathbf{N}} \sum_{i=1}^{n} 1_{Q_i}\right) \\ &= \sup_{n\in\mathbf{N}} I^*\left(\sum_{i=1}^{n} 1_{Q_i}\right) = \sup_{n\in\mathbf{N}} \sum_{i=1}^{n} I^*(1_{Q_i}) \\ &= \sum_{n=1}^{\infty} I^*(1_{Q_n}) \qquad = \sum_{n=1}^{\infty} \mu^*(Q_n). \end{aligned}$$

4. The outer measure μ^* can also be defined by the equality

$$\mu^*(Q) = \begin{cases} \inf\limits_{\substack{f \in \mathfrak{F}_+^* \\ 1_Q \leqq f}} I^*(f), & \text{if } 1_Q \leqq f \text{ for some } f \in \mathfrak{F}_+^* \\ +\infty, & \text{if } 1_Q \leqq f \text{ for no } f \in \mathfrak{F}_+^* \end{cases} \tag{7.1.16}$$

for arbitrary sets $Q \in \mathfrak{P}(\Omega)$. Since, by definition, $1_G \in \mathfrak{F}_+^*$ for all $\mathfrak{F}$-open sets, it suffices to prove that $1_Q \leqq f$ with $f \in \mathfrak{F}_+^*$ implies the existence of an $\mathfrak{F}$-open set $G \supset Q$ as well as $\mu^*(Q) \leqq I^*(f)$. But for every real number $\alpha \in]0,1[$ the set $G_\alpha = \{f > \alpha\}$ is $\mathfrak{F}$-open and satisfies $G_\alpha \supset Q$ and $\alpha\, 1_{G\alpha} \leqq f$. Therefore, $\mu^*(Q) \leqq \mu^*(G_\alpha) = I^*(1_G) \leqq (1/\alpha)I^*(f)$ for all $0 < \alpha < 1$. This implies $\mu^*(Q) \leqq I^*(f)$.

5. By Theorem 1.5.4, the system $\mathfrak{A}^*$ of all μ^*-measurable sets, that is, of all sets $A \subset \Omega$ satisfying

$$\mu^*(Q) \geqq \mu^*(Q \cap A) + \mu^*(Q \setminus A), \qquad \text{for all } Q \in \mathfrak{P}(\Omega), \quad (7.1.17)$$

is a σ-algebra and $\text{rest}_{\mathfrak{A}^*}\, \mu^*$ is a measure. We shall show that $\mathfrak{A}^*$ contains every $\mathfrak{F}$-open set A. Because of (7.1.15) and the isotonicity of μ^*, we need only verify that

$$\mu^*(G) \geqq \mu^*(G \cap A) + \mu^*(G \setminus A)$$

for $\mathfrak{F}$-open sets G with $\mu^*(G) < +\infty$. Since $G \cap A \in \mathfrak{G}$ we can also assume that $A \subset G$, and in particular $\mu^*(A) < +\infty$. For G and A there exist isotone sequences (u_n) and (v_n) in $\mathfrak{F}_+$ such that $1_G = \sup u_n$ and $1_A = \sup v_n$. Since $1_A \leqq 1_G$,

$$f_n \equiv 1_G - v_n = \sup_{k \in \mathbf{N}} (u_k - v_n)^+$$

lies in $\mathfrak{F}_+^*$, and $I^*(f_n) = \mu^*(G) - I(v_n)$. The sequence (f_n) is antitone. Using (7.1.16), we thus obtain

$$\begin{aligned}\mu^*(G \setminus A) &= \inf_{\substack{H \in \mathfrak{G} \\ G \setminus A \subset H}} \mu^*(H) = \inf_{\substack{f \in \mathfrak{F}_+^* \\ 1_{G\setminus A} \leqq f}} I^*(f) \leqq \inf I^*(f_n) \\ &= \lim I^*(f_n) = \mu^*(G) - \mu^*(A)\end{aligned}$$

and hence

$$\mu^*(A) + \mu^*(G \setminus A) = \mu^*(G \cap A) + \mu^*(G \setminus A) \leqq \mu^*(G).$$

Thus $\mathfrak{G} \subset \mathfrak{A}^*$, and hence $\mathfrak{A}(\mathfrak{F}) = \mathfrak{A}(\mathfrak{G}) \subset \mathfrak{A}^*$ has been verified because of Lemma 7.1.3. Then, further, the restriction μ of μ^* to $\mathfrak{A}(\mathfrak{F})$ is a measure which, according to (7.1.15), satisfies the condition (7.1.8).

6. μ also satisfies the conditions (7.1.6) and (7.1.7). For this, let $u \in \mathfrak{F}_+$. We need to show that $I(u) = \int u\, d\mu$, whence follows the μ-integrability of u because $0 \leqq I(u) < +\infty$.

For every natural number n and $i = 1, 2, \ldots, n2^n$, the set

$$G_{in} \equiv \left\{u > \frac{i}{2^n}\right\}$$

is $\mathfrak{F}$-open by Lemma 7.1.3. Therefore,

$$f_n \equiv \frac{1}{2^n} \sum_{i=1}^{n2^n} 1_{G_{in}} \tag{7.1.18}$$

lies in $\mathfrak{F}_+^*$. The sets

$$A_{in} \equiv \left\{u > \frac{i}{2^n}\right\} \cap \left\{u \leqq \frac{i+1}{2^n}\right\}, \qquad i = 1, 2, \ldots, n2^n - 1$$

and

$$A_{n2^n,n} \equiv \{u > n\}$$

are pairwise disjoint and elements of $\mathfrak{A}(\mathfrak{F})$. Thus $G_{in} = \bigcup_{j=1}^{n2^n} A_{jn}$ implies $1_{G_{in}} = \Sigma_{j=i}^{n2^n} 1_{A_{jn}}$ and hence

$$f_n = \sum_{i=1}^{n2^n} \frac{i}{2^n} 1_{A_{in}}.$$

The same reasoning as in the proof of Theorem 2.3.6 then shows the isotonicity of (f_n) and the equality $u = \sup f_n$. By (7.1.13) and the Theorem of Monotone Convergence we need only show that $I^*(f_n) = \int f_n \, d\mu$ for every $n = 1, 2, \ldots$. But this follows directly from (7.1.18) and the definition of μ, and in particular from (7.1.14).

For an arbitrary function $u \in \mathfrak{F}$ we obtain its μ-integrability and (7.1.7) by decomposing u into positive and negative parts.

7. The uniqueness of the measure μ follows from (7.1.8) provided the values $\mu(G)$ for sets $G \in \mathfrak{G}$ are uniquely determined by the remaining conditions. But this can be seen as follows: For $G \in \mathfrak{G}$ there exists an isotone sequence (u_n) in $\mathfrak{F}_+$ with $1_G = \sup u_n$. Therefore for every measure μ on $\mathfrak{A}(\mathfrak{F})$ with the property (7.1.7),

$$\mu(G) = \int 1_G \, d\mu = \sup \int u_n \, d\mu = \sup I(u_n) = I^*(1_G).$$

Therefore, $\mu(G)$ is uniquely determined.

Thus all parts of the theorem have been proved. ⌟

7.1.5. Corollary 1. Under the assumptions of Theorem 7.1.4, for every real number p satisfying $1 \leqq p < +\infty$, $\mathfrak{F} \cap \mathfrak{L}^p(\mu)$ is dense in $\mathfrak{L}^p(\mu)$ with respect to convergence in pth mean.[3]

[3] In particular $\mathfrak{F}$ is dense in $\mathfrak{L}^1(\mu)$ by (7.1.6).

Proof. For every function $f \in \mathfrak{L}^p(\mu)$ and every $\epsilon > 0$ we have to prove the existence of an L^p-function $u \in \mathfrak{F}$ such that

$$N_p(f - u) = (\int |f - u|^p \, d\mu)^{1/p} \leqq \epsilon.$$

We solve this problem by simplifying the function f step by step. Since $|f|$, f^+, and f^- are also in $\mathfrak{L}^p(\mu)$ whenever f is, and N_p is a seminorm on $\mathfrak{L}^p(\mu)$, we can in addition assume $f \geqq 0$. Then there is an isotone sequence (f_p) of $\mathfrak{A}(\mathfrak{F})$-elementary functions such that $f = \sup f_n$. Since $0 \leqq f_n \leqq f$, all f_n lie in $\mathfrak{L}^p(\mu)$. Therefore $\lim\limits_{n \to \infty} N_p(f - f_n) = 0$ by the Theorem of Dominated Convergence. Thus we can also assume that f is an $\mathfrak{A}(\mathfrak{F})$-elementary function and, after another application of the seminorm property of N_p, that it is an indicator function 1_A of a set $A \in \mathfrak{A}(\mathfrak{F})$ with $\mu(A) = [N_p(1_A)]^p < +\infty$. But then (7.1.8) applies. There exists an $\mathfrak{F}$-open set G containing A such that

$$[\mu(G) - \mu(A)]^{1/p} = N_p(1_G - f) \leqq \frac{\epsilon}{2}.$$

Since G is $\mathfrak{F}$-open, there is an isotone sequence (u_n) in $\mathfrak{F}_+$ with $1_G = \sup u_n$. It then follows that $u_n \in \mathfrak{L}^p(\mu)$ for all n and $\lim\limits_{n \to \infty} N_p(1_G - u_n) = 0$. Thus there is an n_0 for which

$$N_p(1_G - u_{n_0}) \leqq \frac{\epsilon}{2}.$$

Hence

$$N_p(f - u_{n_0}) \leqq N_p(f - 1_G) + N_p(1_G - u_{n_0}) \leqq \epsilon.$$

Thus $u = u_{n_0}$ yields the desired result. ┘

7.1.6. Corollary 2. If there exists an isotone sequence (u_n) in $\mathfrak{F}_+$ such that

$$\sup u_n(\omega) > 0, \qquad \text{for all } \omega \in \Omega; \tag{7.1.19}$$

$$\sup I(u_n) < +\infty, \tag{7.1.20}$$

then the measure μ of Theorem 7.1.4 is σ-finite and is uniquely determined by the properties (7.1.6) and (7.1.7) alone.

Proof. The function $f \equiv \sup u_n$ lies in $\mathfrak{F}_+^*$ and is strictly positive. Therefore, $\Omega = \bigcup\limits_{k=1}^{\infty} G_k$ for the sets $G_k \equiv \{f > 1/k\}$, $k \in \mathbf{N}$, which are $\mathfrak{F}$-open according to Lemma 7.1.3. Since $1_{G_k} \leqq kf$, then

$$\mu(G_k) \leqq k \int f \, d\mu = k \sup \int u_n \, d\mu = k \sup I(u_n) < +\infty,$$

for every measure μ on $\mathfrak{A}(\mathfrak{F})$ satisfying the conditions (7.1.6) and (7.1.7). This proves the σ-finiteness of μ. The rest of the assertion follows from the Uniqueness Theorem 1.5.5 since $\mathfrak{G}$ is an $\cap$-stable generator of $\mathfrak{A}(\mathfrak{F})$ on which the values $\mu(G)$ are uniquely determined by the conditions (7.1.6) and (7.1.7). ┘

Note that the conditions of Corollary 7.1.6 are satisfied if the constant function 1 lies in $\mathfrak{F}$. Then we can choose $u_n = 1$ for all $n = 1, 2, \ldots$.

Examples

4. Let Ω be a set, $\mathfrak{R}$ a ring of sets in Ω, and μ a finite content on $\mathfrak{R}$. For every function $u = \sum_{i=1}^n \alpha_i 1_{A_i}$ from the Stone vector lattice $\mathfrak{F} = \mathfrak{F}(\mathfrak{R})$ of Example 1, define

$$I_\mu(u) \equiv \sum_{i=1}^{n} \alpha_i \mu(A_i).$$

Reasoning analogous to that in Section 2.2 (via normal representations) shows that $I_\mu(u)$ is independent of the particular representation of u and hence $I_\mu\colon \mathfrak{F} \to \mathbf{R}$ is a positive linear form on $\mathfrak{F}$. We now have:

I is an abstract integral if and only if μ is σ-additive, that is, is a premeasure.

The second half of the assertion is obvious since for every sequence (A_n) in $\mathfrak{R}$ with $A_n \downarrow \varnothing$, the sequence (1_{A_n}) is antitone and $\inf 1_{A_n} = 0$. Thus μ is σ-additive if I is an abstract integral. For the converse we reason as in the proof of Theorem 2.3.1.

Now we obviously have $\mathfrak{A}(\mathfrak{F}) = \mathfrak{A}(\mathfrak{R})$. Thus Theorem 7.1.4 says: If μ is a premeasure, there is at least one measure $\tilde{\mu}$ on $\mathfrak{A}(\mathfrak{R})$ with $I_\mu(u) = \int u \, d\tilde{\mu}$ for all $u \in \mathfrak{F}$, that is, with

$$\tilde{\mu}(A) = \mu(A)$$

for all $A \in \mathfrak{R}$. Thus we obtain once more the familiar theorem whereby μ can be extended to a measure $\tilde{\mu}$ on $\mathfrak{A}(\mathfrak{R})$. $\tilde{\mu}$ is uniquely determined if we also require (7.1.8). The reader should verify that we obtain the measure constructed in Theorem 1.5.2. The $\mathfrak{F}$-open sets here are indeed all sets of the form $\bigcup_{n=1}^{\infty} A_n$ where (A_n) is a sequence in $\mathfrak{R}$.

5. Let $\mathfrak{F}$ be the Stone vector lattice of real functions on [0,1] defined in Example 2. Now

$$I(u) \equiv u'(0) = \lim_{\substack{x \to 0 \\ x > 0}} \frac{u(x)}{x} \qquad (u \in \mathfrak{F})$$

defines a positive linear form on $\mathfrak{F}$. This is *not* an abstract integral. Indeed, consider the functions $u_n(x) \equiv \inf\,(x,1/n)$, $x \in [0,1]$, $n = 1, 2, \ldots$. Then (u_n) is an antitone sequence in $\mathfrak{F}_+$ with $\inf u_n = 0$, but with $I(u_n) = 1$ for all n.

Remark. Example 4 above can be constructed in such a way that starting with a premeasure μ we immediately obtain the theory of the integral relative to the measure $\tilde{\mu}$ above without having defined the latter in advance. To do this we start with the concept of abstract integral and use the Daniell-Stone Theorem. Then the abstract integral, at first defined only on the system $\mathfrak{F}$ of "elementary functions," is extended to the more comprehensive domain $\mathfrak{L}^1(\tilde{\mu})$ with preservation of its properties. This extension is then the "concrete" integral. For details see Aumann [23], where the presentation corresponds in part to the original work of Daniell [29] and Stone [43].

Condition (7.1.8) is a *regularity condition*. It will appear in the next section in a sharper form. The refinement will be, primarily, that the $\mathfrak{F}$-open sets coincide with the open subsets of Ω for certain Stone vector lattices $\mathfrak{F}$ of *continuous* real functions on a topological space Ω.

PROBLEMS

1. Prove: A set $G \subset \Omega$ is $\mathfrak{F}$-open if and only if it is the union of a sequence of sets of the form $\{u > 0\}$ where $u \in \mathfrak{F}_+$.
2. Let $\mathfrak{F}$ be a Stone vector lattice of real functions on a set Ω. Denote by $\mathfrak{F}_0$ the set of all bounded functions $g \in \mathfrak{F}$ having the following property: There exists a function $h \in \mathfrak{F}_+$ (depending on g) such that $h(\omega) < 1$ implies $g(\omega) = 0$ $(\omega \in \Omega)$. Prove:
   (a) $\mathfrak{F}_0$ is a Stone vector lattice.
   (b) $(f - \alpha)^+ \in \mathfrak{F}_0$ for all bounded functions $f \in \mathfrak{F}$ and all $\alpha > 0$.
   (c) $\mathfrak{F}_0$-open sets and $\mathfrak{F}$-open sets coincide. [*Hint:* Study the proof of 7.1.3, (1) and use Problem 1.]
3. Let $\mathfrak{R}$ be a ring in a set Ω and let $\mathfrak{F}$ be the Stone vector lattice $\mathfrak{F}(\mathfrak{R})$ of Example 1. Prove:
   (a) $\{u > \alpha\} \in \mathfrak{R}$ for all $u \in \mathfrak{F}_+$ and all $\alpha > 0$.
   (b) A set is $\mathfrak{F}$-open if and only if it is the union of a sequence of sets in $\mathfrak{R}$.
4. Let $(\Omega,\mathfrak{A},\mu)$ be a σ-finite measure space. Denote by $\mathfrak{F}$ the vector space of all functions $u = \sum_{i=1}^{n} \alpha_i 1_{A_i}$, where each $A_i \in \mathfrak{A}$ has finite μ-measure $(\alpha_i \in \mathbf{R};\ n \in \mathbf{N})$. Prove: $\mathfrak{F}$ is dense in $\mathfrak{L}^p(\mu)$ with respect to convergence in pth mean $(1 \leqq p < +\infty)$.

7.2 BAIRE AND BOREL SETS AND MEASURES

Let E be a *topological space,* $\mathfrak{O}$ the system of its *open sets* defining the topology, and $\mathfrak{C} = \mathfrak{C}(E) = \mathfrak{C}(E,\mathbf{R})$ [$\mathfrak{C}^b = \mathfrak{C}^b(E) = \mathfrak{C}^b(E,\mathbf{R})$] the Stone vector lattice of all continuous real [continuous real bounded] functions defined on E. Since $1 \in \mathfrak{C}^b \subset \mathfrak{C}$, the Stone condition is trivially satisfied.

7.2.1. Definition. The σ-algebra $\mathfrak{A}(\mathfrak{O})$ generated in E by $\mathfrak{O}$ is called the σ-algebra $\mathfrak{B} = \mathfrak{B}(E)$ of *Borel* sets in E. The smallest σ-algebra $\mathfrak{A}(\mathfrak{C})$ in E with respect to which all functions $f \in \mathfrak{C}(E)$ are $\mathfrak{A}(\mathfrak{C})$-$\mathfrak{B}^1$-measurable is called the σ-algebra $\mathfrak{B}_0 = \mathfrak{B}_0(E)$ of *Baire sets* in E.

Since the closed sets are the complements of open ones, $\mathfrak{B}(E)$ is also generated by the system of *closed* sets of E.

Every function $f \in \mathfrak{C}$ is $\mathfrak{B}(E)$-measurable since $\{f > \alpha\}$ is open for every $\alpha \in \mathbf{R}$. Consequently,

$$\mathfrak{B}_0(E) \subset \mathfrak{B}(E), \tag{7.2.1}$$

that is, every Baire set in E is Borel.

Every function $f \in \mathfrak{C}$ is a pointwise limit of a sequence (f_n) in $\mathfrak{C}^b$, say the sequence $f_n \equiv \inf(\sup(f, -n), n)$. Therefore,

$$\mathfrak{B}_0(E) = \mathfrak{A}(\mathfrak{C}^b(E)). \tag{7.2.2}$$

Examples

1. By Theorem 1.6.4 we have

$$\mathfrak{B}(\mathbf{R}^p) = \mathfrak{B}^p \qquad (p = 1, 2, \ldots). \tag{7.2.3}$$

In Corollary 7.2.4 we shall also obtain $\mathfrak{B}_0(\mathbf{R}^p) = \mathfrak{B}(\mathbf{R}^p)$.

2. Let E be a discrete space, that is, $\mathfrak{O} = \mathfrak{P}(E)$. Then all real functions on E are continuous. Thus $\mathfrak{B}_0(E) = \mathfrak{B}(E) = \mathfrak{P}(E)$. The system $\mathfrak{K}$ of *compact* subsets of E is in general *not* a generator of $\mathfrak{B}_0(E) = \mathfrak{B}(E)$. Obviously $\mathfrak{K}$ consists of all finite subsets of E; therefore, $\mathfrak{A}(\mathfrak{K})$ is the σ-algebra of all sets $A \subset E$ for which either A or $\complement A$ is countable. Thus $\mathfrak{A}(\mathfrak{K}) = \mathfrak{B}_0(E)$ if and only if E is countable.

3. Let Q be a *subspace* of a topological space E. Then $\mathfrak{B}(Q) = Q \cap \mathfrak{B}(E)$, that is, $\mathfrak{B}(Q)$ is the trace of $\mathfrak{B}(E)$ in Q. In fact, $\{Q \cap G\colon G \in \mathfrak{O}\}$ is the system of open sets of Q and thus a generator of $\mathfrak{B}(Q)$. Since $Q \cap \mathfrak{B}(E)$ is a σ-algebra in Q containing this generator, we have

$$\mathfrak{B}(Q) \subset Q \cap \mathfrak{B}(E).$$

The system $\{A \subset E\colon Q \cap A \in \mathfrak{B}(Q)\}$ is obviously a σ-algebra in E, which contains the generator of $\mathfrak{B}(E)$ consisting of all open sets. Therefore we also have $Q \cap \mathfrak{B}(E) \subset \mathfrak{B}(Q)$.[4]

If Q itself is Borel in E, then $\mathfrak{B}(Q)$ consists of all Borel subsets of E which are contained in Q.

A subset A of a topological space E is usually called an *F_σ-set* (*G_δ-set*) if it is the union (intersection) of a sequence of closed (open) sets. We call A a *K_σ-set* if it is the union of a sequence of compact sets.

We say that E is *normal* (see Franz [32], p. 64) if the Hausdorff separation axiom is satisfied[5] and if for any two disjoint closed sets $F_0, F_1 \subset E$ there exists a function $f \in \mathfrak{C}$ satisfying $0 \leqq f \leqq 1$, $f(x) = 0$ for all $x \in F_0$ and $f(x) = 1$ for all $x \in F_1$. Every metric and every compact space is normal.

We shall now connect these notions with the considerations of the preceding section.

7.2.2. Lemma. For every subset G of a normal space E, the following properties are equivalent:

(1) G is $\mathfrak{C}(E)$-open.

(2) G is $\mathfrak{C}^b(E)$-open.

(3) G is an open F_σ-set.

(4) There exists a function $f \in \mathfrak{C}^b(E)$ with $f \geqq 0$ and $G = \{f > 0\}$.

The implications (1) $\Rightarrow$ (2) $\Rightarrow$ (3) and (4) $\Rightarrow$ (1) hold in any topological space E.

Proof. (1) $\Rightarrow$ (2): Every $u \in \mathfrak{C}$ with $1 \leqq u \leqq 1_G$ is bounded.

(2) $\Rightarrow$ (3): There is an isotone sequence (u_n) in $\mathfrak{C}^b_+$ such that $1_G = \sup u_n$. Then

$$G = \bigcup_{n=1}^{\infty} \{u_n > 0\}$$

[4] More generally, this reasoning tells us the following: Let $(\Omega,\mathfrak{A})$ be a measurable space, $\mathfrak{E}$ a generator of $\mathfrak{A}$, $\Omega' \subset \Omega$, and $\mathfrak{E}' \equiv \{\Omega' \cap E\colon E \in \mathfrak{E}\}$. Then $\Omega' \cap \mathfrak{A}$ is generated by $\mathfrak{E}'$ (in Ω').

[5] The reader should note that the proofs below of assertions 7.2.2, 7.2.3, 7.2.5, 7.2.6 for normal spaces do *not* refer to the Hausdorff separation axiom.

is open as the union of open sets and since

$$G = \bigcup_{k,n=1}^{\infty} \left\{ u_n \geqq \frac{1}{k} \right\}$$

it is an F_σ-set.

(3) $\Rightarrow$ (4): For G there is a sequence (F_n) of closed sets such that $G = \bigcup_{n=1}^{\infty} F_n$. Since F_n and $\complement G$ are disjoint closed sets, due to the normality of E there is a function $f_n \in \mathcal{C}^b$ satisfying $0 \leqq f_n \leqq 1$, $f_n(x) = 0$ for all $x \in \complement G$ and $f_n(x) = 1$ for all $x \in F_n$. The series $\Sigma_{n=1}^{\infty} (1/n^2) f_n(x)$ is uniformly convergent on E and thus defines a function $f \in \mathcal{C}_+^b$. For this f we have $G = \{f > 0\}$ by construction.

(4) $\Rightarrow$ (1): This follows directly from Lemma 7.1.3. ⌟

7.2.3. Corollary 1. The σ-algebra of Baire sets of a normal space is generated by the system of open F_σ-sets as well as by the system of closed G_δ-sets.

Proof. This follows from Lemmas 7.1.3 and 7.2.2. We have only to note that the complements of the open F_σ-sets are just the closed G_δ-sets of the space. ⌟

7.2.4. Corollary 2. For every metrizable space E, $\mathfrak{B}_0(E) = \mathfrak{B}(E)$.

Proof. Every closed set F in E is a G_δ-set. For if d denotes a metric defining the topology of E, then

$$G_n \equiv \left\{ x \in E : d(x,F) < \frac{1}{n} \right\}$$

is open for every $n \in \mathbf{N}$ and $F = \bigcap_{n=1}^{\infty} G_n$. But the closed sets in E generate $\mathfrak{B}(E)$. ⌟

Every finite measure defined on $\mathfrak{B}_0(E)$, $[\mathfrak{B}(E)]$ is called a *finite Baire (Borel) measure* on the topological space E. If we note that $\mathcal{C}^b$ contains the constant function 1, then Theorem 7.1.4 together with Corollary 7.1.6 gives us:

7.2.5. Theorem. If E is a topological space, then for every abstract integral I on $\mathcal{C}^b(E)$, there exists exactly one measure μ defined on $\mathfrak{B}_0(E)$

such that $\mathcal{C}^b(E) \subset \mathcal{L}^1(\mu)$ and

$$I(f) = \int f \, d\mu$$

for all $f \in \mathcal{C}^b(E)$. μ is a finite Baire measure on E.

Conversely, for every finite Baire measure μ on a topological space E, the mapping $f \to \int f \, d\mu$ is an abstract integral on $\mathcal{C}^b(E)$. Thus we can consider abstract integrals on $\mathcal{C}^b(E)$ and finite Baire measures on E as one and the same mathematical object for every topological space E. Hence we have further:

7.2.6. Corollary. For every finite Baire measure μ on a topological space E and every Baire set $A \subset E$,

$$\mu(A) = \inf \{\mu(G) : A \subset G,\ G \text{ open},\ G \in \mathfrak{B}_0(E)\}, \tag{7.2.4}$$
$$\mu(A) = \sup \{\mu(F) : F \subset A,\ F \text{ closed},\ F \in \mathfrak{B}_0(E)\}. \tag{7.2.5}$$

Proof. By Corollary 7.1.6, (7.2.4) follows from (7.1.8). If we apply (7.2.4) to $\complement A$ in place of A, then (7.2.5) follows because

$$\mu(A) = \mu(E) - \mu(\complement A). \quad \lrcorner$$

Note that this corollary takes on a particularly simple form for metrizable spaces E since then $\mathfrak{B}_0(E) = \mathfrak{B}(E)$ and (by the proof of Corollary 7.2.4) every open subset of E is an F_σ-set.

Examples

4. Let E be a topological space and ϵ_a the finite Baire measure on E defined by the unit mass in $a \in E$. Then ϵ_a is the only finite Baire measure μ on E satisfying

$$\int f \, d\mu = f(a),$$

for all $f \in \mathcal{C}^b(E)$.

5. Let $E = [a,b]$ be a compact interval in $\mathbf{R}$. The L-B-measure λ_E^1 on E is the only finite Borel measure μ on E such that

$$\int f \, d\mu = \int_a^b f(x) \, dx \qquad \text{for all } f \in \mathcal{C}(E).$$

Remark. In general $\mathfrak{B}_0(E) \neq \mathfrak{B}(E)$. This will be shown by Example 5 in Section 7.5, where E will be even compact (or by the following Problem).

PROBLEM

Let E be an infinite set with the so-called cofinal topology, that is, a set $O \subset E$ is open if and only if $\complement O$ is finite or E. Prove:

(a) Two nonempty open sets of E cannot be disjoint. In particular, E is not a Hausdorff space if it contains more than one point.

(b) $\mathcal{C}(E)$ only consists of the constant real functions on E.

(c) $\mathfrak{B}_0(E) = \{\emptyset, E\}$.

(d) $\mathfrak{B}(E)$ is the σ-algebra (of Section 1.1, Problem 2) consisting of all sets $A \subset E$ for which A or $\complement A$ is countable. Hence, $\mathfrak{B}_0(E) \neq \mathfrak{B}(E)$ if E contains more than one point.

7.3 REGULARITY OF FINITE BOREL MEASURES ON POLISH SPACES

A refined form of property (7.2.5) of a finite Baire measure on a topological space can be established in two special cases, one of which we shall discuss now.

7.3.1. Definition. Let E be a topological space and μ a measure defined on the σ-algebra $\mathfrak{B}(E)$ of Borel sets of E. We say that μ is *regular* if for every set $A \in \mathfrak{B}(E)$,

$$\mu(A) = \inf \{\mu(G) : A \subset G,\ G \text{ open}\} \qquad (\textit{outer regularity}), \qquad (7.3.1)$$
$$\mu(A) = \sup \{\mu(K) : K \subset A,\ K \text{ compact}\} \qquad (\textit{inner regularity}). \qquad (7.3.2)$$

A measure μ defined on $\mathfrak{B}_0(E)$ is said to be regular if for every $A \in \mathfrak{B}_0(E)$,

$$\mu(A) = \inf \{\mu(G) : A \subset G,\ G \text{ open},\quad G \in \mathfrak{B}_0(E)\}, \qquad (7.3.1')$$
$$\mu(A) = \sup \{\mu(K) : K \subset A,\ K \text{ compact},\ K \in \mathfrak{B}_0(E)\}. \qquad (7.3.2')$$

Properties (7.3.1′) and (7.3.2′) are also called outer and inner regularity of μ, respectively.

Every finite Baire measure on a topological space has the property of outer regularity by (7.2.4). By Corollary 7.2.6 every finite Baire measure on a compact space is regular since closed subsets coincide with compact ones in compact spaces.

7.3.2. Definition. A topological space E is called *Polish*[6] if there is a complete metric defining its topology and if E has a countable base.

[6] The name, due to Bourbaki, recalls the achievements of Polish topologists during the developmental years of general topology.

Remember that a metric is said to be *complete* if the associated metric space is complete, that is, if the Cauchy convergence criterion holds in it. *Countable base* means the existence of countably many open sets such that every open set is the union of certain of these sets.[7]

Examples

1. The Euclidean space $\mathbf{R}^p$ is Polish ($p = 1, 2, \ldots$). It suffices to consider the Euclidean metric on $\mathbf{R}^p$.

2. The *product* $E' \times E''$ of two Polish spaces is Polish. If d' and d'' are complete metrics defining the topology of E' and E'', respectively, then

$$d(x,y) \equiv d'(x',y') + d''(x'',y'')$$

is a complete metric defining the topology of $E' \times E''$. Here we let $x = (x',x'')$ and $y = (y',y'')$ be points of $E' \times E''$. Further if $\mathfrak{G}'$ ($\mathfrak{G}''$) is a countable base of E' (E''), then $\{G' \times G'' : G' \in \mathfrak{G}', G'' \in \mathfrak{G}''\}$ is a countable base of $E' \times E''$.

3. Every *closed* subspace F of a Polish space E is Polish. It suffices to restrict a complete metric defining the topology of E to F.

4. Every *open* subspace G of a Polish space E is Polish.

Proof. We can assume $G \neq E$. By Examples 1 and 2, $\mathbf{R} \times E$ is Polish. In this product we consider the set F of all $(\lambda,x) \in \mathbf{R} \times E$ with $\lambda \cdot d(x,E \setminus G) = 1$. As usual, $d(x,A)$ is the distance of a point $x \in E$ from a set $A \subset E$; the mapping $x \to d(x,A)$ is known to be continuous on E. Then, in particular, $(\lambda,x) \to \lambda \cdot d(x,E \setminus G)$ is a continuous real function on $\mathbf{R} \times E$; thus, F is a closed subset of $\mathbf{R} \times E$ and hence a Polish space by Example 3. But F is obviously mapped onto G homeomorphically by $(\lambda,x) \to x$. We need note only that $E \setminus G$ is closed and hence $G = \{x \in E : d(x,E \setminus G) > 0\}$. ┘

5. More generally we have (see Bourbaki [26]): A subspace A of a Polish space E is Polish if and only if A is a G_δ-set in E. Thus, for example, the space $\mathbf{J}$ of all irrational numbers taken as a subspace of $\mathbf{R}$ is Polish. We have

$$J = \bigcap_{x \text{ rational}} (\mathbf{R} \setminus \{x\}).$$

[7] For metrizable spaces E this is equivalent to the existence of a countable, dense subset in E.

6. *Every compact space E with countable basis* is Polish. By a well-known theorem of Urysohn (see Franz [32], p. 100), E is metrizable. Because of the compactness, every metric defining the topology of E is complete.

The significance of Polish spaces for measure theory rests primarily on the following result. Note that Polish spaces are metrizable and thus the Borel sets coincide with the Baire sets.

7.3.3. Theorem. Every finite Borel measure μ on a Polish space E is regular.

Proof. We need to prove only the inner regularity of μ. By (7.2.5) it suffices to treat the case of a closed set $A \subset E$. Since for compact K, the set $A \cap K$ is compact and according to (1.3.5)

$$\mu(A) - \mu(A \cap K) = \mu(A \cup K) - \mu(K) \leqq \mu(E) - \mu(K),$$

it remains to show that for every number $\epsilon > 0$ there is a compact set $K \subset E$ such that

$$\mu(E) - \mu(K) \leqq \epsilon. \tag{7.3.3}$$

Now let d be a complete metric defining the topology of E and let $(x_n)_{n \in \mathbf{N}}$ be a dense sequence of points in E. Let $K_r(x)$ denote the closed sphere with center x and radius r (relative to d). Then $E = \bigcup_{i=1}^{\infty} K_r(x_i)$ for every $r > 0$ since every sphere $K_r(x)$ contains some x_i and thus x lies in $K_r(x_i)$. Since μ is continuous from below,

$$\mu(E) = \lim_{k \to \infty} \mu\Big(\bigcup_{i=1}^{k} K_r(x_i)\Big).$$

Thus for every $\epsilon > 0$ and $n \in \mathbf{N}$ there is a $k_n \in \mathbf{N}$ such that

$$\mu\Big(\bigcup_{i=1}^{k_n} K_{1/n}(x_i)\Big) \geqq \mu(E) - \frac{\epsilon}{2^n}.$$

Each of the sets $B_n \equiv \bigcup_{i=1}^{k_n} K_{1/n}(x_i)$ is closed, and we have

$$\mu(B_1 \cap \cdots \cap B_n) \geqq \mu(E) - \epsilon \sum_{i=1}^{n} \frac{1}{2^i}.$$

By a derivation similar to that in the proof of Theorem 1.4.4 this is shown by means of the inequality

$$\begin{aligned}\mu(B_1 \cap \cdots \cap B_{n+1}) \\ &= \mu(B_1 \cap \cdots \cap B_n) + \mu(B_{n+1}) - \mu((B_1 \cap \cdots \cap B_n) \cup B_{n+1}) \\ &\geqq \mu(B_1 \cap \cdots \cap B_n) + \mu(B_{n+1}) - \mu(E)\end{aligned}$$

using induction. The set $K \equiv \bigcap_{n=1}^{\infty} B_n$ is then closed, and we have

$$\mu(K) = \lim_{n\to\infty} \mu\left(\bigcap_{i=1}^{n} B_i\right) \geqq \mu(E) - \epsilon \sum_{n=1}^{\infty} \frac{1}{2^n} = \mu(E) - \epsilon,$$

and thus (7.3.3).

It remains only to show the compactness of K. But we have

$$K \subset B_n = K_{1/n}(x_1) \cup \cdots \cup K_{1/n}(x_{k_n});$$

further, each of the sets $K_{1/n}(x_i)$ has diameter $\leqq 2/n$. Therefore, K is precompact[8] and closed, and thus compact due to the completeness of the metric space E. ⌟

PROBLEM

Let E be a Polish space. Prove: The measures ϵ_a, $a \in E$, are the only probability measures on $\mathfrak{B}(E)$ which only attain the values 0 and 1. [*Hint:* Consider a probability measure μ on $\mathfrak{B}(E)$ which only attains 0 and 1 as values. Prove that the system $\mathfrak{K}$ of all compact sets K with $\mu(K) = 1$ is $\cap$-stable. Study $\bigcap_{K\in\mathfrak{K}} K$.]

7.4 SOME PROPERTIES OF LOCALLY COMPACT SPACES

A topological space E is said to be *locally compact* if it is Hausdorff and each of its points has at least one compact neighborhood. Examples of such spaces are the Euclidean space $\mathbf{R}^p$, every manifold (that is, every locally Euclidean Hausdorff space), every discrete, and every compact space.

By removing an arbitrary point from a compact space we obtain a locally compact space. This follows from the regularity of compact spaces. In this fashion we indeed obtain *all* locally compact spaces. If E is a locally

[8] Precompact sets are also called totally bounded. See Franz [32], pp. 87 and 91.

compact space, $\mathfrak{O}$ the system of its open sets and ω_0 a point not lying in E, then we can define a topology on $E' \equiv E \cup \{\omega_0\}$ as follows: Let the system $\mathfrak{O}'$ of open sets of E' be the union of $\mathfrak{O}$ with the system of all sets $E' \setminus K$ provided K runs through all compact subsets of E. Thus, E' becomes a compact space and E an open subspace of E'.[9] If E is itself already compact, then ω_0 is an isolated point of E'. If E is not compact, then E is dense in E'. The space E' is called the (Alexandrov) *one point compactification* of E.[10] We call ω_0 the *point at infinity* of E.

We build the further theory of locally compact spaces on the existence of E'. First we study a special class of functions in $\mathfrak{C}^b(E)$.

7.4.1. Definition. Let $f: E \to \mathbf{R}$ be a real function on a topological space E. Then

$$S_f \equiv \overline{\{f \neq 0\}} \tag{7.4.1}$$

is called the *support* of f.

The complement of S_f is thus the largest open set on which f is equal to zero. For a locally compact set E, let

$$\mathfrak{C}^c = \mathfrak{C}^c(E)$$

denote the set of all continuous real functions on E with *compact support.* A function $f \in \mathfrak{C}(E)$ thus lies in $\mathfrak{C}^c(E)$ if and only if f is zero on the complement of a suitable compact subset of E. We have $\mathfrak{C}^c(E) \subset \mathfrak{C}^b(E)$ since f is bounded on the compact support S_f and thus on all of E.

For any n functions $f_1, \ldots, f_n \in \mathfrak{C}^c$, $\varphi(f_1, \ldots, f_n)$ also lies in $\mathfrak{C}^c$ if $\varphi \in \mathfrak{C}(\mathbf{R}^n)$ and $\varphi(0, \ldots, 0) = 0$ ($n = 1, 2, \ldots$). Indeed,

$$S_{\varphi(f_1, \ldots, f_n)} \subset S_{f_1} \cup \cdots \cup S_{f_n}.$$

Then in particular $\mathfrak{C}^c$ is a linear subspace of $\mathfrak{C}^b(E)$ which, when it contains f, g, also contains the functions $|f|$, $\inf(1,f)$, and $f \cdot g$. Therefore, $\mathfrak{C}^c$ is a Stone vector lattice of real functions on E. It is thus meaningful to determine the $\mathfrak{C}^c$-open subsets of E. For this we need:

7.4.2. Lemma. In a locally compact space E let K be a compact set and U a neighborhood of K. Then there is a function $f \in \mathfrak{C}^c(E)$ with the following properties: $0 \leqq f \leqq 1$, $f(x) = 1$ for all $x \in K$, and $S_f \subset U$. In particular S_f is a compact neighborhood of K.

Proof. We can assume that U is open. In the one point compactification E' of E, the sets K and $E' \setminus U$ are then disjoint and closed. From

[9] See Franz [32], p. 75.

[10] E' is often defined only for noncompact E.

the normality of the compact space E' follows the existence of an open neighborhood V of K whose closure $\bar{V}$ is disjoint from $E' \setminus U$. Again the normality of E' yields the existence of a function $f' \in \mathcal{C}(E')$ such that $0 \leqq f' \leqq 1$, $f'(x) = 1$ for all $x \in K$ and $f'(y) = 0$ for all $y \in E' \setminus V$. But then $f \equiv \operatorname{rest}_E f'$ yields the desired result: S_f is a closed subset of $\bar{V}$ and thus is compact in E' and hence also in E because $S_f \subset E$. Since $K \subset \{f > 0\} \subset S_f$, we see that S_f is a compact neighborhood of K in E. ⌋

In analogy to Lemma 7.2.2 we now obtain:

7.4.3. Lemma. A subset G of a locally compact space E is $\mathcal{C}^c(E)$-open if and only if it is an open K_σ-set.

Proof. The step "(2) ⇒ (3)" in the proof of Lemma 7.2.2 (with $u_n \in \mathcal{C}^c_+$) for $\mathcal{C}^c$-open G shows that G is an open K_σ-set. Note that for $f \in \mathcal{C}^c$ and $\alpha > 0$, the set $\{|f| \geqq \alpha\}$ is a closed subset of S_f and is thus compact. Conversely, let G be an open K_σ-set and (K_n) a sequence of compact sets with $G = \bigcup K_n$. By Lemma 7.4.2, for every $n \in \mathbf{N}$ there is an $f_n \in \mathcal{C}^c$ such that $1_{K_n} \leqq f_n \leqq 1_G$. Consequently, $1_G = \sup f_n$. But then $u_n \equiv \sup(f_1, \ldots, f_n)$ yields an isotone sequence in $\mathcal{C}^c_+$ such that $1_G = \sup u_n$. Hence G is $\mathcal{C}^c$-open. ⌋

Example

1. Let E be a discrete space. Open K_σ-sets and thus $\mathcal{C}^c(E)$-open sets in E are just the countable subsets of E. On the other hand, all subsets of E are $\mathcal{C}(E)$-open. (See Section 7.2, Example 2.) The $\mathcal{C}^c(E)$-open sets thus coincide exactly with the $\mathcal{C}(E)$-open ones if E is countable.

The question now arises under which condition the $\mathcal{C}^c(E)$-open sets coincide with the $\mathcal{C}(E)$-open sets. Since $1 \in \mathcal{C}(E)$, the entire space E is always $\mathcal{C}(E)$-open. Therefore, by Lemma 7.4.3, the condition that E be a K_σ-set is necessary for the identity of the two concepts of openness.

7.4.4. Definition. A locally compact space E is said to be *countable at infinity* if E is a K_σ-set.

The following theorem shows that this condition is also sufficient:

7.4.5. Theorem. For every locally compact space countable at infinity,

$$\mathfrak{A}(\mathcal{C}^c(E)) = \mathfrak{B}_0(E). \tag{7.4.2}$$

More precisely, the $\mathcal{C}^c(E)$-open subsets coincide with the $\mathcal{C}(E)$-open subsets.

Proof. By Lemma 7.1.3 the system $\mathfrak{G}_{\mathfrak{C}^c}$ [$\mathfrak{G}_{\mathfrak{C}}$] of $\mathfrak{C}^c$-open [$\mathfrak{C}$-open] sets is a generator of $\mathfrak{A}(\mathfrak{C}^c)$ [$\mathfrak{B}_0(E)$]. $\mathfrak{C}^c \subset \mathfrak{C}$ implies $\mathfrak{G}_{\mathfrak{C}^c} \subset \mathfrak{G}_{\mathfrak{C}}$. Thus we still must show that $\mathfrak{G}_{\mathfrak{C}} \subset \mathfrak{G}_{\mathfrak{C}^c}$. For this, let $G \in \mathfrak{G}_{\mathfrak{C}}$ and let (u_n) be an associated isotone sequence in $\mathfrak{C}_+$ such that $1_G = \sup u_n$. Since E is a K_σ-set, then by Lemma 7.4.3 there is an isotone sequence (k_n) in $\mathfrak{C}_+^c$ with $1_E = 1 = \sup k_n$. But then $f_n \equiv u_n k_n$ $(n = 1, 2, \ldots)$ is an isotone sequence in $\mathfrak{C}_+^c$ such that $1_G = \sup f_n$. Thus G lies in $\mathfrak{G}_{\mathfrak{C}^c}$. ⌟

Examples

2. The following locally compact spaces are countable at infinity:

(a) Every compact space.

(b) The Euclidean $\mathbf{R}^p$, $p = 1, 2, \ldots$. The closed spheres with fixed center and radii $r = 1, 2, \ldots$ are all compact and cover $\mathbf{R}^p$.

(c) Every locally compact space E with a countable base $\mathfrak{Q}$. Then $\mathfrak{Q}^* \equiv \{\bar{Q} \colon Q \in \mathfrak{Q},\ Q \text{ relatively compact}\}$ is a countable system of compact sets which covers E. The latter is obtained as follows: Every point $x \in E$ has a compact neighborhood V. Since $\mathfrak{Q}$ is a base of the topology, there is a $Q \in \mathfrak{Q}$ such that $x \in Q \subset V$. Thus $\bar{Q} \in \mathfrak{Q}^*$ and $x \in \bar{Q} \subset V$.

3. According to Example 1, a discrete space is countable at infinity if and only if it is countable.

The terminology "countable at infinity" is finally justified by the following lemma and the associated remark:

7.4.6. Lemma. Let E be a locally compact space countable at infinity. Then there is an isotone sequence $(L_n)_{n \in \mathbf{N}}$ of compact Baire sets, covering E, with the property

$$L_n \subset \mathring{L}_{n+1} \qquad (n = 1, 2, \ldots).[11] \tag{7.4.3}$$

Every compact set $K \subset E$ is contained in L_n for n sufficiently large.

Proof. By Lemma 7.4.3, E is a $\mathfrak{C}^c$-open set. Thus there is an isotone sequence (u_n) in $\mathfrak{C}_+^c$ such that $1 = \sup u_n$. But then the sequence defined by

$$L_n \equiv \left\{u_n \geqq \frac{1}{n}\right\} \qquad (n = 1, 2, \ldots)$$

yields the desired result. First u_n is $\mathfrak{A}(\mathfrak{C}^c)$-measurable, and thus L_n is a

[11] $\mathring{A}$ denotes the *interior* of a set A.

(closed) Baire set because $\mathfrak{A}(\mathcal{C}^c) = \mathfrak{B}_0$; it is compact because $L_n \subset S_{u_n}$. The isotonicity of (u_n) implies

$$L_n \subset \left\{u_{n+1} \geqq \frac{1}{n}\right\} \subset \left\{u_{n+1} > \frac{1}{n+1}\right\} \subset L_{n+1}$$

and hence obviously $L_n \subset \mathring{L}_{n+1}$. Finally, that every compact $K \subset E$ is contained in one and thus in all L_k for k sufficiently large follows from the remark that $Q_n \equiv K \setminus \mathring{L}_n$, $n = 1, 2, \ldots$ is an antitone sequence of compact sets with empty intersection. Therefore, Q_n must be empty for n sufficiently large. ⌟

In other words, this lemma says the following: A locally compact space E is countable at infinity if and only if the point at infinity ω_0 has a countable fundamental system of neighborhoods in the one point compactification E'.

For an arbitrary locally compact space E, besides $\mathcal{C}^c(E)$, the function space $\mathcal{C}^0(E)$, to be defined in Definition 7.4.7, is also meaningful. For its definition we associate with every *bounded* real function f on E the *norm of uniform convergence* or sup-norm on E defined by

$$\|f\| \equiv \sup_{x \in E} |f(x)|.$$

Then the mapping $(f,g) \to \|f - g\|$ gives $\mathcal{C}^b(E)$, and more generally the vector space of bounded real functions on E, the structure of a metric space. Therefore we also speak of the *metric of uniform convergence.* A sequence (f_n) of bounded real functions on E converges uniformly on E to a bounded real function f if and only if $\lim\limits_{n \to \infty} \|f_n - f\| = 0$.

7.4.7. Definition. A continuous real function f on a locally compact space E is said to *vanish at infinity* if it lies in the closure $\mathcal{C}^0 = \mathcal{C}^0(E)$ of $\mathcal{C}^c(E)$ relative to the metric of uniform convergence on $\mathcal{C}^b(E)$.

Hence we have

$$\mathcal{C}^0(E) = \overline{\mathcal{C}^c(E)} \subset \mathcal{C}^b(E).$$

This teminology is justified by:

7.4.8. Theorem. For every real function f on a locally compact space E, the following statements are equivalent:

(a) $f \in \mathcal{C}^0(E)$.

(b) $f \in \mathcal{C}(E)$ and for every real $\epsilon > 0$, $\{|f| \geqq \epsilon\}$ is compact.

(c) The function

$$f'(x) \equiv \begin{cases} f(x), & x \in E \\ 0, & x = \omega_0 \end{cases}$$

is continuous on the one point compactification E' of E.

Proof. (a) $\Rightarrow$ (b). We need show only the second property. Let $\epsilon > 0$. By definition there is a $g \in \mathcal{C}^c(E)$ such that $\|f - g\| \leqq \epsilon/2$. Due to the inequality $|f(x)| - |g(x)| \leqq |f(x) - g(x)| \leqq \|f - g\|$, we have

$$\{|f| \geqq \epsilon\} \subset \left\{|g| \geqq \frac{\epsilon}{2}\right\} \subset S_g,$$

and thus the set $\{|f| \geqq \epsilon\}$ is relatively compact. But by the continuity of f, the set is also closed and hence compact.

(b) $\Rightarrow$ (c): Condition (b) says that f is continuous and that for every $\epsilon > 0$ there exists a compact set $K \subset E$ with $|f(x)| < \epsilon$ for all $x \in E \setminus K$. The assertion now follows since $\{E' \setminus K\colon K \text{ compact} \subset E\}$ is a fundamental system of neighborhoods of ω_0 in E'.

(c) $\Rightarrow$ (a): For every $\epsilon > 0$ there is a compact set $K \subset E$ with $|f(x)| = |f(x) - f'(\omega_0)| \leqq \epsilon$ for all $x \in E \setminus K$. By Lemma 7.4.2 there exists a $g \in \mathcal{C}^c(E)$ satisfying $0 \leqq g \leqq 1$ and $g(x) = 1$ on K. But then fg lies in $\mathcal{C}^c(E)$ and we have

$$|f(x)g(x) - f(x)| = |f(x)|(1 - g(x)) \leqq \epsilon$$

for all $x \in E$; hence $\|fg - f\| \leqq \epsilon$. This shows that f lies in $\overline{\mathcal{C}^c(E)}$. ┘

From the definition or from property (c) it follows that $\mathcal{C}^0(E)$ is a *linear subspace* of $\mathcal{C}^b(E)$, and it is, like $\mathcal{C}^c(E)$, a *Stone vector lattice.*

Finally, we obtain an analog of Lemma 7.4.2:

7.4.9. Theorem. A locally compact space E is countable at infinity if and only if there exists a function $f \in \mathcal{C}^0(E)$ with $f(x) > 0$ for all $x \in E$.

Proof. Let f be such a function. Then $K_n \equiv \{|f| \geqq 1/n\}$, $n = 1, 2, \ldots$, is an isotone sequence of compact sets such that $E = \bigcup K_n$. If, conversely, E is countable at infinity, then there is an isotone sequence (K_n) of compact sets which covers E. By Lemma 7.4.2 there then exists a sequence (u_n) in $\mathcal{C}^c(E)$ satisfying $0 \leqq u_n \leqq 1$ and $u_n(x) = 1$ for all $x \in K_n$ $(n = 1, 2, \ldots)$. Therefore, the series $\Sigma_{n=1}^{\infty} (1/2^n)u_n(x)$ is uniformly convergent on E. Hence,

$$f \equiv \sum_{n=1}^{\infty} \frac{1}{2^n} u_n$$

yields the desired result. ┘

PROBLEMS

1. Let E be a discrete space. Determine the σ-algebra $\mathfrak{A}(\mathfrak{K})$ which is generated by the system $\mathfrak{K}$ of all compact subsets of E. Prove: $\mathfrak{A}(\mathfrak{K}) = \mathfrak{B}(E)$ $(= \mathfrak{B}_0(E))$ if and only if E is countable.
2. Let E be a locally compact space. Prove: For $\mathfrak{F} \equiv \mathfrak{C}^c(E)$, the space $\mathfrak{F}_0$ defined in Section 7.1, Problem 2 coincides with $\mathfrak{C}^c(E)$.
3. Let K be a compact and U an open subset of a locally compact space E such that $K \subset U$. Prove the existence of a compact G_δ-set K' and an open K_σ-set U' in E such that $K \subset U' \subset K' \subset U$.
4. Let E be a locally compact space countable at infinity. Prove: For every $f \in \mathfrak{C}_+^0(E)$, there exists an increasing sequence (u_n) in $\mathfrak{C}_+^c(E)$ such that $f = \sup u_n$.
5. Let $E' = E \cup \{\omega_0\}$ be the one point compactification of a locally compact space E. Describe the Borel sets of E' by means of the Borel sets of E. Show that this description fits into the following framework: Let $(E,\mathfrak{A})$ be a measurable space, $\omega_0 \notin E$ and $E^{\omega_0} = E \cup \{\omega_0\}$. Prove: The σ-algebra $\mathfrak{A}^{\omega_0}$ in E^{ω_0} generated by $\mathfrak{A}$ and $\{\omega_0\}$ consists of all sets $A' \subset E^{\omega_0}$ such that $A' \cap E \in \mathfrak{A}$.
6. Let $E' = E \cup \{\omega_0\}$ be the one point compactification of a locally compact space E which is countable at infinity. Prove that, in the notation of Problem 5, we have: $\mathfrak{B}_0(E') = \mathfrak{B}_0(E)^{\omega_0}$.

7.5 BAIRE MEASURES ON LOCALLY COMPACT SPACES COUNTABLE AT INFINITY

After the topological preparations of the preceding section, we now proceed to the study of measures on locally compact spaces E. Here we are concerned essentially with an application of the results of Section 7.1 to the following special cases $\Omega = E$, $\mathfrak{F} = \mathfrak{C}^c(E)$. We generally assume E to be countable at infinity, so that the σ-algebra $\mathfrak{A}(\mathfrak{C}^c(E))$ is that of the Baire sets of E (see Theorem 7.4.5). The starting point of our investigations is the noteworthy fact that every positive linear form is an abstract integral on the Stone vector lattice $\mathfrak{C}^c(E)$.

7.5.1. Lemma. If E is a locally compact space and I is a positive linear form on $\mathfrak{C}^c(E)$, then for every compact set $K \subset E$ there exists a real number $\alpha_K \geqq 0$ such that

$$|I(u)| \leqq \alpha_K \|u\|, \tag{7.5.1}$$

for all $u \in \mathfrak{C}^c(E)$ with $S_u \subset K$.

Proof. By Lemma 7.4.2, for K there exists an $f \in \mathfrak{C}^c_+$ with $f(x) = 1$ for all $x \in K$. Therefore, for every $u \in \mathfrak{C}^c$ with $S_u \subset K$, we have $|u| \leqq \|u\| f$, whence $I(u) \leqq \|u\| I(f)$ and $-I(u) = I(-u) \leqq \|u\| I(f)$, and thus $|I(u)| \leqq \|u\| I(f)$. Hence $\alpha_K \equiv I(f)$ yields the desired result. ⌟

7.5.2. Corollary. Every positive linear form I on $\mathfrak{C}^c(E)$ is an abstract integral.

Proof. According to (7.1.4′), we have to show: For every antitone sequence (u_n) in $\mathfrak{C}^c_+$ with $\inf u_n = 0$, we have $\inf I(u_n) = 0$. We show the following sharper result:

For every nonempty set $\mathfrak{J} \subset \mathfrak{C}^c_+$ which is filtering to the left and satisfies $\inf_{t \in \mathfrak{J}} t(x) = 0$ for all $x \in E$, we have $\inf_{t \in \mathfrak{J}} I(t) = 0$. (7.5.2)[12]

We can assume without loss of generality that there exists a largest function t_0 in $\mathfrak{J}$, since if necessary we can choose an arbitrary t_0 in $\mathfrak{J}$ and go over to $\mathfrak{J}_0 = \{t \in \mathfrak{J}: t \leqq t_0\}$. Since $\mathfrak{J}$ is filtering to the left, we then have: $\mathfrak{J}_0$ is filtering to the left and $\inf_{t \in \mathfrak{J}_0} t(x) = 0$ for all $x \in E$.

But if t_0 is the largest element of $\mathfrak{J}$, then $S_t \subset S_{t_0}$ for all $t \in \mathfrak{J}$. Referring to Lemma 7.5.1, we therefore have to show only that

$$\inf_{t \in \mathfrak{J}} \|t\| = 0.$$

But this is just the statement of the so-called Theorem of Dini, which will be proved briefly here: For every $x \in E$ and $\epsilon > 0$, since $\inf_{t \in \mathfrak{J}} t(x) = 0$, there is a $t_x \in \mathfrak{J}$ with $t_x(x) < \epsilon$. t_x is continuous; thus, there is an open neighborhood U_x of x such that $t_x(y) < \epsilon$ for all $y \in U_x$. Now $S_{t_0} \subset \bigcup_{x \in S_{t_0}} U_x$, and the compactness of S_{t_0} implies the existence of finitely many points $x_1, \ldots, x_n \in S_{t_0}$ with the property $S_{t_0} \subset U_{x_1} \cup \cdots \cup U_{x_n}$. Since $\mathfrak{J}$ is filtering to the left, we find a $t \in \mathfrak{J}$ with $t \leqq \inf(t_{x_1}, \ldots, t_{x_n})$. But then $t(y) < \epsilon$ for all $y \in S_{t_0}$, and thus for all $y \in E$ because

[12] Filtering to the left means: For any two functions $t_1, t_2 \in \mathfrak{J}$, there is a $t \in \mathfrak{J}$ with $t \leqq t_1$ and $t \leqq t_2$.

$S_t \subset S_{t_0}$. If we note that $t \geqq 0$, then $\|t\| \leqq \epsilon$ and hence the assertion follows. ┘

By the Daniell-Stone Theorem for I, there exists a measure μ on $\mathfrak{A}(\mathfrak{C}^c)$ with $\mathfrak{C}^c \subset \mathfrak{L}^1(\mu)$ and $I(u) = \int u \, d\mu$ for all $u \in \mathfrak{C}^c$. Since a $u \in \mathfrak{C}^c_+$ with $1_K \leqq u$ exists for every compact set K (see Lemma 7.4.2), then $\mu(K) < +\infty$ for every compact set $K \in \mathfrak{A}(\mathfrak{C}^c)$. This observation leads to the following definition:

7.5.3. Definition. A *Baire measure* [*Borel measure*] on a locally compact space E is any measure μ on $\mathfrak{B}_0(E)$ [$\mathfrak{B}(E)$] such that for all compact Baire [all compact] sets K,

$$\mu(K) < +\infty. \tag{7.5.3}$$

Examples

1. The L-B-measure λ^p is a Borel measure on $\mathbf{R}^p$. It is not finite ($p = 1, 2, \ldots$).

2. Every finite measure on $\mathfrak{B}_0(E)$ [$\mathfrak{B}(E)$] is a Baire [Borel] measure on the locally compact space E. Therefore we can certainly retain the terminology "finite Baire [Borel] measure."

3. The concepts of Baire and Borel measures coincide for a *discrete* space E since $\mathfrak{B}_0(E) = \mathfrak{B}(E) = \mathfrak{P}(E)$ (Section 7.2, Example 2). Since, moreover, the compact subsets of E coincide with the finite ones, a measure μ on $\mathfrak{P}(E)$ is Borel if and only if $m(x) \equiv \mu(\{x\})$ is finite for all points $x \in E$. Therefore every Borel measure μ on E defines the real function $m \geqq 0$ on the discrete space E. If E is countable, μ is uniquely determined by m. Then

$$\mu(A) = \sum_{x \in A} m(x) \qquad (A \in \mathfrak{P}(E)).$$

In this case $\mu \to m$ is a *bijection* of the set of Borel measures on E onto the set of all nonnegative real functions on E.

Now we obtain the important theorem, whose second part is called the *F. Riesz Representation Theorem.*

7.5.4. Theorem. For every locally compact space E which is countable at infinity, we have:

1. For every Baire measure μ, $\mathfrak{C}^c(E) \subset \mathfrak{L}^1(\mu)$ and hence $u \to \int u \, d\mu$ is a positive linear form on $\mathfrak{C}^c(E)$.

2. For every positive linear form I on $\mathfrak{C}^c(E)$, there exists exactly one Baire measure μ on E such that

$$I(u) = \int u \, d\mu, \qquad \text{for all } u \in \mathfrak{C}^c(E). \tag{7.5.4}$$

Proof. 1. S_u is compact for every $u \in \mathfrak{C}^c$, so that by Lemma 7.4.2 there is an $f \in \mathfrak{C}^c_+$ with $f(x) = 1$ for all $x \in S_u$. But then $L = \{f \geqq \frac{1}{2}\}$ is a compact Baire set with $S_u \subset L$, and thus with $|u| \leqq \|u\| 1_L$. Hence the μ-integrability of u follows, since u is $\mathfrak{A}(\mathfrak{C}^c)$-measurable and $\mathfrak{A}(\mathfrak{C}^c) = \mathfrak{B}_0(E)$ by Theorem 7.4.5.[13]

2. By the Daniell-Stone Theorem there exists a measure μ on $\mathfrak{A}(\mathfrak{C}^c) = \mathfrak{B}_0(E)$ with $\mathfrak{C}^c \subset \mathfrak{L}^1(\mu)$ and property (7.5.4). The observation preceding Definition 7.5.3 shows that μ is a Baire measure. The uniqueness of μ is obtained from Corollary 7.1.6, taking into account 7.4.2 and 7.4.6. ⌟

7.5.5. Corollary. Every Baire measure μ on a locally compact space E countable at infinity is regular and σ-finite. For every $p \in [1, +\infty[$, $\mathfrak{C}^c(E)$ is dense in $\mathfrak{L}^p(\mu)$ relative to convergence in pth mean.

Proof. By Corollary 7.1.6 and Lemma 7.4.3, μ is σ-finite. The outer regularity of μ follows from Theorem 7.1.4 and Corollary 7.1.6. Thus for every $A \in \mathfrak{B}_0$ with $\mu(A) < \infty$ and every $\epsilon > 0$ there is an open set $G \in \mathfrak{B}_0$ such that $\mu(G \setminus A) \leqq \epsilon$. This statement also holds for arbitrary $A \in \mathfrak{B}_0$: By the σ-finiteness of μ, A is the union of a sequence (A_n) of Baire sets with $\mu(A_n) < \infty$. Therefore for every $n = 1, 2, \ldots$ there exists an open set $G_n \in \mathfrak{B}_0$ satisfying $A_n \subset G_n$ and $\mu(G_n \setminus A_n) \leqq 2^{-n}\epsilon$. But then $G \equiv \bigcup G_n$ is an open Baire set with $A \subset G$ and

$$\mu(G \setminus A) \leqq \mu(\cup(G_n \setminus A_n)) \leqq \Sigma\mu(G_n \setminus A_n) \leqq \epsilon.$$

Thus for $\complement A$ there is an open set $H \in \mathfrak{B}_0$ with $\complement A \subset H$ and $\mu(H \setminus \complement A) \leqq \epsilon$. Then $F \equiv \complement H$ is a closed Baire set with $F \subset A$ and $\mu(A \setminus F) \leqq \epsilon$. Hence

$$\mu(A) = \sup \{\mu(F) : F \text{ closed} \subset A,\ F \in \mathfrak{B}_0\}. \tag{7.5.5}$$

The existence of an isotone sequence (L_n) of compact Baire sets with $E = \bigcup L_n$ was shown in Lemma 7.4.6. Then for every closed $F \in \mathfrak{B}_0$, $(F \cap L_n)$ is an isotone sequence of compact sets in $\mathfrak{B}_0$ with F as its union

[13] Since $\mathfrak{A}(\mathfrak{C}^c) \subset \mathfrak{B}_0(E)$ holds in general, the proof shows the validity of 1 even without the hypothesis of countability at infinity.

and thus

$$\mu(F) = \sup \mu(F \cap L_n). \tag{7.5.6}$$

Therefore (7.5.5) and (7.5.6) imply the desired inner regularity of μ.

The rest of the assertion follows from Corollary 7.1.5: When u lies in $\mathcal{C}^c$ $|u|^p$ also lies in $\mathcal{C}^c$; moreover, we have $\mathcal{C}^c \subset \mathcal{L}^1(\mu)$. But hence it follows that $\mathcal{C}^c \subset \mathcal{L}^p(\mu)$, and thus we have the denseness of $\mathcal{C}^c = \mathcal{C}^c \cap \mathcal{L}^p(\mu)$ in $\mathcal{L}^p(\mu)$ relative to convergence in pth mean. ┘

Thus we have verified regularity for another extensive class of measures.

Examples

4. Let E be a discrete uncountable space. On $\mathfrak{B}_0(E) = \mathfrak{B}(E) = \mathfrak{P}(E)$, we consider the Borel measure μ with $\mu(A) = 0$ or $= +\infty$ depending on whether A is countable or uncountable. Then μ is *not regular* since $\mu(K) = 0$ holds for all compact (that is, finite, here) subsets K. The regularity of Baire measures proved in Corollary 7.5.5 is thus lost if we eliminate the countability at infinity.

5. Once again let E be a discrete, uncountable space and E' its one point compactification. E is open in E' and hence Borel in E'. But E is not Baire in E'. Otherwise, by (7.2.4) we would have

$$\mu(E) = \inf \{\mu(G) : E \subset G,\ G \text{ open } K_\sigma\text{-set}\}$$

for every Baire measure μ on E'. Since E is uncountable, E cannot be a K_σ-set in E'. Thus

$$\mu(E) = \mu(E').$$

But this is false for the Baire measure defined by the unit mass at ω_0. Thus we have $\mathfrak{B}_0(E') \neq \mathfrak{B}(E')$ for the compact space E'.

Remark. If E is an arbitrary locally compact space and I is a positive linear form on $\mathcal{C}^c(E)$, then by making essential use of (7.5.2) one can still prove the existence, though not the uniqueness, of Borel measures μ on E which represent I in the sense of following two properties: $\mathcal{C}^c \subset \mathcal{L}^1(\mu)$ and $I(u) = \int u\, d\mu$ for all $u \in \mathcal{C}^c$. However, among these measures exists only one inner regular Borel measure $\bar{\mu}$ and only one outer regular Borel measure $\hat{\mu}$. One has $\bar{\mu} \leqq \hat{\mu}$, and $\bar{\mu} \leqq \mu \leqq \hat{\mu}$ holds for every Borel measure μ on E which represents I. The equality $\bar{\mu}(A) = \hat{\mu}(A)$ holds for all Borel sets A in E which can be covered by a sequence of compact sets. This is the reason why some authors (for example, Halmos [3]) only call these sets Borel sets. The two different definitions of Borel sets only coincide when E is countable at infinity. Furthermore, $\bar{\mu}(A) = \hat{\mu}(A)$ holds for a

Borel set $A \in \mathfrak{B}(E)$ whenever $\hat{\mu}(A) < +\infty$. As a consequence one obtains the following result: Assume that I is *bounded*, that is, there exists a real number $\alpha \geqq 0$ such that $|I(u)| \leqq \alpha\|u\|$ holds for all $u \in \mathfrak{C}^c(E)$ (compare Section 7.5, Problem 1). Then there exists one and only one regular Borel measure μ on E such that $\mathfrak{C}^c \subset \mathfrak{L}^1(\mu)$ and $I(u) = \int u\, d\mu$ for all $u \in \mathfrak{C}^c$. This measure μ is finite. The reader interested in details is referred to the pertinent literature; in particular to Bourbaki [1] and Courrège [2].

These results explain why Bourbaki calls every positive linear form on $\mathfrak{C}^c(E)$ a *Radon measure* on E.

PROBLEMS

In what follows let E be a locally compact space countable at infinity.

1. For a Baire measure μ on E the number $\|\mu\| \equiv \mu(E)$ is called the total mass of E (hence $0 \leqq \|\mu\| \leqq +\infty$). It is finite if and only if μ is finite. [For the special case $E = \mathbf{R}^p$ see (3.4.1).]
   Consider a Baire measure μ on E and prove the equivalence of the following conditions:
   (a) μ is finite.
   (b) There exists a real number $\alpha > 0$ such that $|\int u\, d\mu| \leqq \|u\|$ for all $u \in \mathfrak{C}^c(E)$.
   Prove furthermore: The smallest possible number α in (b) is $\|\mu\|$, and one has $\|\mu\| = \sup \{\int u\, d\mu \colon 0 \leqq u \leqq 1,\ u \in \mathfrak{C}^c(E)\}$.
2. Prove: For a Baire measure μ on E the following three conditions are equivalent:
   (a) μ is finite.
   (b) $\mathfrak{C}^0(E) \subset \mathfrak{L}^1(\mu)$.
   (c) $\mathfrak{C}^b(E) \subset \mathfrak{L}^1(\mu)$.
3. Let I be a positive linear form on $\mathfrak{C}^0(E)$. Prove: There exists a unique finite Baire measure μ on E such that $I(f) = \int f\, d\mu$ for all $f \in \mathfrak{C}^0(E)$. [*Hint:* For each $\epsilon > 0$ and $f \in \mathfrak{C}^0_+(E)$, there exists a function $v \in \mathfrak{C}^c(E)$ such that $|f - v| \leqq \epsilon\sqrt{f}$.]
4. Let E_1 and E_2 be locally compact spaces that are countable at infinity. A continuous mapping $T\colon E_1 \to E_2$ is called proper if $T^{-1}(L)$ is compact in E_1 for all compact sets L in E_2. Prove:
   (a) Every continuous mapping $T\colon E_1 \to E_2$ is $\mathfrak{B}(E_1)$-$\mathfrak{B}(E_2)$-measurable.
   (b) For every proper mapping $T\colon E_1 \to E_2$ and every Baire measure μ on E_1, the image measure $T(\mu)$ is a Baire measure on E_2.

7.6 THE SPECIAL CASE OF LOCALLY COMPACT SPACES WITH COUNTABLE BASE

The theory developed in the preceding section is simplified in an essential way when E is a locally compact space with a countable base. By Section 7.4, Example 2(c), every such space is countable at infinity. Examples of such spaces are: the Euclidean $\mathbf{R}^p (p = 1, 2, \ldots)$, every (abstract) Riemann surface (see Ahlfors-Sario [21], p. 144), the countable discrete spaces, the metrizable compact spaces. (By the theorem of Urysohn quoted in Section 7.3, Example 6, a compact space is metrizable if and only if it has a countable base.)

The announced simplification is due above all to the fact that the Baire and Borel sets coincide in a locally compact space with a countable base. We show an even stronger result:

7.6.1. Theorem. Every locally compact space E with a countable base is a Polish space.

Proof. Let $\mathfrak{Q}$ be a countable base for E and let (L_n) be an isotone sequence of compact sets (which exists by Lemma 7.4.6) such that every compact subset of E is contained in an L_n. Then obviously

$$\mathfrak{Q}' = \mathfrak{Q} \cup \{E' \setminus L_n : n = 1, 2, \ldots\}$$

is a countable base for the one point compactification E' of E. Since E' is compact and E is open in E', the assertion now follows from Examples 6 and 4 of Section 7.3. ⌟

7.6.2. Corollary. For every locally compact space E with a countable base, $\mathfrak{B}_0(E) = \mathfrak{B}(E)$. The system of compact subsets of E is a generator of $\mathfrak{B}(E)$.

Proof. The first part of the assertion follows from Corollary 7.2.4. By definition, $\mathfrak{B}(E)$ is generated by the open and thus by the closed subsets of E. Since, in particular, E is countable at infinity, every closed subset is a K_σ-set. Hence $\mathfrak{B}(E)$ is generated by the compact sets. ⌟

Thus, for locally compact spaces E with a countable base, the concepts of Baire and Borel measure also coincide. By Corollary 7.5.5, every Borel measure on E is regular. For finite Borel measures, the regularity also follows from Theorem 7.3.3.

Finally, we show which property of $\mathfrak{C}^c(E)$ reflects the existence of a countable base for E. For this we again use the metric of uniform convergence.

7.6.3. Theorem. For every locally compact space E, the following two properties are equivalent:

(a) E has a countable base.

(b) With respect to uniform convergence there exists a countable, dense subset of $\mathcal{C}^c(E)$.[14]

Proof. (a) $\Rightarrow$ (b): Let $\mathfrak{Q}$ be a countable base for E. The arguments presented in Section 7.4, Example 2(c), show that without loss of generality we can assume all sets in $\mathfrak{Q}$ to be relatively compact. Then the set I of all pairs $(G,H) \in \mathfrak{Q} \times \mathfrak{Q}$ with the property $\bar{G} \subset H$ is countable. With every pair $(G,H) \in I$ we associate a function $f_{(G,H)} \in \mathcal{C}^c$, which exists by Lemma 7.4.2, with the following properties: $0 \leqq f_{(G,H)} \leqq 1$, $f_{(G,H)}(x) = 1$ for all $x \in \bar{G}$, $S_{f_{(G,H)}} \subset H$. Then for every point $x \in E$ or every pair x_1, x_2 of distinct points of E there exists a function f_i, $i \in I$, with $f_i(x) \neq 0$ or $f_i(x_1) \neq f_i(x_2)$, respectively. Indeed there exists a pair $(G,H) \in I$ with $x \in G$ and hence with $f_{(G,H)}(x) = 1$, or, with $x_1 \in G$, and $x_2 \notin H$, and thus with the property $f_{(G,H)}(x_1) = 1$, $f_{(G,H)}(x_2) = 0$.

Now let $\mathcal{P}$ be the set of all real functions, defined in E, of the form $p(f_{i_1}, \ldots, f_{i_n})$, where $p \in \mathbf{R}[z_1, \ldots, z_n]$ is a real polynomial in finitely many variables $z_1, \ldots, z_n$ such that $p(0, \ldots, 0) = 0$ and $i_1, \ldots, i_n$ are elements of I. Then $\mathcal{P}$ obviously satisfies the hypotheses of the Stone-Weierstrass Theorem[15] and thus is dense in $\mathcal{C}^0$. By what was said preceding Lemma 7.4.2, we also have $\mathcal{P} \subset \mathcal{C}^c$. Thus $\mathcal{P}$ is dense in $\mathcal{C}^c$ with respect to uniform convergence. If in the definition of $\mathcal{P}$ we now replace the real polynomials p by ones with only rational coefficients, we obtain a

[14] Since $\mathcal{C}^c(E)$ is a metric space, this is equivalent to the existence of a countable base for $\mathcal{C}^c(E)$. See Franz [32], p. 50.

[15] In the form needed by us this theorem says: Let E be a locally compact space and $\mathcal{P}$ a linear subspace of $\mathcal{C}^0(E)$ which, whenever it contains two functions f, $g \in \mathcal{P}$ it also contains their product fg. Then if, for every pair of distinct points x, $y \in \mathcal{P}$ there exists a function $f \in \mathcal{P}$ with $f(x) \neq f(y)$ and for every $x \in E$ there exists a $g \in \mathfrak{P}$ with $g(x) \neq 0$, it follows that $\mathcal{P}$ is dense in $\mathcal{C}^0(E)$ relative to the metric of uniform convergence on E. This form of the theorem is obtained from the usual formulation for compact spaces (see, for example, Bourbaki [26]) by introducing the one point compactification E' of E and extending every function $f \in \mathcal{P}$ to a continuous function f' on E' (which is equal to zero at the point at infinity). Then $\mathcal{P}_1' \equiv \{f' + \alpha : f \in \mathcal{P}, \alpha \in \mathbf{R}\}$ is a linear subspace of $\mathcal{C}(E')$ with the properties: $f', g' \in \mathcal{P}_1' \Rightarrow f'g' \in \mathcal{P}_1'$; the constant real functions lie in $\mathcal{P}_1'$; for $x \neq y$ in E' there exists an $f' \in \mathcal{P}_1'$ with $f'(x) \neq f'(y)$.

subset $\mathcal{P}'$ of $\mathcal{P}$ which is obviously countable. $\mathcal{P}'$ yields the desired result since $\mathcal{P}'$ is dense in $\mathcal{P}$ and thus also in $\mathcal{C}^c$. We see the former as follows: Let $g \equiv p(f_{i_1}, \ldots, f_{i_n})$ be a function in $\mathcal{P}$. Like all functions from $\mathcal{C}^c$, the functions $f_{i_1}, \ldots, f_{i_n}$ arc bounded. If we approximate all the coefficients of p sufficiently closely by rational numbers, then we obtain an arbitrarily close uniform approximation of g by functions $p'(f_{i_1}, \ldots, f_{i_n})$ from $\mathcal{P}'$.

(b) $\Rightarrow$ (a): Let $\mathfrak{D}$ be a dense subset in $\mathcal{C}^c$. Then the system $\mathfrak{Q}$ of all sets $\{u > \frac{1}{2}\}$ with $u \in \mathfrak{D}$ is a base for E. Indeed for every open set U and every $x_0 \in U$ there exists, by Lemma 7.4.2, an $f \in \mathcal{C}^c$ such that $f(x_0) = 1$ and $S_f \subset U$. For f there is a $u \in \mathfrak{D}$ with $\|f - u\| < \frac{1}{2}$. But then we have

$$x_0 \in \{u > \tfrac{1}{2}\} \subset \{f > 0\} \subset S_f \subset U.$$

This shows that $\mathfrak{Q}$ is a base. But then, since $\mathfrak{D}$ is countable $\mathfrak{Q}$ is also countable. ⌟

PROBLEMS

1. Let μ be a Borel measure on a locally compact space with a countable base. Prove:
   (a) There exists a greatest open set G of measure zero. $S_\mu \equiv \complement G$ is called the *support of* μ.
   (b) A point $x \in E$ is in S_μ if and only if $\mu(U) > 0$ for all open neighborhoods U of x.
   (c) For a function $f \in \mathcal{C}_+(E)$ one has $\int f\, d\mu = 0$ if and only if f vanishes on S_μ. Determine S_{λ^p} for the L-B-measure λ^p on $\mathbf{R}^p$ and S_{ϵ_a} for a point a in a locally compact space with a countable base.
2. Let E be locally compact and $\mathfrak{D} \subset \mathcal{C}^c(E)$. $\mathfrak{D}$ is called *rich* if for every compact set K there exists an open relatively compact neighborhood U of K such that every $f \in \mathcal{C}^c(E)$ with support $S_f \subset K$ can be uniformly approximated on E by functions $d \in \mathfrak{D}$ having their support in U.
   (a) Assume that E is countable at infinity and that $\mathfrak{D} \subset \mathcal{C}^c(E)$ is rich. Prove: Two Baire measures μ and ν coincide if and only if $\int u\, d\mu = \int u\, dv$ for all $u \in \mathfrak{D}$.
   (b) Prove: For every locally compact space E with a countable base there exists a countable set $\mathfrak{D} \subset \mathcal{C}^c(E)$ which is rich. [*Hint:* Modify slightly the proof of Theorem 7.6.3.]
   (c) Prove: In the situation if (b) one can even assume that for every compact set $K \subset E$ there exists a function $d \in \mathfrak{D}$ such that $0 \leqq d \leqq 1$ and $d(x) = 1$ for all $x \in K$.

7.7 CONVERGENCE OF BAIRE MEASURES

Let E be a locally compact space *countable at infinity*. Henceforth, let $\mathfrak{M} = \mathfrak{M}(E)$ denote the *set of all Baire measures* on E. By the Riesz Representation Theorem, $\mathfrak{M}(E)$ can be canonically mapped bijectively onto the set of all positive linear forms on $\mathcal{C}^c = \mathcal{C}^c(E)$. For $\mu, \nu \in \mathfrak{M}(E)$ and any two real numbers $\alpha \geqq 0$, $\beta \geqq 0$, $\alpha\mu + \beta\nu$ also lies in $\mathfrak{M}(E)$. Thus, $\mathfrak{M}(E)$ is a so-called *convex cone*. Besides $\mathfrak{M}(E)$, we often consider the following two subsets:

$$\mathfrak{M}^e = \mathfrak{M}^e(E) \equiv \{\mu \in \mathfrak{M}(E) : \mu(E) < +\infty\},$$
$$\mathfrak{M}^1 = \mathfrak{M}^1(E) \equiv \{\mu \in \mathfrak{M}(E) : \mu(E) = 1\}.$$

These are correspondingly the sets of all finite Baire measures and of all Baire probability measures on E. Obviously, $\mathfrak{M}^1(E) \subset \mathfrak{M}^e(E) \subset \mathfrak{M}(E)$.[16] In particular, all measures ϵ_a defined by the unit mass at $a \in E$ lie in $\mathfrak{M}^1(E)$. $\mathfrak{M}^e(E)$ is a convex sub-cone of $\mathfrak{M}(E)$.

Depending on whether we consider the elements of $\mathfrak{M}(E)$ as measures on $\mathcal{B}_0(E)$ or as positive linear forms on $\mathcal{C}^c(E)$, the following two concepts of convergence in $\mathfrak{M}(E)$ suggest themselves: We can define the convergence of a sequence (μ_n) in $\mathfrak{M}(E)$ to a measure $\mu \in \mathfrak{M}(E)$ either by the requirement $\lim_{n\to\infty} \mu_n(A) = \mu(A)$ for all $A \in \mathfrak{B}_0(A)$ or by $\lim_{n\to\infty} \int f\,d\mu_n = \int f\,d\mu$ for all $f \in \mathcal{C}^c(E)$. It will be seen immediately that the first concept of convergence is of no further interest, while the second is of high importance.

7.7.1. Definition. A sequence $(\mu_n)_{n\in\mathbf{N}}$ in $\mathfrak{M}(E)$ is said to be *vaguely convergent* to a measure $\mu \in \mathfrak{M}(E)$ if

$$\lim_{n\to\infty} \int f\,d\mu_n = \int f\,d\mu \qquad \text{for all } f \in \mathcal{C}^c(E). \tag{7.7.1}$$

A sequence (μ_n) in $\mathfrak{M}(E)$ is *vaguely convergent if and only if* the sequence $(\int f\,d\mu_n)$ of real numbers converges for every $f \in \mathcal{C}^c(E)$. In fact, then $f \to \lim_{n\to\infty} \int f\,d\mu_n$ is a positive linear form on $\mathcal{C}^c$. By the Riesz Representation Theorem there is thus exactly one $\mu \in \mathfrak{M}$ to which (μ_n) converges vaguely. At the same time it also follows that the vague limit of a sequence in $\mathfrak{M}(E)$ is *uniquely determined*.

Examples

1. Let (x_n) be a convergent sequence in E with $\lim x_n = x$. Then (ϵ_{x_n}) converges vaguely to ϵ_x, since we have $\int f\,d\epsilon_a = f(a)$ for all $a \in E$, and $\lim f(x_n) = f(x)$ for arbitrary $f \in \mathcal{C}^c$.

[16] $\mathfrak{M}^e(E)$ was already introduced in Section 3.4 for the case $E = \mathbf{R}^p$.

On the other hand, $\lim \epsilon_{x_n}(A) = \epsilon_x(A)$ does not hold in general for all $A \in \mathfrak{B}_0(E)$. If E has a countable base and if $x_n \neq x$ for all $n = 1, 2, \ldots$, then it suffices to choose $A = \{x\}$. (A is Borel and thus, in this case, Baire.) This example shows why we must attribute little significance to the first of the concepts of convergence above.

2. Let (α_n) be an arbitrary sequence of real numbers $\geqq 0$ and (x_n) a sequence in E such that for every compact set $K \subset E$ there exist at most finitely many $n = 1, 2, \ldots$ with $x_n \in K$. (In other words, E is not compact, and $\lim x_n = \omega_0$ in E'.) Then the sequence of measures $\mu_n \equiv \alpha_n \epsilon_{x_n}$, $n = 1, 2, \ldots$, is vaguely convergent to the zero measure 0. For arbitrary $f \in \mathfrak{C}^c$, we have $\int f\, d\mu_n = \alpha_n f(x_n) = 0$ for all n with $x_n \notin S_f$. Since S_f is compact, we thus have by our hypothesis $\int f\, d\mu_n = 0$ for all n sufficiently large.

This example shows (for $\alpha_n = 1$) that vaguely convergent sequences of measures in $\mathfrak{M}^1(E)$ need not converge to a measure in $\mathfrak{M}^1(E)$.

An important example of vague convergence is provided by the stochastic convergence, introduced in Definition 2.11.2, of a sequence (X_n) of real random variables on a probability space $(\Omega,\mathfrak{A},P)$. Here the sequence (P_{X_n}) of associated distributions is a sequence in $\mathfrak{M}^1(\mathbf{R})$.

7.7.2. Theorem. If a sequence $(X_n)_{n\in\mathbf{N}}$ of real random variables on a probability space $(\Omega,\mathfrak{A},P)$ converges stochastically to the real random variable X, then the sequence $(P_{X_n})_{n\in\mathbf{N}}$ of distributions converges vaguely to the distribution P_X of X. If X is P-almost surely constant, then the converse also holds.

Proof. Every function $f \in \mathfrak{C}^c(\mathbf{R})$ is continuous and vanishes outside its compact support. Therefore f is uniformly continuous on $\mathbf{R}$, so that for every $\epsilon > 0$ there exists a $\delta > 0$ such that $|x' - x''| < \delta$ implies

$$|f(x') - f(x'')| < \epsilon \qquad (x',x'' \in \mathbf{R}).$$

If we now set $A_n \equiv \{|X_n - X| \geqq \delta\}$, then $A_n \in \mathfrak{A}$ and

$$\begin{aligned}\left|\int f\,dP_{X_n} - \int f\,dP_X\right| &= \left|\int f\circ X_n\,dP - \int f\circ X\,dP\right| \\ &\leqq \int_{A_n} |f\circ X_n - f\circ X|\,dP + \int_{\complement A_n} |f\circ X_n - f\circ X|\,dP \\ &\leqq 2\|f\|P(A_n) + \epsilon P(\complement A_n) \leqq 2\|f\|P(A_n) + \epsilon.\end{aligned}$$

Here we used the trivial inequality $|f\circ X_n - f\circ X| \leqq |f\circ X_n| + |f\circ X| \leqq 2\|f\|$. By the definition of stochastic convergence, $\lim_{n\to\infty} P(A_n) = 0$. Thus $\lim_{n\to\infty} \int f\,dP_{X_n} = \int f\,dP_X$ and the asserted vague convergence follows.

For the proof of the converse, let $X = \alpha$ P-almost surely for some $\alpha \in \mathbf{R}$, that is, $P_X = \epsilon_\alpha$. Suppose further that (P_{X_n}) converges vaguely to ϵ_α. For the interval $I \equiv]\alpha - \epsilon, \alpha + \epsilon[$ there are obviously functions $f, g \in \mathcal{C}^c(\mathbf{R})$ satisfying $f \leqq 1_I \leqq g$ and $f(\alpha) = g(\alpha) = 1$. Then $f \circ X_n \leqq 1_I \circ X_n \leqq g \circ X_n$, and consequently

$$\int f\, dP_{X_n} \leqq \int 1_I \circ X_n\, dP \leqq \int g\, dP_{X_n}.$$

Hence

$$\lim_{n\to\infty} P\{X_n \in I\} = 1$$

because of the convergence of the outer terms to $f(\alpha) = g(\alpha) = 1$. If we note that $\{X_n \in I\} = \{|X_n - \alpha| < \epsilon\}$ and thus

$$P\{|X_n - X| \geqq \epsilon\} = P\{|X_n - \alpha| \geqq \epsilon\} = 1 - P\{X_n \in I\},$$

then we obtain $\lim_{n\to\infty} P\{|X_n - X| \geqq \epsilon\} = 0$ for all $\epsilon > 0$. ┘

If X is not almost surely constant, then we cannot in general derive the stochastic convergence of (X_n) to X from the vague convergence of (P_{X_n}) to P_X. This is shown by the following:

Example

3. Let $(\Omega,\mathfrak{A},P)$ be the probability space defined in Section 5.1, Example 3, (A_n) the independent sequence of events given there, and $X_n \equiv 1_{A_n}$. By the example in Section 6.4, every X_n has distribution β_1^p with $p = \frac{1}{2}$, and thus the sequence (P_{X_n}) is constant equal to β_1^p and hence vaguely convergent to P_{X_1}. On the other hand, (X_n) is not stochastically convergent, since for any two natural numbers $m \neq n$ and every δ with $0 < \delta < 1$, we obviously have $P\{|X_m - X_n| \geqq \delta\} = \frac{1}{2}\cdot\frac{1}{2} + \frac{1}{2}\cdot\frac{1}{2} = \frac{1}{2}$.

The vague convergence of sequences in $\mathfrak{M}(E)$ is derived from a topology on $\mathfrak{M}(E)$, the so-called *vague topology*. This is defined as the coarsest topology on $\mathfrak{M}(E)$ relative to which all of the mappings

$$\mu \to \int f\, d\mu \qquad (f \in \mathcal{C}^c(E)) \tag{7.7.2}$$

are continuous. Thus for every measure $\mu_0 \in \mathfrak{M}(E)$ we obtain a fundamental system of neighborhoods of μ_0 in the vague topology in the form of the system of all sets

$$V_{f_1,\ldots,f_n;\epsilon}(\mu_0) \equiv \{\mu \in \mathfrak{M}(E) : |\int f_i\, d\mu - \int f_i\, d\mu_0| < \epsilon,\ i = 1, \ldots, n\}, \tag{7.7.3}$$

where $f_1, \ldots, f_n$ are finitely many functions from $\mathcal{C}^c(E)$ and $\epsilon > 0$ is a real number. The vague topology is *Hausdorff* since, by the Riesz Representation Theorem, for distinct measures $\mu, \nu \in \mathfrak{M}(E)$, there exists an $f \in \mathcal{C}^c(E)$ such that $\int f\, d\mu \neq \int f\, d\nu$.

Hence it is also clear what is meant by the vague convergence of a mapping $t \to \mu_t$ of a subset A of a topological space T in $\mathfrak{M}(E)$ as t converges to a point t_0 in the closure of A. With respect to the vague topology, the convergence

$$\lim_{\substack{t\to t_0\\ t\in A}} \mu_t = \mu \in \mathfrak{M}(E)$$

means just

$$\lim_{\substack{t\to t_0\\ t\in A}} \int f\,d\mu_t = \int f\,d\mu, \qquad \text{for all } f \in \mathcal{C}^c(E). \tag{7.7.4}$$

Example

4. Let $K \geqq 0$ be a λ^p-integrable real function, defined on $E = \mathbf{R}^p$, with $\int K\,d\lambda^p = 1$ (for example, the indicator function of the unit cube $[\mathbf{0},\mathbf{1}]$). For every real $r > 0$, we set

$$K_r(x) \equiv r^p K(rx) \qquad (x \in \mathbf{R}^p).$$

Then $K_r \geqq 0$ and K_r is λ^p-integrable, and $\int K_r\,d\lambda^p = 1$. We need only take note of (1.7.9), where for the homothetic mapping $x \to H_r(x) \equiv rx$ of $\mathbf{R}^p$ onto itself we have $H_r(\lambda^p) = r^{-p}\lambda^p$. But then

$$\int K_r\,d\lambda^p = r^p\int K \circ H_r\,d\lambda^p = r^p \int K\,dH_r(\lambda^p) = 1.$$

In the sense of the vague topology we have

$$\lim_{r\to+\infty} K_r\lambda^p = \epsilon_0.$$[17]

Here $r \to K_r\lambda^p$ is a mapping of $]0,+\infty[$ into $\mathfrak{M}^1(\mathbf{R}^p)$. For the proof we note that for every $f \in \mathcal{C}^c(\mathbf{R}^p)$, we have the equality

$$\begin{aligned}\int fK_r\,d\lambda^p &= r^p \int f\cdot(K\circ H_r)\,d\lambda^p = r^p\int (f\circ H_r^{-1})K\,dH_r(\lambda^p)\\ &= \int (f\circ H_r^{-1})K\,d\lambda^p = \int f\left(\frac{x}{r}\right)K(x)\lambda^p(dx).\end{aligned}$$

This implies the assertion, using the Lebesgue Convergence Theorem, since, on the one hand, $\lim_{r\to+\infty} f(x/r)K(x) = f(0)K(x)$ for all $x \in \mathbf{R}^p$ and, on the other hand, for all $r > 0$,

$$|(f\circ H_r^{-1})K| \leqq \|f\|\cdot K.$$

In particular, all *discrete* Baire measures on E belong to $\mathfrak{M}^e(E)$. These are the measures δ which can be represented in the form

$$\delta = \sum_{i=1}^{k} \alpha_i \epsilon_{x_i}$$

[17] We shall make essential use of this "approximation of the unit ϵ_0" in Section 8.2.

by means of finitely many points $x_1, \ldots, x_k \in E$ and real numbers $\alpha_1 \geqq 0, \ldots, \alpha_k \geqq 0$. Using the vague topology, we can now show that every Baire measure on E can be approximated by discrete measures.

7.7.3. Theorem. For every locally compact space E countable at infinity, the set of discrete Baire measures on E is dense in $\mathfrak{M}(E)$ relative to the vague topology.

Proof. Suppose we are given a measure $\mu_0 \in \mathfrak{M}$ and a vague neighborhood

$$V_{f_1,\ldots,f_n;\epsilon}(\mu_0)$$

of μ_0 in the form described in (7.7.3) ($f_1, \ldots, f_n \in \mathfrak{C}^c$; $\epsilon > 0$). We want to show the existence of a discrete measure δ lying in $V_{f_1,\ldots,f_n;\epsilon}(\mu_0)$. For this we consider a compact Baire set K such that

$$\bigcup_{i=1}^{n} S_{f_i} \subset K$$

and an $\eta > 0$ with $\eta\mu_0(K) \leqq \epsilon$. For every $y \in K$ there exists a compact neighborhood $U_y \in \mathfrak{B}_0(E)$ of y in E such that $|f_i(y') - f_i(y'')| \leqq \eta$ for any two points $y', y'' \in U_y$ and arbitrary $i = 1, \ldots, n$.[18] Finitely many of these U_y, say $U_{y_1}, \ldots, U_{y_k}$ cover K. If we set

$$A_1 \equiv K \cap U_{y_1},\ A_2 \equiv K \cap U_{y_2} \setminus A_1,\ \ldots,\quad A_k \equiv K \cap U_{y_k} \setminus A_1 \cup \cdots \cup A_{k-1},$$

then $A_1, \ldots, A_k$ are relatively compact, pairwise disjoint Baire sets such that

$$K = A_1 \cup \cdots \cup A_k$$

and $|f_i(y') - f_i(y'')| \leqq \eta$ for all $i = 1, \ldots, n$, $j = 1, \ldots, k$ and arbitrary $y', y'' \in A_j$. Since only these properties will be used below, we can assume that no $A_1, \ldots, A_k$ is empty. If we now arbitrarily choose $x_1 \in A_1, \ldots, x_k \in A_k$, then the discrete measure

$$\delta \equiv \sum_{j=1}^{k} \mu_0(A_j)\epsilon_{x_j}$$

yields the desired result. [Note that the measure $\mu_0(A_j)$ of the relatively compact sets A_j is finite.] The proof is derived from the following inequali-

[18] By Lemma 7.4.2 we can choose U_y, for example, of the form $\{h \geqq \beta\}$ with suitable $h \in \mathfrak{C}^c$ and $\beta > 0$.

ties, valid for $i = 1, \ldots, n$,

$$\left| \int f_i \, d\mu_0 - \int f_i \, d\delta \right| = \left| \sum_{j=1}^{k} \int_{A_j} f_i \, d\mu_0 - \sum_{j=1}^{k} \mu_0(A_j) f_i(x_j) \right|$$
$$= \left| \sum_{j=1}^{k} \int_{A_j} (f_i - f_i(x_j)) \, d\mu_0 \right| \leqq \sum_{j=1}^{k} \int_{A_j} |f_i - f_i(x_j)| \, d\mu_0$$
$$\leqq \eta \sum_{j=1}^{k} \mu_0(A_j) = \eta \mu_0(K) \leqq \epsilon.$$

We have only to recall that $|f_i(x) - f_i(x_j)| \leqq \eta$ for all $x \in A_j$. ⌟

7.7.4. Corollary. The discrete probability measures on E are dense in $\mathfrak{M}^1(E)$ relative to the vague topology.

Proof. Now suppose in particular that μ_0 is a measure in $\mathfrak{M}^1(E)$. We consider the measure $\delta = \Sigma \mu_0(A_j)\epsilon_{x_j}$ in $V_{f_1, \ldots, f_n; \epsilon}(\mu_0)$ constructed above, and set $\alpha_j \equiv \mu_0(A_j)$ for $j = 1, \ldots, k$. When $K = E$, $\alpha_1 + \cdots + \alpha_k = 1$ and there is nothing more to prove. When $K \neq E$, by construction we have $\alpha_1 + \cdots + \alpha_k = \mu_0(K) \leqq \mu_0(E) = 1$. Thus it suffices to set

$$\alpha_{k+1} \equiv 1 - (\alpha_1 + \cdots + \alpha_k)$$

and to choose an $x_{k+1} \in E \setminus K$. Then

$$\delta' \equiv \sum_{j=1}^{k+1} \alpha_j \epsilon_{x_j}$$

is a discrete probability measure satisfying $\int f_i \, d\delta = \int f_i \, d\delta'$ for all $i = 1, \ldots, n$, since x_{k+1} does not lie in $S_{f_1} \cup \cdots \cup S_{f_n}$. Consequently, when δ lies in $V_{f_1, \ldots, f_n; \epsilon}(\mu_0)$, so does δ'. ⌟

Now we investigate the question of whether equality (7.7.1) or (7.7.4) also holds for more general continuous functions. Here we can immediately observe that for a measure $\mu \in \mathfrak{M}^e(E)$ every function $f \in \mathcal{C}^b(E)$ is μ-integrable since it is $\mathfrak{B}_0(E)$-measurable and its absolute value is bounded by a constant and therefore by a μ-integrable function. We formulate the relevant results only for sequences; their generalization to mappings $t \to \mu_t$ is obvious.

7.7.5. Theorem. If a sequence $(\mu_n)_{n \in \mathbf{N}}$ in $\mathfrak{M}(E)$ is vaguely convergent to $\mu \in \mathfrak{M}(E)$ and the sequence $(\mu_n(E))_{n \in \mathbf{N}}$ is bounded,[19] then μ is a finite measure, and for every function $f \in \mathcal{C}^0(E)$,

[19] In particular, every measure μ_n is then finite.

$$\lim_{n\to\infty} \int f\, d\mu_n = \int f\, d\mu.$$

Proof. For every $\mu \in \mathfrak{M}(E)$,

$$\mu(E) = \sup_{\substack{u\in\mathcal{C}^c \\ 0\leqq u\leqq 1}} \int u\, d\mu \tag{7.7.5}$$

We need note only that for every compact set $K \subset E$, by Lemma 7.4.2 there exists a $u \in \mathcal{C}^c$ with $0 \leqq u \leqq 1$ and $1_K \leqq u$ and that $u \leqq 1_{S_u}$ for every $u \in \mathcal{C}^c$ with $0 \leqq u \leqq 1$. Hence

$$\sup_{\substack{u\in\mathcal{C}^c \\ 0\leqq u\leqq 1}} \int u\, d\mu = \sup_{\substack{K \text{ compact} \\ K\in\mathfrak{B}_0}} \mu(K) = \mu(E),$$

where the second equality follows from the regularity of μ.

Now let μ be the vague limit of (μ_n) and $\alpha \equiv \sup_{n\in\mathbf{N}} \mu_n(E)$. Then by (7.7.5), $\int u\, d\mu_n \leqq \mu_n(E) \leqq \alpha$ and hence $\int u\, d\mu \leqq \alpha$ for all $u \in \mathcal{C}^c$ with $0 \leqq u \leqq 1$. Another application of (7.7.5) then yields $\mu(E) \leqq \alpha$, that is, the finiteness of μ. Now let $f \in \mathcal{C}^0(E)$. By Definition 7.4.7, for every $\epsilon > 0$ there exists a $g \in \mathcal{C}^c$ with $\|f - g\| \leqq \epsilon$. Therefore,

$$\left|\int f\, d\mu_n - \int g\, d\mu_n\right| \leqq \|f - g\|\mu_n(E) \leqq \alpha\epsilon \qquad (n = 1, 2, \ldots)$$

and $\left|\int f\, d\mu - \int g\, d\mu\right| \leqq \alpha\epsilon$. Using the triangle inequality, we obtain

$$\left|\int f\, d\mu_n - \int f\, d\mu\right| \leqq 2\alpha\epsilon + \left|\int g\, d\mu_n - \int g\, d\mu\right|$$

for all n, and hence the assertion since $\lim_{n\to\infty} \int g\, d\mu_n = \int g\, d\mu$. ┘

Remark 1. Even when all measures μ_n ($n \in \mathbf{N}$) and μ are finite, we cannot do without the requirement $\sup \mu_n(E) < \infty$ in general. This is shown by Example 2 in the special case $E = \mathbf{R}$, $x_n = n$ and $\alpha_n = n$ for all $n = 1, 2, \ldots$. Indeed, the function $f(x) \equiv \inf(1, 1/|x|)$ for $x \neq 0$ and $f(x) \equiv 1$ for $x = 0$ lies in $\mathcal{C}^0(\mathbf{R})$. But $\int f\, d\mu_n = 1$ for all n and $\int f\, d\mu = 0$.

We emphasize the transition from $\mathcal{C}^0(E)$ to $\mathcal{C}^b(E)$ by the following:

7.7.6. Definition. Let $\mu, \mu_1, \mu_2, \ldots$ be measures from $\mathfrak{M}^e(E)$. The sequence $(\mu_n)_{n\in\mathbf{N}}$ is said to be *weakly* or *Bernoulli convergent* to μ if

$$\lim_{n\to\infty} \int f\, d\mu_n = \int f\, d\mu \qquad \text{for all } f \in \mathcal{C}^b(E). \tag{7.7.6}$$

7.7.7. Theorem. A sequence $(\mu_n)_{n\in\mathbf{N}}$ in $\mathfrak{M}^e(E)$ is weakly convergent to a measure $\mu \in \mathfrak{M}^e(E)$ if and only if $(\mu_n)_{n\in\mathbf{N}}$ converges vaguely to μ and

$$\lim_{n\to\infty} \mu_n(E) = \mu(E). \tag{7.7.7}$$

Proof. The given conditions are necessary for weak convergence since

$\mathcal{C}^c \subset \mathcal{C}^b$ and $1 \in \mathcal{C}^b$. Now we show that these conditions are also sufficient: By (7.7.5), for every $\epsilon > 0$ there is a $u \in \mathcal{C}^c$ such that $0 \leqq u \leqq 1$ and

$$\mu(E) - \int u\, d\mu = \int (1 - u)\, d\mu < \epsilon.$$

By hypothesis,

$$\lim_{n\to\infty} \int u\, d\mu_n = \int u\, d\mu \qquad \text{and} \qquad \lim_{n\to\infty} \int 1\, d\mu_n = \int 1\, d\mu.$$

Thus for n sufficiently large,

$$\int (1 - u)\, d\mu_n < \epsilon.$$

For these n and for all $f \in \mathcal{C}^b$,

$$\left|\int f(1 - u)\, d\mu_n\right| \leqq \|f\| \int (1 - u)\, d\mu_n \leqq \|f\|\epsilon.$$

Analogously, $|\int f(1 - u)\, d\mu| \leqq \|f\|\epsilon$. As in the preceding proof, we use the triangle inequality to obtain

$$\left|\int f\, d\mu_n - \int f\, d\mu\right| \leqq 2\|f\|\epsilon + \left|\int fu\, d\mu_n - \int fu\, d\mu\right|$$

for n sufficiently large. Since fu lies in $\mathcal{C}^c$ and thus $(\int fu\, d\mu_n)$ converges to $\int fu\, d\mu$, the assertion now follows. ┘

7.7.8. Corollary. Every sequence $(\mu_n)_{n\in\mathbf{N}}$ in $\mathfrak{M}^1(E)$ which converges vaguely to a measure $\mu \in \mathfrak{M}^1(E)$ is weakly convergent (and conversely).

Remark 2. The concept of *weak convergence* obviously can also be introduced in the following more general situation: Let E be a topological space, $\mathfrak{B}_0(E)$ the σ-algebra of its Baire sets, and $\mu, \mu_1, \mu_2, \ldots$ a sequence of finite Baire measures on E. Since the functions in $\mathcal{C}^b(E)$ are $\mathfrak{B}_0(E)$-measurable, we can define as above: The sequence $(\mu_n)_{n\in\mathbf{N}}$ is said to be weakly convergent to μ if (7.7.6) holds. We shall shortly make use of this possibility for generalization. Moreover, weak convergence can be derived from a topology in the same way as vague convergence. In (7.7.3), $\mathcal{C}^b(E)$ takes over the role of $\mathcal{C}^c(E)$. We speak of the *weak topology* on the set of finite Baire measures on E.

Example 1 of this section shows that weak convergence of a sequence (μ_n) in $\mathfrak{M}^e(E)$ to $\mu \in \mathfrak{M}^e(E)$ does not imply the convergence $\lim_{n\to\infty} \int f\, d\mu_n = \int f\, d\mu$ for not necessarily continuous, bounded, $\mathfrak{B}_0(E)$-measurable functions f. Nonetheless, we can weaken the continuity of the functions f for weak convergence. We restrict discussion to the case in which E has a countable base. Referring to Theorem 7.6.6, we therefore carry out our reasoning immediately for Polish spaces.

Now let E be a *Polish space* and $\mathfrak{M}^e(E)$ the set of its finite Borel measures. We consider Borel-measurable bounded real functions f on E which are

μ-almost everywhere continuous relative to a measure $\mu \in \mathfrak{M}^e(E)$. These are functions which are continuous in all points $x \in \complement N$ where $N \in \mathfrak{B}(E)$ is a μ-null set. Important examples of such functions are the indicator functions 1_Q of null-boundary Borel sets Q, which are defined as follows:

7.7.9. Definition. A Borel subset Q of a Polish space E is said to be a *null-boundary set* with respect to a measure $\mu \in \mathfrak{M}^e(E)$, or a μ-null-boundary set, if $\mu(Q^*) = 0$ for the topological boundary $Q^* \equiv \bar{Q}\backslash\mathring{Q}$ of Q.[20]

Examples

5. Every interval I of the number line $\mathbf{R}$ is a λ^1-null-boundary set.

6. A set $Q \in \mathfrak{B}(E)$ is a null-boundary set with respect to a unit mass ϵ_a $(a \in E)$ if and only if $a \notin Q^*$.

7.7.10. Theorem. Let E be a Polish space. Suppose a sequence $(\mu_n)_{n\in\mathbf{N}}$ in $\mathfrak{M}^e(E)$ converges weakly to a measure $\mu \in \mathfrak{M}^e(E)$. Then

$$\lim_{n\to\infty} \int f\, d\mu_n = \int f\, d\mu$$

for every Borel-measurable, bounded, μ-almost everywhere continuous function f. In particular,

$$\lim_{n\to\infty} \mu_n(Q) = \mu(Q)$$

for every μ-null-boundary set $Q \in \mathfrak{B}(E)$.

Proof. Let d denote a metric defining the topology of E and $K_r(x) \equiv \{y \in E: d(x,y) < r\}$ be the open ball of center x and of radius r. By hypothesis there is a Borel set E_0 in E such that $\mu(E_0) = \mu(E)$ and such that the bounded Borel-measurable function f is continuous at all points $x \in E_0$. By Theorem 7.3.3, μ is regular. Therefore for every $\epsilon > 0$ there is a compact set $K \subset E_0$ such that $\mu(E_0 \setminus K) \leqq \epsilon$. Every point $x \in K$ is the midpoint of a sphere $U'_x \equiv K_{r_x}(x)$ such that the variation of f in $U'(x)$ is at most ϵ, that is, $|f(y') - f(y'')| \leqq \epsilon$ for all $y', y'' \in U'_x$. Because of the compactness of the set K, we have $K \subset U_{x_1} \cup \cdots \cup U_{x_n}$ for a suitable finite subset $\{x_1, \ldots, x_n\}$ of K if we set $U_x \equiv K_{r_x/2}(x)$. If we define

$$\alpha \equiv \inf f(E),\ \beta \equiv \sup f(E),\ \alpha_\nu \equiv \inf f(U'_{x_\nu}),\ \beta_\nu \equiv \sup f(U'_{x_\nu})$$

for all $\nu = 1, \ldots, n$, then, due to the normality of E, for every $\nu = 1, \ldots, n$ there are functions $g_\nu, h_\nu \in \mathfrak{C}^b(E)$ satisfying $g_\nu(x) = \alpha_\nu$ and

[20] μ-null-boundary sets are also called *μ-squareable.*

$h_\nu(x) = \beta_\nu$ for all $x \in \overline{U_{x_\nu}}$, $g_\nu(x) = \alpha$, and $h_\nu(x) = \beta$ for all $x \in E \setminus U'_{x_\nu}$, and $\alpha \leqq g_\nu \leqq \alpha_\nu \leqq \beta_\nu \leqq h_\nu \leqq \beta$. Then obviously, in particular, $g_\nu \leqq f \leqq h_\nu$. Let

$$g \equiv \sup (g_1, \ldots, g_n),\ h \equiv \inf (h_1, \ldots, h_n);$$

then g and h are functions in $\mathfrak{C}^b(E)$ with $\alpha \leqq g \leqq f \leqq h \leqq \beta$. For all points $x \in K$, $h(x) - g(x) \leqq \epsilon$. Indeed, every point $x \in K$ lies in a U_{x_ν}; consequently, $h(x) - g(x) \leqq h_\nu(x) - g_\nu(x) = \beta_\nu - \alpha_\nu \leqq \epsilon$. Now we can complete the proof as follows: We have

$$\begin{aligned} \int (h - g)\, d\mu &= \int_K (h - g)\, d\mu + \int_{\complement K} (h - g)\, d\mu \\ &\leqq \epsilon\mu(K) + (\beta - \alpha)\mu(E \setminus K) \leqq \epsilon(\mu(E) + \beta - \alpha) \end{aligned}$$

and since $g \leqq f \leqq h$, $g, h \in \mathfrak{C}^b(E)$, we also have

$$\begin{aligned} \int g\, d\mu &= \lim_{n\to\infty} \int g\, d\mu_n \leqq \liminf_{n\to\infty} \int f\, d\mu_n \leqq \limsup_{n\to\infty} \int f\, d\mu_n \leqq \lim_{n\to\infty} \int h\, d\mu_n \\ &= \int h\, d\mu \end{aligned}$$

and

$$\int g\, d\mu \leqq \int f\, d\mu \leqq \int h\, d\mu.$$

Therefore, $\int f\, d\mu$, $\liminf \int f\, d\mu_n$, and $\limsup \int f\, d\mu_n$ differ at most by $\epsilon(\mu(E) + \beta - \alpha)$. Since $\epsilon > 0$ was chosen arbitrarily, the assertion now follows. ⌋

7.7.11. Theorem. Let $\mu, \mu_1, \mu_2, \ldots$ be probability measures on $\mathfrak{B}^1 = \mathfrak{B}(\mathbf{R})$ and $F, F_1, F_2, \ldots$ be the associated distribution functions. Then if the sequence $(\mu_n)_{n\in\mathbf{N}}$ is vaguely convergent to μ,

$$\lim_{n\to\infty} F_n(x) = F(x) \tag{7.7.8}$$

for all points $x \in \mathbf{R}$ at which F is continuous. If F is continuous on all of $\mathbf{R}$, then the sequence $(F_n)_{n\in\mathbf{N}}$ converges uniformly to F.

Proof. By Theorem 7.7.7 we have weak convergence to μ. Thus $\lim \mu_n(Q) = \mu(Q)$ for every μ-null-boundary set $Q \in \mathfrak{B}^1$ and hence by the definition of distribution functions, $\lim_{n\to\infty} F_n(x) = F(x)$ for all $x \in \mathbf{R}$ for which the interval $Q_x \equiv]-\infty, x[$ in a μ-null-boundary set. Now $]-\infty, x] = \overline{Q_x} = \bigcap_{k=1}^{\infty} Q_{x+1/k}$ and therefore

$$\mu(\overline{Q_x}) = \lim_{k\to\infty} \mu\,(Q_{x+1/k}) = \lim_{k\to\infty} F\left(x + \frac{1}{k}\right).$$

Consequently, Q_x is a μ-null-boundary set if and only if F is right-continuous in x and therefore (due to the left-continuity of distribution functions) is continuous in x. This proves the first part of the assertion.

For the second part, we assume that F is continuous. For every $\epsilon > 0$ there are numbers $a < b$ with $F(a) < \epsilon$ and $1 - F(b) < \epsilon$. Further, we can choose finitely many $x_0, \ldots, x_k \in \mathbf{R}$ such that $a = x_0 < x_1 < \cdots < x_k = b$ and $F(x_i) - F(x_{i-1}) < \epsilon$ for $i = 1, \ldots, k$. By the first part of the assertion we can determine n so that $|F_n(x_i) - F(x_i)| < \epsilon$ for $i = 0, 1, \ldots, k$. But then we have $|F_n(x) - F(x)| < 2\epsilon$ for all $x \in \mathbf{R}$, that is, the convergence to F is uniform. In fact, for $x < x_0$, $0 \leqq F(x) \leqq F(x_0) < \epsilon$ and $0 \leqq F_n(x) \leqq F_n(x_0) < F(x_0) + \epsilon < 2\epsilon$, and thus $|F_n(x) - F(x)| < 2\epsilon$. We conclude analogously for $x \geqq x_k$. When $x_{i-1} \leqq x < x_i$ for some $i = 1, \ldots, k$, then $F(x_{i-1}) \leqq F(x) \leqq F(x_i) < F(x_{i-1}) + \epsilon$ and $F(x_{i-1}) - \epsilon < F_n(x_{i-1}) \leqq F_n(x) \leqq F_n(x_i) < F(x_i) + \epsilon < F(x_{i-1}) + 2\epsilon$, hence again, $|F_n(x) - F(x)| < 2\epsilon$. ⌟

Remark 3. The second condition in Theorem 7.7.10 and the first in Theorem 7.7.11 are also sufficient for weak convergence of the sequence (μ_n) to μ. See the following Problems 4 and 5.

PROBLEMS

1. Let E be a locally compact space countable at infinity, and let $\mathfrak{D}$ be a rich subset of $\mathcal{C}^c(E)$ (see Section 7.6, Problem 2). Assume that for every compact set $K \subset E$ there exists a function d in $\mathfrak{D}$ such that $0 \leqq d \leqq 1$ and $d(x) = 1$ for all $x \in K$. Prove:
   (a) A sequence (μ_n) in $\mathfrak{M}(E)$ is vaguely converging to $\mu \in \mathfrak{M}(E)$ if and only if $\lim\limits_{n\to\infty} \mu_n(d) = \mu(d)$ for all $d \in \mathfrak{D}$.
   (b) For $E = \mathbf{R}$ the set $\mathfrak{D}$ of all continuously differentiable functions $d \in \mathcal{C}^c(\mathbf{R})$ has all the properties mentioned above.
2. Use Problem 1 in order to prove that the sequence $\mu_n \equiv f_n\lambda^1$, $n \in \mathbf{N}$, where f_n is the function $x \to 1 - \sin nx$ on $\mathbf{R}$, converges vaguely to λ^1.
3. Let μ be a finite Borel measure on a Polish space E. Prove:
   (a) The system $\mathfrak{Q}_\mu$ of all μ-null-boundary sets is an algebra in Ω.
   (b) For every $f \in \mathcal{C}^b(E)$ there exists a countable set $D_f \subset \mathbf{R}$ such that $\{f > \alpha\} \in \mathfrak{Q}_\mu$ for all $\alpha \in \mathbf{R} \setminus D_f$. [*Hint:* For every finite subset $\{\alpha_1, \ldots, \alpha_n\}$ of $\mathbf{R}$ one has $\sum_{i=1}^n \mu(\{f = \alpha_i\}) \leqq \mu(E) < +\infty$.]
4. Let $\mu, \mu_1, \mu_2, \ldots$ be finite Borel measures on a Polish space. Prove: The condition of Theorem 7.7.10 is not only necessary but also sufficient for weak convergence, that is, (μ_n) is weakly convergent to μ if $\lim \mu_n(Q) = \mu(Q)$ for all μ-null-boundary sets Q. [*Hint:* Imitate the proof of Theorem 2.3.6 and use Problem 3 in order to prove that every $f \in \mathcal{C}_+^b(E)$ is the uniform limit of an increasing sequence (u_n) in $\mathfrak{F}(\mathfrak{Q}_\mu)$ (where $\mathfrak{F}(\mathfrak{Q}_\mu)$ is defined as in Section 7.1, Example 1).]

5. Prove that the condition (7.7.8) is also sufficient for weak convergence of (μ_n) to μ.
6. Let $(\alpha_n)_{n \in \mathbf{N}}$ be a sequence of numbers in the interval $]0,1[$. Remove from $[0,1]$ an open interval I_{11} of length α_1 containing the center $\frac{1}{2}$ of $[0,1]$. This leaves two disjoint closed intervals J_{11} and J_{12}. Remove from J_{1i} an open interval I_{2i} of length $\alpha_2\lambda^1(J_{1i})$ containing the center of J_{1i} $(i = 1, 2)$. This leaves four pairwise disjoint closed intervals $J_{21}, \ldots, J_{24}$. Remove from J_{2i} an open interval I_{3i} of length $\alpha_3\lambda^1(J_{2i})$ containing the center of J_{2i} $(i = 1, 2, 3, 4)$. This leaves $8 = 2^3$, pairwise disjoint closed intervals J_{3i}, $i = 1, \ldots, 8$. Continuing this way one obtains for each $n \in \mathbf{N}$ pairwise disjoint closed intervals J_{ni}, $i = 1, \ldots, 2^n$. The set

$$C \equiv \bigcap_{n=1}^{\infty} (J_{n1} \cup \cdots \cup J_{n2^n})$$

is known as a *Cantor-like* set, and for $\alpha_n = (1/n)(n \in \mathbf{N})$ as *Cantor* (ternary) *set*. Prove:
(a) C is compact and nonempty, and has empty interior.
(b) $\lambda^1(C) = \lim_n \prod_{i=1}^{n} (1 - \alpha_i)$.
(c) $\lambda^1(C) = 0$ if and only if $\sum_{n=1}^{\infty} \alpha_n = +\infty$. [*Hint:* Use (6.2.4) and the theory of absolutely convergent infinite products.]
(d) If $\sum_{n=1}^{\infty} \alpha_n < +\infty$, $U \equiv]0,1[\setminus C$ is an open set in $\mathbf{R}$ whose boundary $U^* = \bar{U} \setminus U$ is not a λ^1-null set.
7. Construct an open subset of $]0,1[\times]0,1[$ whose boundary in $\mathbf{R}^2$ has positive λ^2-measure.

7.8 VAGUELY COMPACT SETS OF MEASURES

We again consider a *locally compact space* E which is *countable at infinity*, and the space $\mathfrak{M} = \mathfrak{M}(E)$, of all Baire measures on E, equipped with the vague topology. We are interested in the compact (relatively compact) subsets of $\mathfrak{M}$ in this topology, which we designate as vaguely compact (vaguely relatively compact).

We can immediately give a necessary condition for vague relative compactness of a set $H \subset \mathfrak{M}$. By the definition of vague topology, the real function $\mu \to \int f\, d\mu$ is continuous on $\mathfrak{M}$ for every function $f \in \mathcal{C}^c(E)$. Consequently, the image of H under this mapping is a relatively compact subset of $\mathbf{R}$, and hence is bounded. This observation leads to the following definition:

7.8.1. Definition. A set $H \subset \mathfrak{M}(E)$ is said to be *bounded* if

$$\sup_{\mu \in H} |\int f\, d\mu| < +\infty, \qquad \text{for all } f \in \mathfrak{C}^c(E). \tag{7.8.1}$$

Thus, boundedness is necessary for vague relative compactness of a set $H \subset \mathfrak{M}$. We shall show that this condition is also sufficient.

7.8.2. Theorem. A set $H \subset \mathfrak{M}(E)$ is vaguely relatively compact if and only if it is bounded.

Proof. We need only show that boundedness of H implies vague relative compactness. Thus let α_f, for all $f \in \mathfrak{C}^c$, denote the supremum occurring in (7.8.1), I_f the compact interval $[-\alpha_f, \alpha_f]$ in $\mathbf{R}$, and $\bar{H}$ the closed hull of H in $\mathfrak{M}$. Then $\int f\, d\mu \in I_f$ for all $f \in \mathfrak{C}^c$ and all $\mu \in \bar{H}$. In fact, for every $f \in \mathfrak{C}^c$ and every $\epsilon > 0$

$$V_{f;\epsilon} = V_{f;\epsilon}(\mu) \equiv \{\nu \in \mathfrak{M}: |\int f\, d\nu - \int f\, d\mu| < \epsilon\}$$

is a vague neighborhood of μ, that is, $H \cap V_{f;\epsilon} \neq \emptyset$. But $\int f\, d\nu$ lies in I_f for every $\nu \in H \cap V_{f;\epsilon}$, and thus

$$|\int f\, d\mu| \leqq \alpha_f + |\int f\, d\nu - \int f\, d\mu| < \alpha_f + \epsilon$$

for every $\epsilon > 0$. This implies $\int f\, d\mu \in I_f$.

Now we consider the Hausdorff product space $\mathbf{R}^{\mathfrak{C}^c} = \prod_{f \in \mathfrak{C}^c} \mathbf{R}_f$, which we obtain by associating a copy $\mathbf{R}_f = \mathbf{R}$ of the real line with each f. The product space $J \equiv \prod_{f \in \mathfrak{C}^c} I_f$ is a subspace of $\mathbf{R}^{\mathfrak{C}^c}$ which, according to the familiar theorem of Tychonov, is compact as the product of compact spaces. We can map $\mathfrak{M}$ into $\mathbf{R}^{\mathfrak{C}^c}$ by associating with each measure μ the real function $f \to \int f\, d\mu$ on $\mathfrak{C}^c$ which lies in $\mathbf{R}^{\mathfrak{C}^c}$. Thus we have a mapping $\Phi: \mathfrak{M} \to \mathbf{R}^{\mathfrak{C}^c}$ which is injective by Theorem 7.5.4. According to what was shown in the introduction, $\Phi(\bar{H}) \subset J$. Hence the assertion is proved if we can show: (a) Φ maps $\mathfrak{M}$ homeomorphically onto $\Phi(\mathfrak{M})$. (b) $\Phi(\mathfrak{M})$ is closed in $\mathbf{R}^{\mathfrak{C}^c}$. Then $\Phi(\bar{H})$, as a closed subset of $\Phi(\mathfrak{M})$, is also closed in $\mathbf{R}^{\mathfrak{C}^c}$, and $\Phi(\bar{H}) \subset J$ implies the compactness of $\Phi(\bar{H})$ and thus of $\bar{H}$.

To prove (a): Continuity of Φ is equivalent to continuity of every "component" of Φ, that is, every mapping $\mu \to \int f\, d\mu$ with $f \in \mathfrak{C}^c$. But this continuity follows from the definition of the vague topology. Continuity of the inverse mapping Ψ of Φ is equivalent to the continuity of $\Phi(\mu) \to \int f\, d\Psi(\Phi(\mu)) = \int f\, d\mu$ on $\Phi(\mathfrak{M})$ for each $f \in \mathfrak{C}^c$. But this mapping is the restriction to $\Phi(\mathfrak{M})$ of the projection of $\mathbf{R}^{\mathfrak{C}^c}$ on the fth coordinate axis. To prove (b): Let $I \in \mathbf{R}^{\mathfrak{C}^c}$ be a point in the closure of $\Phi(\mathfrak{M})$ in $\mathbf{R}^{\mathfrak{C}^c}$. Then I is a positive linear form on $\mathfrak{C}^c$. Indeed, let f, g be functions from $\mathfrak{C}^c$

and $\epsilon > 0$. The set

$$\{I' \in \mathbf{R}^{\mathcal{C}^c}: |I'(f+g) - I(f+g)| < \epsilon, \\ |I'(f) - I(f)| < \epsilon, |I'(g) - I(g)| < \epsilon\}$$

is a neighborhood of I in $\mathbf{R}^{\mathcal{C}^c}$ and thus contains a point $I' = \Phi(\mu) \in \Phi(\mathfrak{M})$. Hence I' is the positive linear form $q \to \int q\, d\mu$ on $\mathcal{C}^c$; hence, we obtain

$$|I(f+g) - I(f) - I(g)| \leqq |I(f+g) - I'(f+g)| \\ + |I(f) - I'(f)| + |I(g) - I'(g)| < 3\epsilon.$$

Since $\epsilon > 0$ was chosen arbitrarily, $I(f+g) = I(f) + I(g)$. Analogously we see that $I(\alpha f) = \alpha I(f)$ ($f \in \mathcal{C}^c$, $\alpha \in \mathbf{R}$) and $I(f) \geqq 0$ for $f \in \mathcal{C}^c_+$. Thus by Theorem 7.5.4 there is exactly one $\nu \in \mathfrak{M}$ with $\Phi(\nu) = I$; that is, I lies in $\Phi(\mathfrak{M})$. Hence $\Phi(\mathfrak{M})$ is closed.

This proves the theorem. ⌟

7.8.3. Corollary. For every real number $\alpha \geqq 0$, the set

$$\mathcal{E}_\alpha \equiv \{\mu \in \mathfrak{M}^e(E): \mu(E) \leqq \alpha\}$$

is vaguely compact.

Proof. For each $f \in \mathcal{C}^c$ and all $\mu \in \mathcal{E}_\alpha$, $|\int f\, d\mu| \leqq \int |f|\, d\mu \leqq \alpha\|f\|$. Therefore $\mathcal{E}_\alpha$ is bounded and thus vaguely relatively compact. Now we still have to show that $\mathcal{E}_\alpha$ is closed in $\mathfrak{M}$. By (7.7.5), $\mathcal{E}_\alpha$ is the set of all $\mu \in \mathfrak{M}$ such that $\int u\, d\mu \leqq \alpha$ for all $u \in \mathcal{C}^c$ with values between 0 and 1. Since $\mu \to \int u\, d\mu$ is continuous, the set $A_u \equiv \{\mu \in \mathfrak{M}: \int u\, d\mu \leqq \alpha\}$ is closed for each $u \in \mathcal{C}^c$. Since

$$\mathcal{E}_\alpha = \bigcap_{\substack{u \in \mathcal{C}^c \\ 0 \leqq u \leqq 1}} A_u,$$

$\mathcal{E}_\alpha$ is also vaguely closed. ⌟

Remark 1. The set of all measures $\mu \in \mathfrak{M}^e(E)$ such that $\mu(E) = \alpha$ is in general *not* closed and therefore not compact. This can be seen from Example 2 of Section 7.7 if we choose the numbers α_n there, all equal to α. More precisely, the example shows that (for $\alpha = 1$) for noncompact E, the set $\mathfrak{M}^1(E)$ is not closed. $\mathfrak{M}^1(E)$ is closed for compact E, since the constant function 1 lies in $\mathcal{C}^c(E) = \mathcal{C}(E)$.

In order to be able to use sequences in investigating the vague topology on $\mathfrak{M}(E)$, we still need to know when $\mathfrak{M}(E)$ is metrizable. Therefore we show:

7.8.4. Theorem. Let E be a locally compact space countable at infinity. The vague topology on $\mathfrak{M}(E)$ is metrizable if and only if E has a countable base.

Proof. The mapping $x \to \varphi(x) \equiv \epsilon_x$ of E into $\mathfrak{M}(E)$ is injective and a homeomorphism of E onto $\varphi(E)$. The latter can be seen as follows: For every point $x \in E$, the sets

$$M_{f_1, \ldots, f_n; \eta}(x) \equiv \{y \in E: |f_i(x) - f_i(y)| < \eta, i = 1, \ldots, n\}$$

$(f_1, \ldots, f_n \in \mathcal{C}^c; \eta > 0; n = 1, 2, \ldots)$ form a fundamental system of neighborhoods of x. (Indeed, if U is a neighborhood of $x \in E$, then by Lemma 7.4.2 there is an $f \in \mathcal{C}^c$ satisfying $0 \leqq f \leqq 1$, $f(x) = 1$, and $S_f \subset U$. But then $M_{f; 1/2}(x) \subset U$.) Using the notation of (7.7.3), we have

$$\varphi(M_{f_1, \ldots, f_n; \eta}(x)) = \varphi(E) \cap V_{f_1, \ldots, f_n; \eta}(\epsilon_x)$$

for arbitrary $f_1, \ldots, f_n \in \mathcal{C}^c$ and $\eta > 0$, which proves that φ is a homeomorphism of E onto $\varphi(E)$. Therefore, the metrizability of $\mathfrak{M}(E)$ and hence of $\varphi(E)$ implies that of E. But for a locally compact space countable at infinity, metrizability is equivalent to the existence of a countable base.[21]

Now assume the existence of a countable base for E. Then by Theorem 7.6.3 there is a countable set $\mathfrak{D}_0 \subset \mathcal{C}^c$ which is dense in $\mathcal{C}^c$.

Moreover, we can choose a sequence (K_n) of compact sets such that $K_n \subset \mathring{K}_{n+1}$ for all $n = 1, 2, \ldots$ and $\bigcup K_n = E$, and a sequence (e_n) in $\mathcal{C}^c$ such that $0 \leqq e_n \leqq 1$ and $e_n(x) = 1$ for all $x \in K_n$ $(n = 1, 2, \ldots)$. Then

$$\mathfrak{D} = \mathfrak{D}_0 \cup \{de_n: d \in \mathfrak{D}_0, n \in \mathbf{N}\} \cup \{e_n: n \in \mathbf{N}\}$$

is again a countable subset of $\mathcal{C}^c$ which is dense in $\mathcal{C}^c$.

As in the proof of Theorem 7.8.2, we consider the mapping $\Phi: \mathfrak{M} \to \mathbf{R}^{\mathfrak{D}}$ which, with every measure $\mu \in \mathfrak{M}$, associates the real function $d \to \int d\, d\mu$ defined on $\mathfrak{D}$. Then Φ is injective for the following reason: Let $\mu, \nu \in \mathfrak{M}$ with $\int d\, d\mu = \int d\, d\nu$ for all $d \in \mathfrak{D}$, and suppose we are given $f \in \mathcal{C}^c$. The support S_f is contained in a K_{n_0}. For every $\epsilon > 0$ there is a $d \in \mathfrak{D}_0$ with $\|f - d\| \leqq \epsilon$. But then $f = fe_{n_0}$ and thus $|f - de_{n_0}| \leqq \epsilon e_{n_0}$, that is, $|\int f\, d\mu - \int de_{n_0}\, d\mu| \leqq \epsilon \int e_{n_0}\, d\mu$ and $|\int f\, d\nu - \int de_{n_0}\, d\nu| \leqq \epsilon \int e_{n_0}\, d\nu$. Since $de_{n_0} \in \mathfrak{D}$, this implies $|\int f\, d\mu - \int f\, d\nu| \leqq 2\epsilon \int e_{n_0}\, d\mu$ for each $\epsilon > 0$. Thus $\int f\, d\mu = \int f\, d\nu$ for every $f \in \mathcal{C}^c$, that is, $\mu = \nu$. We show the continuity of Φ as in the proof of Theorem 7.8.2. Again Φ is a homeomorphism of $\mathfrak{M}$ onto $\Phi(\mathfrak{M})$. For this we need show only (see the proof of Theorem 7.8.2)

[21] See N. Bourbaki [26] as well as Section 7.3, Example 6 and the proof of Theorem 7.6.1.

that for every $f \in \mathfrak{C}^c$, the mapping $\Phi(\mu) \to \int f\,d\mu$ is continuous on $\Phi(\mathfrak{M})$. If f lies in $\mathfrak{D}$, then this mapping is the restriction to $\Phi(\mathfrak{M})$ of the projection of $\mathbf{R}^{\mathfrak{D}}$ onto the fth coordinate axis, whence follows the desired continuity. For an arbitrary $f \in \mathfrak{C}^c$, as above, for $\epsilon > 0$ we determine an e_{n_0} and a $d \in \mathfrak{D}_0$ with $|f - de_{n_0}| \leqq \epsilon e_{n_0}$. Then

$$\left|\int f\,d\mu - \int de_{n_0}\,d\mu\right| \leqq \epsilon \int e_{n_0}\,d\mu,$$

where, according to what has just been shown, $\Phi(\mu) \to \int de_{n_0}\,d\mu$ and $\Phi(\mu) \to \int e_{n_0}\,d\mu$ are continuous on $\Phi(\mathfrak{M})$. Then, in particular, $\Phi(\mu) \to \int e_{n_0}\,d\mu$ is locally bounded on $\Phi(\mathfrak{M})$, that is, every point $\Phi(\mu_0) \in \Phi(\mathfrak{M})$ has an open neighborhood U_{μ_0} in $\Phi(\mathfrak{M})$ on which $\Phi(\mu) \to \int e_{n_0}\,d\mu$ is bounded. The above inequality then implies that $\Phi(\mu) \to \int f\,d\mu$ can be approximated uniformly on U_{μ_0} by continuous functions of the form $\Phi(\mu) \to \int de_{n_0}\,d\mu$ with $d \in \mathfrak{D}$. Therefore $\Phi(\mu) \to \int f\,d\mu$ is continuous on every U_{μ_0}, hence continuous on $\Phi(\mathfrak{M})$.

Now we need only recall the familiar theorem[22] according to which every countable product of metrizable spaces, in particular $\mathbf{R}^{\mathfrak{D}}$, is metrizable. Then the subspace $\Phi(\mathfrak{M})$ and the space $\mathfrak{M}$ homeomorphic to it are also metrizable. ┘

Remark 2. Similarly, but more simply, it can be shown that for every Polish space E, the weak topology on the set of finite Baire measures on E is metrizable. The details are left to the reader since we do not make explicit use of this result.

PROBLEMS

1. Let ν be a Baire measure on a locally compact space E countable at infinity. Prove: The set H of all $\mu \in M(E)$ satisfying $0 \leqq \int u\,d\mu \leqq \int u\,d\nu$ for all $u \in \mathfrak{C}^c_+(E)$ is vaguely compact.
2. Let E be a locally compact space with a countable base and let $(d_n)_{n \in \mathbf{N}}$ be a sequence in $\mathfrak{C}^c(E)$ such that the set $\mathfrak{D}$ of all d_n has the properties mentioned in Section 7.6, Problem 2 (b) and (c). Define

$$\rho(\mu,\nu) := \sum_{n=1}^{\infty} 2^{-n} \min\left(1, \left|\int d_n\,d\mu - \int d_n\,d\nu\right|\right)$$

for Borel measures μ, ν on E. Prove: ρ is a metric on $\mathfrak{M}(E)$. The corresponding topology is the vague topology on $\mathfrak{M}(E)$.

[22] See N. Bourbaki [26].

8

FOURIER ANALYSIS

Below we present the main features of the theory of Fourier transforms. This theory is one of the most powerful tools of analysis available in probability theory. It leads to elegant solutions of many of the problems involving convolution of measures. The importance of the convolution product for probability theory was introduced to the reader mainly in the discussions in Section 5.3. We restrict consideration to the case of measures on Euclidean space $\mathbf{R}^p$, although the theory can be fully developed on locally compact Abelian groups. (See Rudin [42] and Hewitt-Ross [36].)

8.1 FOURIER TRANSFORMS OF MEASURES AND FUNCTIONS

(a) Integration of complex-valued functions

The integration theory developed for real and numerical functions can be extended to complex-valued functions with a few comments. Let $\mathbf{C}$ denote the field of complex numbers $z = x + iy$ (with real part $x = \Re z$ and imaginary part $y = \Im z$). As a topological space, $\mathbf{C}$ is equal to $\mathbf{R}^2$, and thus is equipped with the σ-algebra $\mathfrak{B}^2$ of Borel sets. *Measurability* of complex functions will henceforth always mean measurability relative to $\mathfrak{B}^2$.

Now let $f\colon \Omega \to \mathbf{C}$ be a complex function on a measure space $(\Omega,\mathfrak{A},\mu)$ and $f = u + iv$ its decomposition into the real part $u = \Re f$ and the imaginary part $v = \Im f$. Since f is just the product mapping $u \otimes v$ of Ω into $\mathbf{R}^2 = \mathbf{C}$ (see Section 5.2), the $(\mathfrak{A}\text{-}\mathfrak{B}^2)$-measurability of f is equivalent to the measurability of the real functions u and v. We now say that f is

(μ-) *integrable* if u and v are μ-integrable. The complex number

$$\int f\,d\mu \equiv \int u\,d\mu + i\int v\,d\mu \tag{8.1.1}$$

is then called the *integral* of f with respect to μ.

Many important properties of the integral carry over at once. The complex μ-integrable functions form a vector space (over $\mathbf{C}$), which we denote by $\mathcal{L}^1(\mu,\mathbf{C})$. Then $f \to \int f\,d\mu$ is a linear mapping of $\mathcal{L}^1(\mu,\mathbf{C})$ into $\mathbf{C}$. The transition to the complex conjugate is accomplished by

$$\int \bar{f}\,d\mu = \overline{\int f\,d\mu}. \tag{8.1.2}$$

Here, $\bar{f}$ is always μ-integrable whenever f is.

The following theorem gives us information about the absolute value of functions $f \in \mathcal{L}^1(\mu,\mathbf{C})$.

8.1.1. Theorem. A measurable complex function f on Ω is μ-integrable if and only if $|f|$ is μ-integrable. Then

$$\left|\int f\,d\mu\right| \leqq \int |f|\,d\mu. \tag{8.1.3}$$

Proof. First, when f is measurable, $|f|$ is also measurable. We need only compose the mapping $\omega \to f(\omega)$ with $z \to |z|$ and observe that $z \to |z|$ is continuous and hence $\mathfrak{B}^2$-measurable. The first part of the assertion now follows by Theorem 2.4.2 from the inequalities

$$|f| \leqq |u| + |v|, \qquad |u| \leqq |f|, \qquad \text{and} \qquad |v| \leqq |f|$$

for the real and imaginary part of f.

Inequality (8.1.3) is obtained as follows for integrable f. For every $z \in \mathbf{C}$, $\mathcal{R}z \leqq |z|$ and hence

$$\mathcal{R}\left(\overline{\int f\,d\mu}\cdot f\right) \leqq \left|\int f\,d\mu\right|\,|f|.$$

By integrating, we obtain

$$\begin{aligned}\left|\int f\,d\mu\right|^2 &= \mathcal{R}\left(\overline{\int f\,d\mu}\cdot \int f\,d\mu\right) = \int \mathcal{R}\left(\overline{\int f\,d\mu}\cdot f\right)d\mu \\ &\leqq \left|\int f\,d\mu\right| \cdot \int |f|\,d\mu\end{aligned}$$

and thus the desired inequality. ┘

Now also the spaces $\mathcal{L}^p(\mu,\mathbf{C})$ for $1 \leqq p < \infty$ can be defined in analogy to the real case: $\mathcal{L}^p(\mu,\mathbf{C})$ is the set of all measurable complex functions f on Ω for which $|f|^p$ is μ-integrable. We define $\mathcal{L}^\infty(\mu,\mathbf{C})$ as in the real case. We likewise extend the definition of integrability to the case of measurable complex functions defined μ-almost everywhere on Ω. The reader should verify that the properties proved in the real case for these concepts carry over easily to the complex case.

If $(\Omega,\mathfrak{A},P)$ is a probability space and X is a complex random variable, that is, an $\mathfrak{A}$-$\mathfrak{B}^2$-measurable mapping $X: \Omega \to \mathbf{C} = \mathbf{R}^2$, then

$$E(X) \equiv \int X\, dP$$

is again called the *expected value* of X provided X is P-integrable.

For any two functions $f_1, f_2 \in \mathcal{L}^1(\lambda^p,\mathbf{C})$, where λ^p again denotes the L-B-measure on $\mathbf{R}^p$, we can now define the convolution product $f_1 * f_2$ by decomposing f_j into real and imaginary parts $f_j = u_j + iv_j$ $(j = 1, 2)$:

$$f_1 * f_2 \equiv u_1 * u_2 - v_1 * v_2 + i(u_1 * v_2 + u_2 * v_1).$$

Thus obviously, as in the real case, $f_1 * f_2$ is defined λ^p-almost everywhere on Ω and λ^p-integrable. The definition is formulated in such a way that the properties (3.4.15)–(3.4.17) and (3.4.18), for arbitrary $\alpha \in \mathbf{C}$, are preserved.

Finally, if E is a locally compact space, then we use $\mathcal{C}(E,\mathbf{C})$ $[\mathcal{C}^b(E,\mathbf{C})]$ to denote the vector space of all continuous [continuous, bounded] complex functions on E. By $\mathcal{C}^c(E,\mathbf{C})$ we denote the vector space of all $f \in \mathcal{C}(E,\mathbf{C})$ whose support S_f is compact. Definition 7.4.1 of the support carries over verbatim to complex functions. On $\mathcal{C}^b(E,\mathbf{C})$ we again use

$$\|f\| \equiv \sup_{x \in E} |f(x)|$$

to define the norm of uniform convergence. We then let $\mathcal{C}^0(E,\mathbf{C})$ correspond to the closure of $\mathcal{C}^c(E,\mathbf{C})$ in $\mathcal{C}^b(E,\mathbf{C})$ with respect to the metric of uniform convergence. The functions in $\mathcal{C}^0(E,\mathbf{C})$ are again said to vanish at infinity. The characterization of $\mathcal{C}^0(E)$ given in Theorem 7.4.8, including the proof, carries over to $\mathcal{C}^0(E,\mathbf{C})$. In particular, we thus have for $f \in \mathcal{C}(E,\mathbf{C})$

$$f \in \mathcal{C}^0(E,\mathbf{C}) \Leftrightarrow |f| \in \mathcal{C}^0(E).$$

(b) Definition and elementary properties of Fourier transforms

The following considerations involve the set $\mathfrak{M}^e = \mathfrak{M}^e(\mathbf{R}^p)$ of finite Borel measures on the Euclidean space $\mathbf{R}^p$, $p = 1, 2, \ldots$; here $\mathbf{R}^p$ is equipped with the usual Euclidean *scalar product*

$$<x,y> \equiv \sum_{j=1}^{p} x_j y_j \tag{8.1.4}$$

and the Euclidean *norm*

$$|x| \equiv \sqrt{<x,x>}. \tag{8.1.5}$$

Here x, y denote points in $\mathbf{R}^p$ with coordinates $x = (x_1, \ldots, x_p)$ and $y = (y_1, \ldots, y_p)$.

For every measure $\mu \in \mathfrak{M}^e$ we call

$$\|\mu\| \equiv \int d\mu = \mu(\mathbf{R}^p) \tag{8.1.6}$$

the *total mass* of μ. Since $\|\mu\| < +\infty$, every bounded, continuous, complex function f on $\mathbf{R}^p$ is μ-integrable. In particular $y \to e^{i<x,y>}$ is such a function for every $x \in \mathbf{R}^p$, since $|e^{it}| = 1$ for all $t \in \mathbf{R}$.

8.1.2. Definition. The *Fourier transform* of a measure $\mu \in \mathfrak{M}^e(\mathbf{R}^p)$ is the complex function $\hat{\mu}$ on $\mathbf{R}^p$ defined by

$$\hat{\mu}(x) \equiv \int e^{i<x,y>}\mu(dy) \qquad (x \in \mathbf{R}^p). \tag{8.1.7}$$

In probability theory we are interested in Fourier transforms of distributions P_X of random variables X with values in $\mathbf{R}^p$; these are defined since P_X lies in $\mathfrak{M}^1(\mathbf{R}^p)$.

8.1.3. Definition. Let X be an $(\mathbf{R}^p,\mathfrak{B}^p)$-random variable on a probability space $(\Omega,\mathfrak{A},P)$. Then $\hat{P}_X$ is called the *characteristic function* of X. We also denote it by φ_X.

According to the transformation formula (4.3.6),

$$\hat{P}_X(x) = E(e^{i<x,X>}) \qquad (x \in \mathbf{R}^p). \tag{8.1.8}$$

The term "characteristic function" will be justified by the Uniqueness Theorem 8.2.4, according to which the distribution P_X is uniquely determined by the characteristic function $\hat{P}_X$. Moreover, (8.1.7) and (8.1.8) define the same mathematical object, since every measure $\mu \in \mathfrak{M}^1(\mathbf{R}^p)$ is the distribution of a random variable with values in $\mathbf{R}^p$, for example, the identity mapping of the probability space $(\mathbf{R}^p,\mathfrak{B}^p,\mu)$ onto itself.

Examples

1. For the measure $\epsilon_a \in \mathfrak{M}^1(\mathbf{R}^p)$ defined by the unit mass at $a \in \mathbf{R}^p$,

$$\hat{\epsilon}_a(x) = e^{i<x,a>} \qquad (x \in \mathbf{R}^p).$$

This is the characteristic function of a (degenerate) random variable which is almost surely equal to a. In particular, $\hat{\epsilon}_0 = 1$.

2. For every *discrete distribution* $\mu = \Sigma_{n=1}^{\infty} \alpha_n\epsilon_{a_n} \in \mathfrak{M}^1(\mathbf{R}^p)$ (see Section 4.4, Example 2), we thus have

$$\hat{\mu}(x) = \sum_{n=1}^{\infty} \alpha_n e^{i<x,a_n>} \qquad (x \in \mathbf{R}^p).$$

Special cases of this are:

(a) The *binomial distribution* $\beta_n^p = \Sigma_{\nu=0}^n \binom{n}{\nu} p^\nu q^{n-\nu}\epsilon_\nu$ on $\mathbf{R}$ $(0 < p < 1,$

$q = 1 - p$, $n = 1, 2, \ldots$) has Fourier transform

$$\hat{\beta}_n^p(x) = \sum_{\nu=0}^{n} \binom{n}{\nu} p^\nu q^{n-\nu} e^{ix\nu} = (q + pe^{ix})^n \qquad (x \in \mathbf{R}).$$

(b) The *Poisson distribution* $\pi_\alpha = \sum_{n=0}^{\infty} e^{-\alpha} \frac{\alpha^n}{n!} \epsilon_n$ with parameter $\alpha > 0$ has Fourier transform

$$\hat{\pi}_\alpha(x) = e^{-\alpha} \sum_{n=0}^{\infty} \frac{\alpha^n}{n!} e^{ixn} = e^{\alpha(e^{ix}-1)} \qquad (x \in \mathbf{R}).$$

It is not accidental that the Fourier transforms computed so far are all continuous. Indeed, we have:

8.1.4. Theorem. For the Fourier transform $\hat{\mu}$ of every measure $\mu \in \mathfrak{M}^e(\mathbf{R}^p)$, the following properties hold:

(a) $\hat{\mu}$ is uniformly continuous on $\mathbf{R}^p$.

(b) $|\hat{\mu}(x)| \leqq \|\mu\| = \hat{\mu}(0)$ for all $x \in \mathbf{R}^p$.

(c) $\hat{\mu}$ is *positive-definite*, that is, for arbitrary $n \in \mathbf{N}$, points $x_1, \ldots,$ $x_n \in \mathbf{R}^p$, and complex numbers $\lambda_1, \ldots, \lambda_n$,

$$\sum_{s,t=1}^{n} \lambda_s \overline{\lambda_t} \hat{\mu}(x_s - x_t) \geqq 0. \tag{8.1.9}$$

Proof. (a) Since μ is regular, for every $\epsilon > 0$ there exists a compact set $K \subset \mathbf{R}^p$ such that $\mu(\complement K) < \epsilon$. Then

$$\alpha \equiv \sup \{|y| : y \in K\} < +\infty$$

and hence, by the Cauchy-Schwarz inequality,

$$|\langle x_2 - x_1, y\rangle| \leqq |x_2 - x_1|\,|y| \leqq \alpha |x_2 - x_1|$$

for all $y \in K$ and arbitrary $x_1, x_2 \in \mathbf{R}^p$. Since

$$|e^{i\langle x_1,y\rangle} - e^{i\langle x_2,y\rangle}| = |1 - e^{i\langle x_2 - x_1,y\rangle}|,$$

there is a $\delta > 0$ such that

$$|e^{i\langle x_1,y\rangle} - e^{\langle x_2,y\rangle}| \leqq \epsilon$$

for arbitrary $x_1, x_2 \in \mathbf{R}^p$ with $|x_1 - x_2| \leqq \delta$ and for all $y \in K$. But then the following inequalities, that are valid for all such pairs $x_1, x_2 \in \mathbf{R}^p$, show the uniform continuity of $\hat{\mu}$:

$$\begin{aligned} |\hat{\mu}(x_1) - \hat{\mu}(x_2)| &\leqq \int |e^{i<x_1,y>} - e^{i<x_2,y>}|\mu(dy) \\ &= \int_K |e^{i<x_1,y>} - e^{i<x_2,y>}|\mu(dy) + \int_{\complement K} |e^{i<x_1,y>} - e^{i<x_2,y>}|\mu(dy) \\ &\leqq \epsilon\mu(K) + 2\mu(\complement K) \leqq \epsilon(\|\mu\| + 2). \end{aligned}$$

(b) This follows from (8.1.3).

(c) On the left side of inequality (8.1.9) we have a μ-integral with integrand

$$\begin{aligned} f(y) &\equiv \sum_{s,t=1}^{n} \lambda_s \overline{\lambda_t} e^{i<x_s - x_t, y>} \\ &= \sum_{s=1}^{n} \lambda_s e^{i<x_s,y>} \cdot \overline{\sum_{t=1}^{n} \lambda_t e^{i<x_t,y>}} \\ &= \left| \sum_{s=1}^{n} \lambda_s e^{i<x_s,y>} \right|^2. \end{aligned}$$

Hence $f \geqq 0$ and thus $\int f \, d\mu \geqq 0$. ┘

Remark. A well-known theorem of S. Bochner [25] tells us that, conversely, every continuous positive-definite function $p: \mathbf{R}^p \to \mathbf{C}$ is the Fourier transform of a measure $\mu \in \mathfrak{M}^e(\mathbf{R}^p)$. However, we do not make use of this characterization of Fourier transforms.

Now we study the behavior of the mapping $\mu \to \hat{\mu}$ from $\mathfrak{M}^e(\mathbf{R}^p)$ into $\mathcal{C}^b(\mathbf{R}^p, \mathbf{C})$ relative to several operations in $\mathfrak{M}^e(\mathbf{R}^p)$, in particular to the convolution $*$.

8.1.5. Theorem. For arbitrary finite Borel measures μ and ν on $\mathbf{R}^p$.

(a) $$\widehat{\mu + \nu} = \hat{\mu} + \hat{\nu}$$

(b) $$\widehat{\alpha\mu} = \alpha\hat{\mu} \qquad (\alpha \in \mathbf{R}_+).$$

(c) $$\widehat{\mu * \nu} = \hat{\mu} \cdot \hat{\nu}.$$

(d) For every linear mapping T of the vector space $\mathbf{R}^p$ into itself and its transpose mapping T^t,

$$\widehat{T(\mu)} = \hat{\mu} \circ T^t.[1] \qquad (8.1.10)$$

[1] If we represent T (by fixing a coordinate system) by a matrix, then, as is familiar, T^t is described by the transpose matrix.

(e) For the reflection through the origin $x \to S(x) \equiv -x$,

$$\widehat{S(\mu)} = \bar{\hat{\mu}} = \hat{\mu} \circ S. \tag{8.1.11}$$

In particular, along with $\hat{\mu}$, the conjugate function $\bar{\hat{\mu}}$ is also the Fourier transform of a measure from $\mathfrak{M}^e(\mathbf{R}^p)$.
(f) For every translation $T_a(x) \equiv x + a$ in $\mathbf{R}^p(a \in \mathbf{R}^p)$,

$$\widehat{T_a(\mu)} = \hat{\epsilon}_a \cdot \hat{\mu}. \tag{8.1.12}$$

Proof. Equality (a) follows from the observation that for every bounded, measurable, complex function f on $\mathbf{R}^p$: $\int f\, d(\mu + \nu) = \int f\, d\mu + \int f d\nu$. We obtain (b) similarly.

(c) By (3.4.5) we have $\int f\, d(\mu * \nu) = \int\int f(y + z)\mu(dy)\nu(dz)$ for every $\mu * \nu$-integrable, real and hence also for every $\mu * \nu$-integrable, complex function f on $\mathbf{R}^p$. If we choose $f(y) = e^{i\langle x,y\rangle}$ for given $x \in \mathbf{R}^p$, then (c) follows, since $e^{i\langle x,y+z\rangle} = e^{i\langle x,y\rangle}e^{i\langle x,z\rangle}$.

(d) First of all T is continuous, and is thus a measurable mapping; therefore, along with μ, $T(\mu)$ also lies in $\mathfrak{M}^e$. By the Transformation Theorem we have $\int f\, dT(\mu) = \int f \circ T\, d\mu$ for every $T(\mu)$-integrable f, and in particular for $f(y) = e^{i\langle x,y\rangle}(x \in \mathbf{R}^p)$. If we now take note of the equality $\langle x,T(y)\rangle = \langle T^t(x),y\rangle$, then we have the assertion:

$$\widehat{T(\mu)}(x) = \int e^{i\langle x,T(y)\rangle}\mu(dy) = \int e^{i\langle T^t(x),y\rangle}\mu(dy) = \hat{\mu}(T^t(x)).$$

(e) Obviously, $\overline{\hat{\mu}(x)} = \hat{\mu}(-x) = \hat{\mu}(S(x))$. The rest follows from (8.1.10) because $S^t = S$.

(f) By (3.4.10), $T_a(\mu) = \epsilon_a * \mu$. Thus (8.1.12) follows from (c). ┘

For product measures we have:

8.1.6. Theorem. For the product $\mu \otimes \nu$ of measures $\mu \in \mathfrak{M}^e(\mathbf{R}^p)$ and $\nu \in \mathfrak{M}^e(\mathbf{R}^q)$, the Fourier transform is given by

$$\widehat{\mu \otimes \nu}(x,y) = \hat{\mu}(x)\hat{\nu}(y) \qquad ((x,y) \in \mathbf{R}^{p+q}). \tag{8.1.13}$$

Proof. The scalar product of the $(p + q)$-dimensional vectors (x,y) and (z,z') from $\mathbf{R}^p \times \mathbf{R}^q = \mathbf{R}^{p+q}$ is given by $\langle x,z\rangle + \langle y,z'\rangle$, that is, by the sum of the corresponding p- and q-dimensional scalar products. Hence $\widehat{\mu \otimes \nu}(x,y) = \int\int e^{i\langle x,z\rangle}e^{i\langle y,z'\rangle}\mu(dz)\nu(dz') = \hat{\mu}(x)\hat{\nu}(y)$. ┘

In particular, we now have available the following *properties of characteristic functions* of random variables.

1. The characteristic function of the sum of independent $(\mathbf{R}^p,\mathfrak{B}^p)$-random variables $X_1, \ldots, X_n$ is given by

$$\varphi_{X_1+\cdots+X_n} = \varphi_{X_1} \cdot \ldots \cdot \varphi_{X_n}. \tag{8.1.14}$$

2. For every linear mapping T of the vector space $\mathbf{R}^p$ into itself, every $a \in \mathbf{R}^p$, and every $(\mathbf{R}^p,\mathfrak{B}^p)$-random variable X,

$$\varphi_{a+T\circ X}(x) = e^{i<x,a>}\varphi_X(T^t(x)) \qquad (x \in \mathbf{R}^p). \tag{8.1.15}$$

In particular,

$$\varphi_{-X}(x) = \varphi_X(-x) = \overline{\varphi_X(x)} \qquad (x \in \mathbf{R}^p). \tag{8.1.16}$$

Here 1 follows from Theorem 8.1.5 (c) since, by Theorem 5.3.4, $P_{X_1} * \cdots * P_{X_n}$ is the distribution of $X_1 + \cdots + X_n$. (8.1.15) and (8.1.16) follow from (8.1.10)–(8.1.12). We need note only the transitivity of image measures: If T_a denotes the translation $x \to x + a$, then $T_a \circ T \circ X(P) = T_a(T(P_X)) = \epsilon_a * T(P_X)$ is the distribution of the random variables $a + T \circ X$. By Example 1, however, $\hat{\epsilon}_a(x) = e^{i<x,a>}$.

In particular, all measures $\mu = f\lambda^p$ lie in $\mathfrak{M}^e(\mathbf{R}^p)$, where $f \geqq 0$ is an integrable function relative to the L-B-measure λ^p on $\mathbf{R}^p$. By Theorem 2.9.3, the Fourier transform of μ is given by $\hat{\mu}(x) = \int e^{i<x,y>}f(y)\lambda^p(dy)$. We shall soon see that it is useful to study this integral as a function of x also for complex λ^p-integrable functions. Therefore we define:

8.1.7. Definition. For every λ^p-integrable function $f\colon \mathbf{R}^p \to \mathbf{C}$, the function $\hat{f}\colon \mathbf{R}^p \to \mathbf{C}$ defined by

$$\hat{f}(x) \equiv \int e^{i<x,y>}f(y)\lambda^p(dy) \tag{8.1.17}$$

is called the *Fourier transform* of f.

Thus we have

$$\hat{f} = \widehat{u^+\lambda^p} - \widehat{u^-\lambda^p} + i(\widehat{v^+\lambda^p} - \widehat{v^-\lambda^p}), \tag{8.1.18}$$

where $f = u + iv$ is the decomposition of f into real and imaginary parts and $u = u^+ - u^-$, $v = v^+ - v^-$ are the decompositions of the latter into positive and negative parts. From this observation we obtain many properties of the mapping $f \to \hat{f}$ from the corresponding properties of the mapping $\mu \to \hat{\mu}$.

8.1.8. Theorem

1. For any two functions $f, g \in \mathfrak{L}^1(\lambda^p,\mathbf{C})$:

(a) $\hat{f}$ is uniformly continuous.

(b) $f \to \hat{f}$ is a linear mapping (over $\mathbf{C}$).

(c) $\widehat{f * g} = \hat{f} \cdot \hat{g}$.

(d) $\widehat{\bar{f}} = \bar{f}^*$, where f^* denotes the function $f^*(x) = \overline{f(-x)}$.

2. For every pair of functions $f \in \mathfrak{L}^1(\lambda^p,\mathbf{C})$ and $g \in \mathfrak{L}^1(\lambda^q,\mathbf{C})$, $f \otimes g$ lies in $\mathfrak{L}^1(\lambda^{p+q},\mathbf{C})$ and we have

$$\widehat{f \otimes g}(x,y) = \hat{f}(x)\hat{g}(y) \qquad ((x,y) \in \mathbf{R}^{p+q}). \tag{8.1.19}$$

Proof. For 1: (a)–(c) are obtained directly from the above observation and Theorems 8.1.4 and 8.1.5. Property (d) follows from the definition of $\hat{f}$, when we take into account the reflection invariance of λ^p proved in Section 1.7, Example 4.

For 2: Since $\mathfrak{B}^p \otimes \mathfrak{B}^q = \mathfrak{B}^{p+q}$, the function $f \otimes g$ is $\mathfrak{B}^{p+q}$-measurable. By Fubini's Theorem,

$$\int |f \otimes g| \, d\lambda^{p+q} = \int |f| \, d\lambda^p \cdot \int |g| \, d\lambda^q < +\infty,$$

whence follows the λ^{p+q}-integrability of $f \otimes g$. The rest of the assertion follows either by application of Fubini's Theorem to the integral defining $\widehat{f \otimes g}$ or from (8.1.13) via the above observation. ┘

Examples

3. The Fourier transform of the *normal distribution* $\nu_{\alpha,\sigma^2} = g_{\alpha,\sigma^2}\lambda^1$ and thus also of its density g_{α,σ^2} is given by

$$\hat{\nu}_{\alpha,\sigma^2}(x) = \hat{g}_{\alpha,\sigma^2}(x) = e^{i\alpha x - \sigma^2 x^2/2} \qquad (\alpha,x \in \mathbf{R}, \sigma > 0).$$

It suffices to consider the case $\alpha = 0$, $\sigma = 1$ of the standard normal distribution $\nu_{0,1}$, since $\nu_{\alpha,\sigma^2} = T(\nu_{0,1})$ is the image of $\nu_{0,1}$ under the mapping $T(x) = \alpha + \sigma x$ and since we have available the transformation properties of Theorem 8.1.5 (d) and (f). Thus we have to compute the integral

$$\hat{\nu}_{0,1}(x) = \frac{1}{\sqrt{2\pi}} \int_{-\infty}^{+\infty} e^{ixy} e^{-y^2/2} \, dy.$$

Using function theory, we can do this as follows:

If we set $z = y - ix$, then $2ixy - y^2 = -x^2 - z^2$ and thus

$$\hat{\nu}_{0,1}(x) = \frac{1}{\sqrt{2\pi}} e^{-x^2/2} \int_G e^{-z^2/2} \, dz,$$

where we have to integrate along the line $G = \{y - ix\colon y \in \mathbf{R}\}$. But

$$\int_G e^{-z^2/2}\,dz = \int_{-\infty}^{+\infty} e^{-x^2/2}\,dx = \sqrt{2\pi}$$

and hence

$$\hat{\nu}_{0,1}(x) = e^{-x^2/2}.$$

Indeed, since $z \to e^{-z^2/2}$ is an entire holomorphic function, by integrating along the oriented boundary of the rectangle with vertices $y, -y, -y - ix$, $y - ix$, we obtain the equality

$$\int_{-y}^{y} e^{-z^2/2}\,dz = \int_{-y-ix}^{y-ix} e^{-z^2/2}\,dz + \int_{-y}^{-y-ix} e^{-z^2/2}\,dz + \int_{y-ix}^{y} e^{-z^2/2}\,dz.$$

The second and third integral on the right side have absolute values that are bounded by a multiple of $e^{-y^2/2}$ due to the constant length $|x|$ of the integration path, and thus approaches zero as $y \to +\infty$. The asserted equality then follows from the last equality by passing to the limit $y \to +\infty$.

4. For the *Cauchy distribution* $\gamma_\alpha = c_\alpha\lambda^1$ $(\alpha > 0)$, we have

$$\hat{\gamma}_\alpha(x) = \hat{c}_\alpha(x) = e^{-\alpha|x|}.$$

It suffices to consider the case $\alpha = 1$, since $x \to \alpha x$ maps the measure γ_1 onto γ_α. Thus we have to compute the integral

$$\hat{\gamma}_1(x) = \frac{1}{\pi}\int_{-\infty}^{+\infty} \frac{e^{ixy}}{1 + y^2}\,dy.$$

Since $\hat{\gamma}_1(-x) = \hat{\gamma}_1(x)$ and $\hat{\gamma}_1(0) = 1$, we can assume $x > 0$. Again we use function theory. The meromorphic function

$$z \to \frac{e^{ixz}}{1 + z^2} = \frac{e^{ixz}}{(z - i)(z + i)}$$

has a pole (of first order) in $z = i$ with residue $(1/2i)e^{-x}$. Then by the Residue Theorem

$$\int_{-r}^{+r} \frac{e^{ixy}}{1 + y^2}\,dy + \int_{H_r} \frac{e^{ixz}}{1 + z^2}\,dz = \pi e^{-x},$$

if $r > 1$ and H_r denotes the semicircle arc oriented from $+r$ to $-r$ which is intersection of the half-plane $\Re z > 0$ and the circle of radius r and center 0. Then obviously, for all $z \in H_r$,

$$\left|\frac{e^{ixz}}{1 + z^2}\right| \leqq \frac{1}{r^2 - 1}.$$

Therefore, as $r \to +\infty$, the integral over H_r approaches zero, and we obtain $\hat{\gamma}_1(x) = e^{-x}$.

PROBLEMS

1. Let $f\colon \Omega \to \mathbf{R}^p$ be a mapping defined on a measure space $(\Omega,\mathfrak{A},\mu)$. Denote by $f_1, \ldots, f_p$ the p real components of f. The mapping f is called μ-integrable if all components $f_1, \ldots, f_p$ are μ-integrable; the p-dimensional vector $(\int f_1\,d\mu, \ldots, \int f_p\,d\mu)$ is then called the μ-integral of f and denoted by $\int f\,d\mu$. Prove that Theorem 8.1.1 generalizes to such vector-valued functions f if $|f|$ denotes the mapping $\omega \to |f(\omega)|$ where $|x|$ is the Euclidean norm of a vector $x \in \mathbf{R}^p$.
2. Let X be a real random variable satisfying $P\{X = 1\} = \cdots = P\{X = n\} = 1/n$. Prove that
$$x \to \frac{e^{ix}}{n} \cdot \frac{e^{inx} - 1}{e^{ix} - 1}$$
is the characteristic function of X.
3. Prove that
$$(x_1, \ldots, x_k) \to (p_1e^{ix_1} + \cdots + p_ke^{ix_k})^n$$
is the Fourier transform of the multinomial distribution of Section 4.4, Problem 4.
4. Prove that
$$x \to \frac{\sin \alpha x}{\alpha x}$$
is the Fourier transform of the so-called *rectangular* (or uniform) distribution ρ_α of the interval $[-\alpha,\alpha]$:
$$\rho_\alpha \equiv \frac{1}{2\alpha} 1_{[-\alpha,\alpha]}\lambda^1,$$
where $\alpha > 0$.
5. Prove that the following equality is valid for all measures $\mu, \nu \in \mathfrak{M}^e(\mathbf{R}^p)$ and all points $x \in \mathbf{R}^p$:
$$\int e^{-i<x,y>}\hat{\mu}(y)\nu(dy) = \int \hat{\nu}(y - x)\mu(dy).$$
6. Prove that formula (8.1.10) also holds for linear mappings $T\colon \mathbf{R}^p \to \mathbf{R}^q$. Deduce from this (and prove also directly) that
$$\widehat{\mu \otimes \nu}(x,0) = \hat{\mu}(x) \qquad (x \in \mathbf{R}^p)$$
holds for all measures $\mu \in \mathfrak{M}^1(\mathbf{R}^p)$ and $\nu \in \mathfrak{M}^1(\mathbf{R}^q)$.
7. Let X (resp. Y) be a $(\mathbf{R}^p,\mathfrak{B}^p)$-(resp. $(\mathbf{R}^q,\mathfrak{B}^q)$-) random variable on a given probability space. Prove: X and Y are independent if and only if their characteristic functions φ_X, φ_Y and the characteristic function

$\varphi_{X\otimes Y}$ of their joint distribution satisfy

$$\varphi_{X\otimes Y}(x,y) = \varphi_X(x)\varphi_Y(y)$$

for all $x \in \mathbf{R}^p$ and all $y \in \mathbf{R}^q$.

8. Let X and Y be normally distributed real random variables on a probability space. Use Problem 7 in order to prove that $X + Y$ and $X - Y$ are independent random variables.

8.2 UNIQUENESS AND CONTINUITY THEOREMS

The next theorem, due to Riemann-Lebesgue, whose proof is preceded by a lemma, paves the way to all deeper properties of Fourier transforms, and in particular to the two main theorems, namely, to the Uniqueness and Continuity Theorem.

8.2.1. Lemma. For every $f \in \mathfrak{L}^1(\lambda^p,\mathbf{C})$, the function $\Phi: \mathbf{R}^p \to \mathbf{R}_+$ defined by

$$\Phi(t) \equiv \int |f(x + t) - f(x)|\lambda^p(dx)$$

is continuous at the point $t = 0$.

Proof. Since λ^p is translation-invariant, the integrability of f implies that of $x \to f(x + t)$ and hence the existence of the integral $\Phi(t)$. We prove the continuity of Φ at $t = 0$ in two steps.

STEP 1: Let $f \in \mathcal{C}^c(\mathbf{R}^p,\mathbf{C})$ and let S_f be the compact support of f. Since f is uniformly continuous, for every $\epsilon > 0$ there is a $\delta > 0$ such that $|f(x_1) - f(x_2)| < \epsilon$ for all $x_1, x_2 \in \mathbf{R}^p$ with $|x_1 - x_2| < \delta$. Further, if U is an open, relatively compact neighborhood of S_f (say a sufficiently large open sphere), then we can take δ so small that with every point $x \in S_f$, the open ball of center x and radius δ is contained in U. Therefore, for every $t \in \mathbf{R}^p$ with $|t| < \delta$, we have, on the one hand, $|f(x + t) - f(x)| < \epsilon$ for all $x \in \mathbf{R}^p$; on the other hand, the support of $x \to f(x + t)$ is equal to $S_f - t$ and is therefore contained in U. We thus obtain

$$\Phi(t) = \int_U |f(x + t) - f(x)|\lambda^p(dx) \leqq \epsilon\lambda^p(U)$$

for all t satisfying $|t| < \delta$, that is, since $\Phi(0) = 0$ we have the desired continuity at the origin.

STEP 2: Now let $f \in \mathfrak{L}^1(\lambda^p,\mathbf{C})$ be arbitrary. By applying Corollary 7.5.5 to the real and imaginary part of f, for each $\epsilon > 0$ we obtain the existence of a $g \in \mathcal{C}^c(\mathbf{R}^p,\mathbf{C})$ such that $\int |f - g|\, d\lambda^p \leqq \epsilon$. Since λ^p is translation-invariant, we also have

$$\int |f(x+t) - g(x+t)|\lambda^p(dx) = \int |f-g|\,d\lambda^p \leqq \epsilon$$

for all $t \in \mathbf{R}^p$. Using the triangle inequality, we obtain the majorization

$$\begin{aligned}\Phi(t) \leqq \int |f(x+t) - g(x+t)|\lambda^p(dx) + \int |g(x+t) - g(x)|\lambda^p(dx)\\ + \int |g-f|\,d\lambda^p \leqq 2\epsilon + \int |g(x+t) - g(x)|\lambda^p(dx).\end{aligned}$$

According to Step 1, the remaining integral approaches zero as $t \to 0$. Hence it follows that $\lim_{t\to 0} \Phi(t) = \Phi(0) = 0.$ ┘

It is now a simple matter to obtain the next theorem, which represents a sharpening of the uniform continuity of Fourier transforms $\hat{f}$ of *functions*, which we have already shown.

8.2.2. Theorem (Riemann-Lebesgue). The Fourier transform $\hat{f}$ of a function $f \in \mathcal{L}^1(\lambda^p,\mathbf{C})$ vanishes at infinity.

Proof. Since $e^{-i\pi} = -1$ and because of the translation-invariance of λ^p we have for every $x \neq 0$ in $\mathbf{R}^p$:

$$\begin{aligned}\hat{f}(x) &= 2\cdot\frac{1}{2}\int e^{i<x,y>}f(y)\lambda^p(dy)\\ &= \frac{1}{2}\int e^{i<x,y>}f(y)\lambda^p(dy) - \frac{1}{2}\int e^{i<x,y-\frac{\pi}{|x|^2}x>}f(y)\lambda^p(dy)\\ &= \frac{1}{2}\int e^{i<x,y>}f(y)\lambda^p(dy) - \frac{1}{2}\int e^{i<x,y>}f\left(y+\frac{\pi}{|x|^2}x\right)\lambda^p(dy)\\ &= \frac{1}{2}\int e^{i<x,y>}\left[f(y) - f\left(y+\frac{\pi}{|x|^2}x\right)\right]\lambda^p(dy).\end{aligned}$$

Hence,

$$|\hat{f}(x)| \leqq \frac{1}{2}\int \left| f\left(y+\frac{\pi}{|x|^2}x\right) - f(y)\right| \lambda^p(dy),$$

and the assertion follows from the above lemma, since $|(\pi/|x|^2)x| = \pi/|x|$ approaches zero as $|x| \to +\infty$. ┘

Remark. We have $\hat{\epsilon}_0 \equiv 1$. Thus the Fourier transform of a measure $\mu \in \mathfrak{M}^e(\mathbf{R}^p)$ does not vanish at infinity in general.

8.2.3. Corollary. The set $\mathfrak{F}$ of all Fourier transforms $\hat{f}$ of functions $f \in \mathcal{L}^1(\lambda^p,\mathbf{C})$ is dense in $\mathcal{C}^0(\mathbf{R}^p,\mathbf{C})$ relative to uniform convergence.

Proof. By Theorems 8.1.8 and 8.2.2 $\mathfrak{F}$ is a linear subspace of $\mathcal{C}^0(\lambda^p,\mathbf{C})$ which, along with any two functions $\hat{f}$, $\hat{g}$, contains the product $\hat{f}\cdot\hat{g}$ and, along with every $\hat{f}$, contains the conjugate function $\bar{\hat{f}}$. Thus $\mathfrak{F}$ is a self-adjoint subalgebra of $\mathcal{C}^0(\mathbf{R}^p,\mathbf{C})$. By the Stone-Weierstrass[2] Theorem we need to show only the following properties (a) and (b):

(a) For every $x_0 \in \mathbf{R}^p$ there exists an $\hat{f} \in \mathfrak{F}$ with $\hat{f}(x_0) \neq 0$. Indeed, let V be a compact neighborhood of x_0 and let f be the function

$$x \to 1_V(x)e^{-i<x_0,x>}$$

lying in $\mathcal{L}^1(\lambda^p,\mathbf{C})$. Then $\hat{f}(x_0) = \lambda^p(V) > 0$ since V contains, for example, a nonempty half-open interval.

(b) For every pair of points $x_0 \neq y_0$ from $\mathbf{R}^p$ there exists an $\hat{f} \in \mathfrak{F}$ such that $\hat{f}(x_0) \neq \hat{f}(y_0)$. Indeed, choose $\alpha \in \mathbf{R}$ such that $e^{i<x_0-y_0,z_0>} \neq 1$ for $z_0 \equiv \alpha(x_0 - y_0)$. Then there is a compact neighborhood W of z_0 with $e^{i<x_0,y>} \neq e^{i<y_0,y>}$ for all $y \in W$. The function

$$f(x) \equiv 1_W(x)(e^{-i<x_0,x>} - e^{-i<y_0,x>})$$

lying in $\mathcal{L}^1(\lambda^p,\mathbf{C})$, has a Fourier transform for which

$$\hat{f}(x_0) - \hat{f}(y_0) = \int_W |e^{i<x_0,y>} - e^{i<y_0,y>}|^2\lambda^p(dy) > 0. \quad \lrcorner$$

The main theorem of the theory is now easy to prove:

8.2.4. Theorem (Uniqueness Theorem). The mapping $\mu \to \hat{\mu}$ of $\mathfrak{M}^e(\mathbf{R}^p)$ into $\mathcal{C}^b(\mathbf{R}^p,\mathbf{C})$ is injective. Thus every measure $\mu \in \mathfrak{M}^e(\mathbf{R}^p)$ is uniquely determined by its Fourier transform $\hat{\mu}$.

Proof. We have to prove the equality of any two measures μ, $\nu \in \mathfrak{M}^e$ for which $\hat{\mu} = \hat{\nu}$. Let f be arbitrarily chosen in $\mathcal{L}^1(\lambda^p,\mathbf{C})$. Then, by Fubini's Theorem,

$$\int \hat{f}\, d\mu = \int\int e^{i<x,y>}f(y)\lambda^p(dy)\mu(dx) = \int\int e^{i<x,y>}f(y)\mu(dx)\lambda^p(dy) = \int \hat{\mu} f\, d\lambda^p$$

and correspondingly,

$$\int \hat{f}\, d\nu = \int \hat{\nu} f\, d\lambda^p.$$

Thus we obtain

$$\int \hat{f}\, d\mu = \int \hat{f}\, d\nu, \qquad \text{for all } \hat{f} \in \mathfrak{F}.$$

[2] It suffices to note that $\mathfrak{F}$ contains the real part $\mathfrak{R}\hat{f} = \frac{1}{2}(\hat{f} + \bar{\hat{f}})$ and the imaginary part $\mathfrak{I}\hat{f} = (1/2i)(\hat{f} - \bar{\hat{f}})$ of every function $\hat{f} \in \mathfrak{F}$ as elements. Application of the Stone Weierstrass Theorem to the set of *real* functions in $\mathfrak{F}$ then yields the assertion (see footnote 15 in Chapter 7).

Since μ and ν are finite measures, it follows by Corollary 8.2.3 that

$$\int h\,d\mu = \int h\,d\nu, \qquad \text{for all } h \in \mathcal{C}^0(\mathbf{R}^p,\mathbf{C}).$$

We need note only that for any two functions p, $q \in \mathcal{C}^b(\mathbf{R}^p,\mathbf{C})$, the inequalities

$$\left|\int p\,d\mu - \int q\,d\mu\right| \leqq \int |p - q|\,d\mu \leqq \|p - q\| \cdot \|\mu\|$$

and the corresponding ones for ν hold. Since $\mathcal{C}^c(\mathbf{R}^p) \subset \mathcal{C}^c(\mathbf{R}^p,\mathbf{C}) \subset \mathcal{C}^0(\mathbf{R}^p,\mathbf{C})$, it follows that $\int h\,d\mu = \int h\,d\nu$ for all $h \in \mathcal{C}^c(\mathbf{R}^p)$. But then $\mu = \nu$ by Theorem 7.5.4. ⌟

Thus every problem on finite Borel measures on $\mathbf{R}^p$ can be transformed into an equivalent problem with regard to their Fourier transforms. Herein, and in the property of transforming the convolution product into the ordinary product, lies the significance of Fourier transforms. At the same time, this justifies the terminology "characteristic function" for the Fourier transform of the distribution of a random variable. *The distribution of an $(\mathbf{R}^p,\mathfrak{B}^p)$-random variable X is uniquely determined by the characteristic function φ_X.* The examples below are the first illustrations of these statements.

However, we first prove the analog of Theorem 8.2.4 for functions.

8.2.5. Corollary. For any two functions f, $g \in \mathcal{L}^1(\lambda^p,\mathbf{C})$:

$$\hat{f} = \hat{g} \Rightarrow f = g \qquad \lambda^p\text{-almost everywhere.}$$

Proof. First, let f and g be real-valued. Then by (8.1.18) the measures $(f^+ + g^-)\lambda^p$ and $(g^+ + f^-)\lambda^p$ have the same Fourier transform; thus, $(f^+ + g^-)\lambda^p = (g^+ + f^-)\lambda^p$. Then, by Theorem 2.9.4, the functions $f^+ + g^-$ and $g^+ + f^-$, and hence also $f = f^+ - f^-$ and $g = g^+ - g^-$ are λ^p-almost everywhere equal. The general case follows from the remark that $\hat{f} = \hat{g}$ implies the equalities $\widehat{\mathcal{R}f} = \widehat{\mathcal{R}g}$ and $\widehat{\mathcal{I}f} = \widehat{\mathcal{I}g}$. Indeed,

$$\widehat{\mathcal{R}f} = \frac{1}{2}(\hat{f} + \hat{\bar{f}}) \qquad \text{and} \qquad \widehat{\mathcal{I}f} = \frac{1}{2i}(\hat{f} - \hat{\bar{f}})$$

as well as

$$\hat{\bar{f}}(x) = \overline{\hat{f}(-x)}$$

for all $x \in \mathbf{R}^p$ by (8.1.17) and the reflection-invariance of λ^p. Corresponding formulas for g prove this remark. ⌟

Examples

1. The Fourier transform $\hat{\mu}$ of a measure $\mu \in \mathfrak{M}^e(\mathbf{R}^p)$ is real-valued if and only if μ is invariant with respect to the reflection S through the origin. If $\hat{\mu}$ is real, then (8.1.11) tells us that $S(\mu)$ and μ have the same

Fourier transform. But then, by Theorem 8.2.4, $S(\mu) = \mu$. The converse follows directly from (8.1.11).

2. The Fourier transform $\hat{f}$ of a function $f \in \mathcal{L}^1(\lambda^p, \mathbf{C})$ is real-valued if and only if $f(x) = \overline{f(-x)}$ holds λ^p-almost everywhere on $\mathbf{R}^p$. This follows analogously from Theorem 8.1.8(d) and Corollary 8.2.5.

3. In Section 5.3, Remark 3, we mentioned that the sum $X + Y$ of real, independent random variables X and Y with the Cauchy distributions γ_α and γ_β respectively ($\alpha, \beta > 0$) has distribution $\gamma_{\alpha+\beta}$. The proof can now be presented as follows: By (8.1.14) and Section 8.1, Example 4,

$$\varphi_{X+Y}(x) = \varphi_X(x)\varphi_Y(x) = e^{-(\alpha+\beta)|x|} \qquad (x \in \mathbf{R}).$$

But this is the characteristic function of a random variable with distribution $\gamma_{\alpha+\beta}$. Therefore, $\gamma_{\alpha+\beta}$ is the distribution of $X + Y$. In other words, $\gamma_\alpha * \gamma_\beta = \gamma_{\alpha+\beta}$.

4. The reader can prove, analogously to Example 3, the result from Section 5.3 according to which the sum of independent, normally distributed random variables is normally distributed.

5. Example 3 shows that the validity of $P_{X+Y} = P_X * P_Y$ for real random variables X, Y in general does *not* imply their independence. We need only choose $X = Y$ and $P_X = \gamma_1$.

The second main theorem of the theory concerns continuity properties of the mapping $\mu \to \hat{\mu}$. We facilitate the proof by a lemma which sharpens Theorem 8.1.4(a).

8.2.6. Lemma. For every weakly convergent sequence $(\mu_n)_{n\in\mathbf{N}}$ in $\mathfrak{M}^e(\mathbf{R}^p)$, the sequence $(\hat{\mu}_n)_{n\in\mathbf{N}}$ of Fourier transforms is equi-uniformly continuous on $\mathbf{R}^p$.[3]

Proof. It suffices to modify the proof of Theorem 8.1.4(a) slightly. Let μ be the weak limit of (μ_n). By (7.7.5), for every $\epsilon > 0$ there is a $u \in \mathcal{C}^c(\mathbf{R}^p)$ such that $\int(1 - u)\, d\mu < \epsilon$ and $0 \leqq u \leqq 1$. The proof of Theorem 8.1.4(a) tells us that then there is a $\delta > 0$ such that

$$|e^{i<x_1,y>} - e^{i<x_1,y>}| \leqq \epsilon$$

for all $x_1, x_2 \in \mathbf{R}^p$ with $|x_1 - x_2| \leqq \delta$ and all y from the compact support S_u of u. Now $\lim\limits_{n\to\infty} \int(1 - u)\, d\mu_n = \int(1 - u)\, d\mu$, and thus there is a natural

[3] This means: For every $\epsilon > 0$ there is a $\delta > 0$ such that $|\hat{\mu}_n(x_1) - \hat{\mu}_n(x_2)| \leqq \epsilon$, for any two points $x_1, x_2 \in \mathbf{R}^p$ with $|x_1 - x_2| \leqq \delta$ and *every* $n \in \mathbf{N}$.

number n_0 such that $\int(1-u)\,d\mu_n < \epsilon$ for all $n \geqq n_0$. For all pairs $x_1, x_2 \in \mathbf{R}^p$ with $|x_1 - x_2| \leqq \delta$ and all $n \geqq n_0$, it then follows that

$$\begin{aligned} |\hat{\mu}_n(x_1) - \hat{\mu}_n(x_2)| &\leqq \int |e^{i<x_1,y>} - e^{i<x_2,y>}|\mu_n(dy) \\ &= \int |e^{i<x_1,y>} - e^{i<x_2,y>}|u(y)\mu_n(dy) \\ &\qquad + \int |e^{i<x_1,y>} - e^{i<x_2,y>}|(1-u(y))\mu_n(dy) \\ &\leqq \epsilon\mu_n(S_u) + 2\int(1-u)\,d\mu_n < \epsilon(\|\mu_n\| + 2). \end{aligned}$$

But, due to the weak convergence, the sequence $(\|\mu_n\|)$ is convergent and hence bounded. From this and from Theorem 8.1.4(a), applied to the Fourier transforms $\hat{\mu}_1, \ldots, \hat{\mu}_{n_0-1}$, the assertion now follows. ┘

8.2.7. Theorem (Continuity Theorem). For every sequence $(\mu_n)_{n\in\mathbf{N}}$ in $\mathfrak{M}^e(\mathbf{R}^p)$:

1. If (μ_n) converges weakly to $\mu \in \mathfrak{M}^e(\mathbf{R}^p)$, then the sequence of Fourier transforms $(\hat{\mu}_n)$ converges uniformly to $\hat{\mu}$ on every compact subset of $\mathbf{R}^p$.

2. If the sequence $(\hat{\mu}_n)$ of Fourier transforms converges pointwise to a complex function φ on $\mathbf{R}^p$ continuous at $x = 0$, then φ is the Fourier transform of a (uniquely determined) measure $\mu \in \mathfrak{M}^e(\mathbf{R}^p)$ and the sequence (μ_n) converges weakly to μ.

Proof. 1. Each of the functions $y \to e^{i<x,y>}$ lies in $\mathcal{C}^b(\mathbf{R}^p,\mathbf{C})$. For every $x \in \mathbf{R}^p$, $\lim_{n\to\infty} \hat{\mu}_n(x) = \hat{\mu}(x)$ by the definition of weak convergence. Thus, for every $\epsilon > 0$ and $x \in \mathbf{R}^p$ there exists a natural number n_x such that $|\hat{\mu}_n(x) - \hat{\mu}(x)| \leqq \epsilon$ for all $n \geqq n_x$. By Lemma 8.2.6 the sequence $(\hat{\mu}_n - \hat{\mu})_{n\in\mathbf{N}}$ is equi-uniformly continuous on $\mathbf{R}^p$. Thus, there is a $\delta > 0$ such that for $x_1, x_2 \in \mathbf{R}^p$ with $|x_1 - x_2| \leqq \delta$ and all $n = 1, 2, \ldots$

$$|\hat{\mu}_n(x_1) - \hat{\mu}(x_1) - \hat{\mu}_n(x_2) + \hat{\mu}(x_2)| \leqq \epsilon.$$

This now implies $|\hat{\mu}_n(y) - \hat{\mu}(y)| \leqq 2\epsilon$ for all points y in the open ball $K_\delta(x)$ of center x and radius δ and all $n \geqq n_x$. Hence the assertion follows, since every compact set $K \subset \mathbf{R}^p$ is covered by finitely many balls $K_\delta(x^1), \ldots, K_\delta(x^r)$ with centers $x^1, \ldots, x^r \in K$, and thus

$$|\hat{\mu}_n(y) - \hat{\mu}(y)| \leqq 2\epsilon$$

for all $y \in K$ and all $n \geqq \max(n_{x^1}, \ldots, n_{x^r})$.

2. First we reduce the assertion to the following auxiliary assertion:
(H) Let (ν_n) be a sequence in $\mathfrak{M}^e(\mathbf{R}^p)$ converging vaguely to a measure $\nu \in \mathfrak{M}^e(\mathbf{R}^p)$, for which the sequence $(\hat{\nu}_n)$ converges pointwise on $\mathbf{R}^p$ to a function $\psi\colon \mathbf{R}^p \to \mathbf{C}$ continuous at $x = 0$. Then $\lim_{n\to\infty} \|\nu_n\| = \|\nu\|$.

To prove "(H) $\Rightarrow$ (2)," let (μ_n) be a sequence in $\mathfrak{M}^e$ with the properties noted in 2. Then $\varphi(0) = \lim \hat{\mu}_n(0)$, and thus, by Theorem 8.1.4(b), the sequence $(\|\mu_n\|)$ converges and hence is bounded, that is,

$$\alpha \equiv \sup_{n \in \mathbf{N}} \|\mu_n\| < +\infty.$$

By Corollary 7.8.3 and Theorem 7.8.4 there now exists a vaguely convergent subsequence (ν_n) of the sequence (μ_n); let the vague limit be $\nu \in \mathfrak{M}^e$. Since (ν_n) also converges pointwise to φ, (H) yields the convergence $\lim \|\nu_n\| = \|\nu\|$. Thus by Theorem 7.7.7, (ν_n) is weakly convergent to ν, that is, by 1: $\hat{\nu}(x) = \lim_{n \to \infty} \hat{\nu}_n(x) = \varphi(x)$ for all $x \in \mathbf{R}^p$. Thus, $\varphi = \hat{\nu}$ and hence, by the Uniqueness Theorem, ν is uniquely determined. All vaguely convergent subsequences of (μ_n) therefore have the same limit ν. Since (μ_n) is a sequence in the compact metrizable space of all $\mu \in \mathfrak{M}^e$ with $\|\mu\| \leqq \alpha$, we have the vague convergence of (μ_n) to ν and since

$$\lim \|\mu_n\| = \varphi(0) = \hat{\nu}(0) = \|\nu\|,$$

we also have weak convergence to ν. But this is the assertion of 2.

Now we prove (H): The sequence $\hat{\nu}_n(0) = \|\nu_n\|$, $n = 1, 2, \ldots$ is convergent and thus bounded: $\alpha \equiv \sup \|\nu_n\| < +\infty$. Since $|\hat{\nu}_n| \leqq \|\nu_n\|$, the sequence $(\hat{\nu}_n)$ is uniformly bounded on $\mathbf{R}^p$ by α, and thus $|\psi| \leqq \alpha$. As in Section 7.7, Example 4, let $K \geqq 0$ be a function from $\mathfrak{L}^1(\lambda^p)$ satisfying $\int K \, d\lambda^p = 1$. We again set $K_r(x) \equiv r^p K(rx)$ for each $r > 0$ and

$$s_{r,n} \equiv K_r * \hat{\nu}_n, \quad s_r \equiv K_r * \hat{\nu}, \quad S_r \equiv K_r * \varphi$$

for all $r > 0$ and $n = 1, 2, \ldots$. These definitions are meaningful since $\hat{\nu}_n$ and $\hat{\nu}$ are bounded continuous functions and since ψ is bounded and $\mathfrak{B}^p$-measurable as the limit of $(\hat{\nu}_n)$. According to Section 7.7, Example 4, and Theorem 7.7.10, the continuity of $\hat{\nu}$ and ψ at $x = 0$ implies

$$\lim_{r \to +\infty} s_r(0) = \hat{\nu}(0), \quad \lim_{r \to +\infty} S_r(0) = \psi(0).$$

In fact, we have $\lim_{r \to +\infty} K_r \lambda^p = \epsilon_0$ in the sense of the vague topology. Since $K_r \lambda^p$ and ϵ_0 are all probability measures, $K_r \lambda^p$ converges to ϵ_0 relative to the weak topology. For every bounded, Borel-measurable function f continuous at $x = 0$, we have, by Theorem 7.7.10,

$$\lim_{r \to +\infty} \int f(y) K_r(y) \lambda^p(dy) = f(0).$$

By choosing the functions $y \to \hat{\nu}(-y)$ and $y \to \psi(-y)$ for f, we obtain the asserted results.

According to Fubini,

$$\begin{aligned} s_{r,n}(0) &= \int K_r(y)\hat{\nu}_n(-y)\lambda^p(dy) = \int K_r(-y)\hat{\nu}_n(y)\lambda^p(dy) \\ &= \int\int K_r(-y)e^{i<y,z>}\nu_n(dz)\lambda^p(dy) = \int q_r\,d\nu_n \end{aligned}$$

and analogously

$$s_r(0) = \int q_r\,d\nu,$$

where

$$q_r(z) \equiv \int e^{i<y,z>}K_r(-y)\lambda^p(dy) \qquad (z \in \mathbf{R}^p).$$

For every $r > 0$, q_r is then the Fourier transform of the function $y \to K_r(-y)$, that is, by the Riemann-Lebesgue Theorem, $q_r \in \mathcal{C}^0(\mathbf{R}^p,\mathbf{R})$. With the help of Theorem 7.7.5 we now obtain

$$\lim_{n\to\infty} s_{r,n}(0) = s_r(0) \qquad (r > 0).$$

Since $(\hat{\nu}_n)$ converges pointwise to ψ and $|\hat{\nu}_n| \leqq \alpha$ for all n, it follows from

$$s_{r,n}(0) = \int K_r(-y)\hat{\nu}_n(y)\lambda^p(dy)$$

and the Dominated Convergence Theorem that

$$\lim_{n\to\infty} s_{r,n}(0) = \int K_r(-y)\psi(y)\lambda^p(dy) = S_r(0) \qquad (r > 0).$$

Thus $s_r(0) = S_r(0)$ for all $r > 0$; passage to the limit $r \to +\infty$ yields $\hat{\nu}(0) = \psi(0)$. But then

$$\|\nu\| = \hat{\nu}(0) = \psi(0) = \lim_{n\to\infty} \hat{\nu}_n(0) = \lim_{n\to\infty} \|\nu_n\|. \quad \lrcorner$$

Since the weak topology on $\mathfrak{M}^e(\mathbf{R}^p)$ is metrizable by Section 7.8, Remark 2, Theorem 8.2.7 says in particular the following: On the set $\widehat{\mathfrak{M}^e}$ of all Fourier transforms $\hat{\mu}$ of measures $\mu \in \mathfrak{M}^e(\mathbf{R}^p)$, the topology of *pointwise convergence* coincides with the *topology of uniform convergence on compact subsets*. The mapping $\mu \to \hat{\mu}$ is a *homeomorphism* between $\mathfrak{M}^e$ (equipped with the weak topology) and $\widehat{\mathfrak{M}^e}$ (equipped with the topology of pointwise convergence).

PROBLEMS

1. Prove or disprove: For every vaguely convergent sequence (μ_n) in $\mathfrak{M}^e(\mathbf{R}^p)$ the sequence $(\hat{\mu}_n)$ of Fourier transforms converges pointwise on $\mathbf{R}^p$.
2. Consider the sequence $(\rho_n)_{n\in\mathbf{N}}$ of rectangular distributions defined in Section 8.1, Problem 4. Prove: (ρ_n) converges vaguely and $(\hat{\rho}_n)$ con-

verges pointwise. Can the Continuity Theorem be applied to this situation?

3. Define as usual the Gamma function Γ on $]0, +\infty[$ by the integral

$$\Gamma(t) = \int_0^\infty x^{t-1}e^{-x}\,dx.$$

Then

$$f_t(x) \equiv \begin{cases} \dfrac{1}{\Gamma(t)}\, x^{t-1}e^{-x}, & x > 0 \\ 0, & x \leqq 0 \end{cases}$$

is for each $t > 0$ the λ^1-density of a probability measure $\mu_t \equiv f_t\lambda^1$. μ_t is called the *Gamma distribution* with parameter t.

(a) Calculate the Fourier transform of μ_t.

(b) Let $X_1, \ldots, X_n$ be n independent real random variables with $\nu_{0,1}$ as common distribution. Prove that the distribution of $X_1^2 + \cdots + X_n^2$ has the function

$$g_n(x) \equiv \begin{cases} \dfrac{1}{2^{n/2}\Gamma\left(\dfrac{n}{2}\right)}\, x^{(n/2)-1}e^{-x/2}, & x > 0 \\ 0, & x \leqq 0 \end{cases}$$

as λ^1-density. The probability measure $g_n\lambda^1$ is called the χ^2-*distribution* with n degrees of freedom.

4. Let $(X_n)_{n\in\mathbf{N}}$ be an independent sequence of real, centered random variables on a probability space. Prove: The sequence (X_n) satisfies the weak law of large numbers if and only if the sequence of functions

$$x \to \prod_{i=1}^{n} \varphi_{X_i}\left(\frac{x}{n}\right) \qquad (n \in \mathbf{N})$$

converges pointwise to 1. Apply this result to the sequence (X_n) studied in Section 6.4, Problem 1 and prove that for $\lambda \geqq 1$ the weak law fails. This proves again the result of Section 6.5, Problem 2.

5. Let $K \geqq 0$ be a function in $\mathfrak{L}^1(\lambda^p)$ satisfying $\int K\,d\lambda^p = 1$. Define K_r for $r > 0$ as in Section 7.7, Example 4. Prove that

$$\lim_{r\to 0}\,(K_r * \mu)\lambda^p = \mu$$

holds in the weak topology for all $\mu \in \mathfrak{M}^e(\mathbf{R}^p)$. [*Hint:* Section 7.7, Example 4 treats the special case $\mu = \epsilon_0$. Observe that

$$x \to \int f(x+y)\mu(dy)$$

is in $\mathcal{C}^b(\mathbf{R}^p)$ for all $f \in \mathcal{C}^b(\mathbf{R}^p)$.]

6. A function $\delta \geqq 0$ in $\mathfrak{L}^1(\lambda^p)$ is called a *convergence factor* on $\mathbf{R}^p$ if its Fourier transform satisfies $\int \hat{\delta}\, d\lambda^p = 1$.
   Prove that $x \to (2\pi)^{-p}e^{-|x^2|/2}$ is a convergence factor on $\mathbf{R}^p$.
7. Let δ be a convergence factor on $\mathbf{R}^p$, and let μ be in $\mathfrak{M}^e(\mathbf{R}^p)$. Denote by $s_r(\mu)$ the function

$$x \to \int e^{-i<x,y>}\delta\left(\frac{y}{r}\right)\hat{\mu}(y)\lambda^p(dy) \qquad (r > 0).$$

   (a) Prove the following *inversion formula:*

$$\lim_{r \to 0} s_r(\mu)\lambda^p = \mu$$

   in the weak topology. [*Hint:* $x \to \hat{\delta}(-x)$ is a function K with the properties mentioned in Problem 4. The result is a consequence of Section 8.1, Problem 5 and the above Problem 4.]
   (b) Deduce from (a) a new proof of the Uniqueness Theorem 8.2.4.

8.3 DIFFERENTIABILITY OF FOURIER TRANSFORMS

The following lemma is a preliminary to answering the question of differentiability of Fourier transforms:

8.3.1. Lemma. Let $(\Omega,\mathfrak{A},\mu)$ be a measure space, U an open subset of $\mathbf{R}$ or $\mathbf{C}$, and $f\colon U \times \Omega \to \mathbf{C}$ a function with the following properties:

(a) $\omega \to f(t,\omega)$ is μ-integrable for each $t \in U$;

(b) $t \to f(t,\omega)$ is differentiable in $t_0 \in U$ for all $\omega \in \Omega$; the derivative is denoted by $f'(t_0,\omega)$;

(c) there exists a μ-integrable function $h \geqq 0$ on Ω such that

$$\left|\frac{f(t,\omega) - f(t_0,\omega)}{t - t_0}\right| \leqq h(\omega),$$

for all $\omega \in \Omega$ and all $t \in U \setminus \{t_0\}$.

Then the function $\varphi\colon U \to \mathbf{C}$ defined by $t \to \int f(t,\omega)\mu(d\omega)$ is differentiable in t_0, $\omega \to f'(t_0,\omega)$ is μ-integrable, and $\varphi'(t_0) = \int f'(t_0,\omega)\mu(d\omega)$.

Thus under the given conditions we can differentiate under the integral sign.

Proof. Let (t_n) be a sequence in U with $\lim t_n = t_0$ and $t_n \neq t_0$ for all n. Then for every $n \in \mathbf{N}$,

$$g_n(\omega) \equiv \frac{f(t_n,\omega) - f(t_0,\omega)}{t_n - t_0}$$

is μ-integrable and

$$\lim_{n\to\infty} g_n(\omega) = f'(t_0,\omega), \qquad \text{for all } \omega \in \Omega.$$

Moreover, $|g_n| \leqq h$ for all n. By the Dominated Convergence Theorem $\omega \to f'(t_0,\omega)$ is μ-integrable and

$$\lim_{n\to\infty} \int g_n \, d\mu = \int f'(t_0,\omega)\mu(d\omega).$$

This proves the lemma since

$$\int g_n \, d\mu = \frac{\varphi(t_n) - \varphi(t_0)}{t_n - t_0} \qquad (n = 1, 2, \ldots). \quad \lrcorner$$

Remark. 1. Condition (c) is satisfied if $t \to f(t,\omega)$ is differentiable in U for every $\omega \in \Omega$, if for every point $t \in U$ the line connecting t and t_0 lies in U, and if there exists an $h \in \mathcal{L}^1(\mu)$ such that

$$|f'(t,\omega)| \leqq h(\omega), \qquad \text{for all } (t,\omega) \in U \times \Omega.$$

To see this, it suffices to apply the Mean Value Theorem of differential calculus.

Now we return to the study of Fourier transforms.

8.3.2. Definition. Let μ be a (not necessarily finite) Borel measure on $\mathbf{R}^p$ and let $k_1, \ldots, k_p$ be integers $\geqq 0$. Then if

$$x = (x_1, \ldots, x_p) \to x_1^{k_1} \cdot \ldots \cdot x_p^{k_p}$$

is μ-integrable on $\mathbf{R}^p$,

$$M_{k_1,\ldots,k_p} \equiv \int x_1^{k_1} \cdot \ldots \cdot x_p^{k_p} \mu(dx)$$

is called the $(k_1, \ldots, k_p)$th *moment* of μ and $k_1 + \cdots + k_p$ its order. We then say that the $(k_1, \ldots, k_p)$th moment of μ exists.

Examples

1. There is exactly one moment of order 0. It exists if and only if μ is finite, and $M_{0,\ldots,0} = \|\mu\|$.

2. Let $X_1, \ldots, X_p$ be real random variables on a probability space $(\Omega,\mathfrak{A},P)$ with joint distribution $\mu \equiv P_{X_1 \otimes, \ldots, \otimes X_p}$. By the Transformation

Theorem for Integrals, the $(k_1, \ldots, k_p)$th moment $M_{k_1,\ldots,k_p}$ of μ exists if and only if the random variable $X_1^{k_1} \cdot \ldots \cdot X_p^{k_p}$ is P-integrable.

Then we have

$$M_{k_1,\ldots,k_p} = E(X_1^{k_1} \cdot \ldots \cdot X_p^{k_p}).$$

We also call $M_{k_1,\ldots,k_p}$ the $(k_1, \ldots, k_p)$th (mixed) moment of the random variables $X_1, \ldots, X_p$.

3. In the case $p = 1$, the existence of the kth moment M_k of a measure $\mu \in \mathfrak{M}^e(\mathbf{R})$ implies the existence of all moments M_l with $0 \leqq l \leqq k$. Indeed, $|x|^l \leqq 1 + |x|^k$ for all $x \in \mathbf{R}$.

4. All moments M_k, $k = 0, 1, \ldots,$ exist for the standard normal distribution $\nu_{0,1} = g_{0,1}\lambda^1$ (and thus also for ν_{α,σ^2} with arbitrary $\alpha \in \mathbf{R}$, $\sigma > 0$).

For even k, integration by parts (see the derivation of (4.4.11) and (4.4.12) on p. 145) yields the recursion formula $M_{2n} = (2n - 1)M_{2(n-1)}$ and hence

$$M_{2n} = (2n-1)(2n-3) \cdot \ldots \cdot 1 \qquad (n = 0, 1, \ldots).$$

For odd k, taking note of Example 3, we obtain the existence of M_k. Due to the symmetry of $g_{0,1}$,

$$M_{2n+1} = 0 \qquad (n = 0, 1, \ldots).$$

5. In the case $p > 1$, we cannot generally derive the existence of all moments $M_{l_1,\ldots,l_p}$ with $0 \leqq l_j \leqq k_j$ from the existence of a moment $M_{k_1,\ldots,k_p}$. In fact, let $\Omega \equiv]0,1[$, $\mathfrak{A} \equiv \mathfrak{B}(\Omega) = \Omega \cap \mathfrak{B}^1$, and $P \equiv \lambda_\Omega^1$. For numbers α, β satisfying $0 < \alpha \leqq \beta < 1$, we consider the following random variables $X \geqq 0$, $Y \geqq 0$ on $(\Omega,\mathfrak{A},P)$:

$$X(\omega) \equiv \begin{cases} \dfrac{1}{\omega}, & 0 < \omega < \beta \\ 0, & \beta \leqq \omega < 1, \end{cases} \qquad Y(\omega) \equiv \begin{cases} \dfrac{1}{1-\omega}, & \alpha < \omega < 1 \\ 0, & 0 < \omega \leqq \alpha. \end{cases}$$

Then $X \cdot Y$ is integrable, but neither X nor Y is. For the joint distribution $\mu = P_{X \otimes Y}$ on $\mathbf{R}^2$, the (1,1)th moment exists, but $M_{0,1}$ and $M_{1,0}$ do not. If $\alpha < \beta$, then $M_{1,1} > 0$, and if $\alpha = \beta$, then $M_{1,1} = 0$.

8.3.3. Theorem. Let μ be a finite Borel measure on $\mathbf{R}^p$ for which all moments $M_{l_1,\ldots,l_p}$ with $0 \leqq l_j \leqq k_j$ $(j = 1, \ldots, p)$ exist. Then for all such $l_1, \ldots, l_p$, the partial derivative

$$D_{l_1,\ldots,l_p}\hat{\mu} = \frac{\partial^{l_1+\cdots+l_p}\hat{\mu}}{\partial^{l_1}x_1 \cdots \partial^{l_p}x_p}$$

of the Fourier transform $\hat{\mu}$ exists on all of $\mathbf{R}^p$; we have

$$D_{l_1,\ldots,l_p}\hat{\mu}(x) = i^{l_1+\cdots+l_p}\int e^{i<x,y>}y_1^{l_1} \cdot \ldots \cdot y_p^{l_p}\mu(dy) \tag{8.3.1}$$

and in particular

$$D_{l_1,\ldots,l_p}\hat{\mu}(0) = i^{l_1+\cdots+l_p}M_{l_1,\ldots,l_p}. \tag{8.3.2}$$

Each of the derivatives is uniformly continuous and bounded on $\mathbf{R}^p$.

Proof. For every $y = (y_1, \ldots, y_p) \in \mathbf{R}^p$, $x \to f_y(x) \equiv e^{i<x,y>}$ is differentiable arbitrarily often on $\mathbf{R}^p$. We have

$$D_{l_1,\ldots,l_p}f_y(x) = f_y(x)(iy_1)^{l_1} \cdot \ldots \cdot (iy_p)^{l_p}$$

and hence

$$|D_{l_1,\ldots,l_p}f_y(x)| \leqq |y_1|^{l_1} \cdot \ldots \cdot |y_p|^{l_p}.$$

For all $l_1, \ldots, l_p$ with $0 \leqq l_j \leqq k_j$, $y \to D_{l_1,\ldots,l_p}f_y(x)$ is then μ-integrable; moreover, $|y_1|^{l_1} \cdot \ldots \cdot |y_p|^{l_p}$ is a μ-integrable majorant of $D_{l_1,\ldots,l_p}f_y(x)$ and independent of x. Thus, by Lemma 8.3.1 together with the remark following it,

$$x \to \int D_{l_1,\ldots,l_p}f_y(x)\mu(dy)$$

is differentiable with respect to x_j, provided $l_j < k_j$. The partial differentiation can be put under the integral sign. Using induction, we now obtain the existence of all partial derivatives $D_{l_1,\ldots,l_p}\hat{\mu}$, $0 \leqq l_j \leqq k_j$, as well as formulas (8.3.1) and (8.3.2). The rest of the assertion follows from Theorem 8.1.4 and the remark that the integral in (8.3.1) is the difference of the Fourier transforms of measures with densities

$$y \to [y_1^{l_1} \cdot \ldots \cdot y_p^{l_p}]^+ \qquad \text{and} \qquad y \to [y_1^{l_1} \cdot \ldots \cdot y_p^{l_p}]^-$$

with respect to μ. ┘

8.3.4. Corollary. If the moment M_k of kth order exists for a finite Borel measure μ on the real line $\mathbf{R}$, then $\hat{\mu}$ is k times differentiable. The kth derivative $\hat{\mu}^{(k)}$ is uniformly continuous and bounded. We have

$$\hat{\mu}^{(k)}(0) = i^k M_k \qquad (k = 0, 1, \ldots). \tag{8.3.3}$$

This follows directly from Theorem 8.3.3 if we take Example 3 into account. The case $k = 0$ was taken care of by Theorem 8.1.4.

A typical application of this corollary is the following: Let $\mu \in \mathfrak{M}^e(\mathbf{R})$ be such that M_k exists for some $k = 1, 2, \ldots$ and hence M_l exists for all l such that $0 \leqq l \leqq k$. Then $\hat{\mu}$ has the *Taylor expansion*

$$\begin{aligned}\hat{\mu}(x) &= \sum_{l=0}^{k-1} \frac{(ix)^l}{l!} M_l + \int_0^x \frac{(x-t)^{k-1}}{(k-1)!} \hat{\mu}^{(k)}(t)\,dt \\ &= \sum_{l=0}^{k} \frac{(ix)^l}{l!} M_l + \int_0^x \frac{(x-t)^{k-1}}{(k-1)!} [\hat{\mu}^{(k)}(t) - \hat{\mu}^{(k)}(0)]\,dt. \qquad (8.3.4)^{4}\end{aligned}$$

For the remainder term $R_k(x)$ of the above Taylor expansion we then obtain, in particular, the bound

$$|R_k(x)| \leqq \sup_{0 \leqq \vartheta \leqq 1} |\hat{\mu}^{(k)}(\vartheta x) - \hat{\mu}^{(k)}(0)| \cdot \frac{|x|^k}{k!}. \qquad (8.3.5)$$

Finally, we investigate the question under which conditions the Fourier transform (defined on $\mathbf{R}^p$) of a measure $\mu \in \mathfrak{M}^e(\mathbf{R}^p)$ can be extended holomorphically to the complex p-dimensional space $\mathbf{C}^p$. That this is not always possible is shown by the example of the Cauchy distribution γ_α whose Fourier transform, as a function on $\mathbf{R}$, does not have a derivative at $x = 0$.

We decompose every point $z = (z_1, \ldots, z_p) \in \mathbf{C}^p$ into its real and complex components $x \in \mathbf{R}^p$ and $x' \in \mathbf{R}^p$, respectively, so that $z = x + ix'$ and therefore x and x' are the vectors of the real and imaginary parts respectively of all coordinates $z_1, \ldots, z_p$. More precisely, we investigate the question of when $\hat{\mu}$ can be extended holomorphically onto a "strip" of the form

$$S_\alpha \equiv \{z = x + ix' \in \mathbf{C}^p : |x'_1| < \alpha, \ldots, |x'_p| < \alpha\} \qquad (\alpha > 0).$$

We set

$$<z,y> = \sum_{j=1}^{p} z_j y_j$$

for arbitrary $z \in \mathbf{C}^p$ and $y \in \mathbf{R}^p$.

8.3.5. Theorem. Let $\mu \in \mathfrak{M}^e(\mathbf{R}^p)$ and $\alpha > 0$ be such that the function $y \to e^{<x',y>}$ is μ-integrable for all $x' = (x'_1, \ldots, x'_p) \in \mathbf{R}^p$ such that

[4] The reader should note that for complex-valued functions, the remainder term cannot generally be written in Lagrange form, although it can be written in integral form.

$|x'_1| < \alpha, \ldots, |x'_p| < \alpha$. Then the function

$$\varphi(z) \equiv \int e^{i\langle z,y\rangle}\mu(dy) \tag{8.3.6}$$

is holomorphic in the strip S_α and is obviously an extension of $\hat{\mu}$.

Proof. First, the function $y \to e^{i\langle z,y\rangle}$ is μ-integrable for every $z = x + ix' \in S_\alpha$, since $|e^{i\langle z,y\rangle}| = e^{-\langle x',y\rangle} = e^{\langle -x',y\rangle}$ and $|x'_j| < \alpha$ for all $j = 1, \ldots, p$. Further, for each such j, the function $z \to e^{i\langle z,y\rangle}$ is differentiable with respect to z_j for each $y \in \mathbf{R}^p$ and we have

$$\frac{\partial}{\partial z_j} e^{i\langle z,y\rangle} = iy_j e^{i\langle z,y\rangle}.$$

Every point $z^0 \in S_\alpha$ now possesses a strip $S_{\beta'}$ with $0 < \beta' < \alpha$ as a neighborhood in $\mathbf{C}^p$. We choose $\delta > 0$ such that $\beta \equiv \beta' + \delta < \alpha$.

On the one hand, there exists a positive number M such that $|\xi| \leqq Me^{\delta|\xi|}$ holds for all $\xi \in \mathbf{R}$. On the other hand, the function

$$y \longrightarrow e^{\beta(|y_1|+\cdots+|y_p|)}$$

is μ-integrable. Indeed we have

$$e^{\beta(|y_1|+\cdots+|y_p|)} \leqq \sum_{t=1}^{2^p} e^{\langle b_t,y\rangle},$$

if $b_1, \ldots, b_{2^p}$ denotes all vectors in $\mathbf{R}^p$ each of whose coordinates has absolute value β. However, we have a sum of μ-integrable functions on the right by hypothesis. For all points $z = x + ix' \in S_{\beta'}$ and all $y \in \mathbf{R}^p$, we then have the inequality

$$\left|\frac{\partial}{\partial z_j} e^{i\langle z,y\rangle}\right| = |y_j|e^{-\langle x',y\rangle} \leqq Me^{\delta|y_j|}\, e^{|x'_1|\,|y_1|+\cdots+|x'_p|\,|y_p|}$$
$$\leqq Me^{\beta(|y_1|+\cdots+|y_p|)}.$$

By Lemma 8.3.1 and the subsequent remark, φ is differentiable with respect to z_j in $S_{\beta'}$, hence in a neighborhood of z^0. Since this holds for every $j = 1, \ldots, p$, we have, φ is holomorphic in S_α. ⌋

Remark. 2. According to well-known theorems of function theory, φ is the only holomorphic function in S_α which coincides with $\hat{\mu}$ on $\mathbf{R}^p$.

3. The integrability condition given in Theorem 8.3.5 even is necessary in order that $\hat{\mu}$ may be continued holomorphically onto S_α. See Richter [20], p. 311 ff.

PROBLEMS

1. Let μ be a finite Borel measure on $\mathbf{R}$ for which the moment M_k and hence all moments M_l for $l = 0, 1, \ldots, k$ exist ($k \in \mathbf{N}$). Prove the existence of a continuous function $\Theta\colon \mathbf{R} \to \mathbf{C}$ such that

   (a) $\Theta(0) = 0;$

   (b) $$\mu(x) = \sum_{l=0}^{k} \frac{(ix)^l}{l!} M_l + \Theta(x)\frac{x^k}{k!}, \qquad \text{for all } x \in \mathbf{R}.$$

2. Prove (without and with the aid of Fourier transforms) that no moment M_k of order $k \geqq 1$ exists for the Cauchy distribution.

part IV

Further Development of Probability Theory

9

LIMIT DISTRIBUTIONS

The following considerations on the so-called Central Limit Theorem rightly deserve a "central" significance in probability theory. A multitude of applications, especially in mathematical statistics, underscore the importance of the limit distribution problems to be treated here, whose solution was the chief problem in probability theory for a long time. This chapter also provides the reader with the opportunity to realize the fruitfulness of the methods of Fourier analysis.

9.1 EXAMPLES OF LIMIT THEOREMS

Suppose we are given a sequence $(S_n)_{n \in \mathbf{N}}$ of real random variables on a probability space $(\Omega, \mathfrak{A}, P)$. Suppose every variable S_n is the sum of an *independent, finite family* $(X_{nj})_{j=1,\ldots,k_n}$ of real random variables X_{nj} (on the same probability space):

$$S_n \equiv X_{n1} + \cdots + X_{nk_n} \qquad (n = 1, 2, \ldots). \tag{9.1.1}$$

The following examples show that under simple assumptions on the random variables X_{nj}, the distributions (P_{S_n}) converge weakly to certain familiar limit distributions.

Examples

1. Let $(X_n)_{n \in \mathbf{N}}$ be an independent sequence of real, integrable random variables. We choose $k_n = n$ and $X_{nj} = (1/n)(X_j - E(X_j))$, that is,

$$S_n = \frac{1}{n} \sum_{j=1}^{n} (X_j - E(X_j)).$$

By Theorem 7.7.2, the sequence (X_n) satisfies the *weak law of large numbers* if and only if the sequence (P_{S_n}) is weakly convergent to the singular distribution ϵ_0 defined by the unit mass at the origin. The condition

$$\lim_{n\to\infty} \frac{1}{n^2} \sum_{i=1}^{n} V(X_i) = 0$$

formulated in (6.5.2) for the sequence (X_n) thus implies the weak convergence of (P_{S_n}) to ϵ_0.

2. Let $k_n = n$ and suppose each of the (independent) random variables X_{nj}, $j = 1, \ldots, n$ has distribution $\beta_1^{p_n}$ with $0 \leqq p_n \leqq 1$. Further, suppose that the sequence (np_n) converges to some $\alpha \in \mathbf{R}_+$:

$$\lim_{n\to\infty} np_n = \alpha. \tag{9.1.2}$$

Then, in the sense of weak convergence,

$$\lim P_{S_n} = \pi_\alpha,$$

where π_α is the *Poisson distribution* with parameter α (or ϵ_0 when $\alpha = 0$). This result is due to Poisson.

For the proof, we note that S_n has distribution $\beta_n^{p_n}$, and thus for each $f \in \mathcal{C}^c(\mathbf{R})$,

$$\int f\, dP_{S_n} = \sum_{k=0}^{n} \binom{n}{k} p_n^k (1 - p_n)^{n-k} f(k).$$

Since there are only finitely many integers $k \geqq 0$ in the support of f, the asserted convergence of (P_{S_n}) to π_α is equivalent to the validity of

$$\lim_{n\to\infty} \binom{n}{k} p_n^k (1 - p_n)^{n-k} = e^{-\alpha} \frac{\alpha^k}{k!} \qquad (k = 0, 1, \ldots). \tag{9.1.3}$$

But this can be seen directly, since

$$\binom{n}{k} p_n^k (1 - p_n)^{n-k} = \frac{1}{k!} \left(1 - \frac{np_n}{n}\right)^{n-k} \prod_{j=1}^{k} [(n - k + j)p_n]$$

and since $\lim np_n = \alpha$ implies $\lim p_n = 0$.

Remark. Formula (9.1.3) tells us that the probability

$$P\{S_n = k\} = \binom{n}{k} p_n^k (1 - p_n)^{n-k}$$

is approximately equal to $e^{-\alpha}\alpha^k/k!$ for large n, as long as p_n is asymptotically equal to α/n. This explains why the Poisson distribution is also called the *distribution of rare events.*

Formula (9.1.3) is used in many practical applications where a random occurrence can be described by a random variable X, with binomial distribution, which can take on each of the values $0, 1, \ldots, n$ for large n with constant small probability p but whose expected value $E(X) = pn$ is a known number α on the basis of observations. Examples of this kind are the description of rare diseases or accidents in a large "population" (say the set of subscribers of an insurance company), the description of the disintegration of an atomic nucleus, or the number of defectives in the daily production of a factory.

3. Let (X_n) be an independent sequence of real, square integrable, identically distributed random variables such that $V(X_n) > 0$. Suppose we choose each $k_n = n$ and let

$$X_{nj} \equiv \frac{X_j - E(X_j)}{\sigma(X_1 + \cdots + X_n)} \qquad (j = 1, \ldots, n).$$

Then the sequence (P_{S_n}) converges weakly to the *normal distribution* $\nu_{0,1}$.

The proof of this is our first example of the usefulness of Fourier analysis. We can obviously assume that $E(X_j) = 0$. By hypothesis, the distributions P_{X_n} and standard deviations $\sigma(X_n)$ are independent of n; we set $\mu \equiv P_{X_n}$ and $\sigma \equiv \sigma(X_n)$. Then S_n is written in the form

$$S_n = \frac{X_1 + \cdots + X_n}{\sigma\sqrt{n}}.$$

Since $x \to x/\sigma\sqrt{n}$ is a linear mapping of $\mathbf{R}$ into itself, it follows that

$$\varphi_{S_n}(x) = \hat{\mu}\left(\frac{x}{\sigma\sqrt{n}}\right)^n \qquad (x \in \mathbf{R})$$

is the characteristic function of S_n according to the familiar rules for Fourier transforms. By the Continuity Theorem, we need to show that for every $x \in \mathbf{R}$,

$$\lim_{n\to\infty} \hat{\mu}\left(\frac{x}{\sigma\sqrt{n}}\right)^n = e^{-x^2/2}.$$

Due to the square integrability of the X_n, the moment of second order $M_2 = V(X_n) = \sigma^2$ exists for μ; from (8.3.4) and (8.3.5), observing that $M_1 = E(X_n) = 0$, we obtain [see Section 8.3, Problem 1]

$$\hat{\mu}(x) = 1 - \sigma^2 \cdot \frac{x^2}{2} + \Theta(x)\,\frac{x^2}{2},$$

where Θ is a complex function on $\mathbf{R}$ such that

$$|\Theta(x)| \leqq \sup_{0 \leqq \vartheta \leqq 1} |\hat{\mu}''(\vartheta x) - \hat{\mu}''(0)|,$$

that is, such that $\lim_{x \to 0} \Theta(x) = 0$. We have to examine the limit behavior of

$$\hat{\mu}\left(\frac{x}{\sigma \sqrt{n}}\right)^n = \left(1 + \frac{-x^2/2 + \Theta_n x^2}{n}\right)^n$$

as $n \to \infty$, where $\Theta_n \equiv 1/2\sigma^2\, \Theta(x/\sigma\sqrt{n})$. But now $\lim \Theta_n = 0$ and thus $e^{-x^2/2}$ is in fact the limit as $n \to \infty$.

If in particular $P_{X_n} = \beta_1^p$ is the binomial distribution with $0 < p < 1$, then the above result is the so-called De Moivre-Laplace Theorem.

4. Let (X_n) be an independent sequence of real random variables each with Cauchy distribution γ_1. Suppose we choose $k_n = n$ and let $X_{nj} \equiv (1/n)X_j$, that is,

$$S_n = \frac{1}{n}(X_1 + \cdots + X_n).$$

Then by Section 8.2, Example 3, $X_1 + \cdots + X_n$ has distribution γ_n, and thus $P_{S_n} = \gamma_1$ for all n. Therefore, trivially, the sequence (P_{S_n}) converges weakly to the *Cauchy distribution* γ_1.

This example once again emphasizes how essential the integrability of the random variables is for validity of the weak (or strong) law of large numbers. According to Theorem 7.7.2 we are dealing with weak convergence of the sequence (P_{S_n}) to ϵ_0.

The observation that we can actually obtain *every* probability measure ν on $\mathbf{R}$ as the weak limit of a sequence (P_{S_n}) is crucial for what follows. Suppose we choose a real random variable Y with distribution $P_Y = \nu$ and set $k_n = 1$, $X_{n1} = Y$ for all $n = 1, 2, \ldots$. Then the sequence (P_{S_n}) is constantly equal to ν and therefore trivially converges weakly to ν.

This example shows that a summand Y may dominate the sum S_n. But in many theoretical and practical examples, the variables S_n are composed of summands whose influence becomes arbitrarily small for large n. Thus, as we shall show, the consequences for the limits of weakly convergent sequences (P_{S_n}) which are then possible are the more noteworthy.

First, we propose the task of defining when there should not be a predominant influence of individual summands on the behavior of the sums S_n. The following definition will soon prove to be useful.

9.1.1. Definition. A family $(X_{nj})_{\substack{j=1,\ldots,k_n \\ n=1,2,\ldots}}$ of real random variables is said to be *asymptotically negligible* if

$$\lim_{n\to\infty} \max_{1\leqq j\leqq k_n} P\{|X_{nj}| \geqq \epsilon\} = 0, \qquad \text{for all } \epsilon > 0. \tag{9.1.4}$$

Thus it is required that $P\text{-}\lim_{n\to\infty} X_{nj} = 0$ uniformly in j.

Examples

5. If each of the random variables X_{nj} is square integrable with expected value $E(X_{nj}) = 0$ and if

$$\lim_{n\to\infty} \max_{1\leqq j\leqq k_n} V(X_{nj}) = 0, \tag{9.1.5}$$

then the family (X_{nj}) is asymptotically negligible. This follows from the Chebyshev inequality, according to which

$$P\{|X_{nj}| \geqq \epsilon\} \leqq \frac{1}{\epsilon^2} V(X_{nj}) \qquad (\epsilon > 0).$$

6. If the random variables X_{nj} all have the same distribution μ, then the family $((1/n)X_{nj})$ is asymptotically negligible. We have

$$P\left\{\left|\frac{1}{n} X_{nj}\right| \geqq \epsilon\right\} = P\{|X_{nj}| \geqq n\epsilon\} = \mu(A_n),$$

if we set $A_n \equiv \{x \in \mathbf{R}: |x| \geqq n\epsilon\}$. Since $A_n \downarrow \varnothing$, $\lim \mu(A_n) = 0$ and hence (9.1.4) follows.

7. In all four introductory examples, where we have weak convergence to ϵ_0, π_α, $\nu_{0,1}$, or γ_1, the basic families involved (X_{nj}) are asymptotically negligible. In fact, for Example 1, this follows from Example 5 and condition (6.5.2). We need to observe only that $E(X_{nj}) = 0$ and

$$\max_{1\leqq j\leqq n} V(X_{nj}) = \frac{1}{n^2} \max_{1\leqq j\leqq n} V(X_j) \leqq \frac{1}{n^2}\sum_{j=1}^{n} V(X_j).$$

In Example 2, X_{nj} has distribution $\beta_1^{p_n}$ and $\lim p_n = 0$. Thus we have to observe only that

$$P\{|X_{nj}| \geqq \epsilon\} = \begin{cases} 0, & \text{if } \epsilon > 1 \\ p_n, & \text{if } 0 < \epsilon \leqq 1, \end{cases}$$

for all n and j.

Example 3 is settled by Example 5 since obviously

$$E(X_{nj}) = 0 \qquad \text{and} \qquad V(X_{nj}) = \frac{V(X_j)}{V(X_1 + \cdots + X_n)} = \frac{1}{n}$$

for all n and j.

Finally, Example 6 shows that the family (X_{nj}) of Example 4 is also asymptotically negligible.

8. The example preceding Definition 9.1.1 provides us with a family which is not asymptotically negligible. A family $(X_{n1})_{n=1,2,\ldots}$ for which $k_n = 1$ for all n is asymptotically negligible if and only if the sequence (X_{n1}) converges stochastically to zero.

By Theorem 7.7.2 and the Continuity Theorem, a sequence (X_n) of real random variables converges stochastically to zero if and only if the sequence (φ_{X_n}) of characteristic functions converges pointwise to $\hat{\varepsilon}_0 = 1$. The following lemma sharpens part of this statement.

9.1.2. Lemma. If a family $(X_{nj})_{j=1,\ldots,k_n}$ of real random variables is asymptotically negligible, then for the associated family $(\varphi_{X_{nj}})$ of characteristic functions,

$$\lim_{n\to\infty} \max_{1\leqq j\leqq k_n} |\varphi_{X_{nj}}(x) - 1| = 0, \qquad \text{for all } x \in \mathbf{R}. \tag{9.1.6}$$

Thus the pointwise convergence of $(\varphi_{X_{nj}})$ to 1 as $n \to \infty$ is uniform in j.

Proof. For all allowable values of n and j and for all $x \in \mathbf{R}$, we have

$$\begin{aligned} |\varphi_{X_{nj}}(x) - 1| &= \left| \int (e^{ixy} - 1)\, P_{X_{nj}}(dy) \right| \\ &\leqq \int_{|y|<\epsilon} |e^{ixy} - 1|\, P_{X_{nj}}(dy) + \int_{|y|\geqq\epsilon} |e^{ixy} - 1|\, P_{X_{nj}}(dy) \\ &\leqq \epsilon|x| + 2P\{|X_{nj}| \geqq \epsilon\}. \end{aligned}$$

This follows from

$$|e^{it} - 1| = \left| \int_0^t e^{i\tau}\, d\tau \right| \leqq |t|, \qquad \text{for all } t \in \mathbf{R}. \tag{9.1.7}$$

Thus we have

$$\max_{1\leqq j\leqq k_n} |\varphi_{X_{nj}}(x) - 1| \leqq \epsilon|x| + 2 \max_{1\leqq j\leqq k_n} P\{|X_{nj}| \geqq \epsilon\}$$

and hence the assertion. ┘

PROBLEM

Let $(X_n)_{n \in \mathbf{N}}$ be an independent sequence of identically distributed, real, integrable random variables. Use the Fourier transform technique developed in Example 3 in order to prove that the sequence (X_n) satisfies the weak law of large numbers.

9.2 THE CENTRAL LIMIT THEOREM

We now turn back to Example 3 of Section 9.1, which is especially significant in the question of limit distributions. The first reason for this comes from the observation that, under the hypotheses imposed on the sequence (X_n) in Example 3, the sequence (X_n) satisfies the strong law of large numbers (by Theorem 6.4.2), and thus as $n \to \infty$, $(1/n)\Sigma_{j=1}^{n} X_j$ converges P-almost surely to the expected value $\eta = E(X_n)$ which is independent of n. The weak convergence of the sequence (P_{S_n}) with

$$S_n = \frac{1}{\sigma\sqrt{n}} \sum_{j=1}^{n} (X_j - \eta) \qquad (n = 1, 2, \ldots)$$

to $\nu_{0,1}$ can then be used to compute approximately the probability of the error of the deviation of the random variable $(1/n)\Sigma_{j=1}^{n} X_j$ from η. By Theorem 7.7.11 we have

$$\lim_{n\to\infty} P\left\{\alpha \leqq \frac{1}{\sigma\sqrt{n}} \sum_{j=1}^{n} (X_j - \eta) < \beta\right\} = \frac{1}{\sqrt{2\pi}} \int_{\alpha}^{\beta} e^{-x^2/2}\, dx$$

uniformly in α and β. The numbers

$$P\left\{\gamma \leqq \frac{1}{n} \sum_{j=1}^{n} X_j - \eta < \delta\right\} \qquad \text{and} \qquad \frac{1}{\sqrt{2\pi}} \int_{\sqrt{n}(\gamma/\sigma)}^{\sqrt{n}(\delta/\sigma)} e^{-x^2/2}\, dx$$

thus are arbitrarily close to each other for large n, and indeed uniformly in γ and $\delta \in \mathbf{R}$.

A second reason for the significance of Example 3 is seen from the observation that in the domain of applications we often encounter normally distributed random variables if they can be interpreted as the sum of a large number of independent variables. We may have such a situation, for example, with the total error of a measurement.

Thus, in general, we are faced with the problem of deciding when, for an independent sequence (X_n) of real random variables, the sequence (P_{S_n})

introduced in Example 3 converges weakly to the standard normal distribution $\nu_{0,1}$. In the sense of the following definition, this is the question of the validity of the Central Limit Theorem.

9.2.1. Definition. Let $(X_n)_{n\in\mathbf{N}}$ be an independent sequence of real, square integrable random variables with variance $V(X_n) > 0$. We say that the *Central Limit Theorem* holds for the sequence if the sequence (P_{S_n}) of distributions of the so-called standardized sums

$$S_n \equiv \frac{\sum_{j=1}^{n} (X_j - E(X_j))}{\sigma(X_1 + \cdots + X_n)} \tag{9.2.1}$$

converges weakly to $\nu_{0,1}$.

Thus Example 3 tells us that the Central Limit Theorem holds for the sequence (X_n) in particular if the sequence is *identically distributed.*

For a given independent sequence (X_n) of real, square integrable random variables with variances $V(X_n) > 0$, we let

$$\sigma_n \equiv \sigma(X_n), \tag{9.2.2}$$

$$s_n \equiv \sigma(X_1 + \cdots + X_n) = (\sigma_1^2 + \cdots + \sigma_n^2)^{1/2} \tag{9.2.3}$$

$$\eta_n \equiv E(X_n). \tag{9.2.4}$$

The reasoning of the preceding section makes it clear that a sequence (X_n) can satisfy the Central Limit Theorem only when none of the variables X_n exerts a dominating influence on the distribution of the sum S_n. On the basis of the next theorem, we shall see that it is not sufficient to require the family of random variables

$$X_{nj} \equiv \frac{1}{s_n}(X_j - E(X_j)) \qquad (j = 1, \ldots, n;\ n = 1, 2, \ldots) \tag{9.2.5}$$

to be asymptotically negligible. Rather, the following condition will turn out to be crucial:

$$\lim_{n\to\infty} L_n(\epsilon) = 0, \qquad \text{for every } \epsilon > 0, \tag{9.2.6}$$

where

$$L_n(\epsilon) \equiv \frac{1}{s_n^2}\sum_{j=1}^{n} \int_{|x-\eta_j|\geqq \epsilon s_n} (x - \eta_j)^2 P_{X_j}(dx). \tag{9.2.6$'$}$$

This is known as the *Lindeberg condition.* We immediately give several examples.

Examples

1. If the sequence (X_n) is *identically distributed*, that is, $\mu \equiv P_{X_n}$ and then $\eta \equiv \eta_n$, $\sigma \equiv \sigma_n$ are independent of n, then the Lindeberg condition is satisfied. We need note only that $s_n^2 = n\sigma^2$ and thus $\lim s_n = +\infty$ and

$$L_n(\epsilon) = \frac{1}{\sigma^2}\int_{|x-\eta|\geqq \epsilon s_n} (x-\eta)^2\mu(dx)$$

holds.

2. If the sequence (X_n) satisfies the so-called *Lyapunov condition*, that is, there exists a $\delta > 0$ such that

$$\lim_{n\to\infty} \frac{1}{s_n^{2+\delta}}\sum_{j=1}^{n} E(|X_j - \eta_j|^{2+\delta}) = 0, \tag{9.2.7}$$

then the Lindeberg condition is satisfied. Indeed, for every $\epsilon > 0$, $|x - \eta| \geqq \epsilon s_n$ implies the inequality $|x - \eta|^{2+\delta} \geqq |x - \eta|^2(\epsilon s_n)^\delta$. Hence,

$$L_n(\epsilon) \leqq \frac{1}{\epsilon^\delta s_n^{2+\delta}}\sum_{j=1}^{n}\int_{|x-n_j|\geqq \epsilon s_n} |x-\eta_j|^{2+\delta}P_{X_j}(dx) \leqq \frac{1}{\epsilon^\delta s_n^{2+\delta}}\sum_{j=1}^{n} E(|X_j-\eta_j|^{2+\delta})$$

for each $\epsilon > 0$.

3. If the sequence (X_n) is *uniformly bounded* and $\lim s_n = +\infty$, then the Lyapunov condition is satisfied for every $\delta > 0$ and thus the Lindeberg condition is satisfied. Indeed, by hypothesis, there is an $\alpha > 0$ with $|X_n| \leqq \alpha/2$, that is with $|\eta_n| \leqq E(|X_n|) \leqq \alpha/2$. Hence $|X_n - \eta_n| \leqq \alpha$ for all $n = 1, 2, \ldots$. But then for all $\delta > 0$ the Lyapunov condition follows from the inequality

$$\frac{1}{s_n^{2+\delta}}\sum_{j=1}^{n} E(|X_j-\eta_j|^{2+\delta}) \leqq \frac{\alpha^\delta}{s_n^{2+\delta}}\sum_{j=1}^{n} E(|X_j-\eta_j|^2) = \left(\frac{\alpha}{s_n}\right)^\delta.$$

The connection between the Lindeberg condition and the asymptotic negligibility of the family (X_{nj}) defined in (9.2.5) is the following:

9.2.2. Lemma. If the sequence $(X_n)_{n\in\mathbf{N}}$ satisfies the Lindeberg condition, then it also satisfies the following condition, named after W. Feller:

$$\lim_{n\to\infty}\max_{1\leqq j\leqq n}\left(\frac{\sigma_j}{s_n}\right) = 0. \tag{9.2.8}$$

From the *Feller condition* it follows that the family $(X_{nj})_{\substack{j=1,\ldots,n\\ n=1,2,\ldots}}$ defined by (9.2.5) is asymptotically negligible.

Proof. For every $j = 1, \ldots, n$,

$$\sigma_j^2 = \int (x - \eta_j)^2 P_{X_j}(dx) \leqq \epsilon^2 s_n^2 + \int_{|x-\eta_j| \geqq \epsilon s_n} (x - \eta_j)^2 P_{X_j}(dx)$$
$$\leqq \epsilon^2 s_n^2 + \sum_{j=1}^{n} \int_{|x-\eta_j| \geqq \epsilon s_n} (x - \eta_j)^2 P_{X_j}(dx)$$

and hence

$$\max_{1 \leqq j \leqq n} \left(\frac{\sigma_j}{s_n}\right)^2 \leqq \epsilon^2 + L_n(\epsilon),$$

for arbitrary $\epsilon > 0$. Thus the Lindeberg condition implies the Feller condition. Now we have $(\sigma_j/s_n)^2 = V(X_{nj})$, that is,

$$\lim_{n \to \infty} \max_{1 \leqq j \leqq n} V(X_{nj}) = 0$$

by the Feller condition. The rest of the assertion now follows from Section 9.1, Example 5. ⌋

Remark. 1. The Feller condition is equivalent to the simultaneous validity of

$$\lim_{n \to \infty} s_n = +\infty, \qquad (9.2.9)$$

$$\lim_{n \to \infty} \frac{\sigma_n}{s_n} = 0, \qquad (9.2.10)$$

as we can verify immediately.

Now we are in a position to formulate and prove the main result.

9.2.3. Theorem (Lindeberg-Feller). For every independent sequence $(X_n)_{n \in \mathbf{N}}$ of real, square integrable random variables with variances $V(X_n) > 0$, the following statements are equivalent:

(a) The Central Limit Theorem holds and the sequence satisfies the Feller condition.

(b) The Central Limit Theorem holds and the family $(X_{nj})_{\substack{i=1,\ldots,n \\ n=1,2,\ldots}}$ is asymptotically negligible.

(c) The sequence satisfies the Lindeberg condition.

Proof. By Lemma 9.2.2 we need prove only the implications (b) $\Rightarrow$ (c) and (c) $\Rightarrow$ (a), and of the latter, only the part concerning the validity of the Central Limit Theorem. We can assume $\eta_n = E(X_n) = 0$ for all n for both proofs. It suffices to center every variable on its expected value. If we now set

$$\mu_{nj} \equiv P_{X_j/s_n} \qquad \text{and} \qquad \tau_{nj} \equiv \frac{\sigma_j}{s_n} \qquad (j = 1, \ldots, n;\ n = 1, 2, \ldots),$$

then $\mu_{nj} \in \mathfrak{M}^1(\mathbf{R})$ has 0 and τ_{nj}^2 as its first and second moments respectively and $\Sigma_{j=1}^n \tau_{nj}^2 = 1$. The Lindeberg term $L_n(\epsilon)$ can then be written as

$$L_n(\epsilon) = \sum_{j=1}^{n} \int_{|x| \geqq \epsilon} x^2 \mu_{nj}(dx).$$

1. First we prove the remaining part of the implication (c) $\Rightarrow$ (a). By (8.3.4) and (8.3.5) we have

$$\hat{\mu}(x) = 1 - \sigma^2 \frac{x^2}{2} + \Theta(x) \frac{x^2}{2},$$

for the Fourier transform of any measure $\mu \in \mathfrak{M}^1(\mathbf{R})$ with $\int x\mu(dx) = 0$ and $\sigma^2 \equiv \int x^2 \mu(dx) < +\infty$, where the function $\Theta: \mathbf{R} \to \mathbf{C}$ satisfies the condition

$$|\Theta(x)| \leqq \sup_{0 \leqq \vartheta \leqq 1} |\hat{\mu}''(\vartheta x) - \hat{\mu}''(0)| \qquad (x \in \mathbf{R}).$$

Moreover by Theorem 8.3.3

$$\hat{\mu}''(x) = -\int e^{ixy} y^2 \mu(dy),$$

that is,

$$|\Theta(x)| \leqq \sup_{0 \leqq \vartheta \leqq 1} \int |e^{i\vartheta xy} - 1| y^2 \mu(dy).$$

If we note that the convergence of e^{ity} to 1 as $y \to 0$ is uniform on every compact t-interval, then for given $x \in \mathbf{R}$, for every $\epsilon > 0$ there exists a $\delta > 0$ such that $|e^{i\vartheta xy} - 1| < \epsilon$ for all $y \in \mathbf{R}$ with $|y| < \delta$ and all $\vartheta \in [0,1]$. Therefore, for fixed $x \in \mathbf{R}$, we have the inequalities

$$\begin{aligned} |\Theta(x)| &\leqq \sup_{0 \leqq \vartheta \leqq 1} \int_{|y| < \delta} |e^{i\vartheta xy} - 1| y^2 \mu(dy) + \sup_{0 \leqq \vartheta \leqq 1} \int_{|y| \geqq \delta} |e^{i\vartheta xy} - 1| y^2 \mu(dy) \\ &\leqq \epsilon\sigma^2 + 2 \int_{|y| \geqq \delta} y^2 \mu(dy). \end{aligned}$$

We apply this result to both μ_{nj} and the normal distribution ν_{0,τ_{nj}^2}. On the one hand, we obtain

$$\begin{aligned} \hat{\mu}_{nj}(x) &= 1 - \tau_{nj}^2 \frac{x^2}{2} + \Theta_{nj}(x) \frac{x^2}{2}, \\ |\Theta_{nj}(x)| &\leqq \epsilon\tau_{nj}^2 + 2 \int_{|y| \geqq \delta} y^2 \mu_{nj}(dy) \end{aligned} \tag{9.2.11}$$

and, on the other hand,

$$\begin{aligned} \hat{\nu}_{0,\tau_{nj}^2}(x) &= e^{-\tau_{nj}^2(x^2/2)} = 1 - \tau_{nj}^2 \frac{x^2}{2} + \Theta_{nj}^*(x) \frac{x^2}{2}, \\ |\Theta_{nj}^*(x)| &\leqq \epsilon\tau_{nj}^2 + 2 \int_{|y| \geqq \delta} y^2 \nu_{0,\tau_{nj}^2}(dy). \end{aligned} \tag{9.2.12}$$

Here δ depends on ϵ and x in the described manner.

The rest of this part of the proof is based on a simple majorization, which is derived from the inequality

$$\left| \prod_{\varkappa=1}^{k} a_\varkappa - \prod_{\varkappa=1}^{k} b_\varkappa \right| \leqq \sum_{\varkappa=1}^{k} |a_\varkappa - b_\varkappa|,$$

which is valid for complex numbers $a_1, \ldots, a_k; b_1, \ldots, b_k$ of absolute value $\leqq 1$.[1] Using the independence of the sequence (X_n), we obtain

$$\begin{aligned} |\varphi_{S_n}(x) - e^{-x^2/2}| &= \left| \prod_{j=1}^{n} \hat{\mu}_{nj}(x) - \prod_{j=1}^{n} e^{-\tau_{nj}^2(x^2/2)} \right| \\ &\leqq \sum_{j=1}^{n} |\hat{\mu}_{nj}(x) - \hat{\nu}_{0,\tau_{nj}^2}(x)| = \frac{x^2}{2} \sum_{j=1}^{n} |\Theta_{nj}(x) - \Theta_{nj}^*(x)|, \end{aligned}$$

for the characteristic function φ_{S_n} of $S_n = \Sigma_{j=1}^n (X_j/s_n)$, and hence, by (9.2.11) and (9.2.12), the majorization

$$\begin{aligned} |\varphi_{S_n}(x) - \hat{\nu}_{0,1}(x)| \leqq x^2 \Big[\epsilon &+ \sum_{=1}^{n} \int_{|y| \geqq \delta} y^2 \mu_{nj}(dy) \\ &+ \sum_{j=1}^{n} \int_{|y| \geqq \delta} y^2 \nu_{0,\tau_{nj}^2}(dy) \Big], \end{aligned} \tag{9.2.13}$$

which holds for all $n = 1, 2, \ldots$. If we now set $\alpha_n \equiv \max_{1 \leqq j \leqq n} \tau_{nj}$ for the number involved in the Feller condition, then

$$\begin{aligned} \sum_{j=1}^{n} \int_{|y| \geqq \delta} y^2 \nu_{0,\tau_{nj}^2}(dy) &= \sum_{j=1}^{n} \tau_{nj}^2 \int_{|y| \geqq \delta/\tau_{nj}} y^2 \nu_{0,1}(dy) \\ &\leqq \sum_{j=1}^{n} \tau_{nj}^2 \int_{|y| \geqq \delta/\alpha_n} y^2 \nu_{0,1}(dy) = \int_{|y| \geqq \delta/\alpha_n} y^2 \nu_{0,1}(dy) \end{aligned}$$

and hence

$$\lim_{n \to \infty} \sum_{j=1}^{n} \int_{|y| \geqq \delta} y^2 \nu_{0,\tau_{nj}^2}(dy) = 0,$$

[1] This follows from the triangle inequality applied to the identity $\Pi a_\varkappa - \Pi b_\varkappa = (a_1 - b_1) \cdot a_2 \cdot \ldots \cdot a_k + b_1(a_2 - b_2) \cdot a_3 \cdot \ldots \cdot a_k + \cdots + b_1 \cdot \ldots \cdot b_{k-1}(a_k - b_k)$.

since $\lim \alpha_n = 0$ by Lemma 9.2.2. From this, the Lindeberg condition, and (9.2.13), it now follows that

$$\lim_{n\to\infty} \varphi_{S_n}(x) = \hat{\nu}_{0,1}(x) \qquad (x \in \mathbf{R}),$$

and thus, by the Continuity Theorem, we have the weak convergence of (P_{S_n}) to $\nu_{0,1}$.

2. Finally, we prove (b) $\Rightarrow$ (c). Since the family of random variables $X_{nj} = (1/s_n)X_j$ is asymptotically negligible and $P_{X_{nj}} = \mu_{nj}$, then by Lemma 9.1.2 for every fixed $x \in \mathbf{R}$:

$$\lim_{n\to\infty} \max_{1\leqq j\leqq n} |\hat{\mu}_{nj}(x) - 1| = 0. \tag{9.2.14}$$

Hence it follows that

$$\max_{1\leqq j\leqq n} |\hat{\mu}_{nj}(x) - 1| \leqq \tfrac{1}{2} \qquad \text{for } n \text{ sufficiently large.} \tag{9.2.15}$$

For every complex number z with $|z| \leqq \frac{1}{2}$,

$$\log(1+z) = z + \sum_{\nu=2}^{\infty} \frac{(-1)^{\nu-1}}{\nu} z^\nu$$

is the Taylor expansion of the principal branch of the logarithm and

$$\left| \sum_{\nu=2}^{\infty} \frac{(-1)^{\nu-1}}{\nu} z^\nu \right| \leqq \frac{1}{2} \sum_{\nu=2}^{\infty} |z|^\nu = \frac{1}{2} \frac{|z|^2}{1-|z|} \leqq |z|^2.$$

Then by (9.2.15), for all sufficiently large n and all $j = 1, \ldots, n$,

$$\log \hat{\mu}_{nj}(x) = \hat{\mu}_{nj}(x) - 1 + R_{nj}(x), \tag{9.2.16}$$

where $|R_{nj}(x)| \leqq |\hat{\mu}_{nj}(x) - 1|^2$. If, moreover, we take into account the inequalities[2]

$$\begin{aligned} |\hat{\mu}_{nj}(x) - 1| &= \left| \int (e^{ixy} - 1 - ixy)\mu_{nj}(dy) \right| \\ &\leqq \int |e^{ixy} - 1 - ixy| \mu_{nj}(dy) \leqq \frac{x^2}{2} \int y^2 \mu_{nj}(dy) = \tau_{nj}^2 \frac{x^2}{2}, \end{aligned}$$

[2] For all $t \in \mathbf{R}$, $|e^{it} - 1 - it| \leqq t^2/2$. In the proof we can obviously assume that $t \geqq 0$. Then we conclude as follows: $|e^{it} - 1| = \left| \int_0^t e^{i\tau}\, d\tau \right| \leqq t$; $|e^{it} - 1 - it| = \left| \int_0^t (e^{i\tau} - 1)\, d\tau \right| \leqq \int_0^t \tau\, d\tau = \dfrac{t^2}{2}$.

which follow from $0 = E(X_{nj}) = \int y\mu_{nj}(dy)$, then we obtain that $|R_{nj}(x)| \leqq |\hat{\mu}_{nj}(x) - 1|\tau_{nj}^2(x^2/2)$ and hence, since $\Sigma_{j=1}^n \tau_{nj}^2 = 1$, we have

$$\left|\sum_{j=1}^{n} R_{nj}(x)\right| \leqq \frac{x^2}{2} \max_{1\leqq j\leqq n} |\hat{\mu}_{nj}(x) - 1|$$

for sufficiently large n, that is, by (9.2.14),

$$\lim_{n\to\infty} \sum_{j=1}^{n} R_{nj}(x) = 0. \tag{9.2.17}$$

Moreover, since we assume that the Central Limit Theorem holds, we know that

$$\varphi_{S_n}(x) = \prod_{j=1}^{n} \hat{\mu}_{nj}(x)$$

converges to $\hat{\nu}_{0,1}(x) = e^{-x^2/2}$ as $n \to \infty$. Therefore from (9.2.16) and (9.2.17) we obtain

$$\lim_{n\to\infty} e^{\sum_{j=1}^{n} (\hat{\mu}_{nj}(x) - 1)} = e^{-x^2/2}$$

and hence

$$\lim_{n\to\infty} e^{\sum_{j=1}^{n} \Re(\hat{\mu}_{nj}(x) - 1)} = \lim_{n\to\infty} |e^{\sum_{j=1}^{n} (\hat{\mu}_{nj}(x) - 1)}| = e^{-x^2/2},$$

that is, finally,

$$\lim_{n\to\infty} \sum_{j=1}^{n} \Re(\hat{\mu}_{nj}(x) - 1) = -\frac{x^2}{2}. \qquad (9.2.18)^3$$

Simple reasoning now completes the proof: For arbitrary $x \neq 0$, the sequence

$$T_n \equiv \frac{1}{2} + \frac{1}{x^2}\sum_{j=1}^{n} \Re(\hat{\mu}_{nj}(x) - 1) = \frac{1}{2} - \frac{1}{x^2}\sum_{j=1}^{n}\int (1 - \cos xy)\mu_{nj}(dy)$$

approaches zero as $n \to \infty$, by (9.2.18). For arbitrary given $\epsilon > 0$, con-

[3] The reader should note that the equality $\log \varphi_{S_n}(x) = \Sigma \log \hat{\mu}_{nj} = \Sigma(\hat{\mu}_{nj} - 1) + \Sigma R_{nj}$ holds only modulo $2\pi i$ and therefore the sign of the real part cannot simply be omitted.

sider the expression

$$T_n^\epsilon \equiv \frac{1}{x^2}\sum_{j=1}^{n}\int_{|y|\geqq\epsilon}(1 - \cos xy)\mu_{nj}(dy).$$

Using the Chebyshev inequality (6.3.11) we can majorize this as follows:

$$0 \leqq T_n^\epsilon \leqq \frac{2}{x^2}\sum_{j=1}^{n}\int_{|y|\geqq\epsilon} d\mu_{nj} = \frac{2}{x^2}\sum_{j=1}^{n} P\left\{\left|\frac{1}{s_n}X_j\right| \geqq \epsilon\right\} \leqq \frac{2}{\epsilon^2 x^2}\sum_{j=1}^{n}\tau_{nj}^2 = \frac{2}{\epsilon^2 x^2}.$$

This leads to the following bound for $L_n(\epsilon)$:

$$\begin{aligned}\frac{1}{2}L_n(\epsilon) &= \frac{1}{2}\sum_{j=1}^{n}\left(\tau_{nj}^2 - \int_{|y|<\epsilon} y^2\mu_{nj}(dy)\right) = \frac{1}{2} - \frac{1}{x^2}\sum_{j=1}^{n}\int_{|y|<\epsilon}\frac{x^2y^2}{2}\mu_{nj}(dy)\\ &\leqq \frac{1}{2} - \frac{1}{x^2}\sum_{j=1}^{n}\int_{|y|<\epsilon}|1 - e^{ixy} + ixy|\mu_{nj}(dy)\\ &\leqq \frac{1}{2} - \frac{1}{x^2}\sum_{j=1}^{n}\left|\int_{|y|<\epsilon}(1 - e^{ixy} + ixy)\mu_{nj}(dy)\right|\\ &\leqq \frac{1}{2} - \frac{1}{x^2}\sum_{j=1}^{n}\int_{|y|<\epsilon}(1 - \cos xy)\mu_{nj}(dy) = T_n + T_n^\epsilon \leqq T_n + \frac{2}{\epsilon^2x^2}\end{aligned}$$

(for every $x \neq 0$ and $\epsilon > 0$). The Lindeberg condition now follows, since $\lim_{n\to\infty} T_n = 0$ for all $x \neq 0$ and $\lim_{x\to+\infty}\frac{2}{\epsilon^2x^2} = 0$ for each $\epsilon > 0$.

Thus the theorem is completely proved. ┘

Remark. 2. Because of Example 1, the result of Section 9.1, Example 3 and also the De Moivre-Laplace Theorem, are contained in the Lindeberg-Feller Theorem as special cases.

3. In the Lindeberg-Feller Theorem we cannot do without the Feller condition or the asymptotic negligibility of the family (X_{nj}) in showing (a) ⇒ (c) or (b) ⇒ (c), respectively. This can be seen from the following.

Example

4. Let (σ_n) be a sequence of positive numbers with $\Sigma_{n=1}^{\infty}\sigma_n^2 < \infty$ and let (X_n) be an independent sequence (which exists by Corollary 1.5.4) of real random variables with $P_{X_n} = \nu_{0,\sigma_n^2}$ $(n = 1, 2, \ldots)$. Then the Central

Limit Theorem holds since each of the standardized sums S_n has distribution $\nu_{0,1}$. On the other hand, the Feller condition and therefore the Lindeberg condition are violated since $s_n^2 = \sigma_1^2 + \cdots + \sigma_n^2$, $n = 1, 2, \ldots$ is bounded and thus (9.2.9) is not satisfied.

PROBLEMS

1. Let (X_n) be an independent sequence of real random variables with $\nu_{0,1}$ as common distribution. Prove that the squared sequence (X_n^2) satisfies the Central Limit Theorem. Can $\nu_{0,1}$ be replaced by other distributions in order to have the same conclusion?
2. Let (X_n) be an independent sequence of real, square-integrable random variables. Prove: $\lim_{n \to \infty} s_n = +\infty$ if and only if $\sum_{n=1}^{\infty} V(X_n) = +\infty$ [where $s_n \equiv \sigma(X_1 + \cdots + X_n)$].
3. Let (X_n) be an independent sequence of real random variables such that each X_n only attains the values 0 and 1 with positive probability. Prove that (X_n) satisfies the Central Limit Theorem if the series

$$\sum_{n=1}^{\infty} P\{X_n = 0\} \cdot P\{X_n = 1\}$$

diverges.
4. Let λ be a real number, and let $(X_n)_{n \in \mathbf{N}}$ be a corresponding independent sequence of real random variables with the properties stated in part (b) of the Problem of Section 5.4. Prove:
(a) For $\lambda < -\frac{1}{2}$ the sequence (X_n) does not satisfy the Feller condition.
(b) For $\lambda \geqq -\frac{1}{2}$ one has

$$S_n^2 \geqq \begin{cases} \int_1^{n+1} x^{2\lambda}\, dx, & \text{if } \lambda \leqq 0 \\ \int_0^n x^{2\lambda}\, dx, & \text{if } \lambda \geqq 0. \end{cases}$$

(c) Conclude that (X_n) satisfies the Central Limit Theorem for all $\lambda \geqq -\frac{1}{2}$.
5. Assume that the Central Limit Theorem holds for an independent sequence (X_n) of real, square-integrable random variables with variances $V(X_n) > 0$. Prove that

$$P\left\{\left|\frac{1}{n}\sum_{i=1}^{n} (X_i - E(X_i))\right| < \epsilon\right\} - \frac{1}{\sqrt{2\pi}} \int_{-\epsilon(n/s_n)}^{\epsilon(n/s_n)} e^{-(x^2/2)}\, dx$$

converges to zero as $n \to \infty$ uniformly for all $\epsilon > 0$. Conclude: The sequence (X_n) does not satisfy the weak law of large numbers if the sequence (n/s_n) is bounded.

6. Apply the result of Problem 5 to the sequence (X_n) of Problem 4 and prove that (X_n) does not satisfy the weak law of large numbers if $\lambda \geqq \frac{1}{2}$. This completes the discussion of Section 6.4, Problem 1 and Section 6.5, Problem 2.

9.3 INFINITELY DIVISIBLE DISTRIBUTIONS

In the first four examples of Section 9.1, the probability measures ϵ_0, π_α, $\nu_{0,1}$, and γ_1 on $\mathbf{R}$ were obtained as limit distributions of sums $S_n = X_{n1} + \cdots + X_{nk_n}$ of the asymptotically negligible families (X_{nj}) considered there. By Section 5.3, Remark 4, these probability measures are all infinitely divisible. We now show that this is no coincidence. But in order to do this we first establish some properties of such probability measures on $\mathbf{R}$ (or, more generally, on $\mathbf{R}^p$, $p = 1, 2, \ldots$).

9.3.1. Definition. A measure $\mu \in \mathfrak{M}^1(\mathbf{R}^p)$ is said to be *infinitely divisible* if for every natural number n there exists a measure $\mu_n \in \mathfrak{M}^1(\mathbf{R}^p)$ whose n-fold convolution product with itself is equal to μ:

$$\mu = \mu_n * \cdots * \mu_n. \tag{9.3.1}$$

By the Uniqueness Theorem for Fourier transforms this is equivalent to

$$\hat{\mu} = (\hat{\mu}_n)^n. \tag{9.3.2}$$

Example

1. If $\mu^{(1)}, \ldots, \mu^{(q)}$ are finitely many infinitely divisible measures on $\mathbf{R}^p$, then $\nu \equiv \mu^{(1)} \otimes \cdots \otimes \mu^{(q)}$ is an infinitely divisible measure on $\mathbf{R}^{pq}$. In fact, for every $n \in \mathbf{N}$ and $i = 1, \ldots, q$ there are measures $\mu_n^{(i)} \in \mathfrak{M}^1(\mathbf{R}^p)$ whose n-fold convolution product $\mu_n^{(i)} * \cdots * \mu_n^{(i)}$ is equal to $\mu^{(i)}$. But then

$$\nu_n \equiv \mu_n^{(1)} \otimes \cdots \otimes \mu_n^{(q)}$$

is a measure from $\mathfrak{M}^1(\mathbf{R}^{pq})$ with ν as its n-fold convolution product.[4]

[4] If $\mu_1, \ldots, \mu_n$ and $\nu_1, \ldots, \nu_n$ are finite Borel measures on $\mathbf{R}^p$, then

$$(\mu_1 \otimes \cdots \otimes \mu_n) * (\nu_1 \otimes \cdots \otimes \nu_n) = (\mu_1 * \nu_1) \otimes \cdots \otimes (\mu_n * \nu_n).$$

This follows at once by Fourier transforms. By Theorems 8.1.5 and 8.1.6,

$$(x_1, \ldots, x_n) \to \hat{\mu}_1(x_1) \cdot \ldots \cdot \hat{\mu}_n(x_n) \cdot \hat{\nu}(x_1) \cdot \ldots \cdot \hat{\nu}(x_n) = \hat{\mu}_1(x_1)\hat{\nu}_1(x_1) \cdot \ldots \cdot \hat{\mu}_n(x_n)\hat{\nu}_n(x_n)$$

is the Fourier transform of both measures involved in the equality to be proved.

9.3.2. Theorem. If μ is an infinitely divisible measure on $\mathbf{R}^p$, then

$$\hat{\mu}(x) \neq 0, \qquad \text{for all } x \in \mathbf{R}^p. \tag{9.3.3}$$

There is exactly one continuous mapping $\varphi: \mathbf{R}^p \to \mathbf{R}$ such that $\varphi(0) = 0$ and

$$\hat{\mu} = |\hat{\mu}|e^{i\varphi}. \tag{9.3.4}$$

Proof. By (9.3.2) we have $|\hat{\mu}_n| = \sqrt[n]{|\hat{\mu}|}$ for all $n \in \mathbf{N}$ and hence $\lim_{n\to\infty} |\hat{\mu}_n(x)| = 1$ or 0 depending on whether $\hat{\mu}(x) \neq 0$ or $\hat{\mu}(x) = 0$. Since $\hat{\mu}(0) = 1$, we have $\hat{\mu}(x) \neq 0$ and thus $\lim_{n\to\infty} |\hat{\mu}_n(x)| = 1$ in a neighborhood of zero; hence $\lim |\hat{\mu}_n| = \lim |\hat{\mu}_n|^2$ is continuous at $x = 0$. By Theorem 8.1.5 $|\hat{\mu}_n|^2$ is the Fourier transform of the measure $\mu_n * S(\mu_n)$, where S denotes the reflection $x \to -x$. By the Continuity Theorem, $\lim |\hat{\mu}_n|^2$ must itself be the Fourier transform of a measure from $\mathfrak{M}^e(\mathbf{R}^p)$, and in particular must be continuous. Since $\lim |\hat{\mu}_n| = \lim |\hat{\mu}_n|^2$ assumes only the values 1 and 0 and the former value is taken on at $x = 0$, it follows that $\lim |\hat{\mu}_n(x)| = 1$ for all $x \in \mathbf{R}^p$. As a consequence of the introductory statements, we then have $\hat{\mu}(x) \neq 0$ for all $x \in \mathbf{R}^p$. The rest of the assertion follows from Theorem A.2.[5] ┘

9.3.3. Corollary 1. For every infinitely divisible measure μ on $\mathbf{R}^p$, each of the measures $\mu_n \in \mathfrak{M}^1(\mathbf{R}^p)$ is uniquely determined by the equality (9.3.1). With the notation introduced in Theorem 9.3.2, we have

$$\hat{\mu}_n = \sqrt[n]{|\hat{\mu}|}\, e^{i(\varphi/n)} \qquad (n = 1, 2, \ldots). \tag{9.3.5}$$

for its Fourier transform.

Proof. By Corollary A.2, for every $n = 1, 2, \ldots$, there is exactly one continuous mapping $\psi: \mathbf{R}^p \to \mathbf{C}$ such that $\psi(0) = 1$ and $\psi^n = \hat{\mu}$. This is given by

$$\psi = \sqrt[n]{|\hat{\mu}|}\, e^{i(\varphi/n)}.$$

Since $\hat{\mu}_n$ is continuous and $\hat{\mu}_n(0) = 1$, we see that $\psi = \hat{\mu}_n$ by (9.3.2). By the Uniqueness Theorem 8.2.4, μ_n is uniquely determined by ψ and thus also by μ. ┘

9.3.4. Corollary 2. The sequence of measures $\mu_n \in \mathfrak{M}^1(\mathbf{R}^p)$ associated with an infinitely divisible measure μ on $\mathbf{R}^p$ by (9.3.1) is weakly convergent to ϵ_0.

[5] This is the second proposition in the Appendix.

Proof. From (9.3.3) and (9.3.5) it follows that $\lim_{n\to\infty} \hat{\mu}_n(x) = 1$ for all $x \in \mathbf{R}^p$. Since $\hat{\epsilon}_0 = 1$, the assertion now follows from the Uniqueness Theorem. ⌋

Remarks. 1. Formula (9.3.5) does *not* say that $\hat{\mu}_n(x)$ is the principal value of the nth root of $\hat{\mu}(x)$. To see this, it suffices to consider the measure $\mu = \epsilon_1$ on $\mathbf{R}$. Then $\mu_n = \epsilon_{1/n}$, $\hat{\mu}(x) = e^{ix}$, and $\hat{\mu}_n(x) = e^{i(x/n)}$. Hence $\hat{\mu}_n(2\pi) = e^{i(2\pi/n)}$ which for $n = 2, 3, \ldots$ is not the principal value of the nth root of $\hat{\mu}(2\pi) = 1$.

2. Theorem 9.3.2 often makes it possible to verify that a measure $\mu \in \mathfrak{M}^1(\mathbf{R}^p)$ is *not* infinitely divisible. Examples of this are the so-called *rectangular* (or uniform) distribution on $\mathbf{R}$

$$\rho \equiv \tfrac{1}{2}1_{[-1,+1]}\lambda^1$$

with Fourier transform

$$\hat{\rho}(x) = \frac{\sin x}{x} \qquad (x \neq 0)$$

and the binomial distribution β_n^p for $p = \frac{1}{2}$ with Fourier transform $x \to 2^{-n}(1 + e^{ix})^n$. Both Fourier transforms vanish at $x = \pi$.

3. For $0 < p < 1$ and $p \neq \frac{1}{2}$, the Fourier transform of the *binomial distribution* β_n^p has no real zeros. Nonetheless, β_n^p is not infinitely divisible. We can represent β_1^p only in trivial way in the form $\beta_1^p = \mu * \nu$ with measures $\mu, \nu \in \mathfrak{M}^1(\mathbf{R})$.[6] Here trivial means a representation in which either μ or ν is a unit mass ϵ_a. It is easy to see that measures μ and $\nu \in \mathfrak{M}^1(\mathbf{R})$ with $\mu * \nu = \beta_1^p$ must be discrete and that the numbers s and t of points having positive μ- and ν-measures, respectively, must satisfy $s + t - 1 \leqq 2$; that is, $s = 1$ or $t = 1$. By considering such a decomposition into $n + 1$ factors instead of 2, one can prove that for $p \in]0,1[$, β_n^p is not infinitely divisible.

Corollary 1 above motivates us to exhibit a further aspect of the concept of infinitely divisible measures. To this end, we give:

9.3.5. Definition. Let $(\mu_t)_{t\in\mathbf{R}_+}$ be a family of probability measures $\mu_t \in \mathfrak{M}^1(\mathbf{R}^p)$ on $\mathbf{R}^p$ indexed by the elements of $\mathbf{R}_+$. Then if

$$\mu_{s+t} = \mu_s * \mu_t, \qquad \text{for all } s, t \in \mathbf{R}_+, \tag{9.3.6}$$

we call $(\mu_t)_{t\in\mathbf{R}_+}$ a *convolution semigroup* of probability measures on $\mathbf{R}^p$. It is called *continuous* if the mapping $t \to \mu_t$ of $\mathbf{R}_+$ into $\mathfrak{M}^1(\mathbf{R}^p)$ is vaguely, and hence also weakly continuous.

[6] Therefore β_1^p is a so-called *indecomposable* probability measure.

The following theorem now shows the connection with the previous investigation:

9.3.6. Theorem. Each of the measures μ_t of a convolution semigroup $(\mu_t)_{t \in \mathbf{R}_+}$ of probability measures on $\mathbf{R}^p$ is infinitely divisible. Conversely, if μ is an infinitely divisible measure on $\mathbf{R}^p$ and $t_0 > 0$ is a positive real number, then there is exactly one continuous convolution semigroup $(\mu_t)_{t \in \mathbf{R}_+}$ of probability measures on $\mathbf{R}^p$ such that $\mu_{t_0} = \mu$.

Proof. Let $(\mu_t)_{t \in \mathbf{R}_+}$ be a convolution semigroup. Then by (9.3.6), for arbitrary $t \in \mathbf{R}_+$ and $n \in \mathbf{N}$,

$$\mu_t = \mu_{t/n + \cdots + t/n} = \mu_{t/n} * \cdots * \mu_{t/n}$$

(with n summands or factors). Therefore every measure μ_t is infinitely divisible. For the proof of the second assertion, let μ and t_0 be given as above. For μ, by Theorem 9.3.2 there is exactly one continuous mapping $\varphi: \mathbf{R}^p \to \mathbf{R}$ such that $\varphi(0) = 0$ and $\hat{\mu} = |\hat{\mu}|e^{i\varphi}$. By Corollary 9.3.3, for every $n \in \mathbf{N}$ there is exactly one measure $\mu_{1/n} \in \mathfrak{M}^1(\mathbf{R}^p)$ with Fourier transform

$$\hat{\mu}_{1/n} = |\hat{\mu}|^{1/n} e^{i(\varphi/n)}.$$

Therefore, for every rational number $r = m/n \geqq 0$ with $n \in \mathbf{N}$ and $m \in \mathbf{N} \cup \{0\}$, $\mu_r \equiv \mu_{1/n} * \cdots * \mu_{1/n}$ (m factors) is the only probability measure on $\mathbf{R}^p$ such that

$$\hat{\mu}_r = |\hat{\mu}|^r e^{ir\varphi}.$$

For every $t \in \mathbf{R}_+$ there is a sequence (r_n) of rational numbers $\geqq 0$ converging to t. The sequence of functions $|\hat{\mu}|^{r_n} e^{ir_n\varphi}$, $n = 1, 2, \ldots$, converges pointwise to the continuous function $|\hat{\mu}|^t e^{it\varphi}$. By the Continuity Theorem 8.2.7 there is thus exactly one measure $\mu_t \in \mathfrak{M}^1(\mathbf{R}^p)$ such that $\hat{\mu}_t = |\hat{\mu}|^t e^{it\varphi}$. But then $\hat{\mu}_s \hat{\mu}_t = \hat{\mu}_{s+t}$, that is, $\mu_s * \mu_t = \mu_{s+t}$ for all $s, t \in \mathbf{R}_+$, and thus $(\mu_t)_{t \in \mathbf{R}_+}$ is a convolution semigroup with $\mu_1 = \mu$. Consequently, $(\mu_{t/t_0})_{t \in \mathbf{R}_+}$ is a convolution semigroup which associates the measure μ with the number t_0. Since $\hat{\mu}_t = |\hat{\mu}|^t e^{it\varphi}$ it follows from the Continuity Theorem 8.2.7 that this convolution semigroup is continuous. To prove uniqueness, let $(\mu_t)_{t \in \mathbf{R}_+}$ be a continuous convolution semigroup of probability measures where $\mu_{t_0} = \mu$. Then, since $\mu_{t_0} = \mu_{t_0/n} * \cdots * \mu_{t_0/n}$ (n factors),

$$\hat{\mu}_{t_0/n} = |\hat{\mu}|^{1/n} e^{i(\varphi/n)}$$

for all $n \in \mathbf{N}$. As above, it follows from the continuity of the semigroup that

$$\hat{\mu}_{tt_0} = |\hat{\mu}|^t e^{it\varphi} \qquad \text{or} \qquad \hat{\mu}_t = |\hat{\mu}|^{t/t_0} e^{i(t/t_0)\varphi}$$

for all $t \in \mathbf{R}_+$. By the Uniqueness Theorem for Fourier transforms, every μ_t is therefore uniquely determined by μ. ┘

9.3.7. Corollary. For every convolution semigroup $(\mu_t)_{t\in\mathbf{R}_+}$ of probability measures on $\mathbf{R}^p$, $\mu_0 = \epsilon_0$.

Proof. From $\mu_1 = \mu_1 * \mu_0$ it follows that $\hat{\mu}_1 = \hat{\mu}_1\hat{\mu}_0$. Since μ_1 is infinitely divisible, $\hat{\mu}_1(x) \neq 0$ for all $x \in \mathbf{R}^p$ by Theorem 9.3.2. Hence we obtain $\hat{\mu}_0 = 1$, and thus, by the Uniqueness Theorem, $\mu_0 = \epsilon_0$. ⌟

Examples

2. A particular convolution semigroup $(\mu_t)_{t\in\mathbf{R}_+}$ of probability measures on $\mathbf{R}^p$ is obtained by defining

$$\mu_t \equiv \begin{cases} \epsilon_0, & t = 0 \\ \underbrace{\nu_{0,t} \otimes \cdots \otimes \nu_{0,t}}_{p \text{ factors}}, & t > 0. \end{cases} \tag{9.3.7}$$ [7]

For $s > t$, $t > 0$, we have (see p. 287, footnote 4)

$$\begin{aligned} \mu_s * \mu_t &= (\nu_{0,s} * \nu_{0,t}) \otimes \cdots \otimes (\nu_{0,s} * \nu_{0,t}) \\ &= \nu_{0,s+t} \otimes \cdots \otimes \nu_{0,s+t} = \mu_{s+t}. \end{aligned}$$

We call $(\mu_t)_{t\in\mathbf{R}_+}$ the *Brownian convolution semigroup* on $\mathbf{R}^p$.

3. The *Poisson convolution semigroup* on $\mathbf{R}$ is the family $(\pi_t)_{t\in\mathbf{R}_+}$ in which $\pi_0 = \epsilon_0$ and π_t is the Poisson distribution with parameter t for $t > 0$. From Section 5.3, Example 2 we see that this is a convolution semigroup.

4. The *Cauchy convolution semigroup on* $\mathbf{R}$ is the family $(\gamma_t)_{t\in\mathbf{R}_+}$ in which $\gamma_0 = \epsilon_0$ and γ_t is the Cauchy distribution with parameter t for $t > 0$. See Section 8.2, Example 4.

Now we derive further properties of infinitely divisible measures which will be needed later.

9.3.8. Lemma. When μ and ν are infinitely divisible, $\mu * \nu$ and $T(\mu)$ are also infinitely divisible measures on $\mathbf{R}^p$. Here we let T be an arbitrary linear mapping of the vector space $\mathbf{R}^p$ into itself.

Proof. For every natural number n there are measures $\mu_n, \nu_n \in \mathfrak{M}^1(\mathbf{R}^p)$ with $\mu_n * \cdots * \mu_n = \mu$ and $\nu_n * \cdots * \nu_n = \nu$. But then $\mu_n * \nu_n$ is a probability measure and $(\mu_n * \nu_n) * \cdots * (\mu_n * \nu_n) = \mu * \nu$, and thus $\mu * \nu$ is infinitely divisible. The infinite divisibility of $T(\mu)$ is obtained from the following more general assertion: If $\tau_1, \ldots, \tau_n$ are measures in $\mathfrak{M}^e(\mathbf{R}^p)$, $p \geqq 1$, then

$$T(\tau_1 * \cdots * \tau_n) = T(\tau_1) * \cdots * T(\tau_n). \tag{9.3.8}$$

[7] For $t > 0$, μ_t is a particular "p-dimensional normal distribution."

By the Transformation Theorem 2.10.1, for every $\mathfrak{B}^p$-measurable function $f \geqq 0$ on $\mathbf{R}^p$ we have

$$\begin{aligned}\int f\,dT(\tau_1 * \cdots * \tau_n) &= \int f \circ T\,d(\tau_1 * \cdots * \tau_n)\\ &= \int f \circ T(x_1 + \cdots + x_n)\tau_1(dx_1) \cdots \tau_n(dx_n)\\ &= \int f(T(x_1) + \cdots + T(x_n))\tau_1(dx_1) \cdots \tau_n(dx_n)\\ &= \int f(y_1 + \cdots + y_n)\tau_1'(dy_1) \cdots \tau_n'(dy_n)\\ &= \int f d(\tau_1' * \cdots * \tau_n'),\end{aligned}$$

where we let $\tau_j' \equiv T(\tau_j)$, $j = 1, \ldots, n$. If in particular we choose the indicator function of an arbitrary set from $\mathfrak{B}^p$ for f, then the assertion follows. ┘

The following theorem shows that the set of infinitely divisible measures on $\mathbf{R}^p$ is *closed* in $\mathfrak{M}^1(\mathbf{R}^p)$ relative to the vague topology.

9.3.9. Theorem. Suppose a sequence $(\nu_k)_{k \in \mathbf{N}}$ of infinitely divisible measures on $\mathbf{R}^p$ converges vaguely to a measure $\mu \in \mathfrak{M}^1(\mathbf{R}^p)$. Then μ is also infinitely divisible.

Proof. We first show that $\hat{\mu}$ is nowhere zero on $\mathbf{R}^p$. Since all ν_k and μ are probability measures, the sequence (ν_k) converges weakly to μ and thus by the Continuity Theorem $\lim_{k \to \infty} \hat{\nu}_k(x) = \hat{\mu}(x)$ and also

$$\lim_{k \to \infty} \sqrt[n]{|\hat{\nu}_k(x)|^2} = \sqrt[n]{|\hat{\mu}(x)|^2} \tag{9.3.9}$$

for all $x \in \mathbf{R}^p$ and $n = 1, 2, \ldots$. Here $|\hat{\nu}_k|^2$ and $|\hat{\mu}|^2$ are the Fourier transforms of the measures $\nu_k * S(\nu_k)$ and $\mu * S(\mu)$ respectively where S again denotes the reflection $x \to -x$. By Lemma 9.3.8, each of the measures $\nu_k * S(\nu_k)$ is infinitely divisible, that is, by Corollary 9.3.3, $\sqrt[n]{|\hat{\nu}_k|^2}$ is the Fourier transform of the measure $\rho_{k,n} \in \mathfrak{M}^1(\mathbf{R}^p)$ whose n-fold convolution product with itself is equal to $\nu_k * S(\nu_k)$. Then, from (9.3.9) and the second part of the Continuity Theorem, we can conclude that $\sqrt[n]{|\hat{\mu}|^2}$ is the Fourier transform of a measure $\mu_n' \in \mathfrak{M}^1(\mathbf{R}^p)(n = 1, 2, \ldots)$. Hence $(\hat{\mu}_n')^n = |\hat{\mu}|^2$, that is, $\mu * S(\mu)$ is infinitely divisible and thus $|\hat{\mu}(x)|^2 \neq 0$ and $\hat{\mu}(x) \neq 0$ for all $x \in \mathbf{R}^p$.

Now, by Corollary A.2, there is exactly one continuous mapping φ: $\mathbf{R}^p \to \mathbf{R}$ such that $\varphi(0) = 0$ and $\hat{\mu} = |\hat{\mu}|e^{i\varphi}$. Also, by Theorem 9.3.2, for each $k = 1, 2, \ldots$ there is exactly one continuous mapping φ_k: $\mathbf{R}^p \to \mathbf{R}$ such that $\varphi_k(0) = 0$ and $\hat{\nu}_k = |\hat{\nu}_k|e^{i\varphi_k}$. We observed already that for every $x \in \mathbf{R}^p$, the sequence $(\hat{\nu}_k(x))$ converges to $\hat{\mu}(x)$. By the Continuity

Theorem, the convergence is uniform on every compact subset of $\mathbf{R}^p$. But then $(|\hat{\nu}_k|/\hat{\mu})$ converges uniformly to 1 on every compact set, and we conclude that the convergence

$$\lim_{k\to\infty} e^{i(\varphi_k - \varphi)} = 1$$

is uniform on every compact set $\subset \mathbf{R}^p$. Now Corollary A.5 tells us that $(\varphi_k - \varphi)$ approaches zero uniformly on every compact set $\subset \mathbf{R}^p$. For the measures $\nu_{k,n} \in \mathfrak{M}^1(\mathbf{R}^p)$ satisfying $\nu_k = \nu_{k,n} * \cdots * \nu_{k,n}$ (n factors) it thus follows that

$$\lim_{k\to\infty} \hat{\nu}_{k,n}(x) = \lim_{k\to\infty} \sqrt[n]{|\hat{\nu}_k(x)|}\, e^{i(\varphi_k(x)/n)} = \sqrt[n]{|\hat{\mu}(x)|}\, e^{i(\varphi(x)/n)}$$

holds for all $x \in \mathbf{R}^p$, by using (9.3.5). By the Continuity Theorem, the continuous function $\sqrt[n]{|\hat{\mu}|}\, e^{i(\varphi/n)}$ is the Fourier transform of a measure $\mu_n \in \mathfrak{M}^1(\mathbf{R}^p)$. Then $(\hat{\mu}_n)^n = \hat{\mu}$, that is, $\mu_n * \cdots * \mu_n = \mu$. Therefore μ is infinitely divisible. ⌟

9.3.10. Corollary. For every measure $\mu \in \mathfrak{M}^1(\mathbf{R}^p)$, the function

$$e^{\hat{\mu}-1}$$

is the Fourier transform of a (uniquely determined) infinitely divisible measure.

Proof. The assertion is true if μ is a discrete probability measure, that is, if μ is of the form

$$\mu = \sum_{j=1}^{n} \lambda_j \epsilon_{x_j} \qquad \Big(x_j \in \mathbf{R}^p,\ \lambda_j > 0,\ \sum_{j=1}^{n} \lambda_j = 1\Big).$$

Then

$$\hat{\mu}(x) = \sum \lambda_j e^{i<x,x_j>}$$

and hence

$$e^{\hat{\mu}(x)-1} = \prod_{j=1}^{n} e^{\lambda_j(e^{i<x,x_j>}-1)} \qquad \text{for all } x \in \mathbf{R}^p. \tag{9.3.10}$$

Now for every $\lambda > 0$ and $y \in \mathbf{R}^p$, we know from Section 8.1, Example 2 that $x \to e^{\lambda(e^{i<x,y>}-1)}$ is the Fourier transform of the probability measure

$$\pi_{\lambda,y} = \sum_{k=0}^{\infty} e^{-\lambda} \frac{\lambda^k}{k!} \epsilon_{ky}.$$

This is infinitely divisible, as can be seen immediately from the form of the Fourier transforms or from its relationship to the Poisson distribution π_λ. Therefore formula (9.3.10) tells us that $e^{\hat{\mu}-1}$ is the Fourier transform of a convolution product of finitely many infinitely divisible measures. Hence the assertion follows from Lemma 9.3.8 in this case.

The general case is disposed of via Corollary 7.7.4 and Theorem 7.8.4. According to the above, for every $\mu \in \mathfrak{M}^1(\mathbf{R}^p)$ there exists a vaguely and thus weakly convergent sequence (μ_k) of discrete probability measures with limit μ. By the first part of the Continuity Theorem, $(\hat{\mu}_k)$ then converges pointwise to $\hat{\mu}$, and thus $(e^{\hat{\mu}_k-1})$ converges pointwise to the continuous function $e^{\hat{\mu}-1}$. The assertion now follows from the second part of the Continuity Theorem and Theorem 9.3.9. ⌟

The preceding theorems, and in particular the last corollary, make it possible for us to construct new examples of infinitely divisible measures on $\mathbf{R}^p$.

With these tools, we return to the question raised at the beginning of this section, namely, the question of the role of infinitely divisible measures among the limit distributions of sums $S_n = X_{n1} + \cdots + X_{nk_n}$ of asymptotically negligible families (X_{nj}) where for each n, the random variables (X_{nj}) are independent.

It is easy to see that *every infinitely divisible measure* $\mu \in \mathfrak{M}^1(\mathbf{R})$ *occurs as a vague limit of the distributions* P_{S_n} *of such sums* S_n. For this we need only give a probabilistic interpretation to the definition of infinitely divisible measures: For μ and $n = 1, 2, \ldots$, consider the measure $\mu_n \in \mathfrak{M}^1(\mathbf{R})$ uniquely determined from (9.3.3) and temporarily set $\mu_{nj} \equiv \mu_n$ for every $j = 1, \ldots, n$. Then by Corollary 5.4.5 there is an independent family $(X_{nj})_{j=1,2,\ldots,n}$ of real random variables on a suitable probability space[8] $(\Omega,\mathfrak{A},P)$ such that $P_{X_{nj}} = \mu_{nj}$ for all allowable pairs of indices n, j.

Then, in particular, for each $n = 1, 2, \ldots, X_{n1}, \ldots, X_{nn}$ are independent and thus $P_{S_n} = P_{X_{n1}} * \cdots * P_{X_{nn}} = \mu_n * \cdots * \mu_n = \mu$, that is, the sequence (P_{S_n}) is constant, equal to μ. The family (X_{nj}) is asymptotically negligible since $P_{X_{nj}} = \mu_n$ is independent of j and since, by Corollary 9.3.4, the sequence (μ_n) converges vaguely to ϵ_0, that is, by Theorem 7.7.2, the sequence $(X_{nj})_{n=1,2,\ldots}$ converges stochastically to 0 $(j = 1, 2, \ldots, n)$. Since $P\{|X_{nj}| \geqq \epsilon\} = \mu_n(\{x \in \mathbf{R}: |x| \geqq \epsilon\})$, the convergence is uniform in j.

[8] For example, $\bigotimes_{\substack{j=1,\ldots,n \\ n\in\mathbf{N}}} (\mathbf{R}_{nj},\mathfrak{B}_{nj},\mu_{nj})$, where each $\mathbf{R}_{nj} = \mathbf{R}$ and each $\mathfrak{B}_{nj} = \mathfrak{B}^1$, is such a probability space.

Conversely we show:

9.3.11. Theorem. Let $(X_{nj})_{\substack{j=1,\ldots,k_n\\ n=1,2,\ldots}}$ be an asymptotically negligible family of square integrable, real random variables which satisfies the following conditions:

$$X_{n1}, \ldots, X_{nk_n} \text{ are independent for every } n = 1, 2, \ldots; \tag{9.3.11}$$

$$E(X_{nj}) = 0 \qquad (j = 1, \ldots, k_n;\ n = 1, 2, \ldots); \tag{9.3.12}$$

$$\sup_{n=1,2,\ldots} V(X_{n1} + \cdots + X_{nk_n}) < +\infty. \tag{9.3.13}$$

If the sequence (P_{S_n}) of distributions of the sums $S_n \equiv X_{n1} + \cdots + X_{nk_n}$ converges vaguely to a measure $\mu \in \mathfrak{M}^1(\mathbf{R})$, then μ is infinitely divisible.

Proof. We proceed similarly to part 2 of the proof of Theorem 9.2.3. By Lemma 9.1.2, for every $x \in \mathbf{R}$ we have

$$\max_{1 \leqq j \leqq k_n} |\hat{\mu}_{nj}(x) - 1| \leqq \tfrac{1}{2}, \qquad \text{for } n \text{ sufficiently large.}$$

Hence, as before (using the principal branch of the logarithm), it follows that

$$\log \hat{\mu}_{nj}(x) = \hat{\mu}_{nj}(x) - 1 + R_{nj}(x)$$

with the majorization

$$|R_{nj}(x)| \leqq |\hat{\mu}_{nj}(x) - 1|^2 \leqq \frac{x^2}{2} V(X_{nj}) |\hat{\mu}_{nj}(x) - 1|$$

$$\leqq \frac{x^2}{2} V(X_{nj}) \max_{1 \leqq j \leqq k_n} |\hat{\mu}_{nj}(x) - 1|$$

of the remainder term. If we note that $V(S_n) = V(X_{n1}) + \cdots + V(X_{nk_n})$ by the Bienaymé equality and that by hypothesis the sequence $(V(S_n))$ is bounded from above by a number $\alpha > 0$, we have

$$\left| \sum_{j=1}^{k_n} R_{nj}(x) \right| \leqq \frac{x^2}{2} V(S_n) \max_{1 \leqq j \leqq k_n} |\hat{\mu}_{nj}(x) - 1|$$

$$\leqq \alpha \frac{x^2}{2} \max_{1 \leqq j \leqq k_n} |\hat{\mu}_{nj}(x) - 1|.$$

Another application of Lemma 9.1.2 now yields

$$\lim_{n \to \infty} \sum_{j=1}^{k_n} R_{nj}(x) = 0,$$

which as before implies

$$\lim_{n\to\infty} e^{\sum_{j=1}^{k_n}(\hat{\mu}_{nj}-1)} = \hat{\mu}. \tag{9.3.14}$$

But now, by Corollary 9.3.10,

$$e^{\sum_{j=1}^{k_n}(\hat{\mu}_{nj}-1)} = \prod_{j=1}^{k_n} e^{\hat{\mu}_{nj}-1}$$

is the Fourier transform of a convolution product of k_n infinitely divisible measures and thus, by Lemma 9.3.8, it is the Fourier transform of an infinitely divisible measure ν_n ($n = 1, 2, \ldots$). By the Continuity Theorem, (9.3.14) tells us that the sequence (ν_n) converges vaguely to μ. But then μ is infinitely divisible by Theorem 9.3.9. ⌟

Remarks. 4. Theorem 9.3.11 remains valid if we drop the hypothesis (9.3.13) and assume only that the random variables X_{nj} are integrable. The proof is then carried out as above, but a finer majorization is required. We omit the details and refer to Gnedenko-Kolmogorov [33].

5. The hypotheses of Theorem 9.3.11 are satisfied, in particular, for the families (X_{nj}) considered in Examples 1 to 3 of Section 9.1: In Example 1,

$$V(S_n) = \frac{1}{n^2}\sum_{j=1}^{n} V(X_j), \qquad n = 1, 2, \ldots,$$

converges to zero and thus is bounded. In Example 2, $V(S_n) = np_n(1 - p_n)$ converges to α and thus again is bounded. Finally, in Example 3, $V(S_n) = 1$ for all n. On the other hand, the random variables of Example 4 are not even integrable (see Section 4.4, Example 7).

6. Corollary 9.3.10 is closely related to the so-called Lévy-Khinchin formula (see [33]), whereby a measure $\nu \in \mathfrak{M}^1(\mathbf{R})$ is infinitely divisible if and only if there exist a (necessarily uniquely determined) measure $\mu \in \mathfrak{M}^e(\mathbf{R})$ and a (also uniquely determined) $\gamma \in \mathbf{R}$ such that

$$\hat{\nu}(x) = e^{i\gamma x}e^{M(x)}, \qquad \text{for all } x \in \mathbf{R}, \tag{9.3.15}$$

where

$$M(x) = \int \left(e^{ixy} - 1 - \frac{ixy}{1+y^2}\right)\frac{1+y^2}{y^2}\,\mu(dy). \tag{9.3.16}$$

The integrand in (9.3.16) has the value $-x^2/2$ at $y = 0$ since one requires continuity of the integrand.[9]

[9] A proof of the Lévy-Khinchin formula based on the "representation theorem of G. Choquet" can be found in Johansen [37].

PROBLEMS

1. Let μ be an infinitely divisible probability measure on $\mathbf{R}^p$ and T: $\mathbf{R}^p \to \mathbf{R}^q$ a linear mapping of the vector space $\mathbf{R}^p$ into the vector space $\mathbf{R}^q$. Prove that $T(\mu)$ is an infinitely divisible probability measure on $\mathbf{R}^q$.
2. Decide whether the so-called triangular distribution on $\mathbf{R}$, that is, the probability measure $t\lambda^1$ with density

$$t(x) \equiv \begin{cases} 1 - |x|, & |x| \leqq 1 \\ 0, & |x| \geqq 1 \end{cases}$$

is infinitely divisible.
3. Prove that the Gamma distribution $f_t\lambda^1$ with parameter $t > 0$, defined in Section 8.2, Problem 3, is an infinitely divisible probability measure on $\mathbf{R}$. For given $t_0 > 0$, determine the continuous convolution semigroup $(\mu_t)_{t\in\mathbf{R}_+}$ of probability measures on $\mathbf{R}$ such that $\mu_{t_0} = f_{t_0}\lambda^1$.
4. Investigate whether the convolution semigroups of Examples 2 to 4 are continuous.
5. Let $(\mu_t)_{t\in\mathbf{R}_+}$ be a convolution semigroup. Prove (μ_t) is continuous if and only if $\lim\limits_{t\to 0} \mu_t = \epsilon_0$ holds in the vague topology.

9.4 CHARACTERIZATION OF THE NORMAL DISTRIBUTION (STABLE DISTRIBUTIONS)

Because of the Central Limit Theorem, the normal distribution assumes a particularly important role among the infinitely divisible measures on $\mathbf{R}$. It is therefore natural to ask for simple characterizations of the normal distribution in the set of all probability measures on $\mathbf{R}$. One answer to this question will be given in this section.

To this end, we consider the group $\mathfrak{J}$ of all mappings of $\mathbf{R}$ into itself of the form

$$x \to \tau x + \beta \qquad (\tau > 0, \beta \in \mathbf{R}). \tag{9.4.1}$$

For every normal distribution ν_{α,σ^2} with parameters α and σ^2, $\nu_{\alpha\tau+\beta,(\sigma\tau)^2}$ is then obviously the image of ν_{α,σ^2} under the mapping (9.4.1). Therefore the set $\mathfrak{J}(\nu_{\alpha,\sigma^2})$ of all images $T(\nu_{\alpha,\sigma^2})$ of ν_{α,σ^2} under the mapping $T \in \mathfrak{J}$ is the set of all normal distributions and is thus *stable* relative to convolution, that is, for every two measures $\nu_1, \nu_2 \in \mathfrak{J}(\nu_{\alpha,\sigma^2})$, $\nu_1 * \nu_2$ also lies in this set (see Section 5.3, Example 3). Therefore every normal distribution is stable in the sense of the following definition:

9.4.1. Definition. A measure $\mu \in \mathfrak{M}^1(\mathbf{R})$ is said to be *stable* (or a stable distribution) if the set $\mathfrak{J}(\mu)$ of all images $T(\mu)$ with $T \in \mathfrak{J}$ contains the convolution product of any two of its elements.

Since $\mathfrak{J}$ is a group of mappings, then of course, when μ is stable every other element from $\mathfrak{J}(\mu)$ is also a stable probability measure on $\mathbf{R}$.

Examples

1. Every *unit mass* ϵ_a, $a \in \mathbf{R}$, is stable, since $\mathfrak{J}(\epsilon_a)$ is the set of all measures ϵ_x with $x \in \mathbf{R}$.

2. The *Cauchy distribution* γ_1 (and hence every Cauchy distribution γ_α, $\alpha > 0$) is stable. For we have that

$$\mathfrak{J}(\gamma_1) = \{\gamma_\tau * \epsilon_\beta : \tau > 0, \beta \in \mathbf{R}\}$$

is stable with respect to convolution (compare Section 8.2, Example 3).

3. The *Poisson distribution* π_1 is *not* stable. For the mapping $x \to T(x) = \frac{1}{2}x$ we have [see (9.3.8)]

$$T(\pi_1) * T(\pi_1) = T(\pi_1 * \pi_1) = T(\pi_2) = \sum_{n=0}^{\infty} e^{-2} \frac{2^n}{n!} \epsilon_{n/2}.$$

On the other hand, the image π' of π_1 under the mapping $x \to \tau x + \beta$ from $\mathfrak{J}$ equals:

$$\pi' = \sum_{n=0}^{\infty} e^{-1} \frac{1}{n!} \epsilon_{\tau n + \beta}.$$

Therefore $T(\pi_1) * T(\pi_1) \notin \mathfrak{J}(\pi_1)$.

The last example shows that not every infinitely divisible measure on $\mathbf{R}$ is stable. Conversely, however, we have:

9.4.2. Theorem. Every stable measure $\mu \in \mathfrak{M}^1(\mathbf{R})$ is infinitely divisible.

Proof. For every $n = 1, 2, \ldots,$ the n-fold convolution product $\mu * \cdots * \mu$ of μ with itself lies in $\mathfrak{J}(\mu)$. Hence there is a mapping $x \to T_\tau(x) = \tau x$ from $\mathfrak{J}$ and a number $\beta \in \mathbf{R}$ such that $\mu * \cdots * \mu = T_\tau(\mu) * \epsilon_\beta$. By (9.3.8), since $T_{1/\tau} = T_\tau^{-1}$, it now follows that

$$\mu = T_{1/\tau}(\mu) * \cdots * T_{1/\tau}(\mu) * T_{1/\tau}(\epsilon_{-\beta}).$$

Therefore, $\mu_n \equiv T_{1/\tau}(\mu * \epsilon_{-\beta/n})$ is a measure from $\mathfrak{M}^1(\mathbf{R})$ with $\mu = \mu_n * \cdots * \mu_n$. ┘

The normal distributions ν_{α,σ^2} are not the only stable probability measures on $\mathbf{R}$, as is shown by the above examples. Nonetheless, they can be characterized by a simple condition in addition to stability:

9.4.3. Theorem. A measure $\mu \in \mathfrak{M}^1(\mathbf{R})$ is a normal distribution if and only if it is stable and if its variance

$$V(\mu) \equiv \int (x - \int x\mu(dx))^2 \,\mu(dx) \tag{9.4.2}$$

is finite and positive.[10]

Proof. By what was shown in Section 4.4, (6), we need to prove only that a stable measure $\mu \in \mathfrak{M}^1(\mathbf{R})$ with variance $0 < V(\mu) < +\infty$ is a normal distribution. If we set $\alpha = \int x\mu(dx)$ and $\sigma = \sqrt{V(\mu)}$, then $x \to T(x) \equiv \sigma^{-1}(x - \alpha)$ is a mapping from $\mathfrak{I}$, that is, $T(\mu) \in \mathfrak{I}(\mu)$. Obviously, $\int x\, dT(\mu) = 0$ and $V(T(\mu)) = 1$. For the proof we can thus assume that $\int x\mu(dx) = 0$ and $V(\mu) = 1$. Then due to the stability of μ, there is a transformation $x \to T(x) = \tau x + \beta$ from $\mathfrak{I}$ such that $\mu * \mu = T(\mu)$. Since $\int x(\mu * \mu)(dx) = 0$ and $\int x^2(\mu * \mu)(dx) = 2$, we must have $\beta = 0$ and $\tau = \sqrt{2}$. Therefore, for the Fourier transforms of the measures we are considering we have, by (8.1.10),

$$[\hat{\mu}(x)]^2 = \hat{\mu}(\sqrt{2}\, x), \qquad \text{for all } x \in \mathbf{R},$$

or the equivalent functional equation

$$\hat{\mu}(x) = \left[\hat{\mu}\left(\frac{x}{\sqrt{2}}\right)\right]^2 \qquad (x \in \mathbf{R}). \tag{9.4.3}$$

By Theorems 9.3.2 and 9.4.2, $\hat{\mu}(x) \neq 0$ for all $x \in \mathbf{R}$;[11] consequently, by Corollary A.2, there is exactly one continuous mapping $\varphi: \mathbf{R} \to \mathbf{C}$ such that $\varphi(0) = 0$ and $\hat{\mu} = e^{\varphi}$. Since $\hat{\mu}$ is twice continuously differentiable by Theorem 8.3.3, this is also true for φ by Corollary A.3. By Theorem 8.3.3 $\hat{\mu}'(0) = 0$ and $\hat{\mu}''(0) = -1$, that is, $\varphi'(0) = 0$ and $\varphi''(0) = -1$.

From (9.4.3) it follows that

$$\varphi(x) = 2\varphi\left(\frac{x}{\sqrt{2}}\right) = 2^2\varphi\left(\frac{x}{\sqrt{2^2}}\right) = \cdots = 2^n\varphi\left(\frac{x}{\sqrt{2^n}}\right) \tag{9.4.4}$$

[10] The integrability of $x \to x^2$ and the positivity of $V(\mu)$ are equivalent to this.

[11] This also follows directly from (9.4.3). $\hat{\mu}(x) = 0$ would imply $\hat{\mu}(x/\sqrt{2^n}) = 0$ for all $n = 1, 2, \ldots$. For $n \to \infty$ this yields a contradiction in the form $\hat{\mu}(0) = 0$.

for all $x \in \mathbf{R}$ and $n = 1, 2, \ldots$. Therefore, for $x \neq 0$ it follows that

$$\frac{\varphi(x)}{x^2} = \left(\frac{x}{\sqrt{2^n}}\right)^{-2} \varphi\left(\frac{x}{\sqrt{2^n}}\right), \qquad \text{for all } n = 1, 2, \ldots$$

and hence for $n \to \infty$ we have the equality $\varphi(x)/x^2 = \frac{1}{2}\varphi''(0) = -\frac{1}{2}$ by Taylor's formula. Therefore, $\varphi(x) = -x^2/2$, that is, $\mu(x) = e^{-x^2/2}$ for all $x \in \mathbf{R}$, and thus $\mu = \nu_{0,1}$. ⌋

The following corollary is essentially a translation of this theorem into the language of probability theory:

9.4.4. Corollary. Let X and Y be independent real random variables with the same distribution μ, expected value 0 and finite positive variance. Then μ is a normal distribution if and only if μ is also the distribution of $(1/\sqrt{2})(X + Y)$.

Proof. Let μ be the distribution of $(1/\sqrt{2})(X + Y)$. By suitable normalization we can assume that $V(X) = V(Y) = 1$. But then $x \to \hat{\mu}(x/\sqrt{2})^2$ is the characteristic function of $(1/\sqrt{2})(X + Y)$, and thus $\hat{\mu}(x) = \hat{\mu}(x/\sqrt{2})^2$ for all $x \in \mathbf{R}$. Therefore we again arrive at (9.4.3). Hence the above proof shows that $\mu = \nu_{0,1}$. The converse follows immediately from Section 5.3, Example 3. ⌋

Remark. With the help of further theorems from Fourier analysis, it can be shown that for every $\alpha \in]0,2]$, there exists a measure $\mu_\alpha \in \mathfrak{M}^1(\mathbf{R})$ having a continuous density f_α with respect to λ^1 such that

$$\hat{\mu}_\alpha(x) = e^{-|x|^\alpha} \qquad (x \in \mathbf{R}).$$[12]

It follows from this form of the Fourier transform that μ_α is stable. The set of Fourier transforms of all measures from $\mathfrak{Z}(\mu_\alpha)$ is indeed given by

$$x \to e^{i\beta x} e^{-\tau^\alpha |x|^\alpha} \qquad (\tau > 0, \beta \in \mathbf{R})$$

and thus is obviously stable with respect to ordinary multiplication. The functions f_α are called *stable symmetric densities of order* α, $0 < \alpha \leqq 2$. For $\alpha = 1$ ($\alpha = 2$) we obviously have $\mu_1 = \gamma_1$ ($\mu_2 = \nu_{0,2}$), that is, a Cauchy (normal) distribution.

PROBLEM

Let μ be a stable distribution on $\mathbf{R}$. Determine the continuous convolution semigroup $(\mu_t)_{t \in \mathbf{R}_+}$ of probability measures on $\mathbf{R}$ which satisfies $\mu_1 = \mu$.

[12] See Meyer [39] and Blumenthal-Getoor [24].

10

CONDITIONAL EXPECTATIONS

This chapter deals with the concept of conditional expectation which is basic to modern probability theory. The approach to it is based on the Radon-Nikodym Theorem of Section 2.9. Our considerations again refer to an arbitrary probability space $(\Omega,\mathfrak{A},P)$.

10.1 CONDITIONAL EXPECTATIONS AND PROBABILITIES

We consider an integrable real random variable X on Ω. As we know, X can be viewed as a mathematical model for describing an experiment with random outcome in which real measurements are observed. In the sense of this interpretation, X then contains complete information with regard to the observed outcome of the experiment. On the other hand, the expected value $E(X)$ contains only an insignificant amount of information about X. If we also interpret $E(X)$ as a real random variable, namely the constant variable $\omega \rightarrow E(X)$ on Ω, then $E(X)$ is measurable with respect to every σ-subalgebra $\mathfrak{Z}$ of $\mathfrak{A}$,[1] and in particular with respect to the smallest σ-subalgebra which contains only $\varnothing$ and Ω as elements. On the other hand, X is $\mathfrak{A}$-measurable and in general not measurable with respect to smaller σ-subalgebras of $\mathfrak{A}$. We can therefore think of measuring the amount of information about X contained in a real random variable Y by σ-subalgebras $\mathfrak{Z}$ of $\mathfrak{A}$ with respect to which Y is measurable.

In this regard, the concept of conditional probability treated earlier (Section 4.2) now also has a suitable place. For this, let $\Omega = \bigcup_{i\in I} B_i$ be a

[1] By this we mean, of course, every σ-algebra $\mathfrak{Z}$ in Ω with $\mathfrak{Z} \subset \mathfrak{A}$.

decomposition of Ω into finitely many ($I = \{1, \ldots, n\}$) or countably infinitely many ($I = \mathbf{N}$) pairwise disjoint events $B_i \in \mathfrak{A}$ with probabilities $P(B_i) > 0$. Then the conditional probability under the hypothesis B_i is the following probability measure P_{B_i} on $\mathfrak{A}$:

$$P_{B_i}(A) = \frac{P(A \cap B_i)}{P(B_i)} = \frac{1}{P(B_i)} \int 1_A 1_{B_i}\, dP = \frac{1}{P(B_i)} (1_{B_i}P)(A)$$

for all $A \in \mathfrak{A}$; that is, we have

$$P_{B_i} = \frac{1}{P(B_i)} (1_{B_i}P). \tag{10.1.1}$$

The expected value of X with respect to this probability measure, that is, the number

$$E_{B_i}(X) \equiv \int X\, dP_{B_i} = \frac{1}{P(B_i)} \int_{B_i} X\, dP \tag{10.1.2}$$

is called the *expected value of X given B_i* ($i \in I$). Thus we arrive at a new random variable

$$X_0 \equiv \sum_{i \in I} E_{B_i}(X) 1_{B_i}, \tag{10.1.3}$$

which, on every B_i, is constant, equal to $E_{B_i}(X)$. In the case $I = \{1\}$, this is just the constant random variable $\omega \to E(X)$, because $B_1 = \Omega$. The variable X_0 is measurable with respect to the σ-subalgebra $\mathfrak{Z} \equiv \mathfrak{A}(B_i;\ i \in I)$ generated by all the B_i, $i \in I$. It consists of all sets of the form $\bigcup_{i \in J} B_i$ where J runs through all elements of the power set $\mathfrak{P}(I)$. Now we have $\int_{B_i} X_0\, dP = E_{B_i}(X)P(B_i) = \int_{B_i} X\, dP$ for all $i \in I$. Due to the special form of $\mathfrak{Z}$ it then follows that

$$\int_Z X_0\, dP = \int_Z X\, dP, \qquad \text{for all } Z \in \mathfrak{Z}. \tag{10.1.4}$$

This elucidates the close relationship between X_0 and X, to be expected from the construction of X_0 through the σ-subalgebra $\mathfrak{Z}$. If instead of X_0 we choose the random variable $\omega \to E(X)$, then (10.1.4) is satisfied for all events Z of the σ-subalgebra $\{\emptyset, \Omega\}$ relative to which $E(X)$ is measurable. When $X_0 = X$, (10.1.4) holds for all $Z \in \mathfrak{A}$.

We shall now see that in the three given cases, the random variable X_0 is P-almost surely uniquely determined by the measurability relative to the σ-subalgebra $\mathfrak{Z}$ being considered and by the validity of (10.1.4). The special form of $\mathfrak{Z}$ even is without significance for this. Through the consideration of σ-subalgebras $\mathfrak{Z}$ of $\mathfrak{A}$ we thus arrive at random variables X_0 which are more or less closely related to X. We can interpret the σ-algebra $\mathfrak{Z}$ as a measure of the information contained in X_0 with regard to X.

10.1.1. Theorem. Let X be a numerical random variable on $(\Omega,\mathfrak{A},P)$ which is $\geqq 0$ (or integrable). Then for every σ-subalgebra $\mathfrak{Z}$ of $\mathfrak{A}$ there exists, up to almost sure equality, exactly one nonnegative (or integrable) numerical random variable X_0 on Ω which is $\mathfrak{Z}$-measurable and satisfies the condition

$$\int_Z X_0\,dP = \int_Z X\,dP, \qquad \text{for all } Z \in \mathfrak{Z}. \tag{10.1.4}$$

If X is integrable and $\geqq 0$, then $X_0 \geqq 0$ almost surely.

Proof. First suppose $X \geqq 0$. We use P_0 and Q to denote the restrictions of the measures P and XP, respectively, to $\mathfrak{Z}$. Then P_0 and Q are measures on $\mathfrak{Z}$; P_0 is a probability measure and is thus finite. Since

$$Q(Z) = \int_Z X\,dP,$$

$Q(Z) = 0$ for every $Z \in \mathfrak{Z}$ with $P_0(Z) = P(Z) = 0$, that is, Q is P_0-continuous. Therefore, by the Radon-Nikodym Theorem, Q has a density relative to P_0, that is, there exists a $\mathfrak{Z}$-measurable numerical function $X_0 \geqq 0$ on Ω such that $Q = X_0P_0$, that is, such that $\int_Z X_0\,dP_0 = \int_Z X\,dP$ for all $Z \in \mathfrak{Z}$. By the definition of integrals of measurable functions $\geqq 0$ we have $\int f\,dP_0 = \int f\,dP$ for every $\mathfrak{Z}$-measurable function $f \geqq 0$. Therefore $\int_Z X_0\,dP_0 = \int_Z X_0\,dP$ and hence $\int_Z X_0\,dP = \int_Z X\,dP$ for every $Z \in \mathfrak{Z}$. The random variable X_0 thus yields the desired result. Moreover, by Theorem 2.9.9, X_0 is a density and is therefore P_0-almost surely uniquely determined. Since the set $\{X_0 = X_0^*\}$ lies in $\mathfrak{Z}$ for every $\mathfrak{Z}$-measurable numerical random variable X_0^*, it follows that $X_0 = X_0^*$ P-almost surely for every $\mathfrak{Z}$-measurable random variable $X_0^* \geqq 0$ satisfying condition (10.1.4).

If X is integrable and of arbitrary sign, we decompose X into its positive part X^+ and negative part X^-. From what has just been proved, follows the existence of $\mathfrak{Z}$-measurable numerical random variables $X_0^* \geqq 0$ and $X_0^{**} \geqq 0$ such that

$$\int_Z X_0^*\,dP = \int_Z X^+\,dP \qquad \text{and} \qquad \int_Z X_0^{**}\,dP = \int_Z X^-\,dP,$$

for all $Z \in \mathfrak{Z}$. If in particular we set $Z = \Omega$, we see that since X^+ and X^- are integrable, X_0^* and X_0^{**} are also integrable and hence almost surely finite. Thus we can assume X_0^*, X_0^{**} to be real-valued without loss of generality. But then $X_0 \equiv X_0^* - X_0^{**}$ is an integrable solution of our problem. For every other integrable solution Y_0 we have, for arbitrary $Z \in \mathfrak{Z}$,

$$\int_Z (X_0^* + Y_0^-)\,dP = \int_Z (X_0^{**} + Y_0^+)\,dP;$$

hence, $(X_0^* + Y_0^-)P_0 = (X_0^{**} + Y_0^+)P_0$ due to the $\mathfrak{Z}$-measurability of all integrands. By Theorem 2.9.9 again, we thus have $X_0^* + Y_0^- = X_0^{**} + Y_0^+$ P_0-almost surely and therefore also P-almost surely. Hence we obtain $X_0 = Y_0$ P-almost surely. Finally, if X is integrable and $\geqq 0$, then we still have to show that $X_0 \geqq 0$ almost surely. But by (10.1.4), $\int_Z X_0 \, dP \geqq 0$ for every $Z \in \mathfrak{Z}$, and in particular for $Z = \{X_0 < 0\}$. For this Z then, $\int_Z X_0 \, dP = 0$ and hence $P(Z) = 0$ since X_0 is strictly negative on Z. ┘

10.1.2. Definition. Under the conditions of Theorem 10.1.1, the random variable X_0 is called the *conditional expectation of X given $\mathfrak{Z}$*; in symbols;

$$E^{\mathfrak{Z}}(X) = E(X \mid \mathfrak{Z}) \equiv X_0. \tag{10.1.5}$$

Hence $E^{\mathfrak{Z}}(X)$ is a $\mathfrak{Z}$-measurable numerical random variable such that

$$\int_Z E^{\mathfrak{Z}}(X)\, dP = \int_Z X \, dP, \qquad \text{for all } Z \in \mathfrak{Z}. \tag{10.1.4'}$$

$E^{\mathfrak{Z}}(X)$ is only P-almost surely uniquely determined by (10.1.4′). We therefore also speak of different, but P-almost surely equal *versions* of the conditional expectation. Therefore, results on conditional expectations in general hold only P-almost surely. The transition from X to $E^{\mathfrak{Z}}(X)$ can be interpreted as a "smoothing" of X relative to $\mathfrak{Z}$. This interpretation is primarily suggested by properties (10.1.16) and (10.1.17), still to be proved.

If $(Y_i)_{i \in I}$ is a family of random variables Y_i: $(\Omega,\mathfrak{A}) \to (\Omega_i,\mathfrak{A}_i)$ with values in measurable spaces $(\Omega_i,\mathfrak{A}_i)$, $i \in I$, and $\mathfrak{Z} = \mathfrak{A}(Y_i;\, i \in I)$ is the σ-subalgebra generated by these random variables, then we also write

$$E^{(Y_i)_{i \in I}}(X) = E(X \mid Y_i,\, i \in I)$$

for $E^{\mathfrak{Z}}(X)$ and speak of the conditional expectation of X given $(Y_i)_{i \in I}$ or the conditional expectation of X for *given random variables* Y_i, $i \in I$. If I consists of only n elements: $I = \{1, \ldots, n\}$, then we also write

$$E^{Y_1, \ldots, Y_n}(X) \qquad \text{or} \qquad E(X \mid Y_1, \ldots, Y_n).$$

Examples

The three introductory, motivating examples tell us the following in the new terminology:

1. If $\mathfrak{Z}$ is the σ-algebra consisting only of $\varnothing$ and Ω, then $E^{\mathfrak{Z}}(X) = E(X)$ almost surely for every admissible random variable X on $(\Omega,\mathfrak{A},P)$. (We say that X is *admissible* if X is $\geqq 0$ or integrable.)

2. $E^{\mathfrak{A}}(X) = X$ almost surely for every admissible random variable X.

3. If $\mathfrak{Z}$ is generated by a decomposition $\Omega = \bigcup_{i \in I} B_i$ of the sets $B_i \in \mathfrak{A}$ as above (I finite or countably infinite), then

$$E^{\mathfrak{Z}}(X) = \sum_{i \in I} E_{B_i}(X) 1_{B_i}, \qquad P\text{-almost surely}$$

provided $P(B_i) > 0$ for all i. If we drop this condition, then the above equality still holds if we take $E_{B_i}(X)$ to be some arbitrarily chosen number for events B_i with $P(B_i) = 0$. Here X is again an admissible random variable.

4. Let $(\Omega, \mathfrak{A}, P)$ be a probability space and G a group of *finite* order n of measurable mappings $g: (\Omega, \mathfrak{A}) \to (\Omega, \mathfrak{A})$ each of which leaves P invariant, that is, we have $g(P) = P$ for all $g \in G$. The system $\mathfrak{Z}$ of all sets $Z \in \mathfrak{A}$ which are G-invariant, that is, for which $g(Z) = Z$ for arbitrary $g \in G$, is a σ-subalgebra of $\mathfrak{A}$. Then for every admissible random variable,

$$E^{\mathfrak{Z}}(X) = \frac{1}{n} \sum_{g \in G} X \circ g, \qquad P\text{-almost surely}$$

Indeed, $X_0 \equiv (1/n)\Sigma_{g \in G} X \circ g$ is $\mathfrak{Z}$-measurable, since X_0 is $\mathfrak{A}$-measurable and $X_0 \circ h = X_0$ for all $h \in G$. Moreover, for every $Z \in \mathfrak{Z}$,

$$\int_Z X_0 \, dP = \frac{1}{n} \sum_{g \in G} \int_Z X \circ g \, dP = \frac{1}{n} \sum_{g \in G} \int_{g(Z)} X \, dg(P) = \int_Z X \, dP.$$

Remark. 1. For numerical random variables X on $(\Omega, \mathfrak{A}, P)$ for which either X^+ or X^- is integrable, that is, the integral $\int X \, dP$ exists in the sense of the remark following Definition 2.4.1, the conditional expectation can still be defined easily. The details are left to the reader.

The following properties (10.1.6) – (10.1.10) of conditional expectation can be obtained directly from the definition and from Theorem 10.1.1. Therefore we omit the proof. Here X and Y are always numerical random variables on $(\Omega, \mathfrak{A}, P)$ which are both $\geqq 0$ or both integrable.

$$E(E^{\mathfrak{Z}}(X)) = E(X), \tag{10.1.6}$$

$$X \ \mathfrak{Z}\text{-measurable} \Rightarrow E^{\mathfrak{Z}}(X) = X, \text{ almost surely}, \tag{10.1.7}$$

$$X = Y \text{ almost surely} \Rightarrow E^{\mathfrak{Z}}(X) = E^{\mathfrak{Z}}(Y), \text{ almost surely}, \tag{10.1.8}$$

$$X = \text{const} = \alpha \Rightarrow E^{\mathfrak{Z}}(X) = \alpha, \text{ almost surely}, \tag{10.1.9}$$

$$E^{\mathfrak{Z}}(\alpha X + \beta Y) = \alpha E^{\mathfrak{Z}}(X) + \beta E^{\mathfrak{Z}}(Y), \text{ almost surely} \quad (\alpha, \beta \in \mathbf{R}_+ \text{ or } \alpha, \beta \in \mathbf{R}), \tag{10.1.10}$$

$$X \leqq Y \text{ almost surely} \Rightarrow E^{\mathfrak{Z}}(X) \leqq E^{\mathfrak{Z}}(Y), \text{ almost surely}. \tag{10.1.11}$$

Because of (10.1.8), we can assume $X \leqq Y$ on all of Ω. But then there is a random variable $Z \geqq 0$ with $Y = X + Z$, and the desired property follows from (10.1.10) since $E^{\mathfrak{Z}}(Z) \geqq 0$ almost surely.

$$|E^{\mathfrak{Z}}(X)| \leqq E^{\mathfrak{Z}}(|X|), \qquad \text{almost surely.} \tag{10.1.12}$$

This is obvious in the case $X \geqq 0$. For integrable X, we decompose X into positive and negative parts.

For every isotone sequence $(X_n)_{n \in \mathbf{N}}$ of random variables $X_n \geqq 0$,

$$\sup E^{\mathfrak{Z}}(X_n) = E^{\mathfrak{Z}}(\sup X_n), \qquad \text{almost surely.} \tag{10.1.13}$$

By (10.1.8) and (10.1.11), we can assume that the sequence $(E^{\mathfrak{Z}}(X_n))$ is isotone. The assertion then follows by a passage to the limit in (10.1.4) using the Monotone Convergence Theorem.

If a sequence $(X_n)_{n \in \mathbf{N}}$ of numerical random variables converges almost surely to X and there exists an integrable random variable Y with $|X_n| \leqq Y$ for all n, then

$$\lim_{n \to \infty} E^{\mathfrak{Z}}(X_n) = E^{\mathfrak{Z}}(X), \qquad \text{almost surely.} \tag{10.1.14}$$

Indeed, introduce $X_n^* \equiv \sup_{k \geqq n} X_k$ and $X_n^{**} \equiv \inf_{k \geqq n} X_k$; then $-Y \leqq X_n^{**} \leqq X_n \leqq X_n^* \leqq Y$ for all $n \in \mathbf{N}$. Furthermore, $(Y - X_n^*)$ and $(Y + X_n^{**})$ are isotone sequences of integrable random variables with $Y - \limsup_{n \to \infty} X_n$ resp. $Y + \liminf_{n \to \infty} X_n$ as supremum. Since (X_n) converges to X almost surely, the almost sure convergence of $(E^{\mathfrak{Z}}(X_n^*))$ and $(E^{\mathfrak{Z}}(X_n^{**}))$ to $E^{\mathfrak{Z}}(X)$ follows from (10.1.10) and (10.1.13). Hence the inequalities $X_n^{**} \leqq X_n \leqq X_n^*$ together with (10.1.11) imply $\lim E^{\mathfrak{Z}}(X_n) = E^{\mathfrak{Z}}(X)$ almost surely.

A simple reformulation of the definition of $E^{\mathfrak{Z}}(X)$ yields:

10.1.3. Lemma. A $\mathfrak{Z}$-measurable numerical random variable X_0 on Ω is the conditional expectation of a numerical random variable X which is $\geqq 0$ (or integrable) if and only if

$$\int Q X_0 \, dP = \int Q X \, dP, \tag{10.1.15}$$

for all $\mathfrak{Z}$-measurable random variables Q on Ω which are $\geqq 0$ (or almost surely bounded).

Proof. (10.1.15) follows from (10.1.4) when we choose indicator functions of events $Z \in \mathfrak{Z}$ for Q. If $X \geqq 0$, then we obtain (10.1.15) from (10.1.4) for all $\mathfrak{Z}$-elementary functions Q on Ω. A passage to the limit yields the assertion. If X is integrable, we decompose it into its positive and negative part. ⌟

Now we obtain the following *smoothing properties* of conditional expectation.

Let X and Y be both $\geqq 0$ or let X be almost surely bounded and Y integrable. Then

$$X\ \mathfrak{Z}\text{-measurable} \Rightarrow E^{\mathfrak{Z}}(XY) = XE^{\mathfrak{Z}}(Y), \text{ almost surely.} \tag{10.1.16}$$

Suppose we are in the first case. If $Q \geqq 0$ is $\mathfrak{Z}$-measurable, then by the above lemma it follows that $\int QXY\,dP = \int QXE^{\mathfrak{Z}}(Y)\,dP$ since QX is $\mathfrak{Z}$-measurable. But on the other hand, $\int QXY\,dP = \int QE^{\mathfrak{Z}}(XY)\,dP$. Due to the $\mathfrak{Z}$-measurability of $XE^{\mathfrak{Z}}(Y)$, it now follows that $E^{\mathfrak{Z}}(XY) = XE^{\mathfrak{Z}}(Y)$ almost surely. We proceed analogously in the second case.

Under the hypotheses of (10.1.16),

$$E^{\mathfrak{Z}}(YE^{\mathfrak{Z}}(X)) = E^{\mathfrak{Z}}(Y)E^{\mathfrak{Z}}(X), \qquad \text{almost surely.} \tag{10.1.17}$$

This follows from (10.1.16) if $E^{\mathfrak{Z}}(X)$ plays the role of X there. By (10.1.9), (10.1.11), and (10.1.12), when X is almost surely bounded, $E^{\mathfrak{Z}}(X)$ also is.

For σ-algebras $\mathfrak{Z}_1$, $\mathfrak{Z}_2$ in Ω, we have

$$\begin{aligned}\mathfrak{Z}_1 \subset \mathfrak{Z}_2 \subset \mathfrak{A} \Rightarrow E^{\mathfrak{Z}_1}(E^{\mathfrak{Z}_2}(X)) &= E^{\mathfrak{Z}_2}(E^{\mathfrak{Z}_1}(X)) \\ &= E^{\mathfrak{Z}_1}(X), \qquad \text{almost surely.}\end{aligned} \tag{10.1.18}$$

$E^{\mathfrak{Z}_1}(X)$ is $\mathfrak{Z}_1$-and thus $\mathfrak{Z}_2$-measurable; by (10.1.7) it then follows that $E^{\mathfrak{Z}_2}(E^{\mathfrak{Z}_1}(X)) = E^{\mathfrak{Z}_1}(X)$ almost surely. Moreover, $\int_Z E^{\mathfrak{Z}_2}(X)\,dP = \int_Z X\,dP$ for all $Z \in \mathfrak{Z}_2$ and $\int_Z E^{\mathfrak{Z}_1}(X)\,dP = \int_Z X\,dP$ for all $Z \in \mathfrak{Z}_1$, hence, $\int_Z E^{\mathfrak{Z}_2}(X)\,dP = \int_Z E^{\mathfrak{Z}_1}(X)\,dP$ for all $Z \in \mathfrak{Z}_1$. Hence it follows that $E^{\mathfrak{Z}_1}(E^{\mathfrak{Z}_2}(X)) = E^{\mathfrak{Z}_1}(X)$ almost surely.

As a last property, we discuss the behavior of conditional expectation with respect to independence:

10.1.4. Theorem. Let $\mathfrak{Z}_1$ and $\mathfrak{Z}_2$ be σ-subalgebras of $\mathfrak{A}$, $\mathfrak{Z} \equiv \mathfrak{A}(\mathfrak{Z}_1, \mathfrak{Z}_2)$ the σ-algebra generated by $\mathfrak{Z}_1$ and $\mathfrak{Z}_2$, and X an integrable random variable. If X and $\mathfrak{Z}_1$ are independent of $\mathfrak{Z}_2$, that is, the σ-algebra $\mathfrak{A}(\mathfrak{A}(X), \mathfrak{Z}_1)$ generated by X and $\mathfrak{Z}_1$ is independent of $\mathfrak{Z}_2$, then

$$E^{\mathfrak{Z}}(X) = E^{\mathfrak{Z}_1}(X), \qquad \text{almost surely.}$$

Proof. Let X_0 be a version of $E^{\mathfrak{Z}_1}(X)$. This is then $\mathfrak{Z}$-measurable and, as we shall show, is also a version of $E^{\mathfrak{Z}}(X)$. Thus we have to verify that $\int_Z X_0\,dP = \int_Z X\,dP$ for all $Z \in \mathfrak{Z}$. Because of the integrability of X, the system $\mathfrak{D}$ of all $Z \in \mathfrak{Z}$ for which $\int_Z X_0\,dP = \int_Z X\,dP$ is a Dynkin system. By Theorem 1.2.3 it thus suffices to prove the equality for all sets Z of an $\cap$-stable generator of $\mathfrak{Z}$. Such a generator is given by

$$\mathfrak{E} \equiv \{Z_1 \cap Z_2 \colon Z_1 \in \mathfrak{Z}_1, Z_2 \in \mathfrak{Z}_2\}.$$

For every set $Z_1 \cap Z_2 \in \mathfrak{E}$ we now have

$$\int_{Z_1 \cap Z_2} X_0 \, dP = E(1_{Z_1} 1_{Z_2} X_0) = E(1_{Z_2}) E(1_{Z_1} X_0),$$

since $1_{Z_1} X_0$ is $\mathfrak{Z}_1$-measurable, 1_{Z_2} is $\mathfrak{Z}_2$-measurable, and the σ-algebras $\mathfrak{Z}_1$ and $\mathfrak{Z}_2$ are independent by Section 5.1, Consequence 2, p. 151. Further, by the definition of X_0,

$$E(1_{Z_1} X_0) = E(1_{Z_1} X)$$

and hence

$$\int_{Z_1 \cap Z_2} X_0 \, dP = E(1_{Z_2}) E(1_{Z_1} X).$$

But 1_{Z_1} and $1_{Z_2} X$ are independent random variables by hypothesis. Another application of the Multiplication Theorem 5.3.1 now yields

$$\int_{Z_1 \cap Z_2} X_0 \, dP = E(1_{Z_1} 1_{Z_2} X) = \int_{Z_1 \cap Z_2} X \, dP.$$

But this is what we needed to show. ┘

10.1.5. Corollary. If an admissible random variable X is independent of a σ-subalgebra $\mathfrak{Z}$ of $\mathfrak{A}$, then

$$E^{\mathfrak{Z}}(X) = E(X), \qquad \text{almost surely.} \tag{10.1.19}$$

Proof. For integrable X this follows from Theorem 10.1.4 if we choose $\mathfrak{Z}_1 = \{\varnothing, \Omega\}$ and $\mathfrak{Z}_2 = \mathfrak{Z}$. But the assertion also holds for nonintegrable random variables $X \geqq 0$, since, by the Multiplication Theorem, $\int_Z X \, dP = E(1_Z X) = P(Z) E(X) = \int_Z E(X) \, dP$ for all $Z \in \mathfrak{Z}$. ┘

By a simple specialization, we finally arrive at the general concept of conditional probability:

10.1.6. Definition. Let $(\Omega, \mathfrak{A}, P)$ be a probability space, $\mathfrak{Z}$ a σ-subalgebra of $\mathfrak{A}$, and $A \in \mathfrak{A}$ an event. Then

$$P^{\mathfrak{Z}}(A) = P(A \mid \mathfrak{Z}) \equiv E^{\mathfrak{Z}}(1_A) \tag{10.1.20}$$

is called the *conditional probability* of A given $\mathfrak{Z}$.

Thus $P^{\mathfrak{Z}}(A)$ is an integrable $\mathfrak{Z}$-measurable function, $\geqq 0$ almost surely for which

$$\int_Z P^{\mathfrak{Z}}(A) \, dP = \int_Z 1_A \, dP = P(A \cap Z) \tag{10.1.21}$$

for all $Z \in \mathfrak{Z}$. Hence $P^{\mathfrak{Z}}(A)$ is almost surely uniquely determined.

Example

5. Suppose we have the situation of Example 3. Then obviously

$$P^{\mathfrak{Z}}(A) = \sum_{i \in I} P_{B_i}(A) 1_{B_i}, \qquad \text{almost surely.}$$

For events B_i with $P(B_i) = 0$ we can again choose $P_{B_i}(A)$ arbitrarily.

By specializing properties of conditional expectation, we obtain, in particular, the following properties of conditional probability:

$$0 \leqq P^{\mathfrak{Z}}(A) \leqq 1, \qquad \text{almost surely.} \tag{10.1.22}$$

$$P^{\mathfrak{Z}}(\emptyset) = 0, \text{ almost surely; } P^{\mathfrak{Z}}(\Omega) = 1, \text{ almost surely.} \tag{10.1.23}$$

$$A_1 \subset A_2 \ (A_i \in \mathfrak{A}) \Rightarrow P^{\mathfrak{Z}}(A_1) \leqq P^{\mathfrak{Z}}(A_2), \text{ almost surely.} \tag{10.1.24}$$

For every sequence $(A_n)_{n \in \mathbf{N}}$ of pairwise disjoint events from $\mathfrak{A}$,

$$P^{\mathfrak{Z}}\Big(\bigcup_{n=1}^{\infty} A_n\Big) = \sum_{n=1}^{\infty} P^{\mathfrak{Z}}(A_n), \qquad \text{almost surely.} \tag{10.1.25}$$

Here (10.1.22)–(10.1.24) follow from (10.1.9) and (10.1.11). The last property follows from (10.1.14) and (10.1.10).

Remark. 2. The properties given here do *not* say that the function $A \to [P^{\mathfrak{Z}}(A)](\omega)$ is a probability measure on $\mathfrak{A}$ for almost all $\omega \in \Omega$; because in each of the four given properties, we have null sets depending on the given events, for example, in (10.1.25) depending on the sequence (A_n). The union of these generally uncountably many null sets is usually not a null set. (See Problem 2.)

PROBLEMS

1. Consider the probability space $(\mathbf{R}^p, \mathfrak{B}^p, P)$, where $P = \mu \otimes \cdots \otimes \mu$ and where $\mu \in \mathfrak{M}^1(\mathbf{R})$. Let $\mathfrak{Z}$ be the system of all Borel sets $B \in \mathfrak{B}^p$ such that for every permutation $i_1, \ldots, i_p$ of $1, \ldots, p$ and every point $x = (x_1, \ldots, x_p) \in B$ we have $(x_{i_1}, \ldots, x_{i_p}) \in B$. Prove: $\mathfrak{Z}$ is a σ-subalgebra of $\mathfrak{B}^p$, and
$$x \to \frac{1}{p!} \sum_{i_1, \ldots, i_p} X(x_{i_1}, \ldots, x_{i_p})$$
(summation over all permutations of $1, \ldots, p$) is a version of $E(X \mid \mathfrak{Z})$ for every admissible random variable.
2. Let $(\Omega, \mathfrak{A}, P)$ be the probability space $\Omega \equiv [0,1]$, $\mathfrak{A} \equiv \Omega \cap \mathfrak{B}^1$, $P \equiv \lambda^1_\Omega$. Let $f\colon \mathfrak{A} \to \Omega$ be the following mapping: $f(A) \equiv \sup A$ for $A \neq \emptyset$ and $f(\emptyset) = 0$. Define $Q\colon \Omega \times \mathfrak{A} \to \mathbf{R}$ to be
$$Q(\omega, A) := 1_A(\omega) + 1_{\{f(A)\}}(\omega).$$
Prove: $\omega \to Q(\omega, A)$ is a version of $P(A \mid \mathfrak{A})$ for all $A \in \mathfrak{A}$. There exists no P-null set $N \in \mathfrak{A}$ such that $A \to Q(\omega, A)$ is a probability measure for all $\omega \in \complement N$.

3. Let X and Y be real random variables on a probability space $(\Omega,\mathfrak{A},P)$, and let $\mathfrak{Z}$ be a σ-subalgebra of $\mathfrak{A}$. Assume $X \in \mathfrak{L}^p(P)$ and $Y \in \mathfrak{L}^q(P)$, where $1 \leqq p < +\infty$ and $p^{-1} + q^{-1} = 1$. Prove:
   (a) $$|E^{\mathfrak{Z}}(XY)| \leqq E^{\mathfrak{Z}}(|X|^p)^{1/p}E^{\mathfrak{Z}}(|Y|^q)^{1/q}, \quad \text{almost surely}$$
   (b) $$|E^{\mathfrak{Z}}(X)|^p \leqq E^{\mathfrak{Z}}(|X|^p), \quad \text{almost surely}$$
4. Let $\mathfrak{Z}$ and $\mathfrak{Z}'$ be σ-subalgebras of the σ-algebra $\mathfrak{A}$ of a probability space. Prove: $\mathfrak{Z}$ and $\mathfrak{Z}'$ are independent if and only if
$$P(Z' \mid \mathfrak{Z}) = P(Z'), \quad \text{almost surely}$$
for all $Z' \in \mathfrak{Z}'$. Formulate the corresponding result for the independence of two random variables.
5. Let $(\Omega,\mathfrak{A},P)$ be a probability space and let Y be a real, integrable random variable $\geqq 0$ on it. Let $\mathfrak{Z}_1$ and $\mathfrak{Z}_2$ be σ-subalgebras of $\mathfrak{A}$, and denote by $\mathfrak{Z}_3$ the σ-algebra $\mathfrak{A}(\mathfrak{Z}_1,\mathfrak{Z}_2)$ generated by $\mathfrak{Z}_1$ and $\mathfrak{Z}_2$. Prove the equivalence of the following two statements:
   (a) $$E(Y \mid \mathfrak{Z}_3) = E(Y) \mid \mathfrak{Z}_2), \quad \text{almost surely}$$
   (b) $$E(XY \mid \mathfrak{Z}_2) = E(X \mid \mathfrak{Z}_2)E(Y \mid \mathfrak{Z}_2), \quad \text{almost surely}$$
   for all $\mathfrak{Z}$-measurable real random variables $X \geqq 0$.

10.2 FACTORIZATION OF CONDITIONAL EXPECTATION

The connection between conditional expectation relative to a σ-subalgebra $\mathfrak{Z}$ of $\mathfrak{A}$ as just explained and the elementary concept of conditional expectation relative to an event of positive probability is closer than one would first expect. This shows up clearly when $\mathfrak{Z}$ is the σ-algebra $\mathfrak{A}(Y)$ generated by a random variable Y. The reason for this is shown by the following lemma, which gives $\mathfrak{A}(Y)$-measurability a very intuitive meaning.

10.2.1. Lemma. Let $Y: \Omega \to \Omega'$ be a mapping of a set Ω into a measurable space $(\Omega',\mathfrak{A}')$ and let $Z: \Omega \to \bar{\mathbf{R}}$ be a numerical function on Ω. Then Z is measurable with respect to the σ-algebra $\mathfrak{A}(Y)$ in Ω generated by Y if and only if there is a measurable numerical function g on $(\Omega',\mathfrak{A}')$ such that
$$Z = g \circ Y. \tag{10.2.1}$$

Proof. If Z is of the form $Z = g \circ Y$, then Z, as the composition of an $\mathfrak{A}(Y)$-$\mathfrak{A}'$-measurable mapping and an $\mathfrak{A}'$-$\bar{\mathfrak{B}}^1$-measurable mapping, is $\mathfrak{A}(Y)$-$\bar{\mathfrak{B}}^1$-measurable. For the proof of the converse, we distinguish several cases:

1. Let $Z = \sum_{i=1}^n \alpha_i 1_{A_i}$ be an $\mathfrak{A}(Y)$-elementary function, that is, $A_i \in \mathfrak{A}(Y)$ and $\alpha_i \in \mathbf{R}_+$ $(i = 1, \ldots, n)$. For every A_i there is a set $A_i' \in \mathfrak{A}'$ with $A_i = Y^{-1}(A_i')$. Therefore $g \equiv \sum_{i=1}^n \alpha_i 1_{A_i'}$ yields the desired result.

2. Suppose $Z \geqq 0$. Then there is an isotone sequence $(Z_n)_{n\in\mathbf{N}}$ of $\mathfrak{A}(Y)$-elementary functions with $Z = \sup Z_n$ and, according to the first case, for every Z_n there is an $\mathfrak{A}'$-elementary function g_n with $Z_n = g_n \circ Y$. But then $g \equiv \sup g_n$ yields the desired result.

3. In the general case we decompose Z into the positive part Z^+ and the negative part Z^-. By case 2, there are $\mathfrak{A}'$-measurable functions $g_0' \geqq 0$ and $g_0'' \geqq 0$ on Ω' with $Z^+ = g_0' \circ Y$ and $Z^- = g_0'' \circ Y$. The difference $g_0'(\omega') - g_0''(\omega')$ is not defined on the set $U' \equiv \{g_0' = +\infty\} \cap \{g_0'' = +\infty\}$. But the set $Y(\Omega)$ is disjoint from U', since $Z(\omega) = Z^+(\omega) - Z^-(\omega) = g_0'(Y(\omega)) - g_0''(Y(\omega))$ for all $\omega \in \Omega$. Therefore, if we set

$$g' \equiv 1_{\complement U'} g_0' \qquad \text{and} \qquad g'' \equiv 1_{\complement U'} g_0'',$$

then $g \equiv g' - g''$ yields the desired result. ⌟

Remark. The restriction of g to $Y(\Omega)$ is uniquely determined. For every $\omega' \in Y(\Omega)$, we have $g(\omega') = Z(\omega)$ for all $\omega \in \Omega$ with $Y(\omega) = \omega'$. Therefore if $Y(\Omega) = \Omega'$ or at least $Y(\Omega) \in \mathfrak{A}'$, then we can obtain a function g by constructing the uniquely determined restriction on $Y(\Omega)$ according to the introductory statement and defining, say, $g(\omega') = 0$ for all $\omega' \in \complement Y(\Omega)$. Here we can also replace $(\bar{\mathbf{R}},\bar{\mathfrak{B}}^1)$ by an arbitrary measurable space in which the one-point subsets are measurable. The lemma is therefore noteworthy insofar as we can get along without the measurability of $Y(\Omega)$.[2] For this the special structure of $(\bar{\mathbf{R}},\bar{\mathfrak{B}}^1)$ is used in an essential way.

We apply the lemma to the following situation. Let X again be a numerical random variable on $(\Omega,\mathfrak{A},P)$ which is either $\geqq 0$ or integrable and let $Y: (\Omega,\mathfrak{A}) \to (\Omega',\mathfrak{A}')$ be an $(\Omega',\mathfrak{A}')$-random variable on Ω. Since the numerical random variable $E^Y(X)$ is $\mathfrak{A}(Y)$-measurable, then by Lemma 10.2.1 there is a measurable numerical function g on $(\Omega',\mathfrak{A}')$ with

$$E^Y(X) = g \circ Y. \tag{10.2.2}$$

With the aid of the distribution P_Y of Y we can characterize g, uniquely determined only when $Y(\Omega) = \Omega'$, as follows:

10.2.2. Theorem. Every $\mathfrak{A}'$-measurable numerical function g with property (10.2.2) satisfies the equality

$$\int_{A'} g\, dP_Y = \int_{Y^{-1}(A')} X\, dP, \qquad \text{for all } A' \in \mathfrak{A}' \tag{10.2.3}$$

and is therefore P_Y-almost surely uniquely determined.

[2] As an example, for a continuous mapping $f: E \to F$ of a Polish space E into a Polish space F, the image $f(E)$ need not be Borel in F. See the theory of so-called Souslin or analytic sets (Bourbaki [26] and Meyer [39]).

Proof. For every $A' \in \mathfrak{A}'$,

$$\int_{A'} g\, dP_Y = \int 1_{A'} g\, dP_Y = \int (1_{A'} \circ Y)(g \circ Y)\, dP$$
$$= \int_{Y^{-1}(A')} g \circ Y\, dP = \int_{Y^{-1}(A')} E^Y(X)\, dP = \int_{Y^{-1}(A')} X\, dP.$$

The last equality is obtained from the definition of conditional expectation. Now if h is another $\mathfrak{A}'$-measurable function with $E^Y(X) = h \circ Y$, then, by the above, $\int_{A'} g\, dP_Y = \int_{A'} h\, dP_Y$ for all $A' \in \mathfrak{A}'$. Now either $X \geqq 0$ or X is integrable. In the first case, as in the proof of Theorem 10.1.1, (10.2.3) implies that $g \geqq 0$ and $h \geqq 0$ P_Y-almost surely. Therefore, by Theorem 2.9.9, $g = h$ P_Y-almost surely. In the second case, g and h are P_Y-integrable by the Transformation Theorem, since $E^Y(X) = g \circ Y = h \circ Y$ is P-integrable. Decomposition into positive and negative parts then yields $\int_{A'} (g^+ + h^-)\, dP_Y = \int_{A'} (g^- + h^+)\, dP_Y$ for all $A' \in \mathfrak{A}'$, that is, $g^+ + h^- = g^- + h^+$ P_Y-almost surely by Theorem 2.9.9. Since g and h are P_Y-almost surely finite, we have that $g = h$ P_Y-almost surely. ⌟

Now if we assume that the set $\{y\}$ is an event in $\mathfrak{A}'$ for some element $y \in \Omega'$, then from (10.2.3) for $A' = \{y\}$ it follows that

$$g(y) P\{Y = y\} = \int_{\{Y = y\}} X\, dP.$$

Therefore, if $P\{Y = y\} = P_Y(\{y\}) \neq 0$, this equality implies that

$$g(y) = \frac{1}{P\{Y = y\}} \int_{\{Y = y\}} X\, dP = E_{\{Y=y\}}(X) \tag{10.2.4}$$

and hence, by (10.2.2),

$$E^Y(X)(\omega) = E_{\{Y=y\}}(X), \qquad \text{for all } \omega \in \{Y = y\}. \tag{10.2.5}$$

Thus, on $\{Y = y\}$ with $P\{Y = y\} > 0$, *every version of the conditional expectation* $E^Y(X)$ *is constant and equal to the conditional expectation of* X *given* $\{Y = y\}$.

Thus, we see that the situation of Example 3 of Section 10.1 was not as special as it seemed at first. If, there, we let $\Omega' = I$ and $\mathfrak{A}' = \mathfrak{P}(\Omega')$ as well as $Y(\omega) = i$ for $\omega \in B_i$ $(i \in I)$, then Y is an $(\Omega', \mathfrak{A}')$-random variable with $\mathfrak{Z} = \mathfrak{A}(Y)$. Then, further, $g(i) = E_{B_i}(X)$ for all $i \in I$ with $P(B_i) > 0$.

In general, however, $P\{Y = y\}$ will be equal to zero for most points $y \in \Omega'$. For example, we can think of the case of a real random variable Y whose distribution P_Y has a density with respect to Lebesgue measure. Nonetheless, the value of g at an arbitrary point $y \in \Omega'$ is then still at our disposal. For $y \in Y(\Omega)$, this is even uniquely determined by $E^Y(X)$. In the general case, we can interpret $g(y)$ as the mean of the values which X

attains at elementary events $\omega \in \Omega$ satisfying $Y(\omega) = y$. Therefore we define:

10.2.3. Definition. For every $y \in \Omega'$, $g(y)$ is called the *conditional expectation of X given that Y equals y*; in symbols,

$$E^{Y=Y}(X) = E(X|Y = y) \equiv g(y).$$

Thus, while $E^Y(X)$ is a random variable, $E^{Y=y}(X)$ is a number. By (10.2.2), from $y \to E^{Y=y}(X)$ we again obtain the conditional expectation $E^Y(X)$ in the form $\omega \to E^{Y=Y(\omega)}(X)$.

PROBLEMS

1. Let X and Y be random variables on $(\Omega,\mathfrak{A},P)$ with the properties mentioned in connection with (10.2.2). Prove that $E(X \mid Y = y)$ can be defined as follows: Up to equality P_Y-almost everywhere, $y \to E(X \mid Y = y)$ is the only $\mathfrak{A}'$-measurable numerical function on Ω' which is $\geqq 0$ for $X \geqq 0$ or P_Y-integrable for integrable X and satisfies
$$\int_{A'} E(X \mid Y = y)P_Y(dy) = \int_{Y^{-1}(A')} X\,dP,$$
for all $A' \in \mathfrak{A}'$.
2. Let Y be an $(\Omega',\mathfrak{A}')$-random variable on $(\Omega,\mathfrak{A},P)$, and let X_1 and X_2 be numerical random variables on Ω which are both $\geqq 0$ or both integrable. Prove that the following equalities hold for P_Y-almost all $y \in \Omega'$:
   (a) $X_1 = \text{const} \Rightarrow E(X_1 \mid Y = y) = \text{const}$;
   (b) $E(\alpha X_1 + \beta X_2 \mid Y = y) = \alpha E(X_1 \mid Y = y) + \beta E(X_2 \mid Y = y)$ $(\alpha,\ \beta \geqq 0 \text{ or } \alpha,\ \beta \in \mathbf{R})$;
   (c) $X_1 \leqq X_2$, almost surely $\Rightarrow E(X_1 \mid Y = y) \leqq E(X_2 \mid Y = y)$;
   (d) for every increasing sequence (X_n) of random variables $X_n \geqq 0$ one has
   $$\sup_{n\in\mathbf{N}} E(X_n \mid Y = y) = E(\sup X_n \mid Y = y), \qquad P_Y\text{-almost surely.}$$
3. Let Y be an $(\Omega',\mathfrak{A}')$-random variable on $(\Omega,\mathfrak{A},P)$. Then $E(1_A \mid Y = y)$ is called the conditional probability of $A \in \mathfrak{A}$ given that Y equals $y \in \Omega'$. It is denoted by $P(A \mid Y = y) = P^{Y=y}(A)$. Prove:
   (a) $P(A \cap \{Y \in A'\}) = \int_{A'} P(A \mid Y = y)P_Y(dy)$ $(A \in \mathfrak{A},\ A' \in \mathfrak{A}')$;
   (b) $P(\varnothing \mid Y = y) = 0$, P_Y-almost surely;
   (c) $P(\Omega \mid Y = y) = 1$, P_Y-almost surely;

(d) $P\left(\bigcup_{n=1}^{\infty} A_n \mid Y = y\right) = \sum_{n=1}^{\infty} P(A_n \mid Y = y)$, P_Y-almost surely

for every sequence (A_n) of pairwise disjoint events $A_n \in \mathfrak{A}$.

10.3 KERNELS, EXPECTATION KERNELS, AND CONDITIONAL DISTRIBUTIONS

Remark 2 at the end of Section 10.1 raises the question of the assumptions needed on the probability space $(\Omega,\mathfrak{A},P)$ and a given σ-subalgebra $\mathfrak{Z}$ of $\mathfrak{A}$ so that there exists a version $P^{\mathfrak{Z}}(A)$ of the conditional probability for which $A \to [P^{\mathfrak{Z}}(A)](\omega)$ is a probability measure on $\mathfrak{A}$ for every $\omega \in \Omega$. In other words, we are interested in the question of the existence of a function $P_{\mathfrak{Z}}: \Omega \times \mathfrak{A} \to \mathbf{R}$, associated with $\mathfrak{Z}$, such that $\omega \to P_{\mathfrak{Z}}(\omega,A)$ is a version of $P^{\mathfrak{Z}}(A)$ for each $A \in \mathfrak{A}$ and $A \to P_{\mathfrak{Z}}(\omega,A)$ is a probability measure on $\mathfrak{A}$ for each $\omega \in \Omega$. We then call $P_{\mathfrak{Z}}$ an *expectation kernel for* $\mathfrak{Z}$. Here we encounter the important concept of kernel for the first time.

10.3.1. Definition. Suppose we are given two measurable spaces $(\Omega,\mathfrak{A})$ and $(\Omega',\mathfrak{A}')$. A *kernel from* $(\Omega,\mathfrak{A})$ *to* $(\Omega',\mathfrak{A}')$ or, briefly, from Ω to Ω', is a numerical function K on $\Omega \times \mathfrak{A}'$ with the following two properties:

$$\omega \to K(\omega,A') \text{ is } \mathfrak{A}\text{-measurable for every } A' \in \mathfrak{A}'; \tag{10.3.1}$$

$$A' \to K(\omega,A') \text{ is a measure on } \mathfrak{A}' \text{ for every } \omega \in \Omega. \tag{10.3.2}$$

A kernel K is called *Markov* [*sub-Markov*] if $K(\omega,\Omega') = 1$ [$K(\omega,\Omega') \leqq 1$] for all $\omega \in \Omega$.

If $\Omega = \Omega'$ and $\mathfrak{A} = \mathfrak{A}'$, then we also speak of a kernel *on* $(\Omega,\mathfrak{A})$ or, briefly, on Ω.

For a Markov kernel, every measure $A' \to K(\omega,A')$ is thus a probability measure. An expectation kernel for a σ-subalgebra $\mathfrak{Z}$ of $\mathfrak{A}$ [and a given probability space $(\Omega,\mathfrak{A},P)$] is thus a Markov kernel from $(\Omega,\mathfrak{Z})$ to $(\Omega,\mathfrak{A})$. The importance of the concept of kernel is emphasized by the following:

Examples

1. Let $\varphi: (\Omega,\mathfrak{A}) \to (\Omega',\mathfrak{A}')$ be a measurable mapping of a measurable space $(\Omega,\mathfrak{A})$ into a second measurable space $(\Omega',\mathfrak{A}')$. Further, let μ be a measure on $\mathfrak{A}$. Then

$$K(\omega,A') \equiv \varphi(\mu)(A') = \mu(\varphi^{-1}(A')) \qquad (\omega \in \Omega,\ A' \in \mathfrak{A}')$$

defines a kernel (independent of ω) from $(\Omega,\mathfrak{A})$ to $(\Omega',\mathfrak{A}')$. Every *image measure* is a kernel in this sense.

2. Suppose $(\pi_{ij})_{(i,j)\in\mathbf{N}\times\mathbf{N}}$ is a double sequence (or, in view of applications, a matrix with countably infinitely many rows and columns) with numbers $0 \leqq \pi_{ij} \leqq +\infty$ as elements. On the measurable space $(\mathbf{N},\mathfrak{P}(\mathbf{N}))$ of the natural numbers (with discrete topology), this matrix defines the following kernel:

$$K(i,A) = \sum_{j\in A} \pi_{ij} \qquad (i \in \mathbf{N},\ A \in \mathfrak{P}(\mathbf{N})).$$

In this way we obtain a bijection of the $\mathbf{N} \times \mathbf{N}$-matrices onto the kernels on $\mathbf{N}$. The matrix associated with a kernel K on $\mathbf{N}$ is obviously the matrix (π_{ij}) where $\pi_{ij} = K(i, \{j\})$ for all $(i,j) \in \mathbf{N} \times \mathbf{N}$. K is Markov (sub-Markov) if and only if all row sums $\Sigma_{j\in\mathbf{N}}\pi_{ij} = 1$ $(\leqq 1)$.[3]

[There are many possible interpretations of this example: In the Markov case, (π_{ij}) might describe the motion of a particle wandering at random in the set $\mathbf{N}$ of natural numbers. Then π_{ij} is to be interpreted as the probability that the particle passes from "state" $i \in \mathbf{N}$ to state $j \in \mathbf{N}$.]

3. Let $(\Omega,\mathfrak{A},\mu)$ be a measure space and $k \geqq 0$ a numerical, $\mathfrak{A} \otimes \mathfrak{A}$-measurable function defined on $\Omega \times \Omega$. Then

$$K(\omega,A) = \int_A k(\omega,\omega')\mu(d\omega')$$

is a kernel on $(\Omega,\mathfrak{A})$.

This example is basic for *potential theory:* There $\Omega = \mathbf{R}^p$, $\mathfrak{A} = \mathfrak{B}^p$, μ is a Borel measure on $\mathbf{R}^p$, and for dimension $p \geqq 3$, the function k is the Newtonian "kernel"

$$k(x,y) \equiv \begin{cases} \dfrac{1}{|x - y|^{p-2}}, & x \neq y \\ +\infty, & x = y. \end{cases}$$

4. Let $(\Omega,\mathfrak{A},P)$ be a probability space. Then for the σ-algebras

$$\mathfrak{Z} = \mathfrak{Z}_0 \equiv \{\emptyset,\Omega\} \qquad \text{and} \qquad \mathfrak{Z} = \mathfrak{A}$$

there always exists an expectation kernel. Indeed, $P_{\mathfrak{Z}_0}$ is the kernel $(\omega,A) \to P_{\mathfrak{Z}_0}(\omega,A) \equiv P(A)$, independent of ω, and $P_{\mathfrak{A}}$ is the kernel which associates the unit mass ϵ_ω at ω with each $\omega \in \Omega$, that is, $P_{\mathfrak{A}}(\omega,A) \equiv 1_A(\omega)$ for all $(\omega,A) \in \Omega \times \mathfrak{A}$.

The last kernel $(\omega,A) \to 1_A(\omega) = \epsilon_\omega(A)$ is called the *unit kernel* on Ω or $(\Omega,\mathfrak{A})$ and is denoted by I. Thus we have

$$I(\omega,A) \equiv \begin{cases} 1, & \omega \in A \\ 0, & \omega \in \complement A \end{cases} \qquad (\omega,A) \in \Omega \times \mathfrak{A}.$$

[3] Then we also say that the matrix (π_{ij}) is *stochastic* (*sub-stochastic*).

5. Suppose we have the situation of Section 10.1, Examples 3 and 5. Then

$$P_{\mathfrak{Z}}(\omega,A) \equiv \sum_{i\in I} P_{B_i}(A)1_{B_i}(\omega)$$

is an expectation kernel for $\mathfrak{Z}$. Indeed, every $\omega \in \Omega$ lies in exactly one B_i. Therefore $A \to P_{\mathfrak{Z}}(\omega,A) = P_{B_i}(A)$ is a probability measure. For events $B_i \neq \varnothing$ with $P(B_i) = 0$, we have to choose some probability measure on $\mathfrak{A}$ for P_{B_i}, for example, the unit mass ϵ_{ω_i} of an arbitrarily chosen $\omega_i \in B_i$.

6. Suppose we have the situation of Section 10.1, Example 4. Then

$$P_{\mathfrak{Z}}(\omega,A) = \frac{1}{n}\sum_{g\in G} \epsilon_{g(\omega)}(A)$$

is obviously an expectation kernel for $\mathfrak{Z}$.

Example 2 suggests the following interpretation for a kernel K from $(\Omega,\mathfrak{A})$ to $(\Omega',\mathfrak{A}')$: K describes a "diffusion process" which takes the unit mass ϵ_ω at a point $\omega \in \Omega$ into the "mass distribution" on Ω' described by the measure $A' \to K(\omega,A')$.

If μ is an arbitrary measure on $\mathfrak{A}$, then this is taken into a measure μ' on $\mathfrak{A}'$ by K as follows: $\mu'(A') \equiv \int K(\omega,A')\mu(d\omega)$, $A' \in \mathfrak{A}'$. That μ' is a measure on $\mathfrak{A}'$ follows immediately from known properties of the integral. We denote μ' by μK and then write the equation defining μK in the form

$$(\mu K)(A') = \int \mu(d\omega)K(\omega,A') \equiv \int K(\omega,A')\mu(d\omega). \qquad (10.3.3)$$

If $\mathcal{E}^*_{\mathfrak{A}}$ ($\mathcal{E}^*_{\mathfrak{A}'}$) denotes the set of all numerical, nonnegative $\mathfrak{A}$- ($\mathfrak{A}'$-) measurable functions on Ω (Ω'), then K also establishes a mapping from $\mathcal{E}^*_{\mathfrak{A}'}$ into $\mathcal{E}^*_{\mathfrak{A}}$. For every $f' \in \mathcal{E}^*_{\mathfrak{A}'}$, $\omega \to \int f'(\omega')K(\omega,d\omega')$ is a function in $\mathcal{E}^*_{\mathfrak{A}}$. To see this, we need only refer to Theorem 2.3.6, and thus approximate f' by elementary functions. As usual, we denote the new mapping again by K, that is, we define

$$(Kf')(\omega) \equiv \int f'(\omega')K(\omega,d\omega') \qquad (\omega \in \Omega) \qquad (10.3.4)$$

for every $f' \in \mathcal{E}^*_{\mathfrak{A}'}$. In particular, for indicator functions $1_{A'}$ of sets $A' \in \mathfrak{A}'$, we then have

$$K1_{A'}(\omega) = K(\omega,A') \qquad (\omega \in \Omega). \qquad (10.3.5)$$

In this new notation $K1 = 1$ ($K1 \leqq 1$) means precisely that K is Markov (sub-Markov).

Finally, we define Kf' for $\mathfrak{A}'$-measurable numerical functions f' on Ω' by (10.3.4) provided f' is integrable with respect to every measure $A' \to K(\omega,A')$, $\omega \in \Omega$. Then Kf' is a real $\mathfrak{A}$-measurable function on Ω. If K is sub-Markov, then K can always be interpreted as a linear positive mapping of the vector space $\mathcal{E}^b_{\mathfrak{A}'}$ of bounded $\mathfrak{A}'$-measurable functions on Ω' into the vector space $\mathcal{E}^b_{\mathfrak{A}}$ of bounded $\mathfrak{A}$-measurable functions on Ω.

For a kernel K from $(\Omega,\mathfrak{A})$ to $(\Omega',\mathfrak{A}')$, the mapping $K: \mathcal{E}^*_{\mathfrak{A}'} \to \mathcal{E}^*_{\mathfrak{A}}$ is obviously additive, positive-homogeneous, and *Daniell-continuous*, that is,

$$\sup Kf'_n = K \sup f'_n, \tag{10.3.6}$$

for every isotone sequence (f'_n) in $\mathcal{E}^*_{\mathfrak{A}'}$. The latter follows directly from the theorem on monotone convergence. The convention of using K to denote the mappings of $\mathcal{E}^*_{\mathfrak{A}'}$ into $\mathcal{E}^*_{\mathfrak{A}}$ associated with K is justified by the following lemma. It concludes our consideration of general kernels.

10.3.2. Lemma. Let $(\Omega,\mathfrak{A})$ and $(\Omega',\mathfrak{A}')$ be two measurable spaces. Then for every additive, positive-homogeneous, Daniell-continuous mapping $N: \mathcal{E}^*_{\mathfrak{A}'} \to \mathcal{E}^*_{\mathfrak{A}}$ there exists exactly one kernel K from $(\Omega,\mathfrak{A})$ to $(\Omega',\mathfrak{A}')$ such that $N(f') = Kf'$ for all $f' \in \mathcal{E}^*_{\mathfrak{A}'}$.

Proof. Because of (10.3.5), only $K(\omega,A') \equiv [N(1_{A'})](\omega)$ can be considered as a definition of K on $\Omega \times \mathfrak{A}'$. We can verify immediately that K is a kernel which yields the desired result by Theorem 2.3.6. ⌋

Example

7. Let I be the *unit kernel* defined in Example 4 on a measurable space $(\Omega,\mathfrak{A})$. Then obviously

$$If = f \qquad \text{and} \qquad \mu I = \mu$$

for all functions $f \in \mathcal{E}^*_{\mathfrak{A}}$ and all measures μ on $\mathfrak{A}$.

Now we turn back to the study of expectation kernels. If there exists an expectation kernel $P_{\mathfrak{Z}}$ for a probability space $(\Omega,\mathfrak{A},P)$ and a σ-subalgebra $\mathfrak{Z}$ of $\mathfrak{A}$, then this kernel plays the same role relative to conditional expectation as P relative to ordinary expectation in the equality $E(X) = \int X\,dP$. If X is a numerical random variable $\geqq 0$, then for every version of $E^{\mathfrak{Z}}(X)$,

$$E^{\mathfrak{Z}}(X) = P_{\mathfrak{Z}}X, \qquad P\text{-almost surely}, \tag{10.3.7}$$

that is, $\omega \to \int X(\omega')\,P_{\mathfrak{Z}}(\omega,d\omega')$ is a version of $E^{\mathfrak{Z}}(X)$. This is obvious for indicator variables $X = 1_A$ with $A \in \mathfrak{A}$. Using Theorem 2.3.6, (10.3.7) follows in general by using known properties of conditional expectation. (10.3.7) also holds for integrable numerical random variables X. We need only decompose X into its positive and negative part and take into account the integrability of $E^{\mathfrak{Z}}(X^+)$ and $E^{\mathfrak{Z}}(X^-)$.

Finally, we can also give an analog to the concept of distribution of a random variable:

10.3.3. Definition. Let $X: (\Omega, \mathfrak{A}) \to (\Omega', \mathfrak{A}')$ be a random variable on a probability space $(\Omega, \mathfrak{A}, P)$ with values in a measurable space $(\Omega', \mathfrak{A}')$, and let $\mathfrak{Z}$ be a σ-subalgebra of $\mathfrak{A}$. Then every Markov kernel $P_{X|\mathfrak{Z}}$ from $(\Omega, \mathfrak{Z})$ into $(\Omega', \mathfrak{A}')$, such that $\omega \to P_{X|\mathfrak{Z}}(\omega, A')$ is a version of $P^{\mathfrak{Z}}(X^{-1}(A'))$ for each $A' \in \mathfrak{A}'$, is called a *conditional distribution* of X given $\mathfrak{Z}$.

Examples

8. If X is the identity mapping of a probability space $(\Omega, \mathfrak{A}, P)$ onto itself, then every conditional distribution $P_{X|\mathfrak{Z}}$ is an expectation kernel $P_{\mathfrak{Z}}$ and vice versa.

9. For the special σ-subalgebras $\mathfrak{Z} = \mathfrak{Z}_0 = \{\varnothing, \Omega\}$ and $\mathfrak{Z} = \mathfrak{A}$, $P_{X|\mathfrak{Z}}$ exists for every $(\Omega', \mathfrak{A}')$-random variable X on $(\Omega, \mathfrak{A}, P)$: $P_{X|\mathfrak{Z}_0}$ is the kernel $(\omega, A') \to P_X(A')$, independent of ω, where P_X is the distribution of X. With every $\omega \in \Omega$, $P_{X|\mathfrak{A}}$ associates the unit mass $\epsilon_{X(\omega)}$ on $\mathfrak{A}'$, that is,

$$P_{X|\mathfrak{A}}(\omega, A') = \begin{cases} 1, & \text{if } X(\omega) \in A' \\ 0, & \text{if } X(\omega) \notin A' \end{cases} \qquad (\omega, A') \in \Omega \times \mathfrak{A}'.$$

(Compare Example 4.)

10. If we again have the situation of Examples 3 and 5 of Section 10.1 and that of Example 5 of this section, then $P_{X|\mathfrak{Z}}$ exists for every $(\Omega', \mathfrak{A}')$-random variable X. Let $Q_{B_i} = X(P_{B_i})$ be the distribution of the probability measure P_{B_i} $(i \in I)$. Then

$$P_{X|\mathfrak{Z}}(\omega, A') = \sum_{i \in I} Q_{B_i}(A') 1_{B_i}(\omega) \qquad (\omega, A') \in \Omega \times \mathfrak{A}'$$

is a conditional distribution according to Example 5. The reader should determine the conditional distribution also in the case of Example 6.

If we are dealing with the situation described in Definition 10.3.3, then for every $\mathfrak{A}'$-measurable numerical function $f' \geqq 0$ on Ω' and every version of the conditional expectation of $f' \circ X$:

$$E^{\mathfrak{Z}}(f' \circ X) = P_{X|\mathfrak{Z}} f', \qquad P\text{-almost surely.} \tag{10.3.8}$$

It again suffices to treat the case $f' = 1_{A'}$ with $A' \in \mathfrak{A}'$. But then

$$E^{\mathfrak{Z}}(1_{A'} \circ X) = E^{\mathfrak{Z}}(1_{X^{-1}(A')}) = P^{\mathfrak{Z}}(X^{-1}(A'))$$

and

$$P_{X|\mathfrak{Z}} 1_{A'}(\omega) = P_{X|\mathfrak{Z}}(\omega, A') = P^{\mathfrak{Z}}(X^{-1}(A'))(\omega), \qquad P\text{-almost surely.}$$

Formula (10.3.8) remains valid if we require the integrability of $f' \circ X$ instead of $f' \geqq 0$. Then, again, we need only decompose f' into its positive and negative part.

Now we proceed to the question of existence and uniqueness of expectation kernels and conditional distributions. By Example 8, it suffices to discuss these questions for the more general concept of conditional distribution. For the question of uniqueness, we show:

10.3.4. Theorem. Let $P_{X|\mathfrak{Z}}$ and $P^*_{X|\mathfrak{Z}}$ be two conditional distributions of a random variable $X: (\Omega,\mathfrak{A}) \to (\Omega',\mathfrak{A}')$ on a probability space $(\Omega,\mathfrak{A},P)$ given a σ-algebra $\mathfrak{Z} \subset \mathfrak{A}$. Then:

(1) $P_{X|\mathfrak{Z}}(\omega,A') = P^*_{X|\mathfrak{Z}}(\omega,A')$, P-almost surely for every $A' \in \mathfrak{A}'$.

(2) If the σ-algebra $\mathfrak{A}'$ has a countable generator $\mathfrak{E}'$, then there is a P-null set $N \in \mathfrak{A}$ such that

$$P_{X|\mathfrak{Z}}(\omega,A') = P^*_{X|\mathfrak{Z}}(\omega,A'), \quad \text{for all } \omega \in \complement N \text{ and all } A' \in \mathfrak{A}'. \tag{10.3.9}$$

Proof. (1) follows directly from the observation that for fixed $A' \in \mathfrak{A}'$, $\omega \to P_{X|\mathfrak{Z}}(\omega,A')$ is a version of $P^{\mathfrak{Z}}(X^{-1}(A'))$ and that any two such versions coincide P-almost surely.

(2) The countable generator $\mathfrak{E}'$ of $\mathfrak{A}'$ can be assumed $\cap$-stable without loss of generality. If necessary, we replace $\mathfrak{E}'$ by the system of all finite intersections of elements from $\mathfrak{E}'$. By (1), for every $E' \in \mathfrak{E}'$, there is a P-null set $N_{E'} \in \mathfrak{A}$ such that $P_{X|\mathfrak{Z}}(\omega,E') = P^*_{X|\mathfrak{Z}}(\omega,E')$ for all $\omega \in \complement N_{E'}$. The countable union $N \equiv \bigcup_{E' \in \mathfrak{E}'} N_{E'}$ is then a P-null set such that

$$P_{X|\mathfrak{Z}}(\omega,E') = P^*_{X|\mathfrak{Z}}(\omega,E')$$

for all $\omega \in \complement N$ and all $E' \in \mathfrak{E}'$. But then, by the Uniqueness Theorem 5.5, the above equality holds for all $\omega \in \complement N$ and all $E' \in \mathfrak{A}'$. ⌋

In particular, the σ-algebra of Borel sets of a Polish space $\Omega' = E$ is countably generated. If we choose a countable, dense subset D of E, then the system of all open balls in E with rational radii and centers in D is obviously a generator of $\mathfrak{B}(E)$. In this case we can now also prove the existence of a conditional distribution in general. The proof depends primarily on the familiar regularity of finite Borel measures on Polish spaces.

10.3.5. Theorem. Let $X: (\Omega,\mathfrak{A}) \to (E,\mathfrak{B}(E))$ be a random variable on a probability space $(\Omega,\mathfrak{A},P)$ with values in a Polish space E. Then for every σ-algebra $\mathfrak{Z} \subset \mathfrak{A}$ there exists a conditional distribution $P_{X|\mathfrak{Z}}$. This is uniquely determined in the sense of (10.3.9).

Proof. First we show that the σ-algebra $\mathfrak{B} = \mathfrak{B}(E)$ has a countable generator $\mathfrak{R}$ which is an algebra in E. To this end, let $\mathfrak{G}$ be a countable base of E containing E, $\mathfrak{G}_c$ the system of all sets $\complement G$ with $G \in \mathfrak{G}$, and $\mathfrak{G}' \equiv \mathfrak{G} \cup \mathfrak{G}_c$. Then $\mathfrak{G}'$ is also countable; $\mathfrak{G}'$ contains $\emptyset$ and E. The system $\mathfrak{R}$ of all sets of the form $\bigcup_{i=1}^{n} \bigcap_{j=1}^{n} G'_{ij}$ with $G'_{ij} \in \mathfrak{G}'$, $i, j = 1, \ldots, n; n \in \mathbf{N}$, yields the desired result. We only have to observe that by the set-theoretic distributive laws, all sets $\bigcap_{i=1}^{n} \bigcup_{j=1}^{n} H'_{ij}$ with $H'_{ij} \in \mathfrak{G}'$ $(i, j = 1, \ldots, n;$ $n \in \mathbf{N})$ also lie in $\mathfrak{R}$. Suppose $\mathfrak{R} = \{R_i: i \in I\}$ is an enumeration of $\mathfrak{R}$. Here we let I denote the set $\mathbf{N}$ of natural numbers or a finite segment $\{1,2, \ldots ,n\}$ of $\mathbf{N}$.

The distribution $\mu \equiv P_X$ of X is regular by Theorem 7.3.3. Therefore, for each $i \in I$ there is an isotone sequence $(K_{ij})_{j=1,2,\ldots}$ of compact subsets of R_i such that

$$\mu(R_i) = \sup_{j \in \mathbf{N}} \mu(K_{ij}).$$

If $\mathfrak{R}^*$ denotes the algebra generated in E by $\mathfrak{R}$ and all K_{ij}, $i \in I, j \in \mathbf{N}$, then this is again countable. To see this, we repeat the above reasoning with $\mathfrak{R} \cup \{K_{ij}: i \in I, j \in \mathbf{N}\}$ instead of $\mathfrak{G}$. After these preliminaries, for every set $B \in \mathfrak{B}$ we determine a version $P^{\mathfrak{Z}}(X^{-1}(B))$ of the conditional probability of $X^{-1}(B)$ given $\mathfrak{Z}$ and set

$$P(\omega,B) \equiv P^{\mathfrak{Z}}(X^{-1}(B))(\omega) \qquad (\omega,B) \in \Omega \times \mathfrak{B}.$$

Now, by (10.1.25),

$$P(\omega, \bigcup_{i=1}^{k} B_i) = \sum_{i=1}^{k} P(\omega,B_i), \qquad P\text{-almost surely}$$

on Ω for every finite family of pairwise disjoint sets $B_1, \ldots, B_k \in \mathfrak{B}$. Because $\mathfrak{R}^*$ is countable, there are only countably many such finite families $B_1, \ldots, B_k$ of sets in $\mathfrak{R}^*$. Hence there is a P-null set N_0 such that $B \to P(\omega,B)$ is finitely additive on $\mathfrak{R}^*$ for every $\omega \in \complement N_0$. Due to the $\mathfrak{Z}$-measurability of $\omega \to P(\omega,B)$, we can choose N_0 in $\mathfrak{Z}$. For each $i \in I$, the isotone sequence $(1_{K_{ij}})_{j=1,2,\ldots}$ converges to 1_{R_i} μ-almost surely, and thus the sequence $(1_{X^{-1}(K_{ij})})_{j=1,2,\ldots}$ converges to $1_{X^{-1}(R_i)}$ P-almost surely. By (10.1.8) and (10.1.13) we then have

$$\begin{aligned}\sup_{j \in \mathbf{N}} P^{\mathfrak{Z}}(X^{-1}(K_{ij})) &= \sup_{j \in \mathbf{N}} E^{\mathfrak{Z}}(1_{X^{-1}(K_{ij})}) = E^{\mathfrak{Z}}(1_{X^{-1}(R_i)}) \\ &= P^{\mathfrak{Z}}(X^{-1}(R_i)), \qquad P\text{-almost surely.}\end{aligned}$$

Thus for every $i \in I$ there is a P-null set $N_i \in \mathfrak{Z}$ such that

$$\sup_{j \in \mathbf{N}} P(\omega,K_{ij}) = P(\omega,R_i), \qquad \text{for all } \omega \in \complement N_i.$$

Finally, by (10.1.23) there is another P-null set $N_\infty \in \mathfrak{Z}$ such that

$$P(\omega,E) = 1, \qquad \text{for all } \omega \in \complement N_\infty.$$

But then $N \equiv N_0 \cup \bigcup_{i \in I} N_i \cup N_\infty$ is a P-null set in $\mathfrak{Z}$ such that $B \to P(\omega,B)$ is a content on $\mathfrak{R}^*$ with $P(\omega,E) = 1$ for all $\omega \in \complement N$. The restriction of this content to $\mathfrak{R}$ is, as we shall show, $\emptyset$-continuous and hence a premeasure on $\mathfrak{R}$. For this, let (B_n) be a sequence in $\mathfrak{R}$ with $B_n \downarrow \emptyset$ and let $\omega \in \complement N$. Since ω does not lie in any N_i $(i \in I)$, then for each $\epsilon > 0$ and $n = 1, 2, \ldots$, there is a compact set K_n (namely, a suitable K_{ij}) such that $K_n \subset B_n$ and

$$P(\omega,B_n) - P(\omega,K_n) = P(\omega,B_n \setminus K_n) \leqq 2^{-n}\epsilon.$$

Since $B_n \downarrow \emptyset$, we have $\bigcap_{n=1}^{\infty} K_n = \emptyset$. The compactness of all K_n then yields the existence of an $n_0 \in \mathbf{N}$ with $K_1 \cap \cdots \cap K_{n_0} = \emptyset$, that is, with $B_{n_0} \subset \bigcup_{i=1}^{n_0} (B_i \setminus K_i)$. But since $B \to P(\omega,B)$ is a finite content even on $\mathfrak{R}^*$, it follows that

$$P(\omega,B_{n_0}) \leqq \sum_{i=1}^{n_0} P(\omega,B_i \setminus K_i) \leqq \sum_{i=1}^{\infty} 2^{-i}\epsilon = \epsilon.$$

Therefore $\inf_{n \in \mathbf{N}} P(\omega,B_n) = 0$, as asserted.

The premeasure $B \to P(\omega,B)$ on $\mathfrak{R}$ can now be extended by Theorem 1.5.7 (in exactly one way) to a probability measure $B \to Q(\omega,B)$ on $\mathfrak{B}$ $(\omega \in \complement N)$. We also define $Q(\omega,B)$ for elements $\omega \in N$ by arbitrarily choosing (in the case $N \neq \emptyset$) an $\omega_N \in N$ and setting

$$Q(\omega,B) = Q(\omega_N,B) \equiv \begin{cases} 1, & \text{if } X(\omega_N) \in B \\ 0, & \text{if } X(\omega_N) \notin B \end{cases} \qquad (B \in \mathfrak{B}).$$

Thus, for every $\omega \in N$, the measure $B \to Q(\omega,B)$ is equal to $\epsilon_{X(\omega_N)}$. Therefore, Q has the property (10.3.2) of a Markov kernel from $(\Omega,\mathfrak{Z})$ to $(E,\mathfrak{B})$, that is, $B \to Q(\omega,B)$ is a probability measure on $\mathfrak{B}$ for all $\omega \in \Omega$. But the function $\omega \to Q(\omega,B)$ is also $\mathfrak{Z}$-measurable for every $B \in \mathfrak{B}$. By the above construction, since $N \in \mathfrak{Z}$, this holds at least for all $B \in \mathfrak{R}$. The system $\mathfrak{D}$ of all $D \in \mathfrak{B}$ for which $\omega \to Q(\omega,D)$ is $\mathfrak{Z}$-measurable is obviously a Dynkin system. Since $\mathfrak{R} \subset \mathfrak{D}$ and because of the $\cap$-stability of $\mathfrak{R}$, $\mathfrak{B} = \mathfrak{A}(\mathfrak{R}) = \mathfrak{D}(\mathfrak{R}) \subset \mathfrak{D} \subset \mathfrak{B}$ by Theorem 1.2.3, that is, $\mathfrak{D} = \mathfrak{B}$. Thus, Q is a Markov kernel from $(\Omega,\mathfrak{Z})$ to $(E,\mathfrak{B})$. Q also has the last property of a conditional distribution of X given $\mathfrak{Z}$: According to the construction of Q, for every $B \in \mathfrak{R}$, $\omega \to P(\omega,B)$ and hence $\omega \to Q(\omega,B)$ is a version of

$P^{\mathfrak{Z}}(X^{-1}(B))$, and thus

$$\int_Z Q(\omega,B)P(d\omega) = P(Z \cap X^{-1}(B))$$

for all $B \in \mathfrak{R}$ and $Z \in \mathfrak{Z}$. For every $Z \in \mathfrak{Z}$, $B \to \int_Z Q(\omega,B)P(d\omega)$ and $B \to P(Z \cap X^{-1}(B))$ are finite measures on $\mathfrak{B}$ which coincide on $\mathfrak{R}$. Then, by the Uniqueness Theorem, the two measures coincide on $\mathfrak{B}$. The last equality thus also holds for all $B \in \mathfrak{B}$ and $\mathfrak{Z} \in \mathfrak{Z}$. Therefore $\omega \to Q(\omega,B)$ is a version of $P^{\mathfrak{Z}}(X^{-1}(B))$ for all $B \in \mathfrak{B}$. As asserted, $P_{X|\mathfrak{Z}} \equiv Q$ yields the desired result. The uniqueness follows from Theorem 10.3.4 as mentioned. ┘

In order to demonstrate the usefulness of conditional distributions, we now prove the so-called *Jensen inequality* for ordinary and conditional expectations.

First we recall the following well-known notion: A real function q defined on an interval $J \subset \mathbf{R}$ is said to be *concave* if

$$\alpha q(x) + (1 - \alpha)q(y) \leqq q(\alpha x + (1 - \alpha)y)$$

for any two points $x, y \in J$ and all numbers $\alpha \in [0,1]$. Using induction, we can prove easily that this is equivalent to the following requirement: For any finite number of points $x_1, \ldots, x_k \in J$ and numbers $\alpha_1, \ldots, \alpha_k \in \mathbf{R}_+$ with $\Sigma_{i=1}^k \alpha_i = 1$,

$$\sum_{i=1}^{k} \alpha_i q(x_i) \leqq q\left(\sum_{i=1}^{k} \alpha_i x_i\right).$$

We say that q is *convex* if $-q$ is concave.

10.3.6. Theorem. Let X be a real, integrable random variable on a probability space $(\Omega,\mathfrak{A},P)$ with values in an arbitrary interval $J \subset \mathbf{R}$. Then $E(X)$ lies in J, and for every continuous concave function q on J,

$$E(q \circ X) \leqq q(E(X)), \tag{10.3.10}$$

provided $q \circ X$ is integrable.

Proof. 1. $E(X)$ lies in J. This follows from the following two remarks: (i) For every real number α, the relation $X \leqq \alpha$ (or $\alpha \leqq X$) implies $E(X) \leqq \alpha$ (or $\alpha \leqq E(X)$); (ii) $X(\omega) < \alpha$ (or $\alpha < X(\omega)$) for all $\omega \in \Omega$ implies $E(X) < \alpha$ (or $\alpha < E(X)$). Remark (i) is an immediate consequence of the isotonicity of integrals. From $X(\omega) < \alpha$ for all $\omega \in \Omega$, we conclude that $E(X) = \alpha$ is impossible, and hence $E(X) < \alpha$. Indeed, $E(X) = \alpha$ would imply $E(\alpha - X) = 0$; hence, the integrand $\alpha - X \geqq 0$ would vanish P-almost surely. But $\{X < \alpha\} = \Omega$ has probability 1. The same argument proves the second part of (ii).

2. Next we prove (10.3.10) for the case of a compact interval J. We consider X as a random variable with values in the compact subspace J of $\mathbf{R}$ and correspondingly consider the distribution P_X as a probability measure on J. By Corollary 7.7.4 and Theorem 7.8.4, there exists a sequence (μ_n) of discrete probability measures on J which converges vaguely and hence weakly to P_X. Consequently, we have, on one hand,

$$\text{(a)} \qquad \lim_{n\to\infty} \int x\mu_n(dx) = \int x\, dP_X = E(X)$$

and on the other

$$\text{(b)} \qquad \lim_{n\to\infty} \int q\, d\mu_n = \int q\, dP_X = E(q \circ X).$$

Every measure μ_n is of the form $\Sigma_{i=1}^k \alpha_i \epsilon_{a_i}$, where $a_1, \ldots, a_k \in J$, $\alpha_1, \ldots, \alpha_k \in \mathbf{R}_+$, and $\Sigma\alpha_i = 1$. (Here all the a_i and α_i depend on both k and n.) Therefore, $m_n \equiv \int x\mu_n(dx) = \Sigma_{i=1}^k \alpha_i a_i \in J$ since J is a convex set. From this and (a) it follows that $E(X) \in J$ because J is closed. Moreover,

$$\int q\, d\mu_n = \sum_{i=1}^{k} \alpha_i q(a_i) \leqq q\Big(\sum_{i=1}^{k} \alpha_i a_i\Big) = q(m_n)$$

due to the concavity of q. From $\lim m_n = E(X)$, the continuity of q, and from (b), it follows that $E(q \circ X) \leqq q(E(X))$, that is, we have the Jensen inequality.

3. The general case can now be derived from 2: There exists an isotone sequence $(J_n)_{n\in\mathbf{N}}$ of compact intervals in R such that $J_n \uparrow J$. Then $\Omega_n \equiv \{X \in I_n\}$ defines a sequence in the σ-algebra $\mathfrak{A}$ such that $\Omega_n \uparrow \Omega$. Thus $\lim P(\Omega_n) = 1$, and we can assume $P(\Omega_n) > 0$ for all $n \in \mathbf{N}$. Since the restriction P_n of $[1/P(\Omega_n)]P$ to $\Omega_n \cap \mathfrak{A}$ is a probability measure and since the continuous function q is bounded on the compact interval I_n, the restriction of $q \circ X$ to Ω_n is an integrable random variable on the probability space $(\Omega_n, \Omega_n \cap \mathfrak{A}, P_n)$. Thus we know from 2

$$\frac{1}{P(\Omega_n)} \int_{\Omega_n} q \circ X\, dP \leqq q\left(\frac{1}{P(\Omega_n)} \int_{\Omega_n} X\, dP\right).$$

Now $\lim P(\Omega_n) = 1$, $\lim \int_{\Omega_n} q \circ X\, dP = E(q \circ X)$, and $\lim \int_{\Omega_n} X\, dP = E(X)$. Since q is continuous on J, we obtain (10.3.10) as $n \to \infty$. ┘

10.3.7. Corollary. If X and q are as above, then for every σ-subalgebra $\mathfrak{Z}$ of $\mathfrak{A}$:

$$E^{\mathfrak{Z}}(X) \in J, \qquad P\text{-almost surely}; \qquad (10.3.11)$$

$$E^{\mathfrak{Z}}(q \circ X) \leqq q(E^{\mathfrak{Z}}(X)), \qquad P\text{-almost surely}. \qquad (10.3.12)$$

Proof. By Theorem 10.3.5 there exists a conditional distribution $P_{X|\mathfrak{Z}}$, considered as a kernel from $(\Omega,\mathfrak{Z})$ to $(J,\mathfrak{B}(J))$. Let Q_ω denote the probability measure $A \to P_{X|\mathfrak{Z}}(\omega,A)$ on $\mathfrak{B}(J)$ $(\omega \in \Omega)$. Then by (10.3.8) we have

$$[E^{\mathfrak{Z}}(X)](\omega) = \int x Q_\omega(dx), \qquad P\text{-almost surely}$$

and also

$$[E^{\mathfrak{Z}}(q \circ X)](\omega) = \int q(x) Q_\omega(dx), \qquad P\text{-almost surely}.$$

Since $(J,\mathfrak{B}(J),Q_\omega)$ is a probability space for every $\omega \in \Omega$, and $x \to x$ is a random variable on it, then by Theorem 10.3.6

$$\int x Q_\omega(dx) \in J$$

and

$$\int q(x) Q_\omega(dx) \leqq q\left(\int x Q_\omega(dx)\right), \qquad P\text{-almost surely}.$$

But these four results imply (10.3.11) and (10.3.12). ⌟

Remark. The function $x \to |x|$ is convex on $\mathbf{R}$. Therefore, we again obtain the familiar inequalities $|E(X)| \leqq E(|X|)$ and $|E^{\mathfrak{Z}}(X)| \leqq E^{\mathfrak{Z}}(|X|)$ almost surely.

PROBLEMS

1. Let $X: (\Omega,\mathfrak{A}) \to (E,\mathfrak{B}(E))$ and $Y: (\Omega,\mathfrak{A}) \to (\Omega',\mathfrak{A}')$ be random variables on a probability space $(\Omega,\mathfrak{A},P)$ with values in a Polish space E or an arbitrary measurable space $(\Omega',\mathfrak{A}')$, respectively. Imitate the proof of Theorem 10.3.5 and prove the existence of a kernel Q from $(\Omega',\mathfrak{A}')$ to $(E,\mathfrak{B}(E))$ such that $y \to Q(y,B)$ is a version of $y \to P(X^{-1}(B) \mid Y = y)$ for all $B \in \mathfrak{B}(E)$. For each $y \in \Omega'$, the probability measure $B \to Q(y,B)$ on $\mathfrak{B}(E)$ is then called a *conditional distribution of X given that Y equals y*. It is denoted by $P_{X|Y=y}$; hence, for a given version of $y \to P(X^{-1}(B) \mid Y = y)$, we have

$$P_{X|Y=y}(B) = P(X^{-1}(B) \mid Y = y), \qquad P_Y\text{-almost surely}.$$

   Why is then $(\omega,B) \to P_{X|Y=Y(\omega)}(B)$ a conditional distribution $P_{X|Y}$?
2. Consider the situation of Problem 1 and denote by μ_y the probability measure $P_{X|Y=y}$. Prove that the joint distribution of $P_{X \otimes Y}$ of X and Y is given by the formula

$$P_{X\otimes Y}(M) = \int[\int 1_M(x,y)\mu_y(dx)]P_Y(dy), \qquad (M \in \mathfrak{B}(E) \otimes \mathfrak{A}').$$

Observe that $y \to \int 1_M(x,y)\mu_y(dx)$ is $\mathfrak{A}'$-measurable. Express $\int f\, dP_{X\otimes Y}$ for nonnegative measurable functions and $P_{X\otimes Y}$-integrable functions f on $E \times \Omega'$ in terms of μ_y and P_Y.

3. Consider again the situation of Problem 1. Prove: X and Y are independent if and only if

$$P_{X|Y=y} = P_X, \qquad P_Y\text{-almost surely.}$$

4. Let X and Y be real random variables on $(\Omega,\mathfrak{A},P)$. Assume that X has ν_{0,σ^2} as distribution and that the conditional distribution $P_{X|Y}$ equals ν_{y,τ^2}. Prove that the joint distribution of X and Y has the following density with respect to λ^2:

$$(x,y) \to \frac{1}{2\pi\sigma\tau}\, e^{-\frac{1}{2}\left(\frac{x^2}{\sigma^2}+\frac{(y-x)^2}{\tau^2}\right)}.$$

5. Let $X_1, \ldots, X_n$ be n real random variables on $(\Omega,\mathfrak{A},P)$ such that their joint distribution has density d with respect to λ^n. Introduce $X \equiv X_1 \otimes \cdots \otimes X_p$ and $Y \equiv X_{p+1} \otimes \cdots \otimes X_n$ for some $p \in \{1, \ldots, n-1\}$, and denote by d_Y the function

$$(x_{p+1}, \ldots, x_n) \to \int d(x_1, \ldots, x_n)\lambda^p(dx_1, \ldots, dx_p)$$

on $\mathbf{R}^{n-p}$. Prove:

(a) $d_Y\lambda^{n-p}$ is a probability measure on $\mathfrak{B}^{n-p}$.

(b) $d_Y \neq 0, \qquad P_Y$-almost surely.

(c) For P_Y-almost all $y = (x_{p+1}, \ldots, x_n) \in \mathbf{R}^{n-p}$ we have

$$P_{X|Y=y} = d_y\lambda^p,$$

where

$$d_y(x_1, \ldots, x_p) = \frac{d(x_1, \ldots, x_n)}{d_Y(x_{p+1}, \ldots, x_n)}.$$

6. Deduce from Minkowski's inequality (2.6.4) that $x \to |x|^p$ is convex on $\mathbf{R}$ for all $p \geqq 1$. Prove that for every random variable $X \in \mathfrak{L}^p(P)$, defined on a probability space $(\Omega,\mathfrak{A},P)$, and every σ-subalgebra $\mathfrak{Z}$ of $\mathfrak{A}$ the inequality

$$|E^{\mathfrak{Z}}(X)|^p \leqq E^{\mathfrak{Z}}(|X|^p)$$

holds almost surely.

11

MARTINGALES

The theory of martingales established by J. L. Doob (see [11] and [39]) is one of the principal tools of the theory of stochastic processes. It provides a unified method for dealing with various limit theorems of probability theory. In particular, the concept of martingale helps us to study theorems connected with the law of large numbers in a new light. We are more or less inevitably led to the concept of martingale through the study of the convergence behavior of conditional expectations $E^{\mathfrak{A}_n}(X)$ relative to an isotone sequence of σ-subalgebras $\mathfrak{A}_n$ of $\mathfrak{A}$.

11.1 DEFINITION AND EXAMPLES

Suppose we are given a probability space $(\Omega,\mathfrak{A},P)$, a partially ordered set I,[1] a family $(\mathfrak{A}_t)_{t\in I}$ of σ-subalgebras $\mathfrak{A}_t$ of $\mathfrak{A}$, and a family $(X_t)_{t\in I}$ of random variables X_t: $(\Omega,\mathfrak{A}) \to (\Omega',\mathfrak{A}')$ with values in a measurable space $(\Omega',\mathfrak{A}')$.[2] We agree on the following terminology: The family $(\mathfrak{A}_t)_{t\in I}$ is said to be *isotone* (or *increasing*) if

$$s \leqq t \Rightarrow \mathfrak{A}_s \subset \mathfrak{A}_t \qquad (s, t \in I). \tag{11.1.1}$$

The family $(X_t)_{t\in I}$ is said to be *adapted to the family* $(\mathfrak{A}_t)_{t\in I}$ if X_t is $\mathfrak{A}_t$-$\mathfrak{A}'$-measurable for every $t \in I$.

[1] Thus a relation $\leqq$ is defined in I which is reflexive ($t \leqq t$ for all $t \in I$), antisymmetric ($s \leqq t, t \leqq s \Rightarrow s = t$), and transitive ($s \leqq t,\ t \leqq u \Rightarrow s \leqq u$). If $s \leqq t$ or $t \leqq s$ holds for every pair $(s,t) \in I \times I$, then I is said to be *totally* ordered.

[2] We denote the families by $(\mathfrak{A}_t)$ and (X_t) when there is no danger of ambiguity.

Example

1. Let $(\Omega,\mathfrak{A},P)$, I and $(X_t)_{t\in I}$ be given as above. For every $t \in I$, let

$$\mathfrak{A}_t^0 \equiv \mathfrak{A}(X_s;\, s \leqq t),$$

be the σ-algebra generated by all X_s with $s \leqq t$ $(s \in I)$. Then the family $(\mathfrak{A}_t^0)_{t\in I}$ is obviously isotone and (X_t) is adapted to the family $(\mathfrak{A}_t^0)$. The family $(X_t)_{t\in I}$ is obviously adapted to a given isotone family $(\mathfrak{A}_t)_{t\in I}$ of σ-subalgebras of $\mathfrak{A}$ if and only if

$$\mathfrak{A}_t^0 \subset \mathfrak{A}_t, \qquad \text{for all } t \in I.$$

11.1.1. Definition. Let $(\Omega,\mathfrak{A},P)$ be a probability space, $(\mathfrak{A}_t)_{t\in I}$ an isotone family of σ-subalgebras of $\mathfrak{A}$, and $(X_t)_{t\in I}$ a family of real, integrable random variables adapted to $(\mathfrak{A}_t)_{t\in I}$. We call (X_t) a *super-martingale* [relative to the family $(\mathfrak{A}_t)$] if one of the following equivalent conditions is satisfied for all pairs $s, t \in I$ with $s \leqq t$:

$$E^{\mathfrak{A}_s}(X_t) \leqq X_s, \qquad P\text{-almost surely}; \tag{11.1.2}$$

$$\int_A X_t\, dP \leqq \int_A X_s\, dP, \qquad \text{for all } A \in \mathfrak{A}_s. \tag{11.1.3}$$

We call (X_t) a *sub-martingale* [relative to $(\mathfrak{A}_t)$] if $(-X_t)$ is a super-martingale relative to $(\mathfrak{A}_t)$. If (X_t) is both a super-martingale and a sub-martingale, then (X_t) is called a *martingale.*[3]

Hence, in the definition of sub-martingale (martingale), the symbols $\leqq$ in (11.1.2) and (11.1.3) should be replaced by $\geqq$ $(=)$. The following remarks serve to clarify the definition further.

Remarks. 1. The equivalence of conditions (11.1.2) and (11.1.3) is easy to see: For all $A \in \mathfrak{A}_s$, $\int_A E^{\mathfrak{A}_s}(X_t)\, dP = \int_A X_t\, dP$. Thus (11.1.3) follows from (11.1.2). Conversely, (11.1.3) implies

$$\int_A (X_s - E^{\mathfrak{A}_s}(X_t))\, dP \geqq 0, \qquad \text{for all } A \in \mathfrak{A}_s.$$

Here the integrand is $\mathfrak{A}_s$-measurable. Therefore, we can choose

$$A = \{X_s - E^{\mathfrak{A}_s}(X_t) < 0\}.$$

The above inequality then shows that $P\{X_s - E^{\mathfrak{A}_s}(X_t) < 0\} = 0$, and thus (11.1.2) holds.

2. Example 1 tells us that a family $(X_t)_{t\in I}$ of real, integrable random variables is always adapted to the isotone family $(\mathfrak{A}_t^0)_{t\in I}$ of the σ-algebras

[3] The strong interrelations of the random variables X_t of a martingale explain the use of the term "martingale." Generally, martingale denotes a part of the bridle of a horse, namely, that part which prevents the horse from tossing his head high.

$\mathfrak{A}_t^0 = \mathfrak{A}(X_s; s \leqq t)$. Therefore we simply call (X_t) a super-martingale, and so on, that is, without reference to a family $(\mathfrak{A}_t)$, if we are dealing with a super-martingale, and so on, relative to $(\mathfrak{A}_t^0)$. Every super-martingale (martingale) relative to $(\mathfrak{A}_t)$ is obviously also a super-martingale (martingale) relative to $(\mathfrak{A}_t^0)$.

3. Properties (11.1.2) and (11.1.3) are obvious (in fact, with equality) in the case $s = t$. In verifying these properties we can therefore always assume that $s < t$, that is, $s \leqq t$ and $s \neq t$.

4. If $(X_n)_{n \in \mathbf{N}}$ is a super-martingale relative to a sequence $(\mathfrak{A}_n)_{n \in \mathbf{N}}$ of σ-algebras (with the set $\mathbf{N}$ of natural numbers, ordered as usual, as index set), then properties (11.1.2) and (11.1.3) follow from the validity of

$$E^{\mathfrak{A}_n}(X_{n+1}) \leqq X_n, \qquad \text{almost surely,}$$

for all $n \in \mathbf{N}$: By (10.1.18),

$$E^{\mathfrak{A}_n}(X_{n+p}) = E^{\mathfrak{A}_n}(E^{\mathfrak{A}_{n+1}}(X_{n+p})) = \cdots = E^{\mathfrak{A}_n}(E^{\mathfrak{A}_{n+1}}(\cdots E^{\mathfrak{A}_{n+p-1}}(X_{n+p})))$$

holds almost surely for every $p = 2, 3, \ldots$, and therefore

$$E^{\mathfrak{A}_n}(X_{n+p}) \leqq X_n, \qquad \text{almost surely}$$

follows from the inequality above. (Corresponding results hold for index sets $I = \{1, \ldots, k\} \subset \mathbf{N}$.)

5. For $A = \Omega$, (11.1.3) yields the assertion that for every super-martingale $(X_t)_{t \in I}$, the expectations $E(X_t)$ depend antitonely on t:

$$s \leqq t \Rightarrow E(X_s) \geqq E(X_t).$$

For a martingale, the expectations $E(X_t)$ are independent of t.

Examples

2. Let $(\Omega, \mathfrak{A}, P)$ be a probability space, $(\mathfrak{A}_t)_{t \in I}$ an isotone family of σ-subalgebras of $\mathfrak{A}$, and X a real, integrable random variable on Ω. We define $X_t \equiv E^{\mathfrak{A}_t}(X)$ for every $t \in I$. Then (X_t) is a martingale relative to $(\mathfrak{A}_t)$. Indeed, every X_t is $\mathfrak{A}_t$-measurable by definition. From $s \leqq t$ it follows that $E^{\mathfrak{A}_s}(X_t) = E^{\mathfrak{A}_s}(E^{\mathfrak{A}_t}(X)) = E^{\mathfrak{A}_s}(X) = X_s$ almost surely according to the smoothing property (10.1.18).

3. Let $(X_n)_{n \in \mathbf{N}}$ be an *independent sequence* of real, integrable random variables on a probability space. Assume each X_n is centered and thus $E(X_n) = 0$ for all $n \in \mathbf{N}$. If $S_n \equiv \sum_{i=1}^{n} X_i$ denotes the nth *partial sum* of the sequence (X_n), then $(S_n)_{n \in \mathbf{N}}$ is a martingale. To see this: For every

natural number n, according to Remark 4, we need to verify the equality

$$E(S_{n+1} \mid S_1, \ldots, S_n) = S_n, \qquad \text{almost surely.}$$

Now obviously $\mathfrak{A}(X_1, \ldots, X_n) = \mathfrak{A}(S_1, \ldots, S_n)$. Therefore, by Corollary 5.1.5, X_{n+1} is independent of $\mathfrak{A}(S_1, \ldots, S_n)$ and hence

$$E(X_{n+1} \mid S_1, \ldots, S_n) = E(X_{n+1}) = 0, \qquad \text{almost surely.}$$

However, for all $j \in \mathbf{N}$ with $1 \leqq j \leqq n$ we have $E(X_j \mid S_1, \ldots, S_n) = X_j$ almost surely. The assertion follows when we add these equalities. The proof further shows the following: If we replace the conditions $E(X_n) = 0$ by $E(X_n) \leqq 0 (\geqq 0)$ for all n, then (S_n) is a super-(sub-) martingale.

4. Martingales allow the following *interpretation:* Let $(X_n)_{n \in \mathbf{N}}$ be a sequence of real, integrable random variables on a probability space. Suppose the variable X_n is interpreted as the amount a player wins on the nth play. Then

$$S_n \equiv X_1 + \cdots + X_n$$

is the fortune of the player after the first n plays. For obvious reasons we call the game *fair* if

$$E(X_1) = 0 \quad \text{and} \quad E(X_{n+1} \mid X_1, \ldots, X_n) = 0, \quad \text{almost surely } (n \in \mathbf{N}).$$

The same reasoning as in Example 3 then shows that $(S_n)_{n \in \mathbf{N}}$ is a martingale. On the other hand, if we had

$$E(X_1) \geqq 0 \quad \text{and} \quad E(X_{n+1} \mid X_1, \ldots, X_n) \geqq 0, \quad \text{almost surely } (n \in \mathbf{N}),$$

then our player would have an advantage over his opponent. (S_n) would then be a sub-martingale.

5. For a probability space $(\Omega, \mathfrak{A}, P)$, let I denote the set of all (finite) decompositions $t = (B_i)_{i=1,\ldots,n_t}$ of Ω into finitely many pairwise disjoint sets $B_i \in \mathfrak{A}$. I is partially ordered as follows: The relationship $t' \leqq t''$ holds between $t' = (B_i')$ and $t'' = (B_j'')$ if and only if t'' is finer than t', that is, if every set B_i' is the union of certain sets B_j''. For every decomposition $t = (B_i) \in I$, let $\mathfrak{A}_t$ denote the σ-algebra in $\mathfrak{A}$ generated by all $B_1, \ldots, B_{n_t}$ (consisting of finitely many sets) (see Section 10.1, p. 302). In addition to P, suppose a finite content Q is defined on $\mathfrak{A}$. If we set

$$X_s \equiv \sum_{i=1}^{n_s} \frac{Q(B_i)}{P(B_i)} 1_{B_i},$$

for every $s = (B_i) \in I$, where the quotient $Q(B_i)/P(B_i)$ is set equal to zero when $P(B_i) = 0$, then $(X_t)_{t \in I}$ is a super-martingale relative to

$(\mathfrak{A}_t)_{t\in I}$. First it is clear that the family $(\mathfrak{A}_t)$ is isotone and (X_t) is adapted to this family. If $s = (B_i)$ and $t = (C_j)$ are decompositions from I such that $s \leqq t$, then every B_i is the union of certain $C_{j_\varkappa}$, $\varkappa = 1, \ldots, k$, and hence

$$\int_{B_i} X_s\,dP = \frac{Q(B_i)}{P(B_i)}\,P(B_i) = \begin{cases} Q(B_i), & \text{if } P(B_i) > 0,\\ 0, & \text{if } P(B_i) = 0,\end{cases}$$

$$\int_{B_i} X_t\,dP = \sum_{\varkappa=1}^{k} \frac{Q(C_{j_\varkappa})}{P(C_{j_\varkappa})}\,P(C_{j_\varkappa}) = {\sum_\varkappa}' Q(C_{j_\varkappa}).$$

The symbol $\Sigma'_\varkappa$ here denotes summation over all $\varkappa = 1, \ldots, k$ with $P(C_{j_\varkappa}) > 0$. Therefore we have $\int_{B_i} X_t\,dP \leqq Q(B_i)$ and $\int_{B_i} X_t\,dP = 0$ when $P(B_i) = 0$. We then obtain $\int_{B_i} X_t\,dP \leqq \int_{B_i} X_s\,dP$ for $i = 1, \ldots, n_s$. Since every element from $\mathfrak{A}_s$ is the union of finitely many B_i, this proves the inequality $E^{\mathfrak{A}_s}(X_t) \leqq X_s$ almost surely.

The above proof tells us that (X_t) is a martingale relative to $(\mathfrak{A}_t)$ if $P(A) = 0$ $(A \in \mathfrak{A})$ implies $Q(A) = 0$, that is, if (in the sense of Definition 2.9.5) the content Q is P-continuous.

The following properties of super-martingales make it possible to construct further examples. Here we let $(X_t)_{t\in I}$ and $(Y_t)_{t\in I}$ be two super-martingales or martingales on a probability space $(\Omega,\mathfrak{A},P)$ relative to the same isotone family $(\mathfrak{A}_t)_{t\in I}$ of σ-subalgebras of $\mathfrak{A}$.

When (X_t) and (Y_t) are super-martingales (or martingales), $(\alpha X_t + \beta Y_t)$ is also a super-martingale (or martingale) relative to $(\mathfrak{A}_t)$ ($\alpha, \beta \in \mathbf{R}_+$ or $\alpha, \beta \in \mathbf{R}$, respectively). (11.1.4)

When (X_t) and (Y_t) are super-martingales, $(\inf(X_t, Y_t))$ is also a super-martingale. (11.1.5)

For every super-martingale (X_t), (X_t^-) is a sub-martingale. (11.1.6)

If (X_t) is a super-martingale (martingale) with values in an interval $J \subset \mathbf{R}$ [that is, $X_t(\Omega) \subset J$ for all $t \in I$] and $q\colon J \to \mathbf{R}$ is a continuous, isotone, concave (continuous concave) function, then $(q \circ X_t)$ is a super-martingale relative to $(\mathfrak{A}_t)$ provided all random variables $q \circ X_t$ are integrable. (11.1.7)

(11.1.4) follows directly from (11.1.2). For (11.1.5) we proceed as follows: For any two elements $s, t \in I$ with $s \leqq t$ we have $E^{\mathfrak{A}_s}(X_t) \leqq X_s$ and $E^{\mathfrak{A}_s}(Y_t) \leqq Y_s$ almost surely. Then $E^{\mathfrak{A}_s}(\inf(X_t,Y_t)) \leqq E^{\mathfrak{A}_s}(X_t) \leqq X_s$ and also $E^{\mathfrak{A}_s}(\inf(X_t,Y_t)) \leqq E^{\mathfrak{A}_s}(Y_t) \leqq Y_s$ almost surely; therefore, $E^{\mathfrak{A}_s}(\inf(X_t,Y_t)) \leqq \inf(X_s,Y_s)$ almost surely. If we choose all Y_t in (11.1.5) equal to 0, then (11.1.6) follows since $X_t^- = -\inf(X_t,0)$. Property (11.1.7) is a direct consequence of the Jensen inequality (10.3.12).

As an example, $x \to x^2$, $x \to \sup(x,0)$, $x \to \sup(-x,0)$, and $x \to |x|$ are continuous convex functions on $\mathbf{R}$. Therefore, for every martingale $(X_t)_{t\in I}$, it is seen that (X_t^2), (X_t^+), (X_t^-), and $(|X_t|)$ are sub-martingales. Of course we must require square integrability of all X_t in the first case.

Remark. 6. Sometimes super-martingales are defined by the requirement (11.1.2) if the random variables X_t are no longer integrable but all $X_t \geqq 0$ or have integrable negative part X_t^-. We do not go into the details of this obvious generalization.

PROBLEMS

1. Consider Polya's urn model of Section 4.2, Problem 6. Let $X_0 \equiv b/(b+w)$ and let X_n be the proportion of black balls attained in the nth drawing. Prove that $(X_n)_{n=0,1,\ldots}$ is a martingale.
2. Let $(X_n)_{n\in\mathbf{N}}$ be a martingale relative to an isotone sequence $(\mathfrak{A}_n)_{n\in\mathbf{N}}$ of σ-subalgebras of $\mathfrak{A}$ where $(\Omega,\mathfrak{A},P)$ is the underlying probability space. Let $(f_n)_{n\in\mathbf{N}}$ be a sequence of bounded real random variables adapted to the sequence $(\mathfrak{A}_n)$. Define by induction

$$Y_1 \equiv X_1,$$
$$Y_{n+1} \equiv Y_n + f_n \cdot (X_{n+1} - X_n) \qquad (n \in \mathbf{N}).$$

   Prove that $(Y_n)_{n\in\mathbf{N}}$ is a martingale relative to $(\mathfrak{A}_n)_{n\in\mathbf{N}}$.
3. Let $((X_t^n)_{t\in I})_{n\in\mathbf{N}}$ be a sequence of super-martingales on a probability space $(\Omega,\mathfrak{A},P)$ with respect to one isotone family $(\mathfrak{A}_t)_{t\in I}$ of σ-subalgebras. Assume that $|X_t^n| \leqq Y_t$ holds almost surely for all $n \in \mathbf{N}$ and $t \in I$, where each Y_t is an integrable random variable. Prove that $(\sup_{n\in\mathbf{N}} X_t^n)_{t\in I}$ resp. $(\inf_{n\in\mathbf{N}} X_t^n)_{t\in I}$ is a super-martingale provided that the sequence $(X_t^n)_{n\in\mathbf{N}}$ is isotone resp. antitone for each $t \in I$.
4. Let $(X_t)_{t\in I}$ be a sub-martingale on $(\Omega,\mathfrak{A},P)$ of random variables $X_t \geqq 0$ in $\mathfrak{L}^p(P)$, $1 \leqq p < +\infty$. Prove that $(X_t^p)_{t\in I}$ is again a sub-martingale.
5. Let $(X_t)_{t\in I}$ be a martingale of random variables $X_t \in \mathfrak{L}^p(P)$, $1 \leqq p < +\infty$. Prove that $(|X_t|^p)_{t\in I}$ is a sub-martingale.

11.2 TRANSFORMATION BY STOPPING TIMES

Example 4 of the preceding section tells us that we can interpret a martingale $(X_n)_{n\in\mathbf{N}}$ as a fair game, where X_n signifies the fortune of the player at time n. The player can now decide to terminate the game at some random point of time T (depending possibly on his mood or on the course of the game up to then). Thus T is a random variable with values in **N**. The run of total wins by the player is then represented by the sequence $(X_n^*)_{n\in\mathbf{N}}$ of random variables

$$X_n^*(\omega) \equiv \begin{cases} X_n(\omega), & \omega \in \{T \geqq n\} \\ X_{T(\omega)}(\omega), & \omega \in \{T < n\} \end{cases}.$$

However, the player may also want to test the course of the game by sampling his fortune at certain times $T_1 \leqq T_2 \leqq \cdots$ (all in **N**) which appear favorable to him and which are again random variables. He may ask himself whether the fair course of the game can be changed to his advantage by passing to (X_{T_n}). Thus he will ask whether a sub-martingale (X_{T_n}) which is not a martingale can be obtained from the martingale (X_n). The second construction is a generalization of the first: We need only set $T_n \equiv \inf(T,n)$ to obtain $X_n^*(\omega) = X_{T_n(\omega)}(\omega)$ for all $n \in \mathbf{N}$ and $\omega \in \Omega$.[4]

Since our player is not supposed to have the gift of prophesy, the random points of time T or T_n contain "no anticipation of the future." The meaning of this will be made precise immediately in the concept of a stopping time. With the help of this clarification, we shall be able to show the invariance of the martingale property in the transition from (X_n) to (X_{T_n}) and thus also to (X_n^*). For the sake of simplicity we shall restrict consideration to the case in which the index set is $\{1, \ldots, k\}$ instead of **N**.

11.2.1. Definition. Let $(\mathfrak{A}_t)_{t\in I}$ be an isotone family of σ-algebras in a set Ω. A *stopping time* (relative to this family) is a mapping $T: \Omega \to I$ of Ω into the partially ordered index set I such that

$$\{T \leqq t\} \in \mathfrak{A}_t, \qquad \text{for all } t \in I.^5 \tag{11.2.1}$$

If I is an interval in $\bar{\mathbf{R}}$ or a subset of the natural numbers, then (11.2.1) implies the measurability of T relative to every σ-algebra $\mathfrak{A}$ in Ω with $\mathfrak{A} \supset \bigcup_{t\in I} \mathfrak{A}_t$. Condition (11.2.1) can be given the interpretation that the random variable T does not contain any knowledge of the future. Here the σ-algebra $\mathfrak{A}_t$ embodies "random events up to time t." Therefore, it is not surprising that one generally determines a stopping time T to be the point of time at which some given random phenomenon is observed for the first time. Example 4 below and the stopping times to be constructed in the proof of Theorem 11.3.2 illuminate this observation.

[4] The transition from (X_n) to (X_n^*) $[(X_{T_n})]$ is called optional stopping [optional sampling].

[5] $\{T \leqq t\}$ of course means the set of all $\omega \in \Omega$ with $T(\omega) \leqq t$. We are often forced (see Example 4) to adjoin an element $+\infty$ to I which, by definition, must satisfy the relationships $t \leqq +\infty$ for all $t \in I$. Then a mapping $T: \Omega \to I \cup \{+\infty\}$ which satisfies (11.2.1) is still called a stopping time. But this is not a real generalization. We need only set $\mathfrak{A}_\infty \equiv \mathfrak{A}(\mathfrak{A}_t;\ t \in I)$ to obtain a stopping time relative to $(\mathfrak{A}_t)_{t\in I\cup\{+\infty\}}$ in the sense of our definition.

Examples

1. Every constant mapping $T: \Omega \to I$ is a stopping time relative to an arbitrary isotone family $(\mathfrak{A}_t)_{t \in I}$ of σ-algebras in Ω. If t_0 is the constant value of T, then

$$\{T \leqq t\} = \begin{cases} \Omega, & \text{if } t_0 \leqq t \\ \varnothing, & \text{otherwise} \end{cases} \qquad (t \in I).$$

2. If I is totally ordered, then when S and T are stopping times, $\inf(S,T)$ and $\sup(S,T)$ are also stopping times (relative to the same amily $(\mathfrak{A}_t)_{t \in I}$). Indeed, we have

$$\{\inf(S,T) \leqq t\} = \{S \leqq t\} \cup \{T \leqq t\}$$

and

$$\{\sup(S,T) \leqq t\} = \{S \leqq t\} \cap \{T \leqq t\}.$$

3. In the case $I = \mathbf{N}$, when S and T are stopping times, $S + T$ is also a stopping time. This follows from

$$\{S + T \leqq n\} = \bigcup_{\substack{k,\, l \in \mathbf{N} \\ k+l \leqq n}} \{S \leqq k\} \cap \{T \leqq l\}.$$

4. Let $I = \mathbf{N} \cup \{+\infty\}$ and let $(X_t)_{t \in I}$ be a family of real random variables on a probability space adapted to the family $(\mathfrak{A}_t)_{t \in I}$; thus, in particular, we have $\mathfrak{A}_t \subset \mathfrak{A}$ for all $t \in I$. For a set $A \in \mathfrak{B}^1$, let T_A denote the first point of time at which X_t lies in A, that is, for each $\omega \in \Omega$ let

$$T_A(\omega) \equiv \begin{cases} \inf\{t \in I: X_t(\omega) \in A\}, & \text{if such } t \text{ exist} \\ +\infty, & \text{otherwise.} \end{cases}$$

Then we call T_A the *first entry time of* A. This is a stopping time since

$$\{T_A \leqq i\} = \bigcup_{\tau=1}^{t} \{X_\tau \in A\}$$

for every $t \in \mathbf{N}$ and $\{T_A \leqq t\} = \Omega$ for $t = +\infty$.

If a mapping $T: \Omega \to I$ takes on only *countably many values* $\leqq t$ for every $t \in I$, then it is a stopping time if and only if

$$\{T = t\} \in \mathfrak{A}_t, \qquad \text{for all } t \in I. \tag{11.2.2}$$

This follows from the equalities

$$\{T = t\} = \{T \leqq t\} \setminus \bigcup_{\substack{s \in I \\ s < t}} \{T \leqq s\}$$

and

$$\{T \leqq t\} = \bigcup_{\substack{s \in I \\ s \leqq t}} \{T = s\},$$

if we note that by hypothesis the right-hand sides above are the unions of countably many sets from $\mathfrak{A}_t$.

We associate with every stopping time T relative to $(\mathfrak{A}_t)_{t\in I}$ a σ-algebra $\mathfrak{A}_T$ by setting

$$\mathfrak{A}_T \equiv \{E \in \mathfrak{P}(\Omega) : E \cap \{T \leqq t\} \in \mathfrak{A}_t \text{ for all } t \in I\}. \quad (11.2.3)$$

It is clear that $\mathfrak{A}_T$ contains the set Ω, and with every sequence of sets, it also contains their union. But when a set E lies in $\mathfrak{A}_T$, also $\complement E$ lies in $\mathfrak{A}_T$; since we have

$$(\Omega \setminus E) \cap \{T \leqq t\} = \{T \leqq t\} \setminus E \cap \{T \leqq t\} \in \mathfrak{A}_t,$$

for all $t \in I$. Thus $\mathfrak{A}_T$ is a σ-algebra. It is called the σ-algebra of *events up to time* T. If T is a constant t_0 (see Example 1), then obviously $\mathfrak{A}_T = \mathfrak{A}_{t_0}$.

11.2.2. Lemma. If T and T^* are stopping times relative to an isotone family $(\mathfrak{A}_t)_{t\in I}$ of σ-subalgebras of a σ-algebra $\mathfrak{A}$ in Ω, then the following statements hold:

$$T \leqq T^* \Rightarrow \mathfrak{A}_T \subset \mathfrak{A}_{T^*}. \quad (11.2.4)$$

If I contains a countable subset D such that for each $t \in I$ there exists a $d \in D$ with $t \leqq d$, then

$$\mathfrak{A}_T \subset \mathfrak{A}. \quad (11.2.5)$$

Proof. For every $\omega \in \Omega$, let $T(\omega) \leqq T^*(\omega)$, that is, $\{T^* \leqq t\} \subset \{T \leqq t\}$ for each $t \in I$. Hence, $E \cap \{T^* \leqq t\} = E \cap \{T \leqq t\} \cap \{T^* \leqq t\}$ for every set $E \subset \Omega$ and every $t \in I$. Therefore, $E \in \mathfrak{A}_T$ implies $E \in \mathfrak{A}_{T^*}$. Under the condition given in (11.2.5), we have

$$\Omega = \bigcup_{d\in D} \{T \leqq d\}$$

and thus

$$E = E \cap \Omega = \bigcup_{d\in D} E \cap \{T \leqq d\},$$

for every set $E \subset \Omega$. Since D is countable, it now follows that $E \in \mathfrak{A}$ for every $E \in \mathfrak{A}_T$. ⌟

Now let $(X_t)_{t\in I}$ be a family of random variables on a probability space $(\Omega,\mathfrak{A},P)$ with values in an arbitrary measurable space $(\Omega',\mathfrak{A}')$; assume that $(X_t)_{t\in I}$ is adapted to an isotone family $(\mathfrak{A}_t)_{t\in I}$ of σ-subalgebras. With every stopping time T relative to $(\mathfrak{A}_t)$ we can then associate a mapping $X_T: \Omega \to \Omega'$ by defining

$$X_T(\omega) \equiv X_{T(\omega)}(\omega).^6 \quad (11.2.6)$$

Under additional hypotheses X_T is then a random variable. We show:

[6] If T is constant, equal to t_0, then $X_T = X_{t_0}$.

11.2.3. Lemma. X_T is $\mathfrak{A}_T$-measurable if the set of all values t' of T with $t' \leqq t$ is countable for every $t \in I$.

Proof. We must verify that $E \equiv \{X_T \in A'\}$ is an element of $\mathfrak{A}_T$ for every set $A' \in \mathfrak{A}'$. But for every $t \in I$,

$$E \cap \{T \leqq t\} = \bigcup_{\substack{t' \in I \\ t' \leqq t}} E \cap \{T = t'\} = \bigcup_{\substack{t' \in I \\ t' \leqq t}} \{X_{t'} \in A'\} \cap \{T = t'\}.$$

Hence it follows that $E \cap \{T \leqq t\} \in \mathfrak{A}_t$ and thus $E \in \mathfrak{A}_T$ from (11.2.3) and the countability hypothesis on T. For this hypothesis implies that at most countably many of the sets $\{T = t'\}$ with $t' \leqq t$ are nonempty. ┘

If we restrict consideration to super-martingales with *finite, totally ordered index set* $I = \{1, \ldots, k\}$, we can now answer the question raised in the introduction (concerning optional sampling).

11.2.4. Theorem. Let $(X_i)_{i=1,\ldots,k}$ be a super-martingale (martingale) relative to an isotone family $(\mathfrak{A}_i)_{i=1,\ldots,k}$ of σ-algebras and let $(T_j)_{j=1,\ldots,p}$ be an isotone family of stopping times relative to $(\mathfrak{A}_i)$. Then $(X_{T_j})_{j=1,\ldots,p}$ is also a super-martingale (martingale) relative to $(\mathfrak{A}_{T_j})_{j=1,\ldots,p}$.

Proof. By Lemma 11.2.2, $(\mathfrak{A}_{T_j})$ is an isotone family of σ-subalgebras of the σ-algebra $\mathfrak{A}$ of the underlying probability space $(\Omega,\mathfrak{A},P)$. By Lemma 11.2.3, (X_{T_j}) is adapted to the family $(\mathfrak{A}_{T_j})$. Each of the random variables X_{T_j} is integrable since

$$E(|X_{T_j}|) = \sum_{i=1}^{k} \int_{\{T_j = i\}} |X_{T_j}| \, dP \leqq \sum_{i=1}^{k} E(|X_i|) < \infty.$$

Therefore we need only show the inequality for super-martingales, namely,

$$\int_A X_{T_{j+1}} \, dP \leqq \int_A X_{T_j} \, dP$$

for every $j \in \{1, \ldots, p-1\}$ and all $A \in \mathfrak{A}_{T_j}$. For this, we set $S \equiv T_j$ and $T \equiv T_{j+1}$. Then S and T are stopping times such that $S \leqq T$. We need to show that

$$\int_A X_T \, dP \leqq \int_A X_S \, dP,$$

for all $A \in \mathfrak{A}_S$.

We first treat the special case $T - S \leqq 1$. Then for every $A \in \mathfrak{A}_S$,

$$\int_A (X_S - X_T) \, dP = \sum_{i=1}^{k-1} \int_{A_i} (X_S - X_T) \, dP = \sum_{i=1}^{k-1} \int_{A_i} (X_i - X_{i+1}) \, dP,$$

when we introduce $A_i \equiv A \cap \{S = i\} \cap \{T > S\} = A \cap \{S = i\} \cap \{T > i\} = A \cap \{S = i\} \cap \{T = i + 1\}$ $(i = 1, \ldots, k - 1)$. Hence the asserted inequality follows from the inequalities for the super-martingale (X_i) provided $A_i \in \mathfrak{A}_i$ for all $i = 1, \ldots, k - 1$. Now $\{T > i\} = \complement \{T \leqq i\}$ lies in $\mathfrak{A}_i$. From $A \in \mathfrak{A}_S$ it follows that

$$A \cap \{S = i\} = A \cap \{S \leqq i\} \setminus \bigcup_{j=1}^{i-1} A \cap \{S \leqq j\} \in \mathfrak{A}_i.$$

Therefore, indeed, $A_i \in \mathfrak{A}_i$.

We can now proceed as follows in the general case: For every $i = 1, \ldots, k$, $R_i \equiv \inf(T, S + i)$ is a stopping time according to Example 2. We have $S \leqq R_1 \leqq R_2 \leqq \cdots \leqq R_k = T$, $R_1 - S \leqq 1$, and $R_{i+1} - R_i \leqq 1$, for all $i = 1, \ldots, k - 1$. From the special case treated above, since $\mathfrak{A}_S \subset \mathfrak{A}_{R_1} \subset \cdots \subset \mathfrak{A}_T$, it now follows that for every $A \in \mathfrak{A}_S$,

$$\int_A X_T \, dP = \int_A X_{R_k} \, dP \leqq \cdots \leqq \int_A X_{R_1} \, dP \leqq \int_A X_S \, dP,$$

and thus we have the desired inequality.

If (X_i) is a martingale, then (X_i) and $(-X_i)$, and hence also (X_{T_j}) and $(-X_{T_j})$ are super-martingales. But then (X_{T_j}) is a martingale. ┘

11.2.5. Corollary. If $(X_i)_{i=1,\ldots,k}$ is a super-martingale with $I = \{1, \ldots, k\}$ as index set and T is a stopping time [relative to $(\mathfrak{A}_i^0)_{i=1,\ldots,k}$], then the following inequalities hold:

$$E(X_1) \geqq E(X_T) \geqq E(X_k), \tag{11.2.7}$$

$$E(|X_T|) \leqq E(X_1) + 2E(X_k^-). \tag{11.2.8}$$

Proof. We apply (11.2.4) to the isotone family of stopping times $T_1 = 1$, $T_2 = T$, $T_3 = k$. Then $(X_{T_j})_{j=1,2,3}$ is a super-martingale and hence

$$E(X_{T_1}) \geqq E(X_{T_2}) \geqq E(X_{T_3}).$$

But this is inequality (11.2.7).

By (11.2.7), $E(X_T) \leqq E(X_1)$. Since $(-X_i^-)_{i=1,\ldots,k}$ is also a super-martingale, the inequality $E(X_T^-) \leqq E(X_k^-)$ follows again from (11.2.7). Since $|X_T| = X_T^+ + X_T^- = X_T + 2X_T^-$, we now obtain the asserted inequality (11.2.8). ┘

Remark. If $(X_i)_{i=1,\ldots,k}$ is a martingale, then (11.2.7) becomes the equality $E(X_1) = E(X_T) = E(X_k)$. As for our introductory example, this tells us that the expected gain remains unchanged by optional stopping.

PROBLEMS

1. Let $(\mathfrak{A}_t)_{t\in I}$ be an isotone family of σ-algebras in a set Ω with index set $I = [0,+\infty[$, and let S and T be stopping times with respect to $(\mathfrak{A}_t)_{t\in I}$. Prove:
   (a) T is $\mathfrak{A}_T$-measurable.
   (b) Each of the sets $\{S < T\}$, $\{S \leqq T\}$, $\{S = T\}$ is in $\mathfrak{A}_S \cap \mathfrak{A}_T$.
2. Let $(\mathfrak{A}_t)_{t\in I}$ be an isotone family of σ-algebras in a set Ω with index set $I = [0,+\infty[$. Define $\mathfrak{A}_t^+ \equiv \bigcap_{t<s} \mathfrak{A}_s$ for each $t \in I$. Prove:
   (a) $(\mathfrak{A}_t^+)_{t\in I}$ is an isotone family of σ-algebras in Ω.
   (b) A mapping $T: \Omega \to I$ is a stopping time with respect to $(\mathfrak{A}_t^+)$ if and only if $\{T < t\} \in \mathfrak{A}_t$ for all $t \in I$.
   (c) Every stopping time with respect to $(\mathfrak{A}_t)$ is a stopping time with respect to $(\mathfrak{A}_t^+)$ (but not conversely in general).
3. Let $(X_i)_{i=1,\ldots,k}$ be a super-martingale, and $\lambda \in \mathbf{R}_+$. Introduce the random variables $\bar{X} \equiv \sup_{i=1,\ldots,k} X_i$ and $\underline{X} \equiv \inf_{i=1,\ldots,k} X_i$, and prove:
   (a) $\lambda P\{\bar{X} \geqq \lambda\} \leqq E(X_1) - E(X_k^-)$.
   (b) $\lambda P\{\underline{X} \leqq -\lambda\} \leqq E(X_k^-)$.

   [*Hint:* To obtain (a), prove that

$$T(\omega) = \begin{cases} \text{smallest } i \in \{1, \ldots ,k\} \text{ satisfying } X_i(\omega) \geqq \lambda \\ k, \text{ if } X_i(\omega) < \lambda \text{ for all } i \in \{1, \ldots ,k\} \end{cases}$$

   is a stopping time, and apply (11.2.7).]
4. Let $(X_i)_{i=1,\ldots,k}$ be a martingale of square-integrable random variables. Deduce from Problem 3

$$P\{\sup_{i=1,\ldots,k} |X_i| \geqq \lambda\} \leqq \frac{1}{\lambda^2} E(X_k^2)$$

   for $\lambda > 0$. Deduce from this Kolmogorov's inequality (6.3.12).

11.3 THE DOOB INEQUALITIES

Similar to the way in which the Hájek-Rényi inequality plays the decisive role in the proof of Kolmogorov's theorem on the strong law of large numbers, the convergence theorems of the next section are based on two inequalities of J. L. Doob. They are built upon the following combinatorial concepts.

11.3.1. Definition. Let $(x_i)_{i=1,\ldots,k}$ be a finite sequence of real numbers and let $[a,b]$ be a compact interval in $\mathbf{R}$ with $a < b$. Then the integer

$\underline{U}_{[a,b]}$, called the *number of downcrossings* of $[a,b]$ by $(x_i)_{i=1,\ldots,k}$, is defined as follows: If there are indices $i, j \in \{1, \ldots, k\}$ with $i < j$ and $x_i \geqq b$, $x_j \leqq a$, then let $\underline{U}_{[a,b]}$ be the largest natural number l for which there exist indices $i_1 < \cdots < i_{2l}$ from $\{1, \ldots, k\}$ satisfying

$$x_{i_{2\lambda-1}} \geqq b \text{ and } x_{i_{2\lambda}} \leqq a, \qquad \text{for all } \lambda = 1, \ldots, l. \tag{11.3.1}$$

If there are no such indices i, j, then let $\underline{U}_{[a,b]} = 0$.

The number of downcrossings of $[-b,-a]$ by the sequence $(-x_i)_{i=1,\ldots,k}$ is denoted by $\bar{U}_{[a,b]}$ and is called the number of *upcrossings* of $[a,b]$ by $(x_i)_{i=1,\ldots,k}$.

For a given sequence $(x_i)_{i=1,\ldots,k}$, we obviously determine $\underline{U}_{[a,b]}$ by the following procedure: Let i_1 be the smallest index with $x_{i_1} \geqq b$; i_2 the next largest index with $x_{i_2} \leqq a$; i_3 the next largest index with $x_{i_3} \geqq b$; and so on. This process terminates after p steps. Then $\underline{U}_{[a,b]} = [p/2]$ where $[p/2]$ denotes the greatest integer $\leqq p/2$. In particular, $\underline{U}_{[a,b]} \leqq [k/2]$. To determine $\bar{U}_{[a,b]}$, we modify the procedure in the obvious way. Both procedures will be used in the proof of Theorem 11.3.2.

If we consider, say, the sequence

$$0, 1, 0, 0, 1, 1, 0, 0, 0, 1, 1, 1, 0, 0, 0, 0, 1, 1, 1, 1,$$

corresponding to $k = 20$, then $\underline{U}_{[0,1]} = 3$ and $\bar{U}_{[0,1]} = 4$.

Now let $(X_i)_{i=1,\ldots,k}$ be a finite sequence of real random variables on a probability space $(\Omega,\mathfrak{A},P)$. For every $\omega \in \Omega$ we then obtain the numbers $\underline{U}_{[a,b]}(\omega)$ and $\bar{U}_{[a,b]}(\omega)$ of down- and upcrossings respectively of the compact interval $[a,b]$ by the sequence $(X_i(\omega))_{i=1,\ldots,k}$. Then $\omega \to \underline{U}_{[a,b]}(\omega)$ and $\omega \to \bar{U}_{[a,b]}(\omega)$ are random variables. The measurability of $\underline{U}_{[a,b]}$ (and thus also of $\bar{U}_{[a,b]}$) follows from the relationship

$$\{U_{[a,b]} \geqq l\} = \bigcup_{i_1 < \cdots < i_{2l}} \bigcap_{\lambda=1}^{l} (\{X_{i_{2\lambda-1}} \geqq b\} \cup \{X_{i_{2\lambda}} \leqq a\}),$$

which holds for every natural number $l \in \{1, \ldots, [k/2]\}$. Here the union is to be taken over all $(2l)$-tuples of natural numbers $(i_1, \ldots, i_{2l})$ with $i_1 < \cdots < i_{2l}$.

The promised inequalities now relate to the expected values[7] of these random variables $\underline{U}_{[a,b]}$ and $\bar{U}_{[a,b]}$ for the case of a super-martingale (X_i):

11.3.2. Theorem (Doob's Inequalities). For every super-martingale $(X_i)_{i=1,\ldots,k}$ with index set $I = \{1, \ldots, k\}$ and every compact interval $[a,b]$ with $a < b$, the expected values of $\underline{U}_{[a,b]}$ and $\bar{U}_{[a,b]}$ satisfy

[7] As elementary random variables, $\underline{U}_{[a,t]}$ and $\bar{U}_{[at,b]}$ are of course integrable.

the inequalities

$$E(\underline{U}_{[a,b]}) \leqq \frac{1}{b-a} E[\inf (X_1,b) - \inf (X_k,b)], \tag{11.3.2}$$

$$E(\bar{U}_{[a,b]}) \leqq \frac{1}{b-a} E[(X_k - a)^-]. \tag{11.3.3}$$

Proof. If we set $X_i^* = \inf (X_i,b)$ for every $i = 1, \ldots, k$, then (X_i^*) is a super-martingale by (11.1.5). The number $\underline{U}_{[a,b]}$ ($\bar{U}_{[a,b]}$) of downcrossings (upcrossings) of $[a,b]$ obviously remains unchanged if we replace (X_i) by (X_i^*). Now let p be an even natural number $\geqq k$. By induction we define a sequence $(T_j)_{j=0,1,\ldots,p}$ of stopping times relative to the family $(\mathfrak{A}_i^0)_{i=0,1,\ldots,k}$ of σ-algebras by setting, for every $\omega \in \Omega$:

$$T_0(\omega) \equiv 1;$$

$$T_{2\lambda-1}(\omega) \equiv \begin{cases} \text{the smallest } i \in \{1, \ldots, k\} \\ \qquad\qquad \text{with } i \geqq T_{2\lambda-2}(\omega) \text{ and } X_i^*(\omega) = b, \\ k, \qquad\qquad \text{if no such } i \text{ exists;} \end{cases}$$

$$T_{2\lambda}(\omega) \equiv \begin{cases} \text{the smallest } i \in \{1, \ldots, k\} \\ \qquad\qquad \text{with } i \geqq T_{2\lambda-1}(\omega) \text{ and } X_i^*(\omega) \leqq a, \\ k, \qquad\qquad \text{if no such } i \text{ exists.} \end{cases}$$

Here λ runs through the values $1, 2, \ldots, p/2$ and we let $\mathfrak{A}_0^0 \equiv \{\varnothing,\Omega\}$.

We are dealing with stopping times, since for every $i = 1, \ldots, k-1$ and $\lambda = 1, \ldots, p/2$,

$$\{T_{2\lambda-1} = i\}$$
$$= \{T_{2\lambda-2} \leqq i\} \cap \{X_i^* = b\} \cap \complement \bigcup_{j<i} (\{T_{2\lambda-2} \leqq j\} \cap \{X_j^* = b\}),$$

$$\{T_{2\lambda} = i\}$$
$$= \{T_{2\lambda-1} \leqq i\} \cap \{X_i^* \leqq a\} \cap \complement \bigcup_{j<i} (\{T_{2\lambda-1} \leqq j\} \cup \{X_j^* \leqq a\}),$$

and furthermore,

$$\{T_{2\lambda-1} = k\} = \complement \bigcup_{j=1}^{k} (\{T_{2\lambda-2} \leqq j\} \cap \{X_j^* = b\}),$$

$$\{T_{2\lambda} = k\} = \complement \bigcup_{j=1}^{k} (\{T_{2\lambda-1} \leqq j\} \cap \{X_j^* \leqq a\}).$$

Accordingly, all sets $\{T_j = i\}$ and thus also all sets $\{T_j \leqq i\}$ lie in $\mathfrak{A}_i^0$ ($i = 0, 1, \ldots, k$). By definition the sequence $(T_j)_{j=0,1,\ldots,p}$ is isotone. Therefore, Theorem 11.2.4 applies, according to which $(X_{T_j})_{j=0,1,\ldots,p}$ is a super-martingale and thus the sequence $(E(X_{T_j}))$ is antitone. The sequence (T_j) is in fact isotone in a stronger sense: Since the conditions $X_j^*(\omega) \leqq a$

and $X_j^*(\omega) = b$ cannot be satisfied simultaneously, we have $T_j(\omega) < T_{j+1}(\omega)$ for every $j = 0, 1, \ldots, p$ with $T_j(\omega) < k$. Since $p \geqq k$, we must have $T_p(\omega) = k$ for all $\omega \in \Omega$.

By the construction of the stopping times T_j, we see that $\underline{U}_{[a,b]}(\omega)$ is now precisely the number of downcrossings of $[a,b]$ by the sequence $(X^*_{T_j}(\omega))_{j=1,\ldots,p}$ and therefore, obviously,

$$(b-a)\underline{U}_{[a,b]} \leqq \sum_{\lambda=1}^{p/2} (X^*_{T_{2\lambda-1}} - X^*_{T_{2\lambda}}) = X^*_{T_1} - X^*_{T_p} + \sum_{\lambda=1}^{p/2-1} (X^*_{T_{2\lambda+1}} - X^*_{T_{2\lambda}}).$$

By integration we then have

$$(b-a)E(\underline{U}_{[a,b]}) \leqq E(X^*_{T_1}) - E(X^*_{T_p}) + \sum_{j=1}^{p/2-1} (E(X^*_{T_{2\lambda+1}}) - E(X^*_{T_{2\lambda}})),$$

and thus, due to the antitonicity of the sequence $(E(X^*_{T_j}))_{j=0,1,\ldots,p}$ and to $T_0 = 1$, $T_p = k$, we have

$$(b-a)E(\underline{U}_{[a,b]}) \leqq E(X^*_{T_0}) - E(X^*_{T_p}) = E(X^*_1) - E(X^*_k).$$

But this is the inequality (11.3.2).

To prove (11.3.3), we can proceed quite similarly. We choose p as above, but define $(T_j)_{j=0,1,\ldots,p}$ inductively as follows:

$$T_0(\omega) \equiv 1;$$

$$T_{2\lambda-1}(\omega) \equiv \begin{cases} \text{the smallest } i \in \{1, \ldots, k\} \\ \qquad \text{with } i \geqq T_{2\lambda-2}(\omega) \text{ and } X_i^*(\omega) \leqq a, \\ k, \qquad \text{if no such } i \text{ exists;} \end{cases}$$

$$T_{2\lambda}(\omega) \equiv \begin{cases} \text{the smallest } i \in \{1, \ldots, k\}, \\ \qquad \text{with } i \geqq T_{2\lambda-1}(\omega) \text{ and } X_i^*(\omega) = b, \\ k, \qquad \text{if no such } i \text{ exists.} \end{cases}$$

Then we show, as above, that the sequence $(E(X_{T_j}))_{j=0,1,\ldots,p}$ is antitone. By the construction of the above stopping times, the difference $X_{T_{2\lambda}} - X^*_{T_{2\lambda-1}}$ is $\geqq b - a$ for $\lambda = 1, \ldots, \bar{U}_{[a,b]}$ and $= 0$ for $p/2 \geqq \lambda > \bar{U}_{[a,b]} + 1$. In the case $\lambda = \bar{U}_{[a,b]} + 1$, it is equal to $X_k^* - X^*_{T_{2\lambda-1}}$. Here either $X^*_{T_{2\lambda-1}} = X_k^*$ or $X^*_{T_{2\lambda-1}} \leqq a$ and then also $X_{T_{2\lambda}} = X_k^* < b$, that is, $X_k = X_k^*$. For $\lambda = \bar{U}_{[a,b]} + 1$, in both cases we have

$$X_{T_{2\lambda}} - X_{T_{2\lambda-1}} \geqq \inf(0, X_k - a) = -(X_k - a)^-.$$

Thus we obtain the inequality

$$\sum_{\lambda=1}^{p/2} (X^*_{T_{2\lambda}} - X^*_{T_{2\lambda-1}}) \geqq (b-a) \cdot \bar{U}_{[a,b]} - (X_k - a)^-$$

and hence, by integrating and taking account of the antitonicity of $(E(X_{T_j}))$, we obtain inequality (11.3.3). ⌟

PROBLEMS

1. Let $(X_i)_{i=1,\ldots,k}$ be a sub-martingale, and let $[a,b]$ be a compact interval with end points $a < b$. Prove:

(a) $$E(\underline{U}_{[a,b]}) \leqq \frac{1}{b-a} E[(X_k - b)^+];$$

(b) $$E(\bar{U}_{[a,b]}) = \frac{1}{b-a} E(\sup(X_k,a) - \sup(X_1,a)).$$

2. Let $(X_i)_{i=1,\ldots,k}$ be a super-martingale relative to an isotone family $(\mathfrak{A}_i)_{i=1,\ldots,k}$ of σ-algebras. Imitate the proof of Theorem 11.3.2 and prove the following strengthened form of the inequalities (11.3.2) and (11.3.3):

(a) $$E(\underline{U}_{[a,b]} \mid \mathfrak{A}_1) \leqq \frac{1}{b-a} [\inf(X_1,b) - E(\inf(X_k,b) \mid \mathfrak{A}_1)];$$

(b) $$E(\bar{U}_{[a,b]} \mid \mathfrak{A}_1) \leqq \frac{1}{b-a} E((X_k - a)^- \mid \mathfrak{A}_1).$$

11.4 CONVERGENCE THEOREMS

Example 3 of Section 11.1, which showed that the partial sums of an independent sequence of real, integrable, centered random variables form a martingale, and the close connection of this martingale with the sequence of random variables considered in the strong law of large numbers, raise the question of convergence of martingales and super-martingales. We shall now show that (at least in the discrete case, that is, with $\mathbf{N}$ as index set) the interdependence of the random variables of a super-martingale is so strong that even simple additional hypotheses guarantee convergence almost everywhere.

The following considerations again refer to a probability space $(\Omega,\mathfrak{A},P)$. Let $(\mathfrak{A}_n)_{n\in\mathbf{N}}$ be an isotone sequence of σ-subalgebras of $\mathfrak{A}$.

11.4.1. Theorem. Every super-martingale $(X_n)_{n\in\mathbf{N}}$ relative to $(\mathfrak{A}_n)_{n\in\mathbf{N}}$ which satisfies the condition

$$\sup_{n\in\mathbf{N}} E(X_n^-) < \infty \tag{11.4.1}$$

converges almost surely to an integrable random variable X_∞.

Proof. The sequence (X_n) converges almost surely if the event

$$Q \equiv \{\limsup_{n\to\infty} X_n > \liminf_{n\to\infty} X_n\}$$

has probability zero. To verify this property, we let $\bar{U}^k_{[a,b]}(\omega)$ denote the number of upcrossings of the interval $[a,b]$ (with $a < b$, $a, b \in \mathbf{R}$) by the finite sequence $(X_i(\omega))_{i=1,\ldots,k}$. The sequence $(\bar{U}^k_{[a,b]})_{k=1,2,\ldots}$ is then obviously isotone. We set $\bar{U}_{[a,b]} \equiv \sup_{n\in\mathbf{N}} \bar{U}^k_{[a,b]}$ and thus obtain a random variable with values in $\{0\} \cup \mathbf{N} \cup \{+\infty\}$. For every $\omega \in Q$ and rational numbers a, b with $\liminf_{n\to\infty} X_n(\omega) < a < b < \limsup_{n\to\infty} X_n(\omega)$ we now have $\bar{U}_{[a,b]}(\omega) = \infty$, since there are infinitely many n' with $X_{n'}(\omega) > b$ and infinitely many n'' with $X_{n''}(\omega) < a$. Thus we have

$$Q \subset \bigcup_{\substack{a,\, b \text{ rational} \\ a<b}} \{U_{[a,b]} = \infty\}.$$

Therefore, it will be sufficient to show that $P\{\bar{U}_{[a,b]} = \infty\} = 0$ for every pair a, b of real numbers with $a < b$. For this we use (11.3.3), according to which, for every $k \in \mathbf{N}$,

$$E(\bar{U}^k_{[a,b]}) \leqq \frac{1}{b-a} E((X_k - a)^-),$$

hence, by the Monotone Convergence Theorem,

$$E(\bar{U}_{[a,b]}) \leqq \frac{1}{b-a} \sup_{k\in\mathbf{N}} E((X_k - a)^-).$$

If we note the inequality $(X_k - a)^- = \sup(0, a - X_k) \leqq a^+ + X_k^-$, then from the hypothesis (11.4.1) we obtain

$$\sup E((X_k - a)^-) \leqq a^+ + \sup E(X_k^-) < \infty$$

and thus, from the second Doob inequality $E(\bar{U}_{[a,b]}) < \infty$, and therefore $P\{\bar{U}_{[a,b]} = \infty\} = 0$. This proves the almost sure convergence of (X_n) to a numerical random variable X_∞ on Ω. This is integrable, since by (11.2.8) (for the constant stopping time $T = k$) we have the inequality

$$E(|X_k|) \leqq E(X_1) + 2 \sup_{n\in\mathbf{N}} E(X_n^-) \qquad (k = 1, 2, \ldots),$$

which implies $\sup E(|X_k|) < \infty$ by hypothesis. But then Fatou's Lemma yields

$$E(|X_\infty|) = E(\lim |X_n|) \leqq \liminf E(|X_n|) < \infty$$

and thus the desired integrability of X_∞. ┘

In particular, condition (11.4.1) is obviously satisfied if

$$\sup_{n\in\mathbf{N}} E(|X_n|) < \infty. \tag{11.4.2}$$

But the proof of Theorem 11.4.1 tells us that conditions (11.4.1) and (11.4.2) are actually *equivalent* for a super-martingale $(X_n)_{n\in\mathbf{N}}$. Therefore condition (11.4.1) is satisfied if all $X_n \geqq 0$ or if the super-martingale $(X_n)_{n\in\mathbf{N}}$ is *uniformly integrable*. In this case Theorem 2.12.2 yields the validity of (11.4.2). It now follows further from Theorem 2.12.4 that a uniformly integrable super-martingale $(X_n)_{n\in\mathbf{N}}$ also *converges in mean* to X_∞. We need only take into account Theorem 2.11.4, by which stochastic convergence follows from almost sure convergence.

In conjunction with the above convergence theorem, the question arises of whether the super-martingale $(X_n)_{n\in\mathbf{N}}$ of Theorem 11.4.1 becomes a super-martingale $(X_n)_{n\in\mathbf{N}\cup\{\infty\}}$ by the addition of X_∞. To this end, we enlarge the sequence $(\mathfrak{A}_n)$ by the following σ-algebra, generated by all $\mathfrak{A}_n$ in Ω:

$$\mathfrak{A}_\infty \equiv \mathfrak{A}\left(\bigcup_{n=1}^{\infty} \mathfrak{A}_n\right). \tag{11.4.3}$$

11.4.2. Corollary 1. Let $(X_n)_{n\in\mathbf{N}}$ be a super-martingale relative to the sequence $(\mathfrak{A}_n)_{n\in\mathbf{N}}$ whose random variables X_n are all $\geqq 0$ or uniformly integrable. Then (X_n) converges almost surely to an integrable random variable X_∞ such that $(X_n)_{n\in\mathbf{N}\cup\{\infty\}}$ is a super-martingale relative to $(\mathfrak{A}_n)_{n\in\mathbf{N}\cup\{\infty\}}$.

Proof. By Theorem 11.4.1, (X_n) converges almost surely to an integrable random variable X_∞. Since every X_n is $\mathfrak{A}_n$- and thus $\mathfrak{A}_\infty$-measurable, the set $\{\limsup X_n = \liminf X_n\}$ of all ω for which $\lim X_n(\omega)$ exists, lies in $\mathfrak{A}_\infty$. Therefore we can assume that X_∞ is $\mathfrak{A}_\infty$-measurable and real-valued. Thus all we still have to verify is the inequality

$$\int_A X_n\, dP \geqq \int_A X_\infty\, dP$$

for every $A \in \mathfrak{A}_n$ and every $n \in \mathbf{N}$. To this end, let m be a natural number $>n$. Then $\int_A X_n\, dP \geqq \int_A X_m\, dP$. By the passage to the limit $m \to \infty$ we now obtain the assertion: In the case $X_m \geqq 0$ $(m = 1, 2, \ldots)$ from Fatou's Lemma, and in the case of uniform integrability, because of the convergence in mean of (X_m) to X_∞ just noted. ⌟

11.4.3. Corollary 2. Every uniformly integrable martingale $(X_n)_{n\in\mathbf{N}}$ relative to $(\mathfrak{A}_n)_{n\in\mathbf{N}}$ converges almost surely and in mean to an integrable random variable X_∞ such that $(X_n)_{n\in\mathbf{N}\cup\{\infty\}}$ is also a martingale relative to $(\mathfrak{A}_n)_{n\in\mathbf{N}\cup\{\infty\}}$.

Proof. It suffices to apply Corollary 1 to (X_n) and $(-X_n)$. The reasoning used in the discussion following (11.4.2) yields that (X_n) also converges in mean to X_∞. ⌟

Examples of uniformly integrable martingales are easily found. We show that Example 2 of Section 11.1, is such an example.

11.4.4. Theorem. For every integrable real random variable X on $(\Omega,\mathfrak{A},P)$ and every isotone family $(\mathfrak{A}_t)_{t\in I}$ of σ-subalgebras of $\mathfrak{A}$ (with arbitrary partially ordered index set I), $(E^{\mathfrak{A}_t}(X))_{t\in I}$ is a uniformly integrable martingale relative to $(\mathfrak{A}_t)_{t\in I}$.

Proof. All we need to prove is the uniform integrability of the family $(X_t)_{t\in I}$ of random variables $X_t \equiv E^{\mathfrak{A}_t}(X)$. Since $|X_t| \leqq E^{\mathfrak{A}_t}(|X|)$ holds almost surely for every $t \in I$ and since X_t is $\mathfrak{A}_t$-measurable, we obtain for every $\alpha \geqq 0$,

$$\int_{\{|X_t|\geqq\alpha\}} |X_t|\,dP \leqq \int_{\{|X_t|\geqq\alpha\}} |X|\,dP,$$

and, in particular, for $\alpha = 0$, $E(|X_t|) \leqq E(|X|)$. By the Chebyshev-Markov inequality, for $\alpha > 0$ we then have

$$P\{|X_t| \geqq \alpha\} \leqq \frac{1}{\alpha}E(|X_t|) \leqq \frac{1}{\alpha}E(|X|) \qquad (t \in I).$$

By Theorem 2.9.6, for every $\epsilon > 0$ there is a $\delta > 0$ such that $P(A) \leqq \delta$, $A \in \mathfrak{A}$, implies $\int_A |X|\,dP \leqq \epsilon$. Accordingly,

$$\int_{\{|X_t|\geqq\alpha\}} |X_t|\,dP \leqq \int_{\{|X_t|\geqq\alpha\}} |X|\,dP \leqq \epsilon$$

for all $\alpha \geqq \delta^{-1}E(|X|)$ and every $t \in I$. This proves the uniform integrability of $(X_t)_{t\in I}$. ┘

Under an additional assumption on I which is often satisfied in applications, and in particular for every totally ordered set I and thus especially for $I = \mathbf{N}$, the converse of Theorem 11.4.4 holds. This additional assumption is that the partially ordered set I is *filtering to the right,* that is, for any two indices $t_1, t_2 \in I$ there always exists a $t \in I$ such that $t_1 \leqq t$ and $t_2 \leqq t$. We then obtain the following representation theorem:

11.4.5. Theorem. For every uniformly integrable martingale $(X_t)_{t\in I}$ relative to an isotone family $(\mathfrak{A}_t)_{t\in I}$ of σ-subalgebras of $\mathfrak{A}$ such that I is filtering to the right, there exists an integrable real random variable X such that

$$X_t = E^{\mathfrak{A}_t}(X), \qquad \text{almost surely,} \qquad \text{for all } t \in I. \tag{11.4.4}$$

Proof. First, the assertion is true in the case $I = \mathbf{N}$, since by Corollary 11.4.3 there exists an integrable random variable X_∞ such that $(X_n)_{n\in\mathbf{N}\cup\{\infty\}}$

is a martingale relative to $(\mathfrak{A}_n)_{n\in\mathbf{N}\cup\{\infty\}}$. But then $X_n = E^{\mathfrak{A}_n}(X_\infty)$ almost surely for all $n \in \mathbf{N}$; thus, $X \equiv X_\infty$ yields the desired result.

For arbitrary I filtering to the right, we show that for every $\epsilon > 0$, there exists $t_\epsilon \in I$ such that $E(|X_s - X_t|) < \epsilon$ for all $s, t \in I$ satisfying $s \geqq t_\epsilon$ and $t \geqq t_\epsilon$.[8] If this were not the case, then for some $\epsilon > 0$ there would be an isotone sequence $(t_n)_{n\in\mathbf{N}}$ in I such that $E(|X_{t_{n+1}} - X_{t_n}|) \geqq \epsilon$ for all $n \in \mathbf{N}$. Then $(X_{t_n})_{n\in\mathbf{N}}$ would not be a Cauchy sequence in $\mathfrak{L}^1(P)$. But this is impossible, since (X_{t_n}) is a uniformly integrable martingale relative to $(\mathfrak{A}_{t_n})$ and thus, by the remark made after (11.4.2), converges in mean.

Now consider the sequence $(t_{1/n})_{n\in\mathbf{N}}$. Since I is filtering to the right, we can choose this sequence to be isotone. Then $(X_{t_{1/n}})_{n\in\mathbf{N}}$ is a Cauchy sequence in $\mathfrak{L}^1(P)$, that is, it converges in mean to an integrable random variable X. By the construction of $(t_{1/n})$, we have $E(|X_t - X_{t_{1/n}}|) \leqq 1/m$ for all $t \geqq t_{1/m}$ and all $n \geqq m$ for arbitrary $m \in \mathbf{N}$, whence follows, for $n \to \infty$,

$$E(|X_t - X|) \leqq \frac{1}{m}$$

for all $m \in \mathbf{N}$ and all $t \geqq t_{1/m}$. Thus the martingale converges in mean "along I" to X. As in the proof of Corollary 11.4.2, we now show that X yields the desired result: Let $s \in I$ be arbitrarily chosen. Then

$$\int_A X_s\,dP = \int_A X_t\,dP$$

for all $A \in \mathfrak{A}_s$ and all $t \geqq s$. The passage to the limit in t "along I" yields the desired equality

$$\int_A X_s\,dP = \int_A X\,dP,$$

since $(1_A X_t)_{t\in I}$ converges in mean to $1_A X$. ┘

11.4.6. Corollary. The random variable X of Theorem 11.4.5 can be chosen to be measurable relative to the σ-algebra $\mathfrak{A}_\infty \equiv \mathfrak{A}\left(\bigcup_{t\in I} \mathfrak{A}_t\right)$. It is then almost surely uniquely determined by the property (11.4.4).

Proof. The variable X constructed in the proof of Theorem 11.4.5 is the limit of a sequence $(X_{t_{1/n}})_{n\in\mathbf{N}}$, convergent in mean, consisting of random variables of the martingale. All $X_{t_{1/n}}$ are $\mathfrak{A}_\infty$-measurable; by Theorem 2.7.5, a suitable subsequence $(X_{\tau_k})_{k\in\mathbf{N}}$ converges almost surely to X. Since the

[8] In other words, we show that $(X_t)_{t\in I}$ satisfies the Cauchy criterion relative to convergence in mean (as t "tends to infinity").

set of all $\omega \in \Omega$ for which $\lim_{k \to \infty} X_{\tau_k}(\omega)$ exists also lies in $\mathfrak{A}_\infty$, we can thus assume that X is $\mathfrak{A}_\infty$-measurable.

If X^* is another $\mathfrak{A}_\infty$-measurable, integrable random variable such that $X_t = E^{\mathfrak{A}_t}(X^*)$ almost surely for all $t \in I$, then

$$\int_A X\,dP = \int_A X_t\,dP = \int_A X^*\,dP$$

holds for all $t \in I$ and all $A \in \mathfrak{A}_t$. Thus the system $\mathfrak{D}$ of all sets $A \in \mathfrak{A}_\infty$ satisfying

$$\int_A X\,dP = \int_A X^*\,dP$$

is a Dynkin system containing the system $\mathfrak{E} \equiv \bigcup_{t \in I} \mathfrak{A}_t$. Now I is filtering to the right and therefore $\mathfrak{E}$ is $\cap$-stable. Two sets $A_1 \in \mathfrak{A}_{t_1}$ and $A_2 \in \mathfrak{A}_{t_2}$ from $\mathfrak{E}$ lie in one and the same σ-algebra $\mathfrak{A}_t$ provided $t \geqq t_1$ and $t \geqq t_2$. Thus, by Theorem 1.2.3, $\mathfrak{D} = \mathfrak{A}_\infty$, that is, $\int_A X\,dP = \int_A X^*\,dP$ for all $A \in \mathfrak{A}_\infty$. By choosing A first equal to $\{X \geqq X^*\}$ and then equal to $\{X \leqq X^*\}$, we obtain $X = X^*$ almost surely. ┘

Remark. The proof of Theorem 11.4.5 tells us that the martingale $(X_t)_{t \in I}$ converges in mean along I to X. In the case $I = \mathbf{N}$, we showed in Corollary 11.4.3 that we actually have almost sure convergence to X. A counterexample of J. Dieudonné [30] shows that for arbitrary I the almost sure convergence is lost in general. In this counterexample I is countable but not totally ordered.

Finally, we discuss a second type of convergence theorems. In contrast to the Convergence Theorems 11.4.1–11.4.3, here the set $-\mathbf{N}$ (with the usual ordering) of all *negative* integers plays the role of the index set $\mathbf{N}$ used there. It is remarkable that we shall be able to prove convergence under very weak hypotheses on the super-martingale under consideration.[9] This is due mainly to the following:

[9] It is interesting to introduce the set $\mathbf{N}$ of natural numbers as index set for a super-martingale $(X_n)_{n \in -\mathbf{N}}$ relative to $(\mathfrak{A}_n)_{n \in -\mathbf{N}}$. We set $Y_n \equiv X_{-n}$ and $\mathfrak{B}_n \equiv \mathfrak{A}_{-n}$ for every $n \in \mathbf{N}$. Then the sequence $(\mathfrak{B}_n)_{n \in \mathbf{N}}$ is antitone, and the crucial property (11.1.2) of a super-martingale is written in the form

$$E^{\mathfrak{B}_n}(Y_m) \leqq Y_n, \qquad \text{for all } m, n \in \mathbf{N} \text{ with } m \leqq n.$$

Therefore super-martingales with $\mathbf{N}$ $(-\mathbf{N})$ as index set are often called "*increasing*" ("*decreasing*") super-martingales.

11.4.7. Lemma. For every super-martingale $(X_n)_{n\in-\mathbf{N}}$ relative to an isotone family $(\mathfrak{A}_n)_{n\in-\mathbf{N}}$ of σ-subalgebras of $\mathfrak{A}$, the following statements are equivalent:

(a) $$\sup_{n\in-\mathbf{N}} E(X_n) < +\infty:$$

(b) $$\sup_{n\in-\mathbf{N}} E(|X_n|) < +\infty;$$

(c) $$(X_n)_{n\in-\mathbf{N}} \text{ is uniformly integrable.}$$

Proof. (a) $\Leftrightarrow$ (b): Since $X_n \leqq |X_n|$, (a) follows from (b). Thus we need only show that "(a) $\Rightarrow$ (b)". For every $n \in -\mathbf{N}$, $|X_n| = X_n + 2X_n^-$ and hence

$$E(|X_n|) = E(X_n) + 2E(X_n^-).$$

By (11.1.6), $(X_n^-)_{n\in-\mathbf{N}}$ is a sub-martingale and therefore the family $(E(X_n^-))_{n\in-\mathbf{N}}$ is isotone. Consequently, $E(X_n^-) \leqq E(X_{-1}^-)$ for all $n \in -\mathbf{N}$, and hence

$$\sup_{n\in-\mathbf{N}} E(|X_n|) \leqq \sup_{n\in-\mathbf{N}} E(X_n) + 2E(X_{-1}^-) < \infty.$$

(b) $\Rightarrow$ (c): Due to the equivalence of (a) and (b), for every real number $\epsilon > 0$ there is $k_\epsilon \in -\mathbf{N}$ such that

$$\sup_{n\in-\mathbf{N}} E(X_n) \leqq \epsilon + E(X_{k_\epsilon}).$$

Moreover, for all $n \in -\mathbf{N}$ and all real numbers $\alpha > 0$,

$$\int_{\{|X_n|\geqq\alpha\}} |X_n|\, dP = \int_{\{X_n\geqq a\}} X_n\, dP - \int_{\{X_n\leqq-\alpha\}} X_n\, dP$$

and

$$E(X_n) = \int_{\{X_n\geqq\alpha\}} X_n\, dP + \int_{\{X_n<\alpha\}} X_n\, dP.$$

From the two equalities it follows that

$$\int_{\{|X_n|\geqq\alpha\}} |X_n|\, dP = E(X_n) - \int_{\{X_n<\alpha\}} X_n\, dP - \int_{\{X_n\leqq-\alpha\}} X_n\, dP,$$

and thus, using the defining inequality (11.1.3) for super-martingales,

$$\begin{aligned}\int_{\{|X_n|\geqq\alpha\}} |X_n|\, dP &\leqq \epsilon + E(X_{k_\epsilon}) - \int_{\{X_n<\alpha\}} X_{k_\epsilon}\, dP - \int_{\{X_n\leqq-\alpha\}} X_{k_\epsilon}\, dP \\ &= \epsilon + \int_{\{X_n\geqq\alpha\}} X_{k_\epsilon}\, dP - \int_{\{X_n\leqq-\alpha\}} X_{k_\epsilon}\, dP \\ &\leqq \epsilon + \int_{\{|X_n|\geqq\alpha\}} |X_{k_\epsilon}|\, dP\end{aligned}$$

for all $n \in -\mathbf{N}$ with $n \leqq k_\epsilon$. By the Chebyshev-Markov inequality we have

$$P\{|X_n| \geqq \alpha\} \leqq \frac{1}{\alpha} E(|X_n|) \leqq \frac{1}{\alpha} \sup_{n \in -\mathbf{N}} E(|X_n|).$$

By Theorem 2.9.6 we have uniformly for all $n \in -\mathbf{N}$,

$$\lim_{\alpha \to +\infty} \int_{\{|X_n| \geqq \alpha\}} |X_{k_\epsilon}| \, dP = 0.$$

By the above inequality, however, this implies

$$\lim_{\alpha \to +\infty} \int_{\{|X_n| \geqq \alpha\}} |X_n| \, dP = 0, \qquad \text{uniformly for all } n \in -\mathbf{N}.$$

Therefore, according to Theorem 2.12.7, II, $(X_n)_{n \in -\mathbf{N}}$ is uniformly integrable.

(c) ⇒ (b): This follows directly from Theorem 2.12.7, III. ⌟

Now we can give an analog of Theorem 11.4.1 and its corollaries. For this, we introduce the σ-subalgebra

$$\mathfrak{A}_{-\infty} \equiv \bigcap_{n \in -\mathbf{N}} \mathfrak{A}_n \tag{11.4.5}$$

for every isotone family $(\mathfrak{A}_n)_{n \in -\mathbf{N}}$ of σ-subalgebras of $\mathfrak{A}$.

11.4.8. Theorem. Every super-martingale $(X_n)_{n \in -\mathbf{N}}$ relative to a family $(\mathfrak{A}_n)_{n \in -\mathbf{N}}$ satisfying the condition

$$\sup_{n \in -\mathbf{N}} E(X_n) < +\infty \tag{11.4.6}$$

converges almost surely and in mean to an integrable random variable $X_{-\infty}$. Then $(X_n)_{n \in -\mathbf{N} \cup \{-\infty\}}$ is a super-martingale relative to the family $(\mathfrak{A}_n)_{n \in -\mathbf{N} \cup \{-\infty\}}$.

Proof. The proof of almost sure convergence of the super-martingale to an integrable random variable proceeds analogously to the proof of Theorem 11.4.1. In this case we have to apply the Doob inequality (11.4.3) to the finite families $(X_i)_{i=-k,\ldots,-1}$, $k \in \mathbf{N}$. The integrability of the limit variable $X_{-\infty}$ follows in the same way as in the proof of Theorem 11.4.1 since from Lemma 11.4.7 we have the boundedness of the expected values $E(|X_n|)$, $n \in -\mathbf{N}$. If we take into account Theorem 2.12.4 and the uniform integrability of the super-martingale guaranteed by Lemma 11.4.7, we obtain convergence in mean. The rest of the assertion then follows, analogously to the proof of Corollary 11.4.2. ⌟

The situation in which $(X_n)_{n\in-\mathbf{N}}$ is a martingale is especially convenient. Then the family $(E(X_n))_{n\in-\mathbf{N}}$ of expectations is constant, and thus (11.4.6) is trivially satisfied. Hence we immediately obtain the corollary:

11.4.9. Corollary. Every martingale $(X_n)_{n\in-\mathbf{N}}$ relative to a family $(\mathfrak{A}_n)_{n\in-\mathbf{N}}$ converges almost surely and in mean to an integrable random variable $X_{-\infty}$. Then $(X_n)_{n\in-\mathbf{N}\cup\{-\infty\}}$ is a martingale relative to the family $(\mathfrak{A}_n)_{n\in-\mathbf{N}\cup\{-\infty\}}$.

PROBLEMS

1. Let X be a real, integrable random variable on $(\Omega,\mathfrak{A},P)$, and let $\mathbf{S}$ be the set of all σ-subalgebras $\mathfrak{Z}$ of $\mathfrak{A}$. Prove that the family $(E^{\mathfrak{Z}}(X))_{\mathfrak{Z}\in\mathbf{S}}$ is uniformly integrable.
2. Let $(X_n)_{n\in\mathbf{N}}$ be a super-martingale relative to $(\mathfrak{A}_n)_{n\in\mathbf{N}}$ of random variables $X_n \geqq 0$. Suppose that $\sup_{n\in\mathbf{N}} E(X_n^p) < +\infty$ for some real number $p \geqq 1$. Prove: $(X_n)_{n\in\mathbf{N}}$ converges almost surely and in pth mean to a random variable $X_\infty \geqq 0$ such that $E(X_\infty^p) < +\infty$ and $(X_n)_{n\in\mathbf{N}\cup\{\infty\}}$ is a super-martingale relative to $(\mathfrak{A}_n)_{n\in\mathbf{N}\cup\{\infty\}}$.

11.5 APPLICATIONS

We show the range of the theorems of the preceding section by applying them to several particular situations with which we are already partly familiar.

(a) Convergence of sums of independent random variables

As in Section 11.1, Example 4, let $(X_n)_{n\in\mathbf{N}}$ be a sequence of real, integrable random variables such that

$$E(X_1) \geqq 0 \quad \text{and} \quad E(X_{n+1} \mid X_1, \ldots, X_n) \geqq 0, \quad \text{almost surely, for all } n \in \mathbf{N}. \tag{11.5.1}$$

The sequence of partial sums $S_n \equiv X_1 + \cdots + X_n$ is then a sub-martingale. Hence, by Theorem 11.4.1, (S_n) converges almost surely to an integrable random variable if one of the following equivalent conditions is satisfied:

$$\sup_{n\in\mathbf{N}} E(|S_n|) < \infty, \tag{11.5.2}$$

$$\sup_{n\in\mathbf{N}} E(S_n^+) < \infty. \tag{11.5.3}$$

If the random variables are all centered, then (S_n) is a martingale. Since $E(E(X_{n+1} \mid X_1, \ldots, X_n)) = E(X_{n+1}) = 0$, and because of (11.5.1), we then even have

$$E(X_{n+1} \mid X_1, \ldots, X_n) = 0, \qquad \text{almost surely.}$$

By Section 11.1, Example 3, this is the case in particular when the sequence (X_n) is independent and all X_n are centered. In this case the following condition is sufficient for (11.1.2) to hold and thus for almost sure convergence of (S_n):

$$\sum_{n=1}^{\infty} V(X_n) < \infty, \tag{11.5.4}$$

provided we assume that all X_n are square integrable. By the Hölder inequality (for $p = q = 2$) we then have

$$E(|S_n|)^2 = E(1 \cdot |S_n|)^2 \leqq E(1)E(S_n^2) = V(S_n) = \sum_{i=1}^{n} V(X_i) \leqq \sum_{i=1}^{\infty} V(X_i)$$

and hence

$$\sup_{n \in \mathbf{N}} E(|S_n|) \leqq \left[\sum_{i=1}^{\infty} V(X_i) \right]^{1/2} < \infty.$$

The relationship between the above and our considerations of the strong law of large numbers becomes clear when we recall the following lemma from analysis, the so-called *Kronecker Lemma:*

11.5.1. Lemma. Let $(x_n)_{n \in \mathbf{N}}$ and $(\tau_n)_{n \in \mathbf{N}}$ be two sequences of real numbers where the second is isotone, contains only numbers $\tau_n > 0$, and diverges to $+\infty$. Then

$$\sum_{i=1}^{\infty} \frac{x_i}{\tau_i} \text{ convergent} \Rightarrow \lim_{n \to \infty} \frac{1}{\tau_n} \sum_{i=1}^{n} x_i = 0. \tag{11.5.5}$$

Proof. Let $s_n \equiv \Sigma_{i=1}^{n} (x_i/\tau_i)$ and $s \equiv \lim s_n = \Sigma_{i=1}^{\infty} (x_i/\tau_i)$. Then we know that the sequence

$$\left(\frac{\sigma_1 s_1 + \cdots + \sigma_n s_n}{\sigma_1 + \cdots + \sigma_n} \right)$$

also converges to s if (σ_n) is a sequence of real numbers $\geqq 0$ with $\Sigma \sigma_n$ divergent. Thus we have

$$\lim_{n \to \infty} \frac{\tau_2 s_1 + (\tau_3 - \tau_2) s_2 + \cdots + (\tau_{n+1} - \tau_n) s_n}{\tau_{n+1}} = s,$$

and hence, because $\lim\limits_{n\to\infty} \frac{\tau_1 s_1}{\tau_{n+1}} = 0$,

$$\lim_{n\to\infty} \frac{(\tau_2 - \tau_1)s_1 + (\tau_3 - \tau_2)s_2 + \cdots + (\tau_{n+1} - \tau_n)s_n}{\tau_{n+1}} = s.$$

The assertion now follows since

$$\frac{1}{\tau_{n+1}} \sum_{i=1}^{n+1} x_i = s_{n+1} - \frac{(\tau_2 - \tau_1)s_1 + \cdots + (\tau_{n+1} - \tau_n)s_n}{\tau_{n+1}}. \quad \lrcorner$$

Now let $(X_n)_{n\in\mathbf{N}}$ be a sequence of real, integrable, centered random variables satisfying

$$E(X_1) = 0 \text{ and } E(X_{n+1} \mid X_1, \ldots, X_n) = 0, \quad \text{almost surely, for all } n \in \mathbf{N}, \tag{11.5.6}$$

and let $(\tau_n)_{n\in\mathbf{N}}$ be a sequence as in Lemma 11.5.6 Then obviously

$$E\left(\frac{1}{\tau_1} X_1\right) = 0 \text{ and } E\left(\frac{1}{\tau_{n+1}} X_{n+1} \,\middle|\, \frac{1}{\tau_1} X_1, \ldots, \frac{1}{\tau_n} X_n\right) = 0, \quad \text{almost surely} \qquad (n \in \mathbf{N}),$$

so that $(\Sigma_{i=1}^n (1/\tau_i) X_i)_{n\in\mathbf{N}}$ is a martingale. If this converges almost surely, then by the Kronecker Lemma

$$\lim_{n\to\infty} \frac{1}{\tau_n} \sum_{i=1}^{n} X_i = 0, \qquad \text{almost surely.}$$

Because the X_i are centered, this is the strong law of large numbers for the sequence $\tau_n = n$. Thus condition (11.5.2) for $S_n \equiv \Sigma_{i=1}^n (1/\tau_i) X_i$ is sufficient for almost sure convergence of $(1/\tau_n)\Sigma_{i=1}^n X_i$ to zero. Thus we have derived the strong law of large numbers under weaker independence conditions, namely, those formulated in (11.5.6).

If the sequence (X_n) is independent, then our observation concerning (11.5.4) tells us that also the condition

$$\sum_{n=1}^{\infty} \frac{1}{\tau_n^2} V(X_n) < \infty \tag{11.5.7}$$

is sufficient for almost sure convergence of $(1/\tau_n)\Sigma_{i=1}^n X_i$ to zero. For $\tau_n = n$ this is Theorem 6.4.1 of Kolmogorov.

(b) Sums of independent, identically distributed random variables

We can also derive Kolmogorov's Theorem 6.4.2 from the martingale theorems. As in Theorem 6.4.2, let $(X_n)_{n\in\mathbf{N}}$ be an independent sequence of real, integrable random variables with the same distribution μ. We assume every X_n to be centered and then have to prove almost sure convergence of $(1/n)\Sigma_{i=1}^n X_i$ to 0. We again set $S_n \equiv X_1 + \cdots + X_n$ for all n.

For each $n \in \mathbf{N}$, we use $\mathfrak{A}_{-n}$ to denote the σ-algebra generated by all S_m with $m \geqq n$. Then $(\mathfrak{A}_n)_{n\in -\mathbf{N}}$ is obviously an isotone sequence of σ-subalgebras of the σ-algebra $\mathfrak{A}$ of the probability space $(\Omega,\mathfrak{A},P)$.

Since $S_{m+1} - S_m = X_{m+1}$ $(m \in \mathbf{N})$, we have

$$E(X_1 \mid \mathfrak{A}_{-n}) = E(X_1 \mid S_n, X_{n+1}, X_{n+2}, \ldots), \qquad \text{almost surely.}$$

Moreover, X_1 and S_n are independent of the σ-algebra generated by $X_{n+1}, X_{n+2}, \ldots$. Therefore, by Theorem 10.1.4,

$$E(X_1 \mid S_n, X_{n+1}, X_{n+1}, \ldots) = E(X_1 \mid S_n), \qquad \text{almost surely,}$$

and hence

$$E(X_1 \mid \mathfrak{A}_{-n}) = E(X_1 \mid S_n), \qquad \text{almost surely} \qquad (n \in \mathbf{N}). \tag{11.5.8}$$

As we shall show shortly, we also have

$$E(X_1 \mid S_n) = E(X_j \mid S_n), \qquad \text{almost surely} \qquad (j = 1, \ldots, n) \tag{11.5.9}$$

due to the special hypotheses. From (11.5.8) and (11.5.9) it now follows that

$$\begin{aligned} E(X_1 \mid \mathfrak{A}_{-n}) &= \frac{1}{n}\sum_{j=1}^{n} E(X_j \mid S_n) = E\left(\frac{1}{n}\sum_{j=1}^{n} X_j \mid S_n\right) \\ &= \frac{1}{n} S_n, \qquad \text{almost surely.} \end{aligned}$$

Now Theorem 11.4.4 and Corollary 11.4.9 apply, and thus the martingale $(E(X_1 \mid \mathfrak{A}_n))_{n\in -\mathbf{N}}$ converges almost surely and in mean to an integrable random variable $X_{-\infty}$; hence, we have

$$\lim_{n\to\infty} \frac{1}{n} S_n = X_{-\infty}, \qquad \text{almost surely.}$$

The convergence in mean implies $E(X_{-\infty}) = \lim E((1/n)S_n) = 0$ since the variables X_n are assumed to be all centered. The considerations after the Kolmogorov 0-1 Law (Theorem 6.2.3) yield that $X_{-\infty}$ is measurable relative to the σ-algebra of terminal events determined by (X_n). Therefore, by the 0-1 Law, $X_{-\infty}$ is almost surely constant.[10] Hence and from $E(X_{-\infty}) = 0$, it follows that $X_{-\infty} = 0$ almost surely.

The proof of (11.5.9) still to be given is an application of the Transformation Theorem for integrals. Obviously we need to show only that $\int_A X_1\,dP = \int_A X_j\,dP$ for all $A \in \mathfrak{A}(S_n)$ and $j = 1, \ldots, n$, that is, for all sets $A = S_n^{-1}(B)$ with Borel $B \subset \mathbf{R}$. But now

$$\begin{aligned}\int_A X_j\,dP &= \int (1_B \circ S_n)X_j\,dP\\ &= \int 1_B(s(X_1, \ldots, X_n)) \cdot p_j(X_1, \ldots, X_n)\,dP\\ &= \int (1_B \circ s)p_j\,dP_{X_1 \otimes \cdots \otimes X_n},\end{aligned}$$

if s and p_j denote the mappings $(x_1, \ldots, x_n) \to x_1 + \cdots + x_n$ and $(x_1, \ldots, x_n) \to x_j$ of $\mathbf{R}^n$ into $\mathbf{R}$, respectively. Because of the independence of all X_n, the joint distribution $P_{X_1 \otimes \cdots \otimes X_n}$ is equal to $\mu \otimes \cdots \otimes \mu$. Thus, using Fubini's Theorem, we obtain the value

$$\begin{aligned}\int_A X_j\,dP &= \int \cdots \int 1_B(x_1 + \cdots + x_n) \cdot x_j\,\mu(dx_1) \cdots \mu(dx_n)\\ &= \int \mu^{(n-1)}(B - x_j) \cdot x_j\,\mu(dx_j) = \int \mu^{n-1}(B - x) \cdot x\,\mu(dx),\end{aligned}$$

independent of j, where $\mu^{(n-1)}$ denotes the $(n-1)$-fold convolution product of μ with itself. Thus we have proved Kolmogorov's Theorem 6.4.2.

(c) Connections with differentiation theory

First, we note that our presentation of martingale theory uses the *existence* of conditional expectations only in Theorem 11.4.4. In all other cases, they serve only to shorten the statements of definitions [as in (11.1.2)] or results. Thus, with the exception of Theorem 11.4.4, the convergence theorems of Section 11.4 do not need the Radon-Nikodym Theorem in their proofs. It is therefore noteworthy that martingale theory provides us with a new proof of the Radon-Nikodym Theorem (for the case of finite measures, which was the essential case in our earlier proof.

Thus, let $(\Omega,\mathfrak{A},P)$ be a probability space and Q a finite, P-continuous measure on $\mathfrak{A}$. We refer to Section 11.1, Example 5 and consider the partially ordered set I of all finite decompositions of Ω into measurable

[10] Since $\{X_{-\infty} \leqq n\} \uparrow \Omega$ and $\{X_{-\infty} \leqq -n\} \downarrow \varnothing$ $(n \in \mathbf{N})$, the set of all $\alpha \in \mathbf{R}$ with $P\{X_{-\infty} \leqq \alpha\} = 1$ is nonempty and bounded below. For the infimum α_0 of these α we therefore have $P\{X_{-\infty} \leqq \alpha_0\} = 1$ and $P\{X_{-\infty} \leqq \beta\} = 0$ for all $\beta < \alpha_0$, that is, $P\{X_{-\infty} < \alpha_0\} = 0$ and hence $P\{X_{-\infty} = \alpha_0\} = 1$.

sets, the associated isotone family $(\mathfrak{A}_t)_{t\in I}$ of finite σ-subalgebras of $\mathfrak{A}$, and the martingale $(X_t)_{t\in I}$, defined via Q, relative to $(\mathfrak{A}_t)_{t\in I}$. We note that:

1. I is filtering to the right. If $s = (B_i)_{i=1,\ldots,n_s}$ and $t = (C_j)_{j=1,\ldots,n_t}$ are two partitions from I, then $u \equiv (B_i \cap C_j)_{i=1,\ldots,n_s;\,j=1,\ldots,n_t}$ is another element of I which satisfies $s \leqq u$ and $t \leqq u$.

2. We have $\mathfrak{A} = \bigcup_{t\in I} \mathfrak{A}_t$. Thus, in particular, $\mathfrak{A} = \mathfrak{A}_\infty$ in the sense of the notation of Corollary 11.4.6. Indeed, every $A \in \mathfrak{A}$ lies in the σ-algebra $\mathfrak{A}_{t_A}$ associated with the decomposition $t_A \equiv (A, \complement A)$.

3. For every $\epsilon > 0$, by Theorem 2.9.6 there exists a $\delta > 0$ such that $P(A) \leqq \delta$ implies $Q(A) \leqq \epsilon$ $(A \in \mathfrak{A})$.

4. The martingale $(X_t)_{t\in I}$ is uniformly integrable. For every real $\alpha > 0$ and every $t = (B_i) \in I$, we have

$$\int_{\{|X_t|\geqq\alpha\}} X_t\,dP = \int_{\{X_t\geqq\alpha\}} X_t\,dP = \sum_{\substack{X_t\geqq\alpha \\ \text{on } B_i}} Q(B_i) = Q\{X_t \geqq \alpha\}$$

and

$$P\{X_t \geqq \alpha\} \leqq \frac{1}{\alpha} E(X_t) = \frac{1}{\alpha} Q(\Omega).$$

We need note only that $E(X_t)$ is independent of t. For $t = (\varnothing,\Omega) \in I$, we compute the constant value to be $Q(\Omega)$. Uniform integrability now follows directly from property 3 above.

Using Theorem 11.4.5 and martingale theory, we find that there exists a random variable $X \in \mathfrak{L}^1(P)$ such that $X_t = E^{\mathfrak{A}_t}(X)$ almost surely, that is, such that

$$\int_A X_t\,dP = \int_A X\,dP,$$

for all $A \in \mathfrak{A}_t$ and every $t \in I$. For $A \in \mathfrak{A}$ and the decomposition t_A of property 2 we thus obtain

$$Q(A) = \int_A X_{t_A}\,dP = \int_A X\,dP.$$

But then X is P-almost surely equal to the density of Q relative to P, and thus the Radon-Nikodym Theorem is proved anew.

We can work with the martingale $(X_t)_{t\in J}$ in the above proof, where J is a subset of I filtering to the right such that $\bigcup_{t\in J} \mathfrak{A}_t$ is still a generator of $\mathfrak{A}$. Since this generator is $\cap$-stable, $Q(A) = \int_A X\,dP$ for all $A \in \bigcup_{t\in J} \mathfrak{A}_t$ implies this equality for all $A \in \mathfrak{A}$.

If, in particular, $(X_t)_{t\in J}$ is a *sequence*, then we also have available Corollary 11.4.3. Accordingly, $(X_t)_{t\in J}$ converges almost surely to the density X. This is the case when $\mathfrak{A}$ is *countably generated.* If $(A_n)_{n\in\mathbf{N}}$ is a sequence of events generating $\mathfrak{A}$, then we choose for t_n the decomposition of Ω "generated" by $A_1, \ldots, A_n$. Since $\mathfrak{A}_{t_n}$ contains $A_1, \ldots, A_n$, this proves that $\mathfrak{A}$ is generated by $\bigcup_{n=1}^{\infty} \mathfrak{A}_{t_n}$. Thus we can choose $J = \{t_1, t_2, \ldots\}$. If we make use of these observations, then in certain cases the density X appears not only formally as above, but explicitly as a derivative. Therefore we speak of the *Radon-Nikodym derivative* X. An example of this is the following:

Example

Let $\Omega \equiv [0,1[$, $\mathfrak{A} \equiv \Omega \cap \mathfrak{B}^1$ and $P \equiv \lambda^1_\Omega$, and let Q be a P-continuous measure on $\mathfrak{A}$. We examine it with the help of its distribution function

$$F(x) \equiv Q([0,x[) \qquad (x \in [0,1[),$$

which appears here restricted to $[0,1[$. For every $n = 1, 2, \ldots$ we choose a finite sequence $\mathfrak{F}^{(n)} = (x_0^{(n)}, \ldots, x_{k_n}^{(n)})$ of points

$$0 = x_0^{(n)} < x_1^{(n)} < \cdots < x_{k_n}^{(n)} = 1$$

such that all elements of the sequence $\mathfrak{F}^{(n)}$ occur in $\mathfrak{F}^{(n+1)}$ and such that

$$\lim_{n\to\infty} \max_{0\leq j<k_n} (x_{j+1}^{(n)} - x_j^{(n)}) = 0.$$

Each of the sequences $\mathfrak{F}^{(n)}$ defines a partition t_n of Ω consisting of the sets $[x_j^{(n)}, x_{j+1}^{(n)}[$, $j = 0, \ldots, k_n - 1$. For the associated (finite) σ-algebras $\mathfrak{A}_{t_n}$, we then obviously have $\mathfrak{A}_{t_n} \subset \mathfrak{A}_{t_{n+1}}$ and $\mathfrak{A} = \mathfrak{A}\left(\bigcup_{n=1}^{\infty} \mathfrak{A}_{t_n}\right)$. The density X of Q relative to P is P-almost surely equal to $\lim_{n\to\infty} X_{t_n}$ by Corollary 11.4.3. But now, for every $x \in [0,1[$ and every $n \in \mathbf{N}$,

$$X_{t_n}(x) = \frac{Q([x_{j_x}^{(n)}, x_{j_x+1}^{(n)}[)}{P([x_{j_x}^{(n)}, x_{j_x+1}^{(n)}[)} = \frac{F(x_{j_x+1}^{(n)}) - F(x_{j_x}^{(n)})}{x_{j_x+1}^{(n)} - x_{j_x}^{(n)}},$$

where j_x denotes the uniquely determined number from $\{0, \ldots, k_n - 1\}$ with $x \in [x_{j_x}^{(n)}, x_{j_x+1}^{(n)}[$. It can then be derived from the Convergence Theorem 11.4.3 that F is differentiable at λ^1-almost all points from $[0,1[$ and $X(x)$ is its derivative λ^1-almost everywhere. We will not present a proof of that fact.

This example can also be modified in many ways and can then be used, via Corollary 11.4.3, for the proof of the theorem that every monotone function on an interval in $\mathbf{R}$ is λ^1-almost everywhere differentiable.

PROBLEMS

1. Let $(X_n)_{n \in \mathbf{N}}$ be an independent sequence of real random variables such that $\Sigma_{n=1}^{\infty}|X_n|$ is integrable. Define for each $n \in \mathbf{N}$:

$$\mathfrak{A}_{-n} = \mathfrak{A}(X_n, X_{n+1}, \ldots),$$
$$Y_{-n} = E\left(\sum_{n=1}^{\infty} X_n | \mathfrak{A}_n\right).$$

Prove:

(a) $Y_{-n} = \sum_{i=n}^{\infty} X_i + E(X_1 + \cdots + X_{n-1})$, for all $n \in \mathbf{N}$.
(b) $(Y_n)_{n \in -\mathbf{N}}$ is a martingale relative to $(\mathfrak{A}_n)_{n \in -\mathbf{N}}$.
(c) (Y_n) converges almost surely and in the mean to an integrable random variable. Determine this variable explicitly.

3. Let $(X_n)_{n \in \mathbf{N}}$ be a sequence of real, integrable random variables. Define a new sequence $(Y_n)_{n \in \mathbf{N}}$ as follows:

$$Y_1 \equiv X_1;$$
$$Y_n \equiv X_n - \sum_{i=2}^{n} [E(X_i \mid X_1, \ldots, X_{i-1}) - X_{i-1}] \qquad (n \geqq 2).$$

Prove that $(Y_n)_{n \in \mathbf{N}}$ is a martingale. Give conditions which imply that (Y_n) converges almost surely.

12

STOCHASTIC PROCESSES

The concept of stochastic process is basic to many branches of probability theory. It is at the same time so simple that the reader will wonder why this concept was not introduced earlier. The reason is that the concepts of conditional expectation and of kernel are fundamental in answering the questions characteristic of the theory of stochastic processes. Use of the term "stochastic process" only as an abbreviation would have been of no advantage in the previous material.

12.1 DEFINITION AND CONSTRUCTION OF STOCHASTIC PROCESSES

The problems of probability theory treated so far mostly involved a sequence $(X_n)_{n \in \mathbf{N}}$ of random variables with values in a measurable space $(\Omega', \mathfrak{A}')$. Here we often interpret X_n as the outcome of a random experiment carried out at "points of time" $n = 1, 2, \ldots$. But it is often necessary to handle a random occurrence with continuous (and not only discrete) time with the methods of probability theory. A typical example from physics is that of Brownian motion, of a particle (for example, a molecule) in a liquid or a gas; motion due to molecular collisions. We can describe the position of the particle at time $t \geqq 0$ by a random variable X_t with values in $\mathbf{R}^3$. The family $(X_t)_{t \in \mathbf{R}_+}$ can then be used as a probability-theoretic model for the investigation of Brownian motion. Here it is assumed that the random variables X_t are all defined on one and the same probability space $(\Omega, \mathfrak{A}, P)$. The probability measure P "controls" the motion of the Brownian particle. The various paths that the particle can travel are described by $t \to X_t(\omega)$, that is, by a mapping of $\mathbf{R}_+$ into $\mathbf{R}^3$, where ω is arbitrarily chosen in Ω. Such considerations lead to the following:

12.1.1. Definition. Every quadruple $(\Omega,\mathfrak{A},P,(X_t)_{t\in I})$, where $(\Omega,\mathfrak{A},P)$ is a probability space and $(X_t)_{t\in I}$ is a family of random variables on this space with values all in a measurable space $(E,\mathfrak{B})$, is called a *stochastic process* (or, simply, process). We call I the *parameter-* or *time-*set and E [or, more precisely, $(E,\mathfrak{B})$] the *state space* of the stochastic process. For every $\omega \in \Omega$, the mapping from I into E defined by $t \to X_t(\omega)$ is called a *path*[1] of the process.

Since we generally think of the probability space $(\Omega,\mathfrak{A},P)$ as given, we often denote a stochastic process $(\Omega,\mathfrak{A},P,(X_t)_{t\in I})$ simply by $(X_t)_{t\in I}$ or even just (X_t).

For $I = \mathbf{N}$ we thus obtain the sequences of random variables with values in a fixed measurable space as a special case. Brownian motion is to be considered as a stochastic process with state space $\mathbf{R}^3$ and parameter set $\mathbf{R}_+$. Further, every super- or sub-martingale is a stochastic process. The Doob inequalities contain results about the behavior of the path of a super-martingale.

Now suppose we are given a stochastic process $(\Omega,\mathfrak{A},P,(X_t)_{t\in I})$ with arbitrary parameter set $I \neq \varnothing$ and arbitrary state space $(E,\mathfrak{B})$. For every nonempty subset J of I, let E^J denote the set of all mappings from J into E, that is, the product set $\prod_{t\in J} E_t$ in which every factor E_t is equal to E. Correspondingly, we let $\mathfrak{B}^J$ denote the σ-algebra $\bigotimes_{t\in J} \mathfrak{B}_t$ in E^J, where every $\mathfrak{B}_t = \mathfrak{B}$. Further let

$$X_J \equiv \bigotimes_{t\in J} X_t \tag{12.1.1}$$

be the product mapping from Ω into E^J already introduced in Section 5.4. This is an $(E^J,\mathfrak{B}^J)$-random variable. The joint distribution of the family $(X_t)_{t\in J}$ of random variables is the distribution of X_J, namely,

$$P_J \equiv X_J(P). \tag{12.1.2}$$

There are simple relationships between the probability spaces $(E^J,\mathfrak{B}^J,P_J)$: Let J and H be nonempty subsets of I with $J \subset H$ and let

$$p_J^H\colon E^H \to E^J \tag{12.1.3}$$

[1] The terms "trajectory" and "realization of the processes" are also commonly used. Since to every $\omega \in \Omega$ there corresponds a path and hence a mapping (from I into E), stochastic processes are sometimes also called *random functions* or mappings.

be the $\mathfrak{B}^H$-$\mathfrak{B}^J$-measurable projection mapping. (With every element of E^H, that is, every mapping from H into E, it associates its restriction to J.) For $H = I$ we let

$$p_J \equiv p_J^I. \tag{12.1.4}$$

Since, for any two nonempty subsets J and H of I with $J \subset H$, we see that $X_J = p_J^H \circ X_H$ holds, then in view of (12.1.2) we obtain

$$P_J = p_J^H(P_H) \tag{12.1.5}$$

and for $H = I$, in particular,

$$P_J = p_J(P_I). \tag{12.1.6}$$

These are the relationships announced above between the probability spaces $(E^J,\mathfrak{B}^J,P_J)$. For every finite, nonempty subset $J = \{t_1, \ldots ,t_n\}$ of I, the measure P_J has a simple interpretation. For any n sets $B_1, \ldots , B_n \in \mathfrak{B}$,

$$\begin{aligned} P_J(B_1 \times \cdots \times B_n) &= P(X_J^{-1}(B_1 \times \cdots \times B_n)) \\ &= P\{X_{t_1} \in B_1, \ldots , X_{t_n} \in B_n\}. \end{aligned} \tag{12.1.7}$$ [2]

Thus if we interpret the process $(X_t)_{t \in I}$ as random motion of a particle in E, then P_J is the only probability measure on $\mathfrak{B}^J$ such that for any n sets $B_1, \ldots , B_n \in \mathfrak{B}$, the number $P_J(B_1 \times \cdots \times B_n)$ is equal to the probability that the particle finds itself in B_1 at time t_1, in B_2 at time $t_2, \ldots$, in B_n at time t_n. If $\mathfrak{H} = \mathfrak{H}(I)$ denotes the set of all nonempty, *finite* subsets of I, it thus becomes clear why we call $(P_J)_{J \in \mathfrak{H}(I)}$ or also $(E^J,\mathfrak{B}^J,P_J)_{J \in \mathfrak{H}(I)}$ the *family of finite-dimensional distributions* of the process. This family is *projective* in the sense of the following definition:

12.1.2. Definition. If condition (12.1.5) holds for a family $(P_J)_{J \in \mathfrak{H}(I)}$ of probability measures on the measurable spaces $(E^J,\mathfrak{B}^J)$ for any two sets $J, H \in \mathfrak{H}(I)$ satisfying $J \subset H$, then the family (P_J) is said to be *projective.*

Remark. 1. In this definition, it would be sufficient to require (12.1.5) only for sets $J, H \in \mathfrak{H}$ such that H has exactly one element more than J. If J and H are sets from $\mathfrak{H}$ with $J \subset H$ and if, say, $H \setminus J$ contains exactly n elements, then we determine the sets $H_0, H_1, \ldots , H_n$ such that $H_0 = J \subset H_1 \subset \cdots \subset H_n = H$ and $H_i \setminus H_{i-1}$ always has only one element. The assertion then follows from

$$p_J^H = p_{H_0}^{H_1} \circ p_{H_1}^{H_2} \circ \cdots \circ p_{H_{n-1}}^{H_n}.$$

[2] Here we set $\{X_{t_1} \in B_1, \ldots , X_{t_n} \in B_n\} \equiv \{X_{t_1} \in B_1\} \cap \cdots \cap \{X_{t_n} \in B_n\}$.

A crucial question for the construction of processes is whether, under certain hypotheses on the measurable space $(E,\mathfrak{B})$ and the set I, every projective family $(P_J)_{J\in\mathfrak{H}(I)}$ of probability measures on the measurable spaces $(E^J,\mathfrak{B}^J)$ is the family of finite-dimensional distributions of a stochastic process on a suitable probability space with state space $(E,\mathfrak{B})$ and parameter set I. This question plays the same role in the theory of processes as the question, answered positively by the results in Section 5.4, of whether every family $(Q_i)_{i\in I}$ of probability measures is the family of distributions of an independent family $(Y_i)_{i\in I}$ of random variables.

Our question is answered in the generality suitable to our purposes by the following theorem of Kolmogorov and a corollary.

12.1.3. Theorem. If E is a Polish space, $\mathfrak{B}$ is the σ-algebra of its Borel sets and I is an arbitrary nonempty set, then for every projective family $(P_J)_{J\in\mathfrak{H}(I)}$ of probability measures on $(E^J,\mathfrak{B}^J)$ there exists exactly one probability measure P_I on $(E^I,\mathfrak{B}^I)$ satisfying

$$p_J(P_I) = P_J, \qquad \text{for all } J \in \mathfrak{H}(I). \tag{12.1.8}$$

We call P_I the *projective limit* of the family $(P_J)_{J\in\mathfrak{H}(I)}$ and write

$$\lim_{\substack{\longleftarrow \\ J\in\mathfrak{H}(I)}} P_J = \lim_{\longleftarrow} P_J \equiv P_I. \tag{12.1.9}$$

Proof. Similar to the proof of Theorem 5.4.2, let $\mathfrak{Z}_J \equiv p_J^{-1}(\mathfrak{B}^J)$ be the σ-algebra of the so-called J-cylinders and let $\mathfrak{Z} \equiv \bigcup_{J\in\mathfrak{H}(I)} \mathfrak{Z}_J$. Then $\mathfrak{Z}$ is again an algebra in E^I since $J, H \in \mathfrak{H}(I)$ and $J \subset H$ imply $\mathfrak{Z}_J \subset \mathfrak{Z}_H$ due to the $\mathfrak{B}^H$-$\mathfrak{B}^J$-measurability of p_J^H. Moreover, by the definition of $\mathfrak{B}^I$, $\mathfrak{Z}$ is a generator of $\mathfrak{B}^I$ and thus $\mathfrak{B}^I = \mathfrak{A}(\mathfrak{Z})$. Condition (12.1.8) on P_I tells us that P_I is a probability measure on $\mathfrak{B}^I$ with $P_I(Z) = P_J(B)$ if $Z = p_J^{-1}(B)$ is a cylinder set $(B \in \mathfrak{B}^J)$. It now follows from the Uniqueness Theorem 1.5.5 that there can be at most one such probability measure P_I. The existence of P_I is obtained from the Extension Theorem 1.5.2 if we can show that

$$P_0(Z) \equiv P_J(B) \qquad (Z = p_J^{-1}(B),\ B \in \mathfrak{B}^J,\ J \in \mathfrak{H}(I))$$

defines a premeasure on $\mathfrak{Z}$ with $P_0(E^I) = 1$. Due to the projectivity of the family (P_J), we see that P_0 is well defined on $\mathfrak{Z}$ and is a content. For this, we need only repeat the corresponding reasoning in the proof of Theorem 5.4.2. Since $E^I = p_J^{-1}(E^J)$ for every $J \in \mathfrak{H}(I)$, we have $P_0(E^I) = P_J(E^J) = 1$. Thus all we have left to show is the $\varnothing$-continuity of P_0. Thus we assert: If $(Z_n)_{n\in\mathbf{N}}$ is an antitone sequence in $\mathfrak{Z}$ such that $P_0(Z_n) \geqq \lambda > 0$ for all $n \in \mathbf{N}$, then $\bigcap_{n=1}^{\infty} Z_n \neq \varnothing$. But this can be seen as follows:

Every Z_n is of the form $Z_n = p_{J_n}^{-1}(B_n)$ with $B_n \in B^{J_n}$ and $J_n \in \mathfrak{H}(I)$. Since every J-cylinder is obviously also an H-cylinder for all $H \in \mathfrak{H}(I)$ with $J \subset H$, we can assume $J_n \subset J_{n+1}$ for all $n \in \mathbf{N}$ without loss of generality. By Section 7.3, Example 2, every finite product of Polish spaces is a Polish space and in particular each of the spaces E^{J_n} is Polish. Therefore, by Theorem 7.3.3, the probability measure P_{J_n} is regular on E^{J_n}. Thus for every B_n there exists a set $K_n \subset B_n$, compact in E^{J_n}, satisfying $P_{J_n}(B_n) - P_{J_n}(K_n) = P_{J_n}(B_n \setminus K_n) \leqq 2^{-n}\lambda$. Then $Z_n' \equiv p_{J_n}^{-1}(K_n)$ is a J_n-cylinder satisfying $Z_n' \subset Z_n$ and

$$P_0(Z_n \setminus Z_n') = P_0(p_{J_n}^{-1}(B_n \setminus K_n)) = P_{J_n}(B_n \setminus K_n) \leqq 2^{-n}\lambda.$$

In order to obtain an antitone sequence, we set $Y_n \equiv Z_1' \cap \cdots \cap Z_n'$, so that $Y_n \supset Y_{n+1}$ and $Y_n \subset Z_n' \subset Z_n$ for all n. Then $P_0(Y_n) > 0$ and hence $Y_n \neq \emptyset$. Indeed, because of the finite additivity of P_0,

$$\begin{aligned} P_0(Z_n \setminus Y_n) &= P_0\Big(\bigcup_{i=1}^{n} (Z_n \setminus Z_i')\Big) \leqq P_0\Big(\bigcup_{i=1}^{n} (Z_i \setminus Z_i')\Big) \\ &\leqq \sum_{i=1}^{n} P_0(Z_i \setminus Z_i') \leqq \lambda \sum_{i=1}^{n} 2^{-i} < \lambda, \end{aligned}$$

and thus $P_0(Y_n) = P_0(Z_n \setminus (Z_n \setminus Y_n)) = P_0(Z_n) - P_0(Z_n \setminus Y_n) > \lambda - \lambda = 0$. Now we arbitrarily choose y_n in Y_n. Because of the antitonicity of (Y_n) we have $y_{n+p} \in Y_n$ and thus $y_{n+p} \in Z_n'$ for all $p \in \mathbf{N}$, that is, $p_{J_n}(y_m) \in K_n$ for all $m, n \in \mathbf{N}$ with $m \geqq n$. For every $t \in \cup J_n$ which belongs to, say, J_n, all terms of the sequence $(p_{\{t\}}(y_m))_{m \in \mathbf{N}}$ with the exception of finitely many initial terms then lie in $p_{\{t\}}^{J_n}(K_n)$. Since every projection mapping p_J^H is continuous, this set is compact. Now $\bigcup J_n$ is countable as a countable union of finite sets; let $t_1, t_2, \ldots$ be the elements of this set. Then, due to the compactness of $p_{\{t\}}^{J_n}(K_n)$, $t \in J_n$, there exists a subsequence (y_m') of (y_m) for which $(p_{\{t_1\}}(y_m'))$ converges, a subsequence (y_m'') of (y_m') for which $(p_{\{t_2\}}(y_m''))$ converges, and so on. For the diagonal sequence $(y_m^{(m)})_{m \in \mathbf{N}}$,

$$z_t \equiv \lim_{m \to \infty} p_{\{t\}}(y_m^{(m)}) = \lim_{m \to \infty} y_m^{(m)}(t)$$

then exists for every $t \in \bigcup J_n$. Since for the initial sequence $p_{J_n}(y_m)$ was in K_n for all $m, n \in \mathbf{N}$ with $m \geqq n$, we now have

$$p_{J_n}(y_m^{(m)}) \in K_n \subset E^{J_n}$$

for all $m, n \in \mathbf{N}$ with $m \geqq n$ and thus, because E^{J_n} is a product space,

$$(z_{\tau_1}, \ldots, z_{\tau_{k_n}}) \in K_n, \qquad \text{for all } n,$$

where $\tau_1, \ldots, \tau_{k_n}$ denote the elements of J_n. But then the mapping $z\colon I \to E$ defined by

$$z(t) \equiv \begin{cases} z_t, & \text{if } t \in \bigcup J_n \\ \text{an arbitrary point of } E, & \text{if } t \in I \setminus \bigcup J_n \end{cases}$$

is a point from E^I satisfying

$$p_{J_n}(z) = (z_{\tau_1}, \ldots, z_{\tau_{k_n}}) \in K_n,$$

that is, $z \in Z_n' = p_{J_n}^{-1}(K_n)$ for all n and thus $z \in \bigcap Z_n' \subset \bigcap Z_n$. Hence we have shown that the set $\bigcap Z_n$ is nonempty. ┘

12.1.4. Corollary. If E is a Polish space, $\mathfrak{B}$ is the σ-algebra of its Borel sets, and I is an arbitrary nonempty set, then for every projective family $(P_J)_{J\in\mathfrak{H}(I)}$ of probability measures on $(E^J,\mathfrak{B}^J)$ there exists a stochastic process with state space E and parameter set I such that $(P_J)_{J\in\mathfrak{H}(I)}$ is the family of its finite-dimensional distributions.

Proof. We choose $\Omega \equiv E^I$, $\mathfrak{A} \equiv \mathfrak{B}^I$, and $P \equiv P_I = \underleftarrow{\lim}\, P_J$. Then $(\Omega,\mathfrak{A},P)$ is a probability space. For every $t \in I$, the projection mapping $p_{\{t\}}$ with $p_{\{t\}}(\omega) = \omega(t)$, $\omega \in \Omega$, is an $\mathfrak{A}$-$\mathfrak{B}$-measurable mapping, and thus $X_t \equiv p_{\{t\}}$ is a random variable with values in E, that is, $(\Omega,\mathfrak{A},P,(X_t)_{t\in I})$ is a stochastic process with state space E and parameter set I. This process yields the desired result, since for every $J \in \mathfrak{H}(I)$ and every set $B \in \mathfrak{B}^J$

$$P(X_J^{-1}(B)) = P(p_J^{-1}(B)) = p_J(P_I)(B) = P_J(B). \quad ┘$$

The proof thus tells us that

$$(E^I,\mathfrak{B}^I, \underleftarrow{\lim}\, P_J, (p_{\{t\}})_{t\in I})$$

is a process with the desired properties. We often use it explicitly in constructions and therefore give it a special name: It is called the *canonical process*[3] associated with the projective family $(P_J)_{J\in\mathfrak{H}(I)}$. Since $p_{\{t\}}(\omega) = \omega(t)$ for $\omega \in E^I$, E^I is here equal to the set of all paths of the process. Thus *every* mapping $\omega\colon I \to E$ is a path of the canonical process.

Remark. 2. We still need to clarify the relationship of Theorem 12.1.3 to Theorem 5.4.2 on the existence of infinite products of probability measures. Let $(E,\mathfrak{B})$ be an arbitrary measurable space and P a probability measure on $\mathfrak{B}$. Then by (5.4.8), for an arbitrarily given set $I \neq \varnothing$, the following family $(P_J)_{J\in\mathfrak{H}(I)}$ is projective: If n is the number of elements

[3] The name "*first* canonical process" is also commonly used. See Meyer [39, p. 53].

of J, then let $P_J \equiv P \otimes \cdots \otimes P$ be the n-fold product of P with itself. Then, although E need not be Polish, by Theorem 5.4.2 there exists exactly one probability measure P_I on $(E^I,\mathfrak{B}^I)$ with $p_J(P_I) = P_J$ for all $J \in \mathfrak{H}(I)$.

Thus if the projective family (P_J) is of the special form above, that is, the associated stochastic process $(X_t)_{t\in I}$ is an independent family of random variables, then the measurable space $(E,\mathfrak{B})$ need not have any additional structure.

For an arbitrary projective family, however, we cannot get along without additional hypotheses of a topological type on the measurable space $(E,\mathfrak{H})$, as was shown by E. S. Andersen and B. Jessen [22]. See also Halmos [3, p. 214, Exercise 3].

PROBLEM

Prove that the assertions of Theorem 12.1.3 and Corollary 12.1.4 remain valid if E is a locally compact space countable at infinity and if $\mathfrak{B}$ is the σ-algebra $\mathfrak{B}_0(E)$ of its Baire sets.

12.2 PROCESSES WITH SPECIAL PATHS

We start with a stochastic process $(\Omega,\mathfrak{A},P,(X_t)_{t\in I})$ with a Polish space $(E,\mathfrak{B})$ as state space. Let $(P_J)_{J\in\mathfrak{H}(I)}$ be the projective family of its finite-dimensional distributions and $(E^I,\mathfrak{B}^I,P_I,(Y_t)_{t\in I})$ the canonical process associated with this family. Both processes $(X_t)_{t\in I}$ and $(Y_t)_{t\in I}$ thus have the same finite-dimensional distributions, but are different in this respect: The set $\tilde{\Omega}$ of all paths of (X_t) is in general a proper subset of the path set E^I of (Y_t). The latter is indeed the largest path set of a process with state space E and parameter set I.

Example

1. Let $\Omega = I \equiv \mathbf{R}_+$, $\mathfrak{B} \equiv \mathbf{R}_+ \cap \mathfrak{B}^1$, let P be a λ^1-continuous probability measure on Ω, and let $(\mathbf{R},\mathfrak{B}^1)$ be the state space of the two following stochastic processes $(X_t^*)_{t\in\mathbf{R}_+}$, $(X_t^{**})_{t\in\mathbf{R}_+}$:

$$X_t^*(\omega) \equiv 0, \qquad \text{for all } (t,\omega) \in \mathbf{R}_+ \times \Omega$$

$$X_t^{**}(\omega) \equiv \begin{cases} 0, & \omega \in \Omega \setminus \{t\} \\ 1, & \omega = t \end{cases} \qquad \text{for all } t \in \mathbf{R}_+.$$

We obviously have $X_t^* = X_t^{**}$ P-almost surely for all $t \in \mathbf{R}_+$. Therefore, in particular, both processes have the same finite-dimensional distributions: For every $J \in \mathfrak{H}(\mathbf{R}_+)$, P_J is the unit mass ϵ_0 at the point $0 = (0, \ldots ,0)$ of $\mathbf{R}^J$. (X_t^*) has only the constant path $t \to 0$. The various paths of (X_t^{**}) are given by the indicator functions on $\mathbf{R}_+$ of the one point subsets of $\mathbf{R}_+$, and thus correspond one-to-one to the points of $\mathbf{R}_+$. Hence, in this example, the only path of the first process, and no path of the second, is continuous.

Therefore it seems important to refine the question of the last section so that we ask not only about the existence of a process with given finite-dimensional distributions but also with given path set $\tilde{\Omega}$. For a more convenient formulation of this question and its answer we introduce the following definition.

12.2.1. Definition. Let $(E,\mathfrak{B})$ be a measurable space and $(P_J)_{J\in\mathfrak{H}(I)}$ a projective family of probability measures on $(E^J,\mathfrak{B}^J)$. A set $\tilde{\Omega} \subset E^I$ is said to be *essential* [relative to the family (P_J)] if there is a stochastic process, with state space E, parameter set I, and finite-dimensional distributions P_J, such that $\tilde{\Omega}$ is the set of its paths.

The construction of the canonical process contained in Corollary 12.1.4 says just that E^I is always essential. We now show the following result of J. L. Doob:

12.2.2. Theorem. A set $\tilde{\Omega} \subset E^I$ is essential relative to a projective family $(P_J)_{J\in\mathfrak{H}(I)}$ if and only if

$$P_I^*(\tilde{\Omega}) = 1 \tag{12.2.1}$$

holds for the outer measure P_I^* associated with $P_I \equiv \varprojlim P_J$.

Proof. We first assume that $\tilde{\Omega}$ is essential. Then there exists a process $(\Omega,\mathfrak{A},P,(X_t)_{t\in I})$ with state space E, finite-dimensional distributions P_J, and path set $\tilde{\Omega}$. If we again use the notation introduced after Definition 12.1.1, then we have $\tilde{\Omega} = X_I(\Omega)$ and $P_I = X_I(P)$ by (12.1.1) and (12.1.2). Here the mapping $X_I\colon \Omega \to E^I$ is $\mathfrak{A}$-$\mathfrak{B}$-measurable. Since $\mathfrak{B}^I$ is a σ-algebra, by (1.5.1) we have

$$P_I^*(\tilde{\Omega}) = \inf_{\substack{\tilde{\Omega}\subset Q\\ Q\in\mathfrak{B}^I}} P_I(Q).$$

Thus we need to show that $P_I(Q) = 1$ for all $Q \in \mathfrak{B}^I$ with $\tilde{\Omega} \subset Q$. But $\tilde{\Omega} \subset Q$ implies that $\Omega = X_I^{-1}(\tilde{\Omega}) \subset X_I^{-1}(Q) \subset \Omega$, that is, $\Omega = X_I^{-1}(Q)$, so

$$P_I(Q) = X_I(P)(Q) = P(X_I^{-1}(Q)) = P(\Omega) = 1.$$

Conversely, let $P_I^*(\tilde{\Omega}) = 1$. Then, as we shall show, the process

$$(\tilde{\Omega},\tilde{\Omega} \cap \mathfrak{B}^I,\tilde{P},(\tilde{X}_t)_{t\in I}) \tag{12.2.2}$$

solves our problem, where $\tilde{X}_t$ is the restriction of the projection $p_{\{t\}}$ to $\tilde{\Omega}$, that is,

$$\tilde{X}_t(\omega) = p_{\{t\}}(\omega) = \omega(t), \qquad \text{for all } \omega \in \tilde{\Omega}, \tag{12.2.3}$$

and where for every set $\tilde{\Omega} \cap Q \in \tilde{\Omega} \cap \mathfrak{B}^I$ with $Q \in \mathfrak{B}^I$, we define

$$\tilde{P}(\tilde{\Omega} \cap Q) \equiv P_I(Q). \tag{12.2.4}$$

Now it is clear that $(\tilde{\Omega}, \tilde{\Omega} \cap \mathfrak{B}^I)$ is a measurable space, that $(\tilde{X}_t)_{t \in I}$ is a family of measurable mappings of this measurable space into $(E, \mathfrak{B})$, and that by (12.2.3), the mappings $t \to \tilde{X}_t(\omega)$ of I into E for $\omega \in \tilde{\Omega}$ coincide with the mappings $\omega \in \tilde{\Omega} \subset E^I$. Thus we have a stochastic process with state space E, parameter set I, and path set $\tilde{\Omega}$ provided that $\tilde{P}$ is a probability measure. But this is obvious, provided the Definition 12.2.4 is independent of particular representations of the set $\tilde{\Omega} \cap Q$ by means of $Q \in \mathfrak{B}^I$. Thus if $\tilde{\Omega} \cap Q_1 = \tilde{\Omega} \cap Q_2$ for sets $Q_1, Q_2 \in \mathfrak{B}^I$, then we must show that $P_I(Q_1) = P_I(Q_2)$. We can assume without loss of generality that $Q_1 \subset Q_2$. [Otherwise, we note the relationships $Q_1 \subset Q_1 \cup Q_2$, $Q_2 \subset Q_1 \cup Q_2$, and $\tilde{\Omega} \cap (Q_1 \cup Q_2) = \tilde{\Omega} \cap Q_1 = \tilde{\Omega} \cup Q_2$.] But $Q_1 \subset Q_2$ implies that $\tilde{\Omega} \cap (Q_2 \setminus Q_1) = \emptyset$ and thus $\tilde{\Omega} \subset \complement(Q_2 \setminus Q_1)$; hence, by (12.2.1),

$$1 = P_I(\complement(Q_2 \setminus Q_1)) = 1 - P_I(Q_2 \setminus Q_1).$$

If we now take into account the equality $P_I(Q_2 \setminus Q_1) = P_I(Q_2) - P_I(Q_1)$, we obtain $P_I(Q_1) = P_I(Q_2)$. Thus it remains to be proved that (P_J) is the family of finite-dimensional distributions of the new process, that is, $P_J = \tilde{X}_J(\tilde{P})$ for all $J \in \mathfrak{H}(I)$. This follows from the equalities

$$\tilde{P}(\tilde{X}_J^{-1}(B)) = \tilde{P}(\tilde{\Omega} \cap p_J^{-1}(B)) = P_I(p_J^{-1}(B)) = p_J(P_I)(B) = P_J(B),$$

where B is an arbitrary set in $\mathfrak{B}^J$. We need only take (12.1.8) into account. ⌟

Example

2. Let $(E, \mathfrak{B})$ be an arbitrary measurable space and $I \neq \emptyset$ an arbitrary set. We choose $\omega_0 \in E^I$ arbitrarily. Then for the measures ϵ_{ω_0} $[\epsilon_{\omega_0(t)}, t \in I]$ on $\mathfrak{B}^I$ $[\mathfrak{B}]$ defined by the unit mass at ω_0 $[\omega_0(t)]$, we obviously have

$$\epsilon_{\omega_0} = \bigotimes_{t \in I} \epsilon_{\omega_0(t)} = \lim_{\substack{\longleftarrow \\ J \in \mathfrak{H}(I)}} \bigotimes_{t \in J} \epsilon_{\omega_0(t)}.$$

Hence a set $\tilde{\Omega} \subset E^I$ is essential relative to the projective family $(\bigotimes_{t \in J} \epsilon_{\omega_0(t)})_{J \in \mathfrak{H}(I)}$ if and only if ω_0 lies in $\tilde{\Omega}$.

An essential set $\tilde{\Omega} \subset E^I$ does not generally lie in the σ-algebra $\mathfrak{B}^I$. For example, for a topological space E, the set C of all continuous mappings of $\mathbf{R}_+$ into E is an essential subset of $E^{\mathbf{R}_+}$ relative to the projective family of probability measures which is obtained for some $\omega_0 \in C$ in the preceding example. But C does not generally lie in $\mathfrak{B}^{\mathbf{R}_+}$, as is shown by Corollary 12.2.4 following the lemma below. This lemma expresses the intrinsically interesting fact that every set from the σ-algebra $\mathfrak{B}^I$ is already determined by countably many parameters $t_1, t_2, \ldots \in I$.

12.2.3. Lemma. Let $(E,\mathfrak{B})$ be a measurable space and $I \neq \emptyset$ a set. Then for every set $B \in \mathfrak{B}^I$ there exists a countable set $S \subset I$ such that for every pair of elements $x \in E^I$ and $y \in B$, the following implication holds:

$$x(t) = y(t), \text{ for all } t \in S \Rightarrow x \in B. \tag{12.2.5}$$

Proof. Let $\mathfrak{B}^c$ denote the system of all sets $B \subset E^I$ for which there exists a countable set (depending on B) $S \subset I$ such that the implication (12.2.5) holds for every two elements $x \in E^I$, $y \in B$. Then $\mathfrak{B}^c$ is a σ-algebra in E^I. Of the properties (1.1.1)–(1.1.3) to be proved, we show the second; the proof of the others is obvious. Thus let $B \in \mathfrak{B}^c$ and let S be an appropriate countable subset of I. Then $\complement B$ lies in $\mathfrak{B}^c$. To see this, if $x \in E^I$ and $y \in \complement B$ are elements with $x(t) = y(t)$ for all $t \in S$, then x does not lie in B, and hence lies in $\complement B$. Otherwise $B \in \mathfrak{B}^c$, $x \in B$ would imply $y \in B$. For every set $J \in \mathfrak{H}(I)$ and every set $B_0 \in \mathfrak{B}^J$, the associated J-cylinder $p_J^{-1}(B_0)$ obviously lies in $\mathfrak{B}^c$ since $S = J$ has the desired properties. Hence,

$$\bigcup_{J \in \mathfrak{H}(I)} p_J^{-1}(\mathfrak{B}^J) \subset \mathfrak{B}^c.$$

Since we have a generator of B^I on the left-hand side, $B^I \subset \mathfrak{B}^c$ and thus the assertion follows. ⌟

12.2.4. Corollary. Let E be a Hausdorff space of at least two points, $\mathfrak{B}$ a σ-algebra in E, and $C \subset E^{\mathbf{R}_+}$ the set of all continuous mappings of $\mathbf{R}_+$ into E. Then C does not lie in $\mathfrak{B}^{\mathbf{R}_+}$.

Proof. By Lemma 12.2.3, $C \in \mathfrak{B}^{\mathbf{R}_+}$ implies the existence of a countable set $S \subset \mathbf{R}_+$ such that every mapping $x \in E^{\mathbf{R}_+}$ which coincides on S with a mapping $y \in C$ is continuous. By taking the union of S with the set of rational numbers from $\mathbf{R}_+$ if necessary, we can assume that S is dense in $\mathbf{R}_+$. If we choose a $t_0 \in \mathbf{R}_+ \setminus S$, then

$$y(t_0) = \lim_{\substack{t \to t_0 \\ t \in S}} y(t)$$

for every $y \in C$. Since E is Hausdorff and therefore limits in E are uniquely determined, every function $y \in C$ is uniquely determined by its values on S. Since E contains at least two points, for every $y \in C$ we can choose a mapping $x\colon \mathbf{R}_+ \to E$ such that $x(t) = y(t)$ for all $t \in S$ and $x(t_0) \neq y(t_0)$. Now this mapping x cannot be continuous. But this contradicts the choice of S. ⌟

We now generally seek conditions which imply the continuity of the paths of a process. The following two results of W. Hansen and M. Sieveking,[4] as well as a theorem of Kolmogorov-Prohorov, answer this question.

12.2.5. Lemma. Let $(\Omega,\mathfrak{A},P,(X_t)_{t\in\mathbf{R}_+})$ be a stochastic process with $\mathbf{R}^p$ as state space and $\mathbf{R}_+$ as parameter set. The condition below is then sufficient in order that every random variable X_t can be modified on a null set (depending on t) in such a way that all paths of the process $(\Omega,\mathfrak{A},P,(\tilde{X}_t)_{t\in\mathbf{R}_+})$ thus obtained are continuous. The condition is: There is a countable, dense subset S of $\mathbf{R}_+$ with the properties:
(a) For every pair of positive real numbers $\eta > 0$, $k > 0$,

$$\lim_{\delta\to 0} P\Big(\bigcup_{\substack{|s-t|\leqq\delta\\ s,\,t\in S\\ s,\,t\leqq k}} \{|X_s - X_t| \geqq \eta\}\Big) = 0. \tag{12.2.6}$$

(b) For every $t \in \mathbf{R}_+$, there is a sequence (s_n) in S, converging to t, such that for all $\eta > 0$,

$$\lim_{n\to\infty} P\{|X_{s_n} - X_t| \geqq \eta\} = 0.^5 \tag{12.2.7}$$

Proof. For every triple of positive real numbers δ, η, k, let

$$A(\delta,\eta,k) \equiv \bigcup_{\substack{|s-t|\leqq\delta\\ s,\,t\in S\\ s,\,t\leqq k}} \{|X_s - X_t| \geqq \eta\}$$

denote the event occurring in (12.2.6). If $\epsilon_0 > 0$ is arbitrarily given, then, by (12.2.6), for every $k \in \mathbf{N}$ and for $\eta = 1/k$ there is a $\delta_k > 0$ such that $P(A(\delta_k,k^{-1},k)) \leqq 2^{-k}\epsilon_0$. For

$$A_{\epsilon_0} \equiv \bigcup_{k=1}^{\infty} A(\delta_k,k^{-1},k),$$

we thus have $P(A_{\epsilon_0}) \leqq \epsilon_0$. Moreover, for every $\omega \in \Omega \setminus A_{\epsilon_0}$, every $k \in \mathbf{N}$,

[4] See H. Bauer, *Markov Processes*, Lecture Notes, Universtiy of Hamburg (1963).
[5] If we extend Definition 2.11.2 verbatim to measurable mappings with values in $\mathbf{R}^p$, then (12.2.7) says that (X_{s_n}) converges *stochastically* to X_t.

and every pair of elements $s, t \in S \cap [0,k]$ with $|s - t| \leqq \delta_k$, we have

$$|X_s(\omega) - X_t(\omega)| < \frac{1}{k}.$$

For each $\omega \in \complement A_\epsilon$, and thus in particular for every ω in

$$\Omega_0 \equiv \complement \bigcap_{n=1}^{\infty} A_{1/n},$$

the mapping $s \to X_s(\omega)$ is uniformly continuous on $S \cap [0,\alpha]$ for arbitrary $\alpha > 0$. Since the Euclidean space $\mathbf{R}^p$ is complete, for every $\omega \in \Omega_0$ there exists

$$\tilde{X}_t(\omega) \equiv \lim_{\substack{s \to t \\ s \in S}} X_s(\omega), \qquad \text{for arbitrary } t \in \mathbf{R}_+, \tag{12.2.8}$$

and, moreover, $t \to \tilde{X}_t(\omega)$ is then continuous on $\mathbf{R}_+$. Because $\complement \Omega_0 \subset A_{1/n}$ and

$$P(\complement \Omega_0) \leqq P(A_{1/n}) \leqq \frac{1}{n}$$

for all $n \in \mathbf{N}$, $\complement \Omega_0$ is a P-null set. For every ω from this set we define

$$\tilde{X}_t(\omega) \equiv a \qquad (t \in \mathbf{R}_+), \tag{12.2.9}$$

where $a \in \mathbf{R}^p$ is chosen arbitrarily. Then $(\Omega, \mathfrak{A}, P, (\tilde{X}_t)_{t \in \mathbf{R}_+})$ yields the desired result.

In fact, by (12.2.8) and (12.2.9), and due to the countability of S, each of the mappings $\tilde{X}_t : \Omega \to \mathbf{R}^p$ is the pointwise limit of a sequence of $(\mathbf{R}^p, \mathfrak{B}^p)$-random variables. Hence it follows (when we decompose $\tilde{X}_t$ into its p real components) that $\tilde{X}_t$ is also an $(\mathbf{R}^p, \mathfrak{B}^p)$-random variable. Thus $(\tilde{X}_t)_{t \in \mathbf{R}_+}$ is a stochastic process whose paths are all continuous by construction. Now we still have to show the almost sure equality of X_t and $\tilde{X}_t$ for every $t \in \mathbf{R}_+$. By (12.2.8) and (12.2.9) this is obvious for all $t \in S$. For arbitrary $t \in \mathbf{R}_+$, we need condition (b). Accordingly, there is a sequence (s_n) in S with $\lim s_n = t$ such that the sequence (X_{s_n}) converges stochastically to X_t. By (12.2.8) and (12.2.9), the sequence (X_{s_n}) converges almost surely, and thus, by Theorem 2.11.4 applied to the p component sequences, it converges stochastically to $\tilde{X}_t$. Since stochastic limits are almost surely uniquely determined, it follows that $\tilde{X}_t = X_t$ almost surely. ┘

Remarks. 1. It is easy to show that the condition involved in Lemma 12.2.5 is also necessary for the existence of a process

$$(\Omega, \mathfrak{A}, P, (\tilde{X}_t)_{t \in \mathbf{R}_+})$$

with the given properties. (See Problem 4.)

2. Lemma 12.2.5 remains valid when the state space E is a Polish space with complete metric d. Then we have to replace $|X_s - X_t|$ by $d(X_s,X_t)$. Further, we then have to extend some of the results of Section 2.11 on stochastic convergence to measurable mappings with values in E.

12.2.6. Theorem.[6] The following conditions (c) and (d) are sufficient for being able to obtain a process $(\Omega,\mathfrak{A},P,(\tilde{X}_t)_{t\in\mathbf{R}_+})$ with the properties described in Lemma 12.2.5 from $(\Omega,\mathfrak{A},P,(X_t)_{t\in\mathbf{R}_+})$; to formulate these conditions, we set

$$q_n(\eta) \equiv \sup_{i=0,\ldots,n2^n-1} P\{|X_{(i+1)2^{-n}} - X_{i2^{-n}}| \geqq \eta\} \qquad (12.2.10)$$

for arbitrary $\eta > 0$ and $n \in \mathbf{N}$.

(c) There exists a sequence $(\eta_n)_{n\in\mathbf{N}}$ of positive real numbers satisfying

$$\sum_{n=1}^{\infty} \eta_n < \infty \qquad \text{and} \qquad \sum_{n=1}^{\infty} n2^n q_n(\eta_n) < \infty. \qquad (12.2.11)$$

(d) For every $t \in \mathbf{R}_+$ there exists a sequence (s_n) of dyadic numbers $\geqq 0$ converging to t such that $(X_{s_n})_{n\in\mathbf{N}}$ converges stochastically to X_t [in the sense of (12.2.7)].

Proof. Let $S \equiv \{m2^{-n}: m, n = 0, 1, \ldots\}$ be the set of all nonnegative dyadic numbers. S is a countable, dense subset of $\mathbf{R}_+$. We shall show that condition (a) of Lemma 12.2.5 is satisfied with this S. Since obviously (b) is also satisfied, Lemma 12.2.5 yields the assertion.

Thus, let $\eta > 0$, $k > 0$, and $\epsilon > 0$ be given. Now we choose $n_0 \in \mathbf{N}$ so large that $n_0 > k$ and

$$\sum_{n=n_0}^{\infty} \eta_n < 2^{-1}\eta \qquad \text{and} \qquad \sum_{n=n_0}^{\infty} n2^n q_n(\eta_n) < \epsilon.$$

For the event

$$B \equiv \bigcup_{n=n_0}^{\infty} \bigcup_{i=0}^{n2^n-1} \{|X_{(j+1)2^{-n}} - X_{i2^{-n}}| \geqq \eta_n\},$$

we then have

$$P(B) \leqq \sum_{n=n_0}^{\infty} \sum_{i=0}^{n2^n-1} P\{|X_{(i+1)2^{-n}} - X_{i2^{-n}}| \geqq \eta_n\},$$

[6] See also Courrège [28] and Neveu [18].

that is, by (12.2.10),

$$P(B) \leqq \sum_{n=n_0}^{\infty} n2^n q_n(\eta_n) < \epsilon.$$

Now we consider elements $s, t \in S$ satisfying $s \leqq t \leqq k$ and $t - s \leqq 2^{-n_0}$. Then it will be shown below that one can decompose the interval $[s,t]$ into finitely many intervals of the form

$$J_{i,\nu} \equiv [i2^{-\nu},(i+1)2^{-\nu}]$$

with $i, \nu \in \{0,1, \ldots\}$ in such a way that every $\nu \in \{0,1, \ldots\}$ occurs in the representation of at most two intervals $J_{i,\nu}$ of this decomposition $\mathfrak{z}$. Since every such interval is of length $2^{-\nu} \leqq t - s \leqq 2^{-n_0}$, only natural numbers $\nu \geqq n_0$ are involved. Since $n_0 > k$, we also have

$$i + 1 \leqq t2^\nu \leqq k2^\nu < n_0 2^\nu \leqq \nu 2^\nu$$

for the numbers i and ν of each of the intervals $J_{i,\nu}$ appearing in the decomposition.

The existence of such a decomposition can be seen as follows: There obviously exists a uniquely determined interval $J_0 \equiv J_{i_0,\nu_0} \subset [s,t]$ of maximal length $2^{-\nu_0}$. If $x_0 < y_0$ are the end points of J_0 and $s < x_0$, then we determine an interval $J_{i_1',\nu_1'} \subset [s,t]$ with x_0 as right end point and maximal length $2^{-\nu_1'}$. If the left end point x_1' of $J_{i_1',\nu_1'}$ is still $> s$, we repeat the second step of the construction with x_1' in place of x_0. After finitely many steps we have covered $[s,y_0]$ with $J_0, J_{i_1',\nu_1'}, \ldots, J_{i_p',\nu_p'}$. Here obviously $\nu_0 < \nu_1' < \cdots < \nu_p'$. Analogously, starting with y_0, we obtain finitely many intervals $J_{i_1'',\nu_1''}, \ldots, J_{i_q'',\nu_q''}$, with $\nu_1'' < \cdots < \nu_q''$, which cover $[y_0,t]$. The decomposition of $[s,t]$ by $J_0, J_{i_1',\nu_1'}, \ldots, J_{i_p',\nu_p'}, J_{i_1'',\nu_1''}, \ldots, J_{i_q'',\nu_q''}$ thus yields the desired decomposition.

It now follows by the triangle inequality that

$$|X_s - X_t| \leqq \sum_{i,\nu} |X_{(i+1)2^{-\nu}} - X_{i2^{-\nu}}|,$$

if we sum over all pairs (i,ν) which correspond to an interval $J_{i,\nu}$ appearing in the decomposition $\mathfrak{z}$. Therefore, for every $\omega \in \complement B$,

$$|X_s(\omega) - X_t(\omega)| \leqq \sum_{i,\nu} |X_{(i+1)2^{-\nu}}(\omega) - X_{i2^{-\nu}}(\omega)| < \sum_{(i,\nu)} \eta_\nu$$

if we note that $\nu \geqq n_0$ and $i \leqq \nu 2^\nu - 1$ hold for all admissible (i,ν). Since, furthermore, every $\nu \geqq n_0$ occurs at most twice in an admissible (i,ν), we finally have

$$|X_s(\omega) - X_t(\omega)| < 2 \sum_{\nu=n_0}^{\infty} \eta_\nu < \eta.$$

Hence,

$$A \equiv \bigcup_{\substack{|s-t|\leqq 2^{-n_0} \\ s,t\in S;s,t\leqq k}} \{|X_s - X_t| \geqq \eta\} \subset B,$$

and $P(A) \leqq P(B) < \epsilon$. Now, condition (a) of Lemma 12.2.5 follows. ┘

12.2.7. Corollary (Kolmogorov-Prohorov Theorem).[7] Condition (e) is sufficient for the existence of a process $(\Omega,\mathfrak{A},P,(\tilde{X}_t)_{t\in\mathbf{R}_+})$ derived from $(\Omega,\mathfrak{A},P,(X_t)_{t\in\mathbf{R}_+})$ with the properties described in Lemma 12.2.5. (e) There are real numbers $a > 0$, $b > 1$, $c > 0$ such that

$$E(|X_s - X_t|^a) \leqq c \cdot |s - t|^b \tag{12.2.12}$$

holds for all s, $t \in \mathbf{R}_+$.

Proof. By the Chebyshev-Markov inequality (2.11.1) and by (12.2.12) we have, for every $\eta > 0$,

$$P\{|X_s - X_t| \geqq \eta\} \leqq \eta^{-a}E(|X_s - X_t|^a) \leqq c\eta^{-a}|s - t|^b.$$

Therefore by (12.2.10), for all $n \in \mathbf{N}$,

$$q_n(\eta) \leqq c\eta^{-a}2^{-nb}.$$

Since $b > 1$, we can choose $\delta > 0$ such that $b - a\delta > 1$. But then, for the sequence of positive numbers $\eta_n \equiv 2^{-n\delta}$, $n = 1, 2, \ldots$, the series $\Sigma_{n=1}^{\infty}\eta_n$ and $\Sigma_{n=1}^{\infty}n2^nq_n(\eta_n)$ are convergent. Indeed, the first is a geometric series; the second has as majorant the series $c\Sigma_{n=1}^{\infty}n2^{-n(b-a\delta-1)}$, which is convergent by the ratio test. Therefore condition (c) of Theorem 12.2.6 is satisfied. Moreover, for every convergent sequence (s_n) of real numbers $\geqq 0$ with $t \equiv \lim s_n$ we also have, for arbitrary $\eta > 0$,

$$P\{|X_{s_n} - X_t| \geqq \eta\} \leqq c\eta^{-a}|s_n - t|^b.$$

Hence (X_{s_n}) converges stochastically to X_t, that is, condition (d) is also satisfied. Thus the assertion follows from Theorem 12.2.6. ┘

Finally, we show the connection with the question posed at the beginning of this section.

12.2.8. Lemma. Let $(\Omega,\mathfrak{A},P,(X_t)_{t\in\mathbf{R}_+})$ be a canonical process, with state space $\mathbf{R}^p$, from which we can derive a process $(\Omega,\mathfrak{A},P,(\tilde{X}_t)_{t\in\mathbf{R}_+})$ with the properties described in Lemma 12.2.5. Then the set C of all continuous mappings of $\mathbf{R}_+$ into $\mathbf{R}^p$ is essential relative to the family of finite-dimensional distributions of $(X_t)_{t\in\mathbf{R}_+}$.

[7] See Prohorov [41].

Proof. Since the original process is canonical, $\Omega = (\mathbf{R}^p)^{\mathbf{R}_+}$, $\mathfrak{A} = (\mathfrak{B}^p)^{\mathbf{R}_+}$, and $X_t(\omega) = \omega(t)$ for all $(\omega,t) \in \Omega \times \mathbf{R}_+$. By Theorem 12.2.2, we must show that $P(N) = 0$ for every set $N \in \mathfrak{A}$ which is disjoint from C. By Lemma 12.2.3, for N there is a countable set $S \subset \mathbf{R}_+$ such that $\omega \in \complement N$, $\tilde{\omega} \in \Omega$, and $\omega(t) = \tilde{\omega}(t)$ for all $t \in S$ imply $\tilde{\omega} \in \complement N$. According to the assumed connection between the processes (X_t) and $(\tilde{X}_t)$, for every $t \in \mathbf{R}_+$ there is a P-null set $N_t \in \mathfrak{A}$ such that $X_t(\omega) = \tilde{X}_t(\omega)$ for all $\omega \in \complement N_t$. Moreover, $t \to \tilde{X}_t(\omega)$ is an element $\tilde{\omega} \in C$ for every $\omega \in \Omega$. After these preliminaries we conclude as follows: $N_0 \equiv \bigcup_{t \in S} N_t$ is a P-null set. For the $\tilde{\omega}$ associated with each $\omega \in \complement N_0$, we have $\omega(t) = \tilde{\omega}(t)$ for arbitrary $t \in S$; $\tilde{\omega}$ lies in C and thus in $\complement N$. By the choice of S we now have $\omega \in \complement N$. Hence $\complement N_0 \subset \complement N$, and thus we have shown that $N \subset N_0$ and $P(N) = P(N_0) = 0$. ┘

We shall discuss applications of this result in Section 12.6.

PROBLEMS

1. Let Ω be a set, $(\mathfrak{A}_i)_{i \in I}$ a family of σ-algebras in Ω, and $\mathfrak{A}_0$ the σ-algebra in Ω generated by $\bigcup_{i \in I} \mathfrak{A}_i$. Prove: For every $A \in \mathfrak{A}_0$ there is a countable subset $S \subset I$ such that A is an element of the σ-algebra in Ω generated by $\bigcup_{i \in S} \mathfrak{A}_i$.
2. Let E be a topological space and $\mathfrak{B} = \mathfrak{B}(E)$ the σ-algebra of its Borel sets. Then for every set I, the product space E^I and the σ-algebra $\mathfrak{B}(E^I)$ of its Borel sets are defined. Prove:
   (a) $\mathfrak{B}^I \subset \mathfrak{B}(E^I)$.
   (b) $\mathfrak{B}^I = \mathfrak{B}(E^I)$ for every countable set I.
   (c) For uncountable I, $\mathfrak{B}^I$ and $\mathfrak{B}(E^I)$ do not coincide in general. There may exist even open sets G in E^I such that $G \notin \mathfrak{B}^I$.
3. Let $(\Omega,\mathfrak{A},P,(X_t)_{t \in I})$ be a stochastic process with state space $(E,\mathfrak{B})$. Prove: For each $B \in \mathfrak{B}$ there exists a countable set $S \subset I$ such that the event

$$A_t \equiv \{X_t \notin B\} \cap \bigcap_{s \in S} \{X_s \in B\}$$

has probability zero for all $t \in I$. [*Hint:* Choose $s_1 \in I$ arbitrary. Suppose that $s_1, \ldots, s_n \in I$ are chosen. Define

$$\sigma_n = \sup_{t \in I} P\{X_{s_1} \in B, \ldots, X_{s_n} \in B, X_t \in \complement B\}$$

and choose $s_{n+1} \in I$ such that

$$P\{X_{s_1} \in B, \ . \ . \ . \ , X_{s_n} \in B, X_{s_{n+1}} \in \complement B\} \geqq (1 - 1/n)\sigma_n.$$

Show that $S \equiv \{s_n: n \in \mathbf{N}\}$ has all required properties.]

4. Prove that conditions (a) and (b) of Lemma 12.2.5 are also necessary for the existence of a process $(\Omega,\mathfrak{A},P, (\tilde{X}_t)_{t\in I})$ with the properties mentioned in Lemma 12.2.5. Show that S may be any countable dense subset of $\mathbf{R}_+$. [*Hint:* Note that every continuous path $t \to \tilde{X}_t(\omega)$ is uniformly continuous on each compact interval $[0,k]$.]
5. Decide whether there exists a stochastic process $(X_t)_{t\in I}$ with $\mathbf{R}$ as state space such that the family $(X_t)_{t\in I}$ is independent, all X_t have the same distribution, and all paths of the process are continuous.

12.3 MARKOV SEMIGROUPS

We now study an important method for constructing special projective families of probability measures and thus for constructing stochastic processes. We present a brief consideration of kernels as a preliminary.

Let $(\Omega_i,\mathfrak{A}_i)$, $i = 1, 2, 3$ be three measurable spaces. Let K_i, for $i = 1, 2$ be a kernel from $(\Omega_i,\mathfrak{A}_i)$ to $(\Omega_{i+1},\mathfrak{A}_{i+1})$. By (10.3.6) we can consider K_i as an additive, positive-homogeneous, Daniell-continuous mapping of $\mathcal{E}^*_{\mathfrak{A}_{i+1}}$ into $\mathcal{E}^*_{\mathfrak{A}_i}$. Here $\mathcal{E}^*_{\mathfrak{A}_i}$ again denotes the set of $\mathfrak{A}_i$-measurable numerical functions $\geqq 0$ on Ω_i. Thus we can consider the composed mapping $K_1 \circ K_2$: $E^*_{\mathfrak{A}_3} \to E^*_{\mathfrak{A}_1}$. This is obviously additive, positive-homogeneous, and Daniell-continuous, and thus, by Lemma 10.3.2, is a kernel K_3 from $(\Omega_1,\mathfrak{A}_1)$ to $(\Omega_3,\mathfrak{A}_3)$. We call K_3 the *composition* of K_1 with K_2, and we usually write K_1K_2 rather than $K_1 \circ K_2$. By definition, for every function $f \in \mathcal{E}^*_{\mathfrak{A}_3}$,

$$\begin{aligned}[K_1K_2f](\omega_1) &= [K_1(K_2f)](\omega_1) = \int K_1(\omega_1,d\omega_2)\int K_2(\omega_2,d\omega_3)f(\omega_3) \qquad (12.3.1)\\ &= \int\int K_1(\omega_1,d\omega_2)K_2(\omega_2,d\omega_3)f(\omega_3),\end{aligned}$$

for every $\omega_1 \in \Omega_1$. By choosing an indicator function $f = 1_{A_3}$, we then obtain the kernel $K_3 = K_1K_2$ in the form

$$[K_1K_2](\omega_1,A_3) = \int K_1(\omega_1,d\omega_2)K_2(\omega_2,A_3), \qquad (12.3.2)$$

for arbitrary $(\omega_1,A_3) \in \Omega_1 \times \mathfrak{A}_3$. From (12.3.2) we can see that K_1K_2 is *sub-Markov* (*Markov*) if the kernels K_1 and K_2 are sub-Markov (Markov). In both of these cases we can interpret K_1, K_2, and K_1K_2 as positive linear mappings K_1: $\mathcal{E}^b_{\mathfrak{A}_2} \to \mathcal{E}^b_{\mathfrak{A}_1}$, K_2: $\mathcal{E}^b_{\mathfrak{A}_3} \to \mathcal{E}^b_{\mathfrak{A}_2}$, K_1K_2: $\mathcal{E}^b_{\mathfrak{A}_3} \to \mathcal{E}^b_{\mathfrak{A}_1}$ of the vector spaces $\mathcal{E}^b_{\mathfrak{A}_i}$ of bounded, real, $\mathfrak{A}_i$-measurable functions on Ω_i. K_1K_2 is then again simply the composition of the first mapping with the second.

Now the following definition is meaningful:

12.3.1. Definition. Let $(P_t)_{t \in \mathbf{R}_+}$ be a family of kernels on a measurable space $(E,\mathfrak{B})$ indexed by $\mathbf{R}_+$.[8] If

$$P_{s+t} = P_s P_t, \qquad \text{for all } s,\, t \in \mathbf{R}_+, \tag{12.3.3}$$

then we call $(P_t)_{t \in \mathbf{R}_+}$ a *semigroup of kernels* on E. If in addition all kernels P_t are sub-Markov (Markov), then the semigroup is said to be *sub-Markov* (*Markov*).

Equations (12.3.3) are known as the *Chapman-Kolmogorov equations*. By (12.3.2) they say, in more detail, that for arbitrary $(x,B) \in E \times \mathfrak{B}$ and $s,\, t \in \mathbf{R}_+$,

$$P_{s+t}(x,B) = \int P_s(x,dy) P_t(y,B). \tag{12.3.3$'$}$$

Since $P_{s+t} = P_s P_t$ and $P_{t+s} = P_t P_s$ by (12.3.3), it follows that $P_s P_t = P_t P_s$, that is, every semigroup of kernels is *commutative*.

In applications we often encounter families $(P_t)_{t>0}$ of kernels such that $P_{s+t} = P_s P_t$ for all $s > 0$, $t > 0$. If we enlarge such a family by the unit kernel $P_0 \equiv I$ on E (see Section 10.3, Example 4), we obtain a semigroup of kernels in the above sense.

We now give a probabilistic interpretation of the new concept. Our point of view here is a naive one which, however, will be formally justified below. We refer back to the interpretation of a kernel already developed in Section 10.3, Example 2. Thus, let $(P_t)_{t \in \mathbf{R}_+}$ be a family of Markov kernels on $(E,\mathfrak{B})$ with $P_0 = I$. Then we can interpret $P_t(x,B)$ as the probability that a particle randomly moving in E, which starts at $x \in E$ at time zero, finds itself in the set $B \in \mathfrak{B}$ at time $t \in \mathbf{R}_+$. (This interpretation is also justified for $t = 0$ since $P_0 = I$.) Such a particle will have no "memory"; its further movement is influenced only by chance and not by the "experiences" at times $<t$. Thus if the particle starting at x finds itself in the "volume element" dy at time s, then we compute the probability with which we find it in a set $B \in \mathfrak{B}$ at time $s + t$ as

$$P_{s+t}(x,B) = \int P_s(x,dy) P_t(y,B).$$

But this is just the requirement $P_{s+t} = P_s P_t$. Thus our interpretation automatically leads us to the semigroup property as long as the moving particle has no memory.

Now we finally proceed to examples of semigroups of kernels.

Examples

1. Let $(E,\mathfrak{B})$ be an arbitrary measurable space and $P_t \equiv$ I the unit kernel on E for all $t \in \mathbf{R}_+$. Then $(P_t)_{t \in \mathbf{R}_+}$ is a Markov semigroup.

[8] Thus every P_t is a kernel from $(E,\mathfrak{B})$ to $(E,\mathfrak{B})$.

In our interpretation, it describes a particle that starts at $x \in E$ and never leaves this point.

2. Let $(P_t)_{t\in\mathbf{R}_+}$ be a sub-Markov semigroup on a measurable space $(E,\mathfrak{B})$. Then for every $\lambda \in \mathbf{R}_+$, $(e^{-\lambda t}P_t)_{t\in\mathbf{R}_+}$ is also a sub-Markov semigroup. Here of course $e^{-\lambda t}P_t$ is the kernel

$$(x,B) \to e^{-\lambda t}P_t(x,B).$$

3. Let $(P_t)_{t\in\mathbf{R}_+}$ be a Markov semigroup of kernels on $(\mathbf{R}^p,\mathfrak{B}^p)$. It is said to be *translation-invariant* (or *spatially homogeneous*) if

$$P_t(x,B) = P_t(x + z,\ B + z)$$

holds for all $(x,B) \in \mathbf{R}^p \times \mathfrak{B}^p$, $t \in \mathbf{R}_+$ and all $z \in \mathbf{R}^p$.

Then we define

$$\mu_t(B) \equiv P_t(0,B)$$

and thus obtain a family $(\mu_t)_{t\in\mathbf{R}_+}$ of probability measures on $\mathfrak{B}^p$ such that $P_t(x,B) = \mu_t(B - x)$. The semigroup property for (P_t) implies that

$$\begin{aligned}\mu_{s+t}(B) &= P_{s+t}(0,B) = \int P_s(0,dy)P_t(y,B)\\ &= \int \mu_s(dy)\mu_t(B - y) = (\mu_s * \mu_t)(B),\end{aligned}$$

that is,

$$\mu_{s+t} = \mu_s * \mu_t, \qquad \text{for all } s,\ t \in \mathbf{R}_+ \tag{12.3.4}$$

thus $(\mu_t)_{t\in\mathbf{R}_+}$ is a *convolution semigroup of* probability *measures* on $\mathbf{R}^p$. Conversely if we are given such a convolution semigroup $(\mu_t)_{t\in\mathbf{R}_+}$ and we require the measurability of $x \to \mu_t(B - x)$ for all $t \in \mathbf{R}_+$ and $B \in \mathfrak{B}^p$, then

$$P_t(x,B) \equiv \mu_t(B - x) \tag{12.3.5}$$

yields a translation-invariant Markov semigroup $(P_t)_{t\in\mathbf{R}_+}$ of kernels on $(\mathbf{R}^p,\mathfrak{B}^p)$.

In this special case it is also of interest to determine P_tf for Borel-measurable functions $f \geqq 0$. By definition, for every set $B \in \mathfrak{B}^p$ and every $x \in \mathbf{R}^p$,

$$\begin{aligned}(P_t1_B)(x) &= P_t(x,B) = \mu_t(B - x) = \int 1_{B-x}(y)\mu_t(dy)\\ &= \int 1_B(x + y)\mu_t(dy) = \int 1_B(x - y)\tilde{\mu}_t(dy).\end{aligned}$$

Here $\tilde{\mu}_t$ denotes the image measure $S(\mu_t)$ relative to the reflection $S(x) = -x$ through the origin. When we approximate f by an isotone sequence of elementary functions, we obtain $(P_tf)(x) = \int f(x - y)\tilde{\mu}_t(dy)$. Hence

$$P_tf = f * \tilde{\mu}_t. \tag{12.3.6}$$

In this case, therefore, we call P_t a *convolution kernel*.

Further, we are sure to have measurability of $x \to \mu_t(B - x)$ for a given $t \in \mathbf{R}_+$ if μ_t has a density q_t relative to the L-B-measure λ^p, that is, if $\mu_t = q_t\lambda^p$. Since

$$\mu_t(B - x) = \int_{B-x} q_t \, d\lambda^p = \int_B q_t(y - x)\lambda^p(dy)$$

and due to the $\mathfrak{B}^p \otimes \mathfrak{B}^p$-measurability of $(x,y) \to q_t(y - x)$, this follows from Theorem 3.2.6. For every Borel-measurable function $f \geqq 0$, we then write P_tf in the form

$$P_tf = f * \tilde{q}_t, \tag{12.3.7}$$

where $\tilde{q}_t(x) \equiv q_t(-x)$ for all $x \in \mathbf{R}_+$.

4. Let $(\mu_t)_{t\in\mathbf{R}_+}$ be the Brownian convolution semigroup, defined in Section 9.3, Example 2, on $\mathbf{R}^p$. For every $t > 0$, $\mu_t = \nu_{0,t} \otimes \cdots \otimes \nu_{0,t}$ with p factors; thus,[9] μ_t has the following density g_t relative to the L-B-measure λ^p:

$$g_t(x) = \prod_{i=1}^{p} \left(\frac{1}{2\pi t}\right)^{1/2} e^{-x_i^2/2t} = \left(\frac{1}{2\pi t}\right)^{p/2} e^{-|x|^2/2t}. \tag{12.3.8}$$

Thus, by Example 3, we associate with (μ_t) a translation-invariant Markov semigroup $(P_t)_{t\in\mathbf{R}_+}$ of kernels on $\mathbf{R}^p$. This semigroup is called the *Brownian semigroup* or semigroup of Brownian motion in $\mathbf{R}^p$. (Note that $\tilde{g}_t = g_t$ for all $t > 0$.)

5. On the real line $\mathbf{R}$, let $(\pi_t)_{t\in\mathbf{R}_+}$ be the Poisson convolution semigroup on $\mathbf{R}$ of Section 9.3, Example 3. For every Borel set $B \in \mathfrak{B}^1$ and every $x \in \mathbf{R}$,

$$\pi_t(B - x) = \sum_{\substack{y \geqq 0 \\ x+y\in B}} e^{-t}\frac{t^y}{y!} = \sum_{y=0}^{\infty} e^{-t}\frac{t^y}{y!} 1_{B-y}(x)$$

and hence $x \to \pi_t(B - x)$ is Borel-measurable. Thus, by Example 3, (π_t) is associated with a Markov semigroup $(P_t)_{t\in\mathbf{R}_+}$ of kernels on $\mathbf{R}$. Since $\pi_t(\mathbf{N} \cup \{0\}) = 1$,

$$P_t(x,\mathbf{Z}) = \pi_t(\mathbf{Z} - x) = \pi_t(\mathbf{Z}) = 1$$

holds, for the set $\mathbf{Z}$ of all integers, for all $x \in \mathbf{Z}$, and $t \geqq 0$. Therefore $(P_t)_{t\in\mathbf{R}_+}$ is a Markov semigroup of kernels on $(\mathbf{Z},\mathfrak{P}(\mathbf{Z}))$, since $\mathbf{Z} \cap \mathfrak{B}^1 = \mathfrak{P}(\mathbf{Z})$. We call this Markov semigroup on $\mathbf{Z}$ the *Poisson semigroup* on $\mathbf{Z}$.

[9] If $\sigma_1 = s_1\tau_1$ and $\sigma_2 = s_2\tau_2$ (with measures σ_i, τ_i, and densities s_i), then $(x_1,x_2) \to s_1(x_1)s_2(x_2)$ is the density of $\sigma_1 \otimes \sigma_2$ relative to $\tau_1 \otimes \tau_2$.

6. According to the remark at the end of Section 9.4, for every $\alpha \in]0,2]$ there exists exactly one Lebesgue continuous measure $\mu_\alpha \in \mathfrak{M}^1(\mathbf{R})$ with Fourier transform

$$\hat{\mu}_\alpha(x) = e^{-|x|^\alpha} \qquad (x \in \mathbf{R}).$$

Then for arbitrarily given $c > 0$ and $t > 0$ there is [by (8.1.10)] exactly one measure $\nu_t \in \mathfrak{M}^1(\mathbf{R})$ with Fourier transform

$$\hat{\nu}_t(x) = e^{-ct|x|^\alpha} \qquad (x \in \mathbf{R}).$$

If we now set $\nu_0 = \epsilon_0$, then $(\nu_t)_{t \in \mathbf{R}_+}$ is obviously a convolution semigroup of probability measures which depends on the parameters $\alpha \in]0,2]$ and $c > 0$. The translation-invariant Markov semigroup $(P_t)_{t \in \mathbf{R}_+}$ on $\mathbf{R}$ associated with the above convolution semigroup according to Example 3 is called the *stable symmetric semigroup of order* α (with parameter $c > 0$). For $\alpha = 2$ and $c = \frac{1}{2}$ we obtain the Brownian semigroup on $\mathbf{R}$. When $\alpha = c = 1$, ν_t is the Cauchy distribution γ_t for every $t > 0$. Therefore in this case we call (P_t) the *Cauchy semigroup* on $\mathbf{R}$. Analogs of these semigroups can also be defined in higher dimensions.

7. In a Markov semigroup $(P_t)_{t \in \mathbf{R}_+}$, P_0 is generally *not* the *unit kernel*. In fact, if $(E,\mathfrak{B},\mu)$ is a probability space, then let P_t be the kernel $P_t(x,B) \equiv \mu(B)$ $(x \in E,\ B \in \mathfrak{B},\ t \in \mathbf{R}_+)$ independent of $x \in E$. Then (P_t) is obviously a Markov semigroup and $P_0 \neq I$ provided E does not consist of only one point.

We now show that we can arrive at a projective family of probability measures via a Markov semigroup $(P_t)_{t \in \mathbf{R}_+}$ of kernels on a measurable space $(E,\mathfrak{B})$. The idea for this is suggested by the "randomly moving particle without memory." Let $t_1 < \cdots < t_n$ be n time points and $B_1, \ldots, B_n$ sets from $\mathfrak{B}$. If the particle starts at x_0, then we use the formula

$$\int_{B_1} \cdots \int_{B_{n-1}} \int_{B_n} P_{t_n - t_{n-1}}(x_{n-1},dx_n) P_{t_{n-1}-t_{n-2}}(x_{n-2},dx_{n-1}) \cdots P_{t_1}(x_0,dx_1)$$

to compute the probability that the particle is found successively in $B_1, \ldots, B_n$ at times $t_1, \ldots, t_n$. If the starting point x_0 is chosen at random via a *starting probability*, that is, a probability measure μ on $\mathfrak{B}$, then we have another integration to perform, namely, we have to apply the following formula:

$$\int_E \int_{B_1} \cdots \int_{B_{n-1}} \int_{B_n} P_{t_n-t_{n-1}}(x_{n-1},dx_n) P_{t_{n-1}-t_{n-2}}(x_{n-2},dx_{n-1}) \cdots P_{t_1}(x_0,dx_1)\mu(dx_0) =$$

$$\iint \cdots \int 1_{B_1 \times \cdots \times B_n}(x_1, \ldots, x_n) P_{t_n-t_{n-1}}(x_{n-1},dx_n) \cdots P_{t_1}(x_0,dx_1)\mu(dx_0).$$

After these preliminaries, we now assert:

12.3.2. Theorem. On a measurable space $(E,\mathfrak{B})$, let $(P_t)_{t\in\mathbf{R}_+}$ be a Markov semigroup and μ be a probability measure. For every set $J = \{t_1, \ldots ,t_n\} \in \mathfrak{H}(\mathbf{R}_+)$ with elements $t_1 < t_2 < \cdots < t_n$ and every $B \in \mathfrak{B}^J$ we define

$$P_J(B) \equiv \iint \cdots \int 1_B(x_1, \ldots ,x_n)P_{t_n-t_{n-1}}(x_{n-1},dx_n) \cdots P_{t_1}(x_0,dx_1)\mu(dx_0). \tag{12.3.9}$$

Then $(P_J)_{J\in\mathfrak{H}(\mathbf{R}_+)}$ is a projective family of probability measures on $(E^J,\mathfrak{B}^J)$.

Proof. Every P_J is a measure on $\mathfrak{B}^J$. We need only consider a sequence $(B_j)_{j=1,2,\ldots}$ of pairwise disjoint sets from $\mathfrak{B}^J$ and integrate successively the equality $1_B = \sum_{j=1}^{\infty} 1_{B_j}$. P_J is a probability measure since obviously

$$P_J(E^J) = \int P_{t_1} \circ P_{t_2-t_1} \circ \cdots \circ P_{t_n-t_{n-1}} 1\, d\mu = \int 1\, d\mu = 1.$$

Thus we need show only the projectivity of the family (P_J). In view of the remark after Definition 12.1.2, we need to show that $p_J^H(P_H) = P_J$ for any two sets $J, H \in \mathfrak{H}(\mathbf{R}_+)$ for which $J \subset H$ and $H \setminus J$ is a singleton. Let $t_1 < \cdots < t_n$ be the elements of J and $H \setminus J = \{t'\}$. We have to show that $p_J^H(P_H) = P_J$, that is,

$$P_H((p_J^H)^{-1}(C)) = P_J(C)$$

for all $C \in \mathfrak{B}^J$. Since the sets $B_1 \times \cdots \times B_n$ with $B_1, \ldots , B_n \in \mathfrak{B}$ form an $\cap$-stable generator of $\mathfrak{B}^J$, then by Theorem 1.5.5 it suffices to verify this equality only for sets $C \equiv B_1 \times \cdots \times B_n$ of this generator.

Now we assume that $t_i < t' < t_{i+1}$ for some $i = 1, \ldots , n-1$. But then

$$\begin{aligned} P_H((p_J^H)^{-1}(C)) &= P_H(B_1 \times \cdots \times B_i \times E \times B_{i+1} \times \cdots \times B_n) \\ &= \int_E \int_{B_1} \cdots \int_E \int_{B_{i+1}} \cdots \int_{B_n} P_{t_n-t_{n-1}}(x_{n-1}, dx_n) \\ &\qquad \cdots P_{t_{i+1}-t'}(x',dx_{i+1})P_{t'-t_i}(x_i,dx') \cdots \mu(dx_0) \\ &= \int_E \int_{B_1} \cdots \int_E \int_{B_{i+1}} f(x_{i+1})P_{t_{i+1}-t'}(x',dx_{i+1})P_{t'-t_i}(x_i,dx') \cdots \mu(dx_0), \end{aligned} \tag{12.3.10}$$

where

$$f(x_{i+1}) \equiv \begin{cases} 1, & i = n-1 \\ \displaystyle\int_{B_{i+2}} \cdots \int_{B_n} P_{t_n-t_{n-1}}(x_{n-1},dx_n) \cdots P_{t_{i+2}-t_{i+1}}(x_{i+1},dx_{i+2}), & i < n-1. \end{cases}$$

According to our observations concerning (10.3.4) f is a $\mathfrak{B}$-measurable function $\geqq 0$ on E. As a consequence of the semigroup property,

$$\begin{aligned}\int_E \int_{B_{i+1}} f(x_{i+1}) P_{t_{i+1}-t'}(x',dx_{i+1}) P_{t'-t_i}(x_i,dx') \\ = P_{t'-t_i} P_{t_{i+1}-t'}(1_{B_{i+1}}f)(x_i) \\ = P_{t_{i+1}-t_i}(1_{B_{i+1}}f)(x_i) = \int_{B_{i+1}} f(x_{i+1}) P_{t_{i+1}-t_i}(x_i,dx_{i+1}).\end{aligned} \tag{12.3.11}$$

From (12.3.10) and (12.3.11) we obtain

$$\begin{aligned}P_H((p_J^H)^{-1}(C)) \\ = \int_E \int_{B_1} \cdots \int_{B_{i+1}} f(x_{i+1}) P_{t_{i+1}-t_i}(x_i,dx_{i+1}) P_{t_i-t_{i-1}}(x_{i-1},dx_i) \cdots \mu(dx_0) \\ = \int_E \int_{B_1} \cdots \int_{B_n} P_{t_n-t_{n-1}}(x_{n-1},dx_n) \cdots P_{t_1}(x_0,dx_1)\mu(dx_0).\end{aligned}$$

But this is $P_J(C)$ by definition.

The case $t' < t_1$ can be handled in the same way. The case $t_n < t'$ is trivial. ⌟

In particular, if E is a Polish space and $\mathfrak{B}$ is the σ-algebra of its Borel sets, then Corollary 12.1.4 tells us that the projective family (P_J) just constructed is the family of finite-dimensional distributions of a stochastic process with state space E. For a given Markov semigroup $(P_t)_{t\in\mathbf{R}_+}$, the family (P_J) now depends only on the starting probability μ on $\mathfrak{B}$. If we choose the canonical process belonging to (P_J), then this is a process of the form $(\Omega,\mathfrak{A},P^\mu,(X_t)_{t\in\mathbf{R}_+})$, where only the probability measure P^μ depends on μ (and $\Omega = E^{\mathbf{R}_+}$, $\mathfrak{A} = \mathfrak{B}^{\mathbf{R}_+}$, $X_t = p_{\{t\}}$). Then for every singleton $J = \{t\} \subset \mathbf{R}_+$,

$$P_{\{t\}}(B) = \int P_t(x_0,B)\mu(dx_0) = P^\mu\{X_t \in B\}, \tag{12.3.12}$$

for all $B \in \mathfrak{B}$. In particular, for $\mu = \epsilon_x$, $x \in E$, we obtain

$$P_t(x,B) = P^{\epsilon_x}\{X_t \in B\} \qquad (B \in \mathfrak{B}). \tag{12.3.13}$$

Our introductory remarks, interpreting $P_t(x,B)$ as the probability that a particle which starts at x is found in B at time t, is now fully justified. By (12.3.12) we have

$$P^\mu\{X_0 \in B\} = \int P_0(x,B)\mu(dx) \qquad (B \in \mathfrak{B}).$$

Thus if, in particular, P_0 is the unit kernel, as is the case for many semigroups, then $P^\mu\{X_0 \in B\} = \mu(B)$. This also justifies the previously used terminology "starting probability."

PROBLEMS

1. Prove existence and uniqueness of a Markov semigroup $(H_t)_{t\in\mathbf{R}_+}$ on $\mathbf{R}^{p+1}$ such that, for each Borel-measurable function $f \geqq 0$ on $\mathbf{R}^{p+1}$, we have $H_0 =$ unit kernel and

$$H_tf(x,\tau) = \int_{\mathbf{R}^p} \left(\frac{1}{2\pi t}\right)^{p/2} e^{-|y|^2/2t} f(x + y, \tau - t)\, dy,$$

for all $(x,\tau) \in \mathbf{R}^{p+1} = \mathbf{R}^p \times \mathbf{R}$ and all $t > 0$. $(H_t)_{t\in\mathbf{R}_+}$ is called the *semigroup of the heat equation.*

2. Let $(P_t)_{t\in\mathbf{R}_+}$ and $(P'_t)_{t\in\mathbf{R}_+}$ be sub-Markov semigroups on measurable spaces $(E,\mathfrak{B})$ and $(E',\mathfrak{B}')$, respectively. Prove existence and uniqueness of a sub-Markov semigroup $(Q_t)_{t\in\mathbf{R}_+}$ on $(E \times E', \mathfrak{B} \otimes \mathfrak{B}')$ such that

(a) $$Q_tf(x,x') = \int[\int f(y,y')P_t(x,dy)]P'_t(x',dy')$$

holds for all $(x,x') \in E \times E'$ and all $\mathfrak{B} \otimes \mathfrak{B}'$-measurable functions $f \geqq 0$ on $E \times E'$. Prove that condition (a) is equivalent to

(b) $$\epsilon_{(x,x')}Q_t = \epsilon_xP_t \otimes \epsilon_{x'}P'_t.$$

[In the sense of (10.3.3), ϵ_xP_t is the measure $B \to P_t(x,B)$.] Show that (Q_t) is a Markov semigroup if (P_t) and (P'_t) are Markov semigroups, and that Problem 1 is a special case of Problem 2.

3. Let $(E,\mathfrak{B})$ be a measurable space, and consider the measurable space $(E^{\omega_0}, \mathfrak{B}^{\omega_0})$ of Section 7.4, Problem 5 where $\omega_0 \notin E$.
   (a) Prove: For every sub-Markov kernel K on $(E,\mathfrak{B})$ there exists exactly one Markov kernel K^{ω_0} on $(E^{\omega_0},\mathfrak{B}^{\omega_0})$ which extends K, that is, which satisfies $K^{\omega_0}(x,B) = K(x,B)$ for all $x \in E$ and $B \in \mathfrak{B}$.
   (b) Let $(P_t)_{t\in\mathbf{R}_+}$ be a sub-Markov semigroup on $(E,\mathfrak{B})$. Prove that $(P_t^{\omega_0})_{t\in\mathbf{R}_+}$ is a Markov semigroup on $(E^{\omega_0},\mathfrak{B}^{\omega_0})$.

4. Let $(P_t)_{t\in\mathbf{R}_+}$ be a Markov semigroup on a measurable space $(E,\mathfrak{B})$. A point $a \in E$ is called an *absorbing point* (with respect to the semigroup) if

(a) $$\epsilon_aP_t = \epsilon_a, \qquad \text{for all } t \geqq 0.$$

Prove: Each of the following two conditions is equivalent to this definition:

(b) $P_tf(a) = f(a)$, for all $t \geqq 0$ and all $f \in \mathcal{E}^b_{\mathfrak{B}}$;
(c) $P_tf(a) = 0$, for all $t \geqq 0$ and all $f \in \mathcal{E}^b_{\mathfrak{B}}$ vanishing at a.

Here again $\mathcal{E}^b_{\mathfrak{B}}$ denotes the vector space of all bounded, real, $\mathfrak{B}$-measurable functions on E. Show that the point ω_0 of Problem 3 is absorbing with respect to $(P_t^{\omega_0})_{t\in\mathbf{R}_+}$.

12.4 MARKOV PROCESSES

We have not yet justified the conception of a randomly moving particle "without memory" which we developed in Theorem 12.3.2 and the subsequent discussions after the definition of Markov semigroups. This will now be done.

Let $(\Omega,\mathfrak{A},P,(X_t)_{t\in I})$ be a stochastic process with arbitrary state space $(E,\mathfrak{B})$ and a *totally ordered parameter set* I with order relation $\leqq$. In applications we usually have $I \subset \mathbf{R}$ or even $I \subset \mathbf{R}_+$, and $\leqq$ is the usual order relation of the real numbers. We therefore interpret $s \leqq t$ as: the "time point" $s \in I$ lies before the "time point" $t \in I$. We then consider (as in Section 11.1, Example 1) the σ-algebra

$$\mathfrak{A}_t \equiv \mathfrak{A}(X_s;\, s \leqq t), \tag{12.4.1}$$

generated by all random variables X_s with $s \leqq t$, $s \in I$, and call it the *σ-algebra of events up to time t*.[10] Knowledge of this means information about the development or history of the process up to time t. With this interpretation, it is natural to refine the idea of a process "without memory" by the following definition:

12.4.1. Definition. A stochastic process $(X_t)_{t\in I}$ with totally ordered parameter set I has the *elementary Markov property* if, for every set $B \in \mathfrak{B}$ and every pair s, $t \in I$ with $s < t$,

$$P\{X_t \in B \mid \mathfrak{A}_s\} = P\{X_t \in B \mid X_s\}, \qquad P\text{-almost surely.}^{11} \tag{12.4.2}$$

In the sense of the interpretation developed above, we thus require that, for the "position" of the process at time $t > s$, the information about the development of the process up to time s is equivalent to the information about the position of the process at time s. Since condition (12.4.2) is automatically satisfied for $s = t$ because of the $\mathfrak{A}_t$-measurability of X_t, we can also allow pairs s, $t \in I$ with $s \leqq t$ in the definition.

Example

1. If the process $(X_t)_{t\in I}$ is an *independent* family of random variables, then it has the elementary Markov property. Indeed, Corollary 5.1.5, the σ-algebras $\mathfrak{A}_s$ and $\mathfrak{A}(X_t)$ are independent for any two elements s, $t \in I$

[10] In Section 11.1, $\mathfrak{A}_t$ was denoted by $\mathfrak{A}_t^0$.

[11] Here $P\{X_t \in B|\mathfrak{A}_s\}$, and so on, stands for $P(\{X_t \in B\}|\mathfrak{A}_s)$.

with $s < t$. Therefore by Corollary 10.1.5, for every $B \in \mathfrak{B}$ we have

$$P\{X_t \in B \mid \mathfrak{A}_s\} = P\{X_t \in B\}, \qquad P\text{-almost surely}$$

and

$$P\{X_t \in B \mid X_s\} = P\{X_t \in B\}, \qquad P\text{-almost surely.}$$

From this the assertion follows.

In particular, every independent *sequence* of random variables has the elementary Markov property.

As preparation for our most important example, we give Definition 12.4.1. in an equivalent form.

12.4.2. Lemma. The process $(X_t)_{t\in I}$ has the elementary Markov property if and only if, for every set $B \in \mathfrak{B}$ and any finitely many elements $s_1, \ldots, s_n, t \in I$ with $s_1 < \cdots < s_n < t$,

$$P\{X_t \in B \mid X_{s_1}, \ldots, X_{s_n}\} = P\{X_t \in B \mid X_{s_n}\}, \qquad P\text{-almost surely.} \tag{12.4.3}$$

Proof. Suppose the process has the elementary Markov property. For $B \in \mathfrak{B}$ and elements $s_1 < \cdots < s_n < t$ from I we then have

$$P\{X_t \in B \mid \mathfrak{A}_{s_n}\} = P\{X_t \in B \mid X_{s_n}\}, \qquad P\text{-almost surely.}$$

With the abbreviation $A \equiv \{X_t \in B\}$, this is equivalent to

$$E(1_A \mid \mathfrak{A}_{s_n}) = E(1_A \mid X_{s_n}), \qquad P\text{-almost surely.}$$

Thus we obtain

$$E(E(1_A \mid \mathfrak{A}_{s_1}) \mid X_{s_n}, \ldots, X_{s_n}) = E(E(1_A \mid X_{s_n}) \mid X_{s_1}, \ldots, X_{s_n}), \qquad P\text{-almost surely.}$$

According to (10.1.18), the left (or right) side is almost surely equal to

$$E(1_A \mid X_{s_1}, \ldots, X_{s_n}) \qquad [\text{or } E(1_A \mid X_{s_n})].$$

But the almost sure equality of these conditional expectations is what had to be shown.

Conversely, suppose the condition of the lemma is satisfied. For $B \in \mathfrak{B}$ and elements $s < t$ from I, we have to show that $P\{X_t \in B \mid X_s\}$ is a version of the conditional probability $P\{X_t \in B \mid \mathfrak{A}_s\}$. Now $P\{X_t \in B \mid X_s\}$ is X_s- and thus $\mathfrak{A}_s$-measurable. We must prove that

$$\int_Q P\{X_t \in B \mid X_s\}\, dP = P(\{X_t \in B\} \cap Q) \tag{12.4.4}$$

for all $Q \in \mathfrak{A}_s$. By (12.4.3), this equality holds for all $Q \in \mathfrak{A}(X_{s_1}, \ldots, X_{s_n})$ for arbitrary $s_1, \ldots, s_n \in I$ satisfying $s_1 < \cdots < s_n = s < t$. The system $\mathfrak{E}$ of all these sets Q is obviously an $\cap$-stable generator of $\mathfrak{A}_s$.

Since, on the right- and left-hand sides of (12.4.4), we have finite measures on $\mathfrak{A}$ which coincide on $\mathfrak{E}$ these measures must coincide on $\mathfrak{A}(\mathfrak{E}) = \mathfrak{A}_s$ by the Uniqueness Theorem 1.5.5. Therefore (12.4.4) holds for all $Q \in \mathfrak{A}_s$. ⌟

The problem formulated in the introduction will now be solved in a very satisfactory form by the following theorem.

12.4.3. Theorem. Let $(\Omega,\mathfrak{A},P,(X_t)_{t\in\mathbf{R}_+})$ be a stochastic process with arbitrary state space $(E,\mathfrak{B})$ and $\mathbf{R}_+$ as parameter set, whose finite-dimensional distributions are derived according to (12.3.9) from a Markov semigroup $(P_t)_{t\in\mathbf{R}_+}$ and a starting probability μ on $(E,\mathfrak{B})$. Then the process has the elementary Markov property. Moreover, for arbitrary $B \in \mathfrak{B}$ and $s, t \in \mathbf{R}_+$ with $s < t$,

$$P\{X_t \in B \mid \mathfrak{A}_s\} = P_{t-s}(X_s,B), \qquad P\text{-almost surely.}^{12} \qquad (12.4.5)$$

Since, by (12.3.13), we have already been able to interpret $P_{t-s}(X_s,B)$ as the probability that the particle starting at the point X_s at time 0 is found in B at time $t - s$, our idea of a randomly moving particle without memory is formally justified by (12.4.5).

Proof. We begin with the proof of (12.4.5) by noting that $P_{t-s}(X_s,B)$ is always $\mathfrak{A}(X_s)$- and hence $\mathfrak{A}_s$-measurable. Thus we have to show that

$$\int_Q P_{t-s}(X_s,B)\, dP = P(\{X_t \in B\} \cap Q)$$

for all $Q \in \mathfrak{A}_s$ or, by the Uniqueness Theorem 1.5.5, at least for all sets Q of the generator

$$\mathfrak{E} = \bigcup_{\substack{0\leq s_1<\cdots<s_n \\ s_n=s<t}} \mathfrak{A}(X_{s_1}, \ldots ,X_{s_n})$$

already used in the proof of Lemma 12.4.2. Thus we have only to show that for arbitrary $B \in \mathfrak{B}$ and $s_1 < \cdots < s_n < t$ from $\mathbf{R}_+$,

$$P\{X_t \in B \mid X_{s_1}, \ldots ,X_{s_n}\} = P_{t-s_n}(X_{s_n},B), \qquad P\text{-almost surely.} \qquad (12.4.6)$$

Then in particular

$$P\{X_t \in B \mid X_{s_n}\} = P_{t-s_n}(X_{s_n},B), \qquad P\text{-almost surely.} \qquad (12.4.6')$$

Together, the two equalities yield the elementary Markov property.

Now we give the proof of (12.4.6). Since $Y \equiv P_{t-s_n}(X_{s_n},B)$ is $\mathfrak{A}(X_{s_1}, \ldots ,X_{s_n})$-measurable, we must verify that

$$\int_Q Y\, dP = P(\{X_t \in B\} \cap Q) \qquad (12.4.7)$$

[12] $P_{t-s}(X_s,B)$, of course, denotes the random variable $\omega \to P_{t-s}(X_s(\omega),B)$.

for all $Q \in \mathfrak{A}(X_{s_1}, \ldots, X_{s_n})$. We can take Q in the form $Q = \{X_{s_1} \in B_1\} \cap \cdots \cap \{X_{s_n} \in B_n\}$ $(B_1, \ldots, B_n \in \mathfrak{B})$, since these sets form an $\cap$-stable generator of $\mathfrak{A}(X_{s_1}, \ldots, X_{s_n})$. But then

$$\begin{aligned}\int_Q Y\,dP &= \int 1_Q Y\,dP = \int (1_{B_1} \circ X_{s_1}) \cdot \ldots \cdot (1_{B_n} \circ X_{s_n}) Y\,dP \\ &= \int 1_{B_1}(x_1) \cdot \ldots \cdot 1_{B_n}(x_n) P_{t-s_n}(x_n,B) P_J(dx),\end{aligned}$$

where $J = \{s_1, \ldots, s_n\}$ and P_J is the joint distribution of the random variables $X_{s_1}, \ldots, X_{s_n}$. By (12.3.9) we then have

$$\begin{aligned}\int_Q Y\,dP &= \iint \cdots \int 1_{B_1}(x_1) \cdot \ldots \cdot 1_{B_n}(x_n) P_{t-s_n}(x_n,B) P_{s_n-s_{n-1}}(x_{n-1},dx_n) \\ &\qquad\qquad \cdots P_{s_1}(x_0,dx_1)\mu(dx_0) \\ &= \iint_{B_1} \cdots \int_{B_n} P_{t-s_n}(x_n,B) \cdots P_{s_1}(x_0, dx_1)\mu(dx_0) \\ &= \iint_{B_1} \cdots \int_{B_n} \int_B P_{t-s_n}(x_n,dx_{n+1}) \cdots P_{s_1}(x_0,dx_1)\mu(dx_0) \\ &= P_H(B_1 \times \cdots \times B_n \times B),\end{aligned}$$

where $H = \{s_1, \ldots, s_n, t\}$. P_H is the joint distribution of the variables $X_{s_1}, \ldots, X_{s_n}, X_t$ and therefore, because of the special form of Q,

$$\begin{aligned}P_H(B_1 \times \cdots \times B_n \times B) &= P\{X_{s_1} \in B_1, \ldots, X_{s_n} \in B, X_t \in B\} \\ &= P(\{X_t \in B\} \cap Q).\end{aligned}$$

Thus we have arrived at (12.4.7). ┘

Property (12.4.5) is more closely connected with Markov semigroups than Theorem 12.4.3 indicates. To clarify this, we consider a Markov semigroup $(P_t)_{t \in \mathbf{R}_+}$ on a Polish space E equipped with the σ-algebra $\mathfrak{B} = \mathfrak{B}(E)$ of its Borel sets. After Theorem 12.3.2 we already noted that, for every probability measure μ on $\mathfrak{B}$, there is a canonical stochastic process $(\Omega, \mathfrak{A}, P^\mu, (X_t)_{t \in \mathbf{R}_+})$, with state space E, where only the probability measure P^μ depends on μ and where the finite-dimensional distributions of the process are derived in the sense of Theorem 12.3.2 from the semigroup (P_t) and the starting probability μ. Then, in particular, for $\mu = \epsilon_x$ (with $x \in E$), the probability measure P^{ϵ_x} on $\mathfrak{A}$ is defined, and we henceforth denote it by P^x. By (12.3.9), for arbitrary $t_1 < \cdots < t_n$ in $\mathbf{R}_+$ and sets $B_1, \ldots, B_n \in \mathfrak{B}$,

$$\begin{aligned}&P^x\{X_{t_1} \in B_1, \ldots, X_{t_n} \in B_n\} \\ &\qquad = \int_{B_1} \cdots \int_{B_n} P_{t_n-t_{n-1}}(x_{n-1},dx_n) \cdots P_{t_1}(x,dx_1) \\ &\qquad = P_{t_1}(\cdots (P_{t_{n-1}-t_{n-2}}(P_{t_n-t_{n-1}}1_{B_n})1_{B_{n-1}}) \cdots)1_{B_1}(x),\end{aligned}$$

and hence the mapping

$$x \to P^x\{X_{t_1} \in B_1, \ . \ . \ . \ ,X_{t_n} \in B_n\}$$

is always $\mathfrak{B}$-measurable. Since each of the processes $(\Omega,\mathfrak{A},P^\mu,(X_t)_{t\in\mathbf{R}_+})$ is canonical; that is, $\Omega = E^{\mathbf{R}_+}$, $\mathfrak{A} = \mathfrak{B}^{\mathbf{R}_+}$ and $X_t = p_{\{t\}}$, we see that $\mathfrak{A}$ is generated by the system $\mathfrak{Z}$ of sets $\{X_{t_1} \in B_1\} \cap \cdot \cdot \cdot \cap \{X_{t_n} \in B_n\} = X_{t_1}^{-1}(B_1) \cap \cdot \cdot \cdot \cap X_{t_n}^{-1}(B_n)$ (with $t_1 < \cdot \cdot \cdot < t_n$ from $\mathbf{R}_+$, $B_1, \ . \ . \ . \ , B_n \in \mathfrak{B}$, $n \in \mathbf{N}$). Moreover, the system $\mathfrak{M}$ of all sets A for which $x \to P^x(A)$ is measurable is a Dynkin system. Thus $\mathfrak{E} \subset \mathfrak{M}$ and hence $\mathfrak{A} = \mathfrak{A}(E) = \mathfrak{D}(\mathfrak{E}) \subset \mathfrak{M}$, since $\mathfrak{E}$ is $\cap$-stable. Then $x \to P^x(A)$ is measurable for all $A \in \mathfrak{A}$, that is, $(x,A) \to P^x(A)$ us a *Markov kernel* from $(E,\mathfrak{B})$ to $(\Omega,\mathfrak{A})$.

Finally, the relationships (12.3.13) and (12.4.5) hold, and accordingly, for arbitrary s, $t \in \mathbf{R}_+$, $x \in E$, and $B \in \mathfrak{B}$,

$$P_t(x,B) = P^x\{X_t \in B\},$$
$$P^x\{X_{s+t} \in B \mid \mathfrak{A}_s\} = P_t(X_s,B), \qquad P^x\text{-almost surely.}$$

Consequently,

$$P^x\{X_{s+t} \in B \mid \mathfrak{A}_s\} = P^{X_s}\{X_t \in B\}, \qquad P^x\text{-almost surely.}$$

Here $P^{X_s}(A)$ denotes the random variable $\omega \to P^{X_s(\omega)}(A)$. Thus,

$$(\Omega,\mathfrak{A},(P^x)_{x\in E},\ (X_t)_{t\in\mathbf{R}_+})$$

is a Markov process in the sense of the following definition.

12.4.4. Definition. Let $(E,\mathfrak{A})$ be a measurable space. Then a quadruple

$$(\Omega,\mathfrak{A},(P^x)_{x\in E},\ (X_t)_{t\in\mathbf{R}_+})$$

is called a *Markov process* with state space E [or $(E,\mathfrak{B})$] if it has the following properties:

$(\Omega,\mathfrak{A},P^x,\ (X_t)_{t\in\mathbf{R}_+})$ is a stochastic process with state space E for all $x \in E$; (12.4.8)

$x \to P^x(A)$ is $\mathfrak{B}$-measurable for all $A \in \mathfrak{A}$; (12.4.9)

$P^x\{X_{s+t} \in B \mid \mathfrak{A}_s\} = P^{X_s}\{X_t \in B\}$, P^x-almost surely for all s, $t \in \mathbf{R}_+$, $x \in E$, and $B \in \mathfrak{B}$. (12.4.10)

We call (12.4.10) the *weak Markov property*. Therefore the processes just defined are also called weak Markov processes. We shall not go into the definition of the so-called strong Markov property and additional restrictive properties that are often connected with the concept of a Markov process.[13]

We combine the above into the following theorem:

12.4.5. Theorem. For every Markov semigroup $(P_t)_{t\in\mathbf{R}_+}$ on a Polish space E there exists a Markov process $(\Omega,\mathfrak{A},(P^x)_{x\in E},(X_t)_{t\in\mathbf{R}_+})$ with state space E such that for all $t \in \mathbf{R}_+$, $x \in E$, and $B \in \mathfrak{B}(E)$,

$$P_t(x,B) = P^x\{X_t \in B\}. \tag{12.4.11}$$

It is noteworthy that the following converse also holds:

12.4.6. Theorem. Let $(E,\mathfrak{B})$ be a measurable space and also let $(\Omega,\mathfrak{A},(P^x)_{x\in E},(X_t)_{t\in\mathbf{R}_+})$ be a Markov process with state space E. Then

$$P_t(x,B) \equiv P^x\{X_t \in B\} \qquad (x,B) \in E \times \mathfrak{B} \tag{12.4.11$'$}$$

defines a Markov semigroup $(P_t)_{t\in\mathbf{R}_+}$ on $(E,\mathfrak{B})$.

Proof. Each of the mappings $P_t\colon E \times \mathfrak{B} \to \mathbf{R}_+$ is a Markov kernel on $(E,\mathfrak{B})$, since $x \to P^x(A)$ is $\mathfrak{B}$-measurable for every $A \in \mathfrak{A}$, and thus, in particular, for

$$A \equiv \{X_t \in B\} = X_t^{-1}(B),$$

and since $B \to P_t(x,B)$ is the distribution of the random variable X_t relative to P^x and is thus a probability measure on $\mathfrak{B}$. Thus all we still have to show is the semigroup property. Let $(x,B) \in E \times \mathfrak{B}$ and $s, t \in \mathbf{R}_+$. Then by the weak Markov property and (10.1.6),

$$\begin{aligned} P_{s+t}(x,B) &= P^x\{X_{s+t} \in B\} = E^x(1_{\{X_{s+t}\in B\}}) = E^x(E^x(1_{\{X_{s+t}\in B\}} \mid \mathfrak{A}_s)) \\ &= E^x(P^x\{X_{s+t}\in B \mid \mathfrak{A}_s\}) = E^x(P^{X_s}\{X_t\in B\}).^{14} \end{aligned}$$

In view of (12.4.11$'$), we have

$$\begin{aligned} P_{s+t}(x,B) &= E^x(P_t(X_s,B)) = \int P_t(X_s(\omega),B)P^x(d\omega) \\ &= \int P_t(y,B)P^x_{X_s}(dy). \end{aligned}$$

But according to (12.4.11$'$), the distribution $P^x_{X_s}$ of X_s relative to P^x is the probability measure $B \to P_s(x,B)$. Thus we finally obtain

$$P_{s+t}(x,B) = \int P_s(x,dy)P_t(y,B),$$

and hence (12.3.3$'$). ┘

[13] See Meyer [39], Blumenthal-Getoor [24].

[14] E^x (E^μ) denotes the *expected value* relative to P^x (P^μ).

The circle of reasoning will be completed if we can show that the projective family (P_J^x), according to Theorem 12.4.6 associated with the semigroup (P_t) from Theorem 12.3.2 and with $\mu = \epsilon_x$, is just the family of finite-dimensional distributions of the process $(\Omega,\mathfrak{A},P^x,(X_t)_{t\in\mathbf{R}_+})$.

12.4.7. Corollary. Suppose we have the situation of Theorem 12.4.6. Then for every $x \in E$ and $J \in \mathfrak{H}(\mathbf{R}_+)$, the finite-dimensional distribution P_J^x of the process $(\Omega,\mathfrak{A},P^x,(X_t)_{t\in\mathbf{R}_+})$ is given by

$$P_J^x(B) = \int \cdots \int 1_B(x_1, \ldots, x_n) P_{t_n-t_{n-1}}(x_{n-1},dx_n) \cdots P_{t_1}(x,dx_1) \qquad (B \in \mathfrak{B}^J). \quad (12.4.12)$$

Here $t_1 < \cdots < t_n$ are the elements of J.

Proof. It suffices to prove (12.4.12) for sets $B = B_1 \times \cdots \times B_n$ with $B_1, \ldots, B_n \in \mathfrak{B}$. This is done by induction on n. The case $n = 1$ is contained in the definition (12.4.11′) of the semigroup (P_t). Thus, suppose the assertion has been proved for $n - 1$. Then we can proceed as follows: $Q \equiv \{X_{t_1} \in B_1\} \cap \cdots \cap \{X_{t_{n-1}} \in B_{n-1}\}$ lies in $\mathfrak{A}_{t_{n-1}}$; hence, by the weak Markov property and (12.4.11),

$$\begin{aligned} P_J^x(B_1 \times \cdots \times B_n) &= P^x(\{X_{t_n} \in B_n\} \cap Q) \\ &= \int_Q P^x\{X_{t_n} \in B_n \mid \mathfrak{A}_{t_{n-1}}\}\, dP^x \\ &= \int_Q P^{X_{t_{n-1}}}\{X_{t_n-t_{n-1}} \in B_n\}\, dP^x \\ &= \int 1_Q P^{X_{t_{n-1}}}\{X_{t_n-t_{n-1}} \in B_n\}\, dP^x \\ &= \int (1_{B_1} \circ X_{t_1}) \cdot \ldots \cdot (1_{B_{n-1}} \circ X_{t_{n-1}})\, P_{t_n-t_{n-1}}(X_{t_{n-1}},B_n)\, dP^x. \end{aligned}$$

Now we use the transformation theorem and the induction hypothesis, according to which the joint distribution of $X_{t_1}, \ldots, X_{t_{n-1}}$ relative to P^x is of the form (12.4.12).

Then we have

$$\begin{aligned} P_J^x(B_1 \times \cdots \times B_n) &= \int_{B_1} \cdots \int_{B_{n-1}} P_{t_n-t_{n-1}}(x_{n-1},B_n)\, P_{t_{n-1}-t_{n-2}}(x_{n-2}, dx_{n-1}) \cdots P_{t_1}(x, dx_1) \\ &= \int_{B_1} \cdots \int_{B_n} P_{t_n-t_{n-1}}(x_{n-1}, dx_n) \cdots P_{t_1}(x, dx_1). \end{aligned}$$

But this is what we wanted to show. ┘

Combining our results, we can establish that Markov semigroups and Markov processes can be interpreted as the same mathematical object. Experience tells us that the more complicated concept of a Markov process provides us with a powerful probability-theoretic tool for studying Markov semigroups.

Because of (12.4.11′), we call the Markov semigroup associated with a Markov process the semigroup of *transition probabilities* of the process. By Corollary 12.4.7 we can rederive the measure P^μ associated with an arbitrary starting probability μ on $\mathfrak{B}$ from the measures $(P^x)_{x \in E}$ of a Markov process. Obviously,

$$P^\mu = \int P^x \mu(dx). \tag{12.4.13}$$

We conclude with one of the simplest examples for a Markov process.

Example

2. Let $\Omega = E = \mathbf{R}^p$ $(p = 1, 2, \ldots)$, $\mathfrak{A} = \mathfrak{B} = \mathfrak{B}^p$; let P^x be the measure ϵ_x on $\mathfrak{A}$ defined by the unit mass at x for every $x \in E = \Omega$; and let $v \in E$ be chosen arbitrarily. If we define $X_t \colon \Omega \to E$ by

$$X_t(\omega) \equiv \omega + vt,$$

then $(\Omega, \mathfrak{A}, P^x, (X_t)_{t \in \mathbf{R}_+})$ is a stochastic process with state space E whose paths are all of the form $t \to y + vt$ with $y \in E$. The process thus describes the *motion* of a particle starting at x with *constant* (directed) *velocity* v. Since every X_t is a translation in E, we have $\mathfrak{A}(X_t) = \mathfrak{B}^p = \mathfrak{A}$ and hence $\mathfrak{A}_t = \mathfrak{A}$ for all $t \in \mathbf{R}_+$.

Then $(\Omega, \mathfrak{A}, (P^x)_{x \in E}, (X_t)_{t \in \mathbf{R}_+})$ is a Markov process. Indeed, for every $A \in \mathfrak{A}$, we see that $x \to P^x(A) = 1_A(x)$ is $\mathfrak{B}$-measurable. Since $\{X_t \in B\} = B - vt$ and $\mathfrak{A}_t = \mathfrak{A}$ for all $t \in \mathbf{R}_+$ and $B \in \mathfrak{B}$, we also immediately obtain the weak Markov property: On the one hand, we have

$$P^x\{X_{s+t} \in B \mid \mathfrak{A}_s\} = P^x\{X_{s+t} \in B\} = 1_{B - v(s+t)}(x), \qquad P^x\text{-almost surely;}$$

on the other hand, for all $\omega \in \Omega$,

$$P^{X_s(\omega)}\{X_t \in B\} = 1_{B-vt}(\omega + vs) = 1_{B-v(s+t)}(\omega)$$

and thus

$$P^{X_s}\{X_t \in B\} = 1_{B-v(s+t)}, \qquad P^x\text{-almost surely}$$

$(s, t \in \mathbf{R}_+, x \in E, B \in \mathfrak{B})$.

The associated Markov semigroup $(P_t)_{t \in \mathbf{R}_+}$ is given by

$$P_t(x,B) = 1_{B-vt}(x) = \epsilon_x(B - vt).$$

PROBLEMS

1. Let $(E,\mathfrak{B})$ be a measurable space, and let $(\Omega,\mathfrak{A},(P^x)_{x\in E}(X_t)_{t\in\mathbf{R}_+})$ be a quadruple with properties (12.4.8) and (12.4.9). Prove that the weak Markov property is equivalent to the following property:

$$P^x\{X_t \in B \mid X_{t_1}, \ldots, X_{t_n}\} = P^{X_{t_n}}\{X_{t-t_n} \in B\}, \quad P^x\text{-almost surely.}$$

for all $x \in E$, $0 \leqq t_1 < \cdots < t_n < t$, and all $B \in \mathfrak{B}$. [*Hint:* Imitate the proof of Lemma 12.4.2.]

2. Consider the following subset of $\mathbf{R}^2$:

$$\Omega = \{0\} \times \mathbf{R}_+ \cup]0,+\infty[\times \{0\}.$$

Let $\mathfrak{A}$ be the trace in Ω of the Borel σ-algebra $\mathfrak{B}^2$, and define for each $x \in \mathbf{R}_+$ the following probability measure P^x on $\mathfrak{A}$: for $x > 0$, $P^x = \epsilon_{(x,0)}$; for $x = 0$, $P^0(A) = \int 1_A(0,y)e^{-y}\,dy$. Furthermore, for $t \in \mathbf{R}_+$, we define $X_t\colon \Omega \to \mathbf{R}_+$ as follows:

$$X_t(x,0) = x + t \qquad (x \geqq 0),$$
$$X_t(0,y) = \begin{cases} 0, & \text{if } t \leqq y \\ t - y, & \text{if } t \geqq y \end{cases} \qquad (y \geqq 0).$$

Use the result of Problem 1 and prove that $(\Omega,\mathfrak{A},(P^x)_{x\in\mathbf{R}_+}, (X_t)_{t\in\mathbf{R}_+})$ is a Markov process with $\mathbf{R}_+$ as state space. Determine the semigroup (P_t) of transition probabilities, and prove in particular $P^0\{X_t = 0\} = e^{-t}$. (This Markov process describes the movement of a particle in $\mathbf{R}_+$ with constant velocity 1 but with the additional property that it leaves the origin 0 only after an "exponential holding time.")

3. Let $(\Omega,\mathfrak{A},(P^x)_{x\in E},(X_t)_{t\in\mathbf{R}_+})$ be a Markov process with a measurable space $(E,\mathfrak{B})$ as state space. A set A of the σ-algebra $\mathfrak{B}$ is called *absorbing* if $P^x\{X_t \in A\} = 1$ for all $x \in A$ and all $t \in \mathbf{R}_+$, that is, if the process starting at any $x \in A$ does not leave A P^x-almost surely. In particular, a point $a \in E$ is called absorbing (with respect to the process) if $\{a\} \in \mathfrak{B}$ and $\{a\}$ is an absorbing set.
   (a) Prove: A point $a \in E$ with $\{a\} \in \mathfrak{B}$ is absorbing with respect to a Markov process if and only if a is an absorbing point with respect to the semigroup (P_t) of transition probabilities (see Section 12.3, Problem 4).
   (b) Let $(\Omega,\mathfrak{A},(P^x)_{x\in\mathbf{R}^{p+1}},(X_t)_{t\in\mathbf{R}_+})$ be a Markov process with the semigroup (H_t) of the heat equation (Section 12.3, Problem 1) as semigroup of transition probabilities. For each $t_0 \in \mathbf{R}$, denote by A_{t_0} the closed half space

$$A_{t_0} \equiv \{(x,\tau) \in \mathbf{R}^p \times \mathbf{R}\colon \tau \leqq t_0\}$$

of $\mathbf{R}^{p+1} = \mathbf{R}^p \times \mathbf{R}$. Prove that A_{t_0} is an absorbing set with respect to this Markov process.

4. Let E be a locally compact space with countable base, $E' = E \cup \{\omega_0\}$ its one point compactification, and let $(P_t)_{t \in \mathbf{R}_+}$ be a sub-Markov semigroup of kernels on $(E, \mathfrak{B}(E))$. Prove the existence of a Markov process $(\Omega, \mathfrak{A}, (P^x)_{x \in E'}, (X_t)_{t \in \mathbf{R}_+})$ with $(E', \mathfrak{B}(E'))$ as state space and with the following properties: (a) $P_t(x,B) = P^x\{X_t \in B\}$ for all $x \in E$ and $B \in \mathfrak{B}(E)$; (b) ω_0 is an absorbing point. [*Hint:* Use Section 12.3, Problem 3.]

12.5 PROCESSES WITH STATIONARY AND INDEPENDENT INCREMENTS

In Section 12.3, Example 3, we became acquainted with the concept of a *translation-invariant* Markov semigroup $(P_t)_{t \in \mathbf{R}_+}$ on $\mathbf{R}^p$. If $(\Omega, \mathfrak{A}, (P^x)_{x \in \mathbf{R}^p}, (X_t)_{t \in \mathbf{R}_+})$ is a Markov process with $\mathbf{R}^p$ as state space and $(P_t)_{t \in \mathbf{R}_+}$ as semigroup of its transition probabilities, then (12.4.11) tells us that (P_t) is translation-invariant if and only if

$$P^{x+z}\{X_t \in B + z\} = P^x\{X_t \in B\} \tag{12.5.1}$$

for all x, $z \in \mathbf{R}^p$, $B \in \mathfrak{B}^p$, and $t \in \mathbf{R}_+$. Naturally, this property is said to define the class of *translation-invariant Markov processes* in $\mathbf{R}^p$. Two of its properties will be stressed here. Thus we define:

12.5.1. Definition. Let $(\Omega, \mathfrak{A}, P, (X_t)_{t \in \mathbf{R}_+})$ be a stochastic process with $\mathbf{R}^p$ as state space.

(a) We call $(X_t)_{t \in \mathbf{R}_+}$ a process *with stationary increments* if there exists a family $(\mu_t)_{t \in \mathbf{R}_+}$ of probability measures on $\mathfrak{B}^p$ such that

$$P_{X_t - X_s} = \mu_{t-s} \tag{12.5.2}$$

for arbitrary s, $t \in \mathbf{R}_+$ with $s \leqq t$.[15]

(b) We call $(X_t)_{t \in \mathbf{R}_+}$ a process *with independent increments* if for every finitely many $t_0, t_1, \ldots, t_n \in \mathbf{R}$ with $0 = t_0 < t_1 < \cdots < t_n$, the random variables

$$X_{t_0}, X_{t_1} - X_{t_0}, \ldots, X_{t_n} - X_{t_{n-1}}$$

are independent.

[15] In somewhat unprecise formulation this means that the distribution $P_{X_t - X_s}$ for $0 \leqq s \leqq t < +\infty$ depends only on the difference $t - s$. Obviously, we always have $\mu_0 = \epsilon_0$.

These concepts are of interest first of all because of the following two theorems:

12.5.2. Theorem. Let $(\Omega,\mathfrak{A},(P^x)_{x\in\mathbf{R}^p},(X_t)_{t\in\mathbf{R}_+})$ be a translation-invariant Markov process with state space $\mathbf{R}^p$ and a semigroup $(P_t)_{t\in\mathbf{R}_+}$ of transition probabilities. Then for every $x \in \mathbf{R}^p$, $(\Omega,\mathfrak{A},P^x,(X_t)_{t\in\mathbf{R}_+})$ is a process with stationary increments. For arbitrary s, $t \in \mathbf{R}_+$ with $s \leqq t$, the probability measure $B \to P_{t-s}(0,B)$ on $\mathfrak{B}^p$ is the distribution of $X_t - X_s$ relative to P^x.

Proof. As in Section 12.3, Example 3, we denote the probability measure $B \to P_t(0,B)$ by μ_t ($t \in \mathbf{R}_+$). Then we have to show that

$$P^x_{X_t-X_s} = \mu_{t-s}$$

for arbitrary s, $t \in \mathbf{R}_+$ with $s \leqq t$. The case $s = t$ is disposed of by means of Corollary 9.3.7: According to Section 12.3, Example 3, $(\mu_t)_{t\in\mathbf{R}_+}$ is a convolution semigroup of probability measures and thus $\mu_0 = \epsilon_0 = P^x_{X_t-X_t}$.[16] When $s < t$, we set $Y \equiv X_s \otimes X_t$ and let d: $\mathbf{R}^p \times \mathbf{R}^p \to \mathbf{R}^p$ denote the continuous and thus Borel-measurable mapping $(x_1,x_2) \to x_2 - x_1$. Then $P^x_{X_t-X_s}$ is the distribution of $d \circ Y$ relative to P^x. Therefore, for every $B \in \mathfrak{B}^p$, we have

$$P^x\{X_t - X_s \in B\} = P^x\{d \circ Y \in B] = P^x\{Y \in d^{-1}(B)\},$$

and thus, by (12.4.12),

$$\begin{aligned} P^x\{X_t - X_s \in B\} &= \int\int 1_{d^{-1}(B)}(x_1,x_2)P_{t-s}(x_1,dx_2)P_s(x,dx_1) \\ &= \int P_{t-s}(x_1,x_1 + B)P_s(x,dx_1) \\ &= P_{t-s}(0,B)\int P_s(x,dx_1) = \mu_{t-s}(B). \end{aligned}$$

Here the translation-invariance of P_{t-s}, that is, the equality

$$P_{t-s}(x_1,x_1 + B) = \mu_{t-s}(B)$$

was used for all $x_1 \in \mathbf{R}^p$. ┘

12.5.3. Theorem. Let $(\Omega,\mathfrak{A},(P^x)_{x\in\mathbf{R}^p},(X_t)_{t\in\mathbf{R}_+})$ be a translation-invariant Markov process with $\mathbf{R}^p$ as state space. Then for every $x \in \mathbf{R}^p$,

$$(\Omega,\mathfrak{A},P^x,(X_t)_{t\in\mathbf{R}_+})$$

is a process with independent increments.

[16] Thus P_0 is the unit kernel I in every translation-invariant Markov semigroup $(P_t)_{t\in\mathbf{R}_+}$ on $(\mathbf{R}^p,\mathfrak{B}^p)$. (However, see Section 12.3, Example 7.)

Proof. Let $(P_t)_{t\in\mathbf{R}_+}$ be the Markov semigroup associated with the given Markov process. This is translation-invariant and thus $P_t(x,B) = P_t(x + z, B + z)$, that is,

$$\int 1_B(y)P_t(x,dy) = \int 1_{B+z}(y)P_t(x + z,dy) = \int 1_B(y - z)P_t(x + z,dy).$$

By Theorem 2.3.6 this is equivalent to

$$\int f(y)P_t(x,dy) = \int f(y - z)P_t(x + z,dy) \qquad (x,z \in \mathbf{R}^p)$$

for all $\mathfrak{B}^p$-measurable functions f which are $\geqq 0$ or bounded. The independence of

$$Y_0 \equiv X_{t_0}, Y_1 \equiv X_{t_1} - X_{t_0}, \ldots, Y_n \equiv X_{t_n} - X_{t_{n-1}},$$

to be proved (where $0 = t_0 < t_1 < \cdots < t_n$) is equivalent, by Theorems 5.2.4, 8.1.6, and the Uniqueness Theorem 8.2.4 for Fourier transforms, to the validity of

$$\varphi_{Y_0\otimes\ldots\otimes Y_n}(y_0, \ldots ,y_n) = \varphi_{Y_0}(y_0) \cdot \ldots \cdot \varphi_{Y_n}(y_n) \qquad (y_0, \ldots, y_n \in \mathbf{R}^p),$$

where φ_X denotes the characteristic function of the random variable X. Thus we have to show that

$$E^x\left(e^{i\sum_{j=0}^{n} <y_j, Y_j>}\right) = \prod_{j=0}^{n} E^x(e^{i<y_j,Y_j>})$$

for all $x, y_0, \ldots, y_n \in \mathbf{R}^p$. Here E^x denotes the expected value relative to P^x. By (12.4.12), we know the joint distribution of the random variables

$$X_{t_0}, \ldots, X_{t_n}$$

relative to P^x. Therefore,

$$\begin{aligned} E^x(e^{i\Sigma<y_j,Y_j>}) &= E^x(e^{i\Sigma<y_j,X_{t_j}-X_{t_{j-1}}>}) \\ &= \int \cdots \int e^{i\Sigma<y_j,z_j-z_{j-1}>}P_{t_n-t_{n-1}}(z_{n-1},dz_n) \cdots P_{t_0}(x,dz_0), \end{aligned}$$

where we agree that $X_{t_{-1}} \equiv 0$ and $z_{-1} \equiv 0$ (because of the summation from $j = 0$ to $j = n$). In the first integration we have to compute

$$q_n \equiv \int e^{i<y_n,z_n-z_{n-1}>}P_{t_n-t_{n-1}}(z_{n-1},dz_n),$$

and thus, according to the preliminaries, to compute

$$q_n = \int e^{i<y_n,z_n>}P_{t_n-t_{n-1}}(0,dz_n).$$

Since, by Theorem 12.5.2, $B \rightarrow P_{t_n-t_{n-1}}(0,B)$ is the distribution of

$$Y_n = X_{t_n} - X_{t_{n-1}}$$

relative to P^x, we obtain

$$q_n = E^x(e^{i<y_n,Y_n>}).$$

But this expression is independent of $z_1, \ldots, z_{n-1}$, so that the next integration can be carried out analogously, and we finally obtain the desired equality. ┘

Remark. Of course the random variables X_s and X_t corresponding to times $s < t$ of a translation-invariant Markov process with state space $\mathbf{R}^p$ are in general *not* independent relative to P^x. Independence would mean $P^x_{X_s \otimes X_t} = P^x_{X_s} \otimes P^x_{X_t}$, that is,

$$\iint 1_B(x_1,x_2) P_{t-s}(x_1,dx_2) P_s(x,dx_1) = P^x_{X_s} \otimes P^x_{X_t}(B)$$

for all $B \in \mathfrak{B}^p \otimes \mathfrak{B}^p$, and thus

$$\int_{B_1} P_{t-s}(x_1,B_2) P_s(x,dx_1) = P_s(x,B_1) \cdot P_t(x,B_2)$$

for all $B_1, B_2 \in \mathfrak{B}^p$. We can easily verify that this equality fails, for example, for a Markov process corresponding to a Poisson or Brownian semigroup

We discuss another interesting connection with the convolution semigroup of probability measures considered in Section 9.3 and thus with infinitely divisible distributions.

12.5.4. Theorem. For every stochastic process $(\Omega,\mathfrak{A},P,(X_t)_{t\in\mathbf{R}_+})$ with stationary and independent increments (and with $\mathbf{R}^p$ as state space), $(P_{X_t-X_0})_{t\in\mathbf{R}_+}$ is a convolution semigroup of probability measures on $\mathbf{R}^p$. Conversely, for every convolution semigroup $(\mu_t)_{t\in\mathbf{R}_+}$ of probability measures on $\mathbf{R}^p$, there exists a process $(\Omega,\mathfrak{A},P,(X_t)_{t\in\mathbf{R}_+})$ with stationary and independent increments such that $P_{X_t-X_s} = \mu_{t-s}$ for all $s, t \in \mathbf{R}_+$ with $s \leqq t$.

Proof. Let $(X_t)_{t\in\mathbf{R}_+}$ be a process with stationary and independent increments and $\mu_t \equiv P_{X_t-X_0}$. For every two numbers $s, t \in \mathbf{R}_+$, $X_{s+t} - X_t$, and $X_t - X_0$ are then independent random variables, and thus, by Theorem 5.3.4,

$$\mu_{s+t} = P_{X_{s+t}-X_0} = P_{X_{s+t}-X_t} * P_{X_t-X_0} = \mu_s * \mu_t.$$

Conversely, let $(\mu_t)_{t\in\mathbf{R}_+}$ be a convolution semigroup of probability measures on $\mathbf{R}^p$. We shall construct the desired stochastic process with the help of Kolmogorov's Theorem 12.1.3 and therefore begin with the construction of a projective family $(P_J)_{J\in\mathfrak{H}(\mathbf{R}_+)}$. We use the abbreviation $E \equiv \mathbf{R}^p$. For every set $J \in \mathfrak{H}(\mathbf{R}_+)$ with elements $t_1 < \cdots < t_n$, we define the following linear (and therefore also continuous) mapping $T_J \equiv E^J \to E^J$. For every point $(x_1, \ldots, x_n) \in E^J$, let

$$T_J(x_1, \ldots, x_n) \equiv (x_1, x_1 + x_2, \ldots, x_1 + \cdots + x_n).$$

Then T_J is bijective. For the inverse mapping we obviously have

$$T_J^{-1}(x_1, \ldots, x_n) = (x_1, x_2 - x_1, \ldots, x_n - x_{n-1}).$$

If we set

$$P_J \equiv T_J(\mu_{t_1} \otimes \mu_{t_2 - t_1} \otimes \cdots \otimes \mu_{t_n - t_{n-1}}), \tag{12.5.3}$$

then $(P_J)_{J \in \mathfrak{H}(\mathbf{R}_+)}$ is a family of probability measures on the spaces E^J. To verify the projectivity, let H be another set from $\mathfrak{H}(\mathbf{R}_+)$ such that $H \setminus J$ contains exactly one element t'. Then, according to Remark 1 of Section 12.1 we need only verify that $p_J^H(P_H) = P_J$.[17] We prove this for the case $t_{n-1} < t' < t_n$. The remaining cases can be taken care of analogously (only with somewhat more writing); the case $t_n < t'$ is even trivial.[18] For every point $(x_1, \ldots, x_{n-1}, x', x_n) \in E^H$,

$$\begin{aligned} p_J^H \circ T_H\,(x_1, &\ldots, x_{n-1}, x', x_n) \\ &= (x_1, \ldots, x_1 + \cdots + x_{n-1}, x_1 + \cdots + x_{n-1} + x' + x_n). \end{aligned}$$

For every Borel-measurable function $f \geqq 0$ on E^J we therefore have, when we take into account the semigroup property of (μ_t),

$$\begin{aligned} \int f \circ p_J^H \circ T_H d(\mu_{t_1} \otimes &\cdots \otimes \mu_{t' - t_{n-1}} \otimes \mu_{t_n - t'}) \\ = \int f(x_1, \ldots, x_1 + &\cdots + x_{n-1} + x' + x_n)\mu_{t_n - t'}(dx_n)\mu_{t' - t_{n-1}}(dx') \cdots \mu_{t_1}(dx_1) \\ = \int f(x_1, \ldots, x_1 + &\cdots + x_{n-1} + y)(\mu_{t_n - t'} * \mu_{t' - t_{n-1}})(dy) \cdots \mu_{t_1}(dx_1) \\ = \int\int f(x_1, \ldots, x_1 + &\cdots + x_{n-1}, x_1 + \cdots + x_n)\mu_{t_n - t_{n-1}}(dx_n) \cdots \mu_{t_1}(dx_1) \\ = \int f\, dP_J. & \end{aligned}$$

Consequently, $p_J^H(P_H) = p_J^H \circ T_H(\mu_{t_1} \otimes \cdots \otimes \mu_{t_n - t'}) = P_J$, and thus the family (P_J) is indeed projective.

Let $(\Omega, \mathfrak{A}, P, (X_t)_{t \in \mathbf{R}_+})$ be a stochastic process (which exists by Theorem 12.1.3) whose family of finite-dimensional distributions is just (P_J): Then for every set $J \in \mathfrak{H}(\mathbf{R}_+)$ with elements $t_1 < \cdots < t_n$,

$$P_{X_{t_1} \otimes \cdots \otimes X_{t_n}} = P_J.$$

By using T_J^{-1}, we now obtain (because of the transitivity of image measures)

$$\begin{aligned} P_{X_{t_1} \otimes (X_{t_2} - X_{t_1}) \otimes \cdots \otimes (X_{t_n} - X_{t_{n-1}})} \\ = T_J^{-1}(P_J) = \mu_{t_1} \otimes \mu_{t_2 - t_1} \otimes \cdots \otimes \mu_{t_n - t_{n-1}}. \end{aligned} \tag{12.5.4}$$

If we now apply the natural projections of E^J on the n factors E, it follows that

$$P_{X_{t_1}} = \mu_{t_1}, P_{X_{t_2} - X_{t_1}} = \mu_{t_2 - t_1}, \ldots, P_{X_{t_n} - X_{t_{n-1}}} = \mu_{t_n - t_{n-1}}. \tag{12.5.5}$$

[17] p_J^H again denotes the natural projection of E^H onto E^J [see (12.1.3)].

[18] See also the proof of Theorem 12.3.2, which is carried out similarly.

The equalities (12.5.5) tell us that we are dealing with a process with stationary increments. We need only take account of Corollary 9.3.7, whereby $\mu_0 = \epsilon_0$, and thus $P_{X_t - X_s} = \mu_{t-s}$ also holds for $t = s$. By Theorem 5.2.4, the independence of the increments follows from (12.5.4) and (12.5.5). ⌟

Finally, real-valued stochastic processes with independent increments provide us with further important examples for *martingales*. We have:

12.5.5. Theorem. Let $(\Omega,\mathfrak{A},P,(X_t)_{t\in\mathbf{R}_+})$ be a stochastic process with independent increments and the real line $\mathbf{R}$ as state space. If all the random variables X_t are integrable, then

$$E(X_t \mid \mathfrak{A}_s) = E(X_t - X_s) + X_s, \qquad P\text{-almost surely} \qquad (12.5.6)$$

for arbitrary $s, t \in \mathbf{R}_+$ with $s < t$.

Proof. As in the proof of Lemma 12.4.2, we only have to prove

$$E(X_t \mid X_{t_0}, \ldots ,X_{t_n}) = E(X_t - X_{t_n}) + X_{t_n}, \qquad \text{almost surely}$$

for arbitrary $t_0, \ldots , t_n \in \mathbf{R}_+$ with $0 = t_0 < \cdots < t_n < t$. But now

$$\begin{aligned} E(X_t \mid X_{t_0}, \ldots ,X_{t_n}) &= E(X_t \mid X_{t_0}, X_{t_1} - X_{t_0}, \ldots ,X_{t_n} - X_{t_{n-1}}) \\ &= E(X_t - X_{t_n} \mid X_{t_0}, X_{t_1} - X_{t_0}, \ldots , X_{t_n} - X_{t_{n-1}}) \\ &\qquad + E(X_{t_n} \mid X_{t_0}, \ldots ,X_{t_n}). \end{aligned}$$

Hence the assertion follows, by Corollary 10.1.5 and by (10.1.7). ⌟

This proves that the stochastic process from Theorem 12.5.5 is a super- (or sub-) martingale if and only if

$$E(X_t - X_s) \leqq 0 \qquad (\text{or } \geqq 0),$$

for arbitrary numbers $s, t \in \mathbf{R}_+$ with $s < t$. Moreover, if the process is stationary and $(\mu_t)_{t\in\mathbf{R}_+}$ is the associated convolution semigroup according to Theorem 12.5.4, then

$$E(X_t - X_s) = \int x\mu_{t-s}(dx) \qquad (s,t \in \mathbf{R}_+; s < t).$$

PROBLEMS

1. Let $(\Omega,\mathfrak{A},(P^x)_{x\in\mathbf{R}},(X_t)_{t\in\mathbf{R}_+})$ be a Markov process associated with the Brownian semigroup on $\mathbf{R}$ (in the sense of Theorem 12.4.5). Prove that each of the stochastic processes $(\Omega,\mathfrak{A},P^x,(X_t)_{t\in\mathbf{R}_+})$ with $x \in \mathbf{R}$ is a martingale.
2. Let $(\Omega,\mathfrak{A},(P^x)_{x\in\mathbf{Z}},(X_t)_{t\in\mathbf{R}_+})$ be a Markov process associated with the Poisson semigroup on $\mathbf{Z}$. Prove that each of the stochastic processes $(\Omega,\mathfrak{A},P^x,(X_t)_{t\in\mathbf{R}_+})$ with $x \in \mathbf{Z}$ is a sub-martingale.

12.6 THE BROWNIAN MOTION PROCESS

According to Theorem 12.4.5, there is a Markov process associated with the Brownian semigroup on $\mathbf{R}^p$ (Section 12.3, Example 4). This section is devoted to a careful study of this process, and in particular, of the behavior of its paths.

12.6.1. Definition. *Brownian motion process* (or Brownian motion) in $\mathbf{R}^p$ is called any Markov process $(\Omega,\mathfrak{A},(P^x)_{x\in\mathbf{R}^p},(X_t)_{t\in\mathbf{R}_+})$ with state space $\mathbf{R}^p$ whose semigroup of transition probabilities is the Brownian semigroup on $\mathbf{R}^p$ and whose paths are all continuous.

The existence of such a process is given by the following theorem.

12.6.2. Theorem. There exists a Brownian motion process $(\Omega,\mathfrak{A},(P^x)_{x\in\mathbf{R}^p},(X_t)_{t\in\mathbf{R}_+})$. In particular, Ω can be chosen equal to the set $C \subset (\mathbf{R}^p)^{\mathbf{R}_+}$ of all continuous mappings of $\mathbf{R}_+$ into $\mathbf{R}^p$ and $\mathfrak{A}$ equal to the trace of $(\mathfrak{B}^p)^{\mathbf{R}_+}$ in C.

Proof. The existence of a Markov process $(\Omega,\mathfrak{A},(P^x)_{x\in\mathbf{R}^p},(X_t)_{t\in\mathbf{R}_+})$ with the Brownian semigroup (P_t) as semigroup of transition probabilities is given by Theorem 12.4.5. Since this is a combination of Theorems 12.3.2 and 12.4.3, we can assume that each of the processes $(\Omega,\mathfrak{A},P^x,(X_t)_{t\in\mathbf{R}_+})$ is canonical, and thus in particular that $\Omega = (\mathbf{R}^p)^{\mathbf{R}_+}$ and $\mathfrak{A} = (\mathfrak{B}^p)^{\mathbf{R}_+}$ and that every P^x is the projective limit of the projective family $(P^x_J)_{J\in\mathfrak{H}(\mathbf{R}_+)}$ given by (12.4.12). We have to show that C is essential relative to this family (P^x_J). By Lemma 12.2.8, a verification of conditions (c) and (d) of Theorem 12.2.6 for the process $(\Omega,\mathfrak{A},P^x,(X_t)_{t\in\mathbf{R}_+})$ will suffice. But this can be done as follows:

Since the Brownian semigroup is translation-invariant, by Theorem 12.5.2 we know the distribution of $X_t - X_s$ relative to P^x for all $s, t \in \mathbf{R}_+$ with $s < t$. It is

$$P^x_{X_t-X_s} = g_{t-s}\lambda^p, \tag{12.6.1}$$

where g_{t-s} is the density from (12.3.8). Therefore, for arbitrary $t \in \mathbf{R}_+$, $\eta > 0$, and $\delta > 0$,

$$P^x\{|X_{t+\delta} - X_t| \geqq \eta\} = \int_{A_\eta} g_\delta(x)\lambda^p(dx),$$

where

$$A_\eta \equiv \{x \in \mathbf{R}^p\colon |x| \geqq \eta\}.$$

Now obviously

$$A_\eta \subset \bigcup_{i=1}^{p} \left\{x \in \mathbf{R}^p\colon |x_i| \geqq \frac{\eta}{\sqrt{p}}\right\}$$

and thus

$$P^x\{|X_{t+\delta} - X_t| \geqq \eta\} \leqq \sum_{i=1}^{p} \int_{|x_i| \geqq \frac{\eta}{\sqrt{p}}} g_\delta(x)\lambda^p(dx)$$
$$= 2p(2\pi\delta)^{-1/2} \int_{\eta/\sqrt{p}}^{+\infty} e^{-\xi^2/2\delta}\, d\xi$$
$$\leqq 2p(2\pi\delta)^{-1/2} \frac{\delta\sqrt{p}}{\eta} \int_{\eta/\sqrt{p}}^{+\infty} \frac{\xi}{\delta} e^{-\xi^2/2\delta}\, d\xi.$$

A primitive of the last integrand is $\xi \to -e^{-\xi^2/(2\delta)}$. Thus we finally obtain the inequality

$$P^x\{|X_{t+\delta} - X_t| \geqq \eta\} \leqq \frac{p}{\eta}\sqrt{\frac{2p\delta}{\pi}}\, e^{-\eta^2/2p\delta}. \tag{12.6.2}$$

Since $\lim_{\delta\to 0} \sqrt{\delta}\, e^{-\eta^2/(2p\delta)} = 0$, we have $\lim_{\delta\to 0} P^x\{|X_{t+\delta} - X_t| \geqq \eta\} = 0$ for every $\eta > 0$, and hence condition (d) holds. From (12.6.2) we now obtain a bound for the number $q_n(\eta)$ defined in (12.2.10):

$$q_n(\eta) \leqq \frac{p}{\eta}\sqrt{\frac{2p}{\pi}}\, 2^{-n/2} e^{-\eta^2 2^n/2p}. \tag{12.6.3}$$

Thus if we choose $\eta_n \equiv 2^{-n/4}$ $(n = 1, 2, \ldots)$, then $\Sigma_{n=1}^{\infty} \eta_n < \infty$ and

$$\sum_{n=1}^{\infty} n2^n q_n(\eta_n) \leqq \text{const.} \sum n 2^{(3/4)n} e^{-\frac{2^{n/2}}{2p}} < \infty,$$

as we see from a simple application of the root test. Thus condition (c) has also been verified. ┘

Remarks. 1. In the above proof we could have used Theorem 12.2.7 of Kolmogorov-Prohorov instead of Theorem 12.2.6. We show this for the case $p = 1$ and leave the somewhat more complicated computations for $p > 1$ to the reader: For every pair $s, t \in \mathbf{R}_+$ with $s < t$, we know the distribution $P^x_{X_t - X_s}$ by (12.6.1). Therefore,

$$E^x(|X_t - X_s|^4) = (2\pi(t-s))^{-1/2} \int_{-\infty}^{+\infty} y^4 e^{-y^2/2(t-s)}\, dy$$
$$= \frac{(t-s)^2}{\sqrt{2\pi}} \int_{-\infty}^{+\infty} z^4 e^{-z^2/2}\, dz.$$

But now $(2\pi)^{-1/2} \int_{-\infty}^{+\infty} z^4 e^{-z^2/2}\, dz$ is the fourth moment of the standard normal distribution $\nu_{0,1}$ and thus, by Section 8.3, Example 4, equals 3. Hence,

$$E^x(|X_t - X_s|^4) = 3(t-s)^2.$$

Therefore condition (e) of Theorem 12.2.7 is satisfied, namely, with $a = 4$, $c = 3$, and $b = 2$.

2. For a Brownian motion process $(\Omega,\mathfrak{A},(P^x)_{x\in\mathbf{R}^p},(X_t)_{t\in\mathbf{R}_+})$, the measurable space $(\Omega,\mathfrak{A})$ is *not* uniquely determined. For example, we can replace Ω by $\hat{\Omega} \equiv \Omega \cup \Omega_0$, where Ω_0 is an arbitrary set disjoint from Ω, choose for $\hat{\mathfrak{A}}$ the σ-algebra of all sets A or $A \cup \Omega_0$ with $A \in \mathfrak{A}$ generated by $\mathfrak{A}$ and Ω_0 in $\hat{\Omega}$, and set $\hat{P}^x(A) = \hat{P}^x(A \cup \Omega_0) = P^x(A)$ for every $A \in \mathfrak{A}$ and $x \in \mathbf{R}^p$ and define

$$\hat{X}_t(\omega) = \begin{cases} X_t(\omega), & \omega \in \Omega \\ 0, & \omega \in \Omega_0, \end{cases} \qquad (t \in \mathbf{R}_+).$$

Then $(\hat{\Omega},\hat{\mathfrak{A}},(\hat{P}^x)_{x\in\mathbf{R}^p},(\hat{X}_t)_{t\in\mathbf{R}_+})$ is obviously again a Brownian motion process. Nevertheless, we speak of *the Brownian motion process,* since the same family of finite-dimensional distributions is associated with each $x \in \mathbf{R}^p$.

3. If we choose $\Omega = C$ and $\mathfrak{A} = C \cap (\mathfrak{B}^p)^{\mathbf{R}_+}$, then P^x is called the *Wiener measure* associated with the starting point $x \in \mathbf{R}^p$

The name "Brownian motion process" is justified by the fact that this Markov process provides us with a far-reaching, convenient mathematical model for the Brownian motion already discussed in Section 12.1. This type of description of Brownian motion is based on results of Einstein and Smoluchowski. If we interpret Brownian motion as a stochastic process $(X_t)_{t\in\mathbf{R}_+}$, then we can first of all assume that we are dealing with a process with stationary and independent increments. Secondly, we can consider the increment $X_t - X_s$ as the result of many irregular successions of molecular collisions. The Central Limit Theorem therefore makes it natural to consider $X_t - X_s$ as normally distributed. Theoretical considerations led Einstein to determine the variance of this normal distribution. It turned out to be $2D^2(t - s)$, where D is the diffusion constant of fluid. It is no loss of generality to set $D = 1$.

A theorem of N. Wiener, by which P^x-almost surely every Brownian path $t \to X_t(\omega)$ is nowhere differentiable (that is, the Brownian particle has no velocity), shows that this model is not sufficient, from the physicists' viewpoint, to describe all phenomena of Brownian motion. We shall be content with a weaker result:

12.6.3. Theorem. Let $(\Omega,\mathfrak{A},(P^x)_{x\in\mathbf{R}^p},(X_t)_{t\in\mathbf{R}_+})$ be the Brownian motion process. Then, for every $x \in \mathbf{R}^p$ P^x-almost surely, every path $t \to X_t(\omega)$ is differentiable only at the points of a Borel set $D_\omega \subset \mathbf{R}_+$ of Lebesgue measure zero.

Proof. The lemma below shows that the mapping $(\omega,t) \to X_t(\omega)$ is $\mathfrak{A} \otimes \mathfrak{B}^1_+$-$\mathfrak{B}^p$-measurable ($\mathfrak{B}^1_+ \equiv \mathbf{R}_+ \cap \mathfrak{B}^1$). Therefore, if $X^1_t, \ldots, X^p_t$ denote the p coordinates of X_t, the mappings

$$\begin{aligned}(\omega,t) &\to D_+X^i_t(\omega) \equiv \limsup_{\substack{h\to 0\\ h \text{ rational}}} \frac{1}{h}[X^i_{t+h}(\omega) - X^i_t(\omega)],\\ (\omega,t) &\to D_-X^i_t(\omega) \equiv \liminf_{\substack{h\to 0\\ h \text{ rational}}} \frac{1}{h}[X^i_{t+h}(\omega) - X^i_t(\omega)]\end{aligned} \qquad (i = 1, \ldots, p)$$

of $\Omega \times \mathbf{R}_+$ into $\bar{\mathbf{R}}$ are all $\mathfrak{A} \otimes \mathfrak{B}^1_+$-measurable. Hence the set D^i of all (ω,t) for which $\tau \to X^i_\tau(\omega)$ is differentiable at the point $t \in \mathbf{R}_+$ lies in $\mathfrak{A} \otimes \mathfrak{B}^1_+$. Consequently, $D \equiv \bigcap_{i=1}^{p} D^i$, the set of all (ω,t) for which $\tau \to X_\tau(\omega)$ is differentiable at $t \in \mathbf{R}_+$, also lies in $\mathfrak{A} \otimes \mathfrak{B}^1_+$. Thus by Lemma 3.2.1 every ω-section D_ω (t-section D_t) of D lies in $\mathfrak{B}^1_+$ ($\mathfrak{A}$). Fubini's Theorem applied to the indicator function of D [or (3.2.3)] then yields, for every $x \in \mathbf{R}^p$,

$$E^x\left(\int_0^\infty 1_D(\omega,t)\,dt\right) = \int_0^\infty P^x(D_t)\,dt.$$

Therefore, if $P^x(D_t) = 0$ for all $t \in \mathbf{R}_+$, then we have

$$\lambda^1(D_\omega) = \int_0^\infty 1_D(\omega,t)\,dt = 0, \qquad P^x\text{-almost surely}$$

and thus the assertion. But we can see that $P^x(D_t) = 0$ always holds, as follows: D_t is the set of all $\omega \in \Omega$ for which the path $\tau \to X_\tau(\omega)$ is differentiable at t. The boundedness of $(1/h)[X_{t+h}(\omega) - X_t(\omega)]$ as $h \to 0$ ($h > 0$) is necessary for the differentiability of $\tau \to X_\tau(\omega)$ at t. Therefore, if we define

$$A^M_h \equiv \left\{\frac{1}{h}|X_{t+h} - X_t| \leqq M\right\}$$

for arbitrary numbers $h > 0$, $M > 0$, then

$$D_t \subset \bigcup_{M=1}^{\infty} \liminf_{n\to\infty} A^M_{1/n},$$

and it suffices to show that $P^x\{\liminf_{n\to\infty} A^M_{1/n}\} = 0$ for every $M > 0$. By Fatou's Lemma, the following is thus sufficient:[19]

$$\lim_{h\to 0} P^x(A^M_h) = 0, \qquad \text{for every } M > 0. \tag{12.6.4}$$

[19] If (B_n) is a sequence of subsets of a set and (1_{B_n}) is the corresponding sequence of indicator functions, then $\liminf_{n\to\infty} 1_{B_n}$ is the indicator function of $\liminf_{n\to\infty} B_n = \bigcup_{m=1}^{\infty} \bigcap_{n=m}^{\infty} B_n$.

By Theorem 12.5.2 for every $h > 0$, the distribution of $(1/h)(X_{t+h} - X_t)$ relative to P^x has $y \to h^p g_h(hy)$ as density relative to λ^p. Hence,

$$P^x(A_h^M) = h^p \int_{|y| \leqq M} g_h(hy)\lambda^p(dy)$$
$$\leqq \left(\frac{h}{2\pi}\right)^{p/2} \int_{|y| \leqq M} \lambda^p(dy) \leqq \left(\frac{h}{2\pi}\right)^{p/2} (2M)^p,$$

and (12.6.4) obviously follows. ┘

We still need:

12.6.4. Lemma. Let $(\Omega,\mathfrak{A},P,(X_t)_{t \in \mathbf{R}_+})$ be a stochastic process with $\mathbf{R}^p$ as state space such that all paths are right continuous. Then $(\omega,t) \to X_t(\omega)$ is $\mathfrak{A} \otimes \mathfrak{B}_+^1$-$\mathfrak{B}^p$-measurable.

Proof. For every $n \in \mathbf{N}$, we define $Y_n\colon \Omega \times \mathbf{R}^p \to \mathbf{R}^p$ by

$$Y_n(\omega,t) \equiv X_{(j+1)/2^n}(\omega), \qquad \text{provided } j2^{-n} \leqq t < (j+1)2^{-n} \quad (j = 0, 1, \ldots).$$

Due to the right continuity of all paths, $\lim_{n \to \infty} Y_n(\omega,t) = X_t(\omega)$ for all (ω,t). Since we can pass to the p coordinate functions and since Theorem 2.1.5 holds, we need prove only the $\mathfrak{B}_+^1$-measurability of Y_n. But for every $B \in \mathfrak{B}^p$,

$$Y_n^{-1}(B) = \bigcup_{j=0}^{\infty} \{X_{(j+1)/2^n} \in B\} \times [j2^{-n},(j+1)2^{-n}[\; \in \mathfrak{A} \otimes \mathfrak{B}_+^1. \quad ┘$$ [20]

The Brownian motion process has found many applications in mathematics during the last decades. We mention only the many connections with *potential theory* and the so-called *invariance principles.* The interested reader is referred to the special literature in these areas; above all, to Blumenthal-Getoor [24], Meyer [39], and Prohorov [41].

PROBLEMS

1. Let $(\Omega,\mathfrak{A},(P^x)_{x \in \mathbf{R}^p},(X_t)_{t \in \mathbf{R}_+})$ be the Brownian motion process in $\mathbf{R}^p$. Prove: For each $x \in \mathbf{R}^p$, and any four real numbers $0 \leqq t_0 < t_1 \leqq t_2 < t_3$, one has

$$E^x(<X_{t_3} - X_{t_2}, X_{t_1} - X_{t_0}>) = 0,$$

[20] The lemma remains valid if $\mathbf{R}^p$ is replaced by a Polish space E. Then we have to check that the pointwise limit of a sequence of E-valued random variables is itself again a random variable.

that is, for each $x \in \mathbf{R}^p$, the stochastic process $(\Omega,\mathfrak{A},P^x,(X_t)_{t\in\mathbf{R}_+})$ has *orthogonal increments.*

2. Let $(\Omega,\mathfrak{A},(P^x)_{x\in\mathbf{R}},(X_t)_{t\in\mathbf{R}_+})$ be the Brownian motion process in $\mathbf{R}$, and choose $a \in \mathbf{R}$ and $T > 0$ arbitrarily. For every finite subdivision $t = (t_0, \ldots, t_n)$ of the interval $[0,T]$, that is, $0 = t_0 < \cdots < t_n = T$, define

$$\delta(t) = \max (t_i - t_{i-1}; i = 1, \ldots, n).$$

$$V_1(t) = \sum_{i=1}^{n} |X_{t_i} - X_{t_{i-1}}|,$$

$$V_2(t) = \sum_{i=1}^{n} |X_{t_i} - X_{t_{i-1}}|^2.$$

Prove:

(a) $V_2(t) \leqq \sup (|X_{t_i} - X_{t_{i-1}}|; i = 1, \ldots, n) \cdot V_1(t);$

(b) $\lim\limits_{\delta(t)\to 0} E^a(|V_2(t) - (b - a)|^2) = 0;$

(c) $\sup\limits_{t} V_1(t) = +\infty,$ $\quad P^a$-almost surely,

where the supremum is taken over all finite subdivisions t of $[0,T]$. Hence the paths of the Brownian motion process P^a-almost surely are *not of bounded variation* on $[0,T]$.

12.7 THE POISSON PROCESS

In formal analogy to the treatment of the Brownian motion processes, we now study the Markov processes associated with the Poisson semigroup (Section 12.3, Example 5), and in particular their paths.

We first agree on a useful convention: Let $f: \mathbf{R}_+ \to \mathbf{R}$ be an isotone real function on $\mathbf{R}_+$. Then the following limits exist for all $t \in \mathbf{R}_+$ by the monotonicity criterion:

$$f_t^+ \equiv \lim_{\substack{s\to t\\ s>t}} f(s) \qquad \text{and} \qquad f_t^- \equiv \begin{cases} \lim\limits_{\substack{s\to t\\ s<t}} f(s), & \text{if } t > 0 \\ f(0), & \text{if } t = 0. \end{cases}$$

Then we call the difference

$$s_t \equiv f_t^+ - f_t^-$$

the *jump* of f at $t \in \mathbf{R}_+$. We say that f has (only) *jumps of size* 1 if, for all $t \in \mathbf{R}_+$, either $s_t = 1$ or $s_t = 0$.

12.7.1. Definition. *Poisson process* is called any Markov process $(\Omega,\mathfrak{A},(P^x)_{x\in\mathbf{Z}},(X_t)_{t\in\mathbf{R}_+})$ with the set $\mathbf{Z}$ of integers as state space, whose semigroup of transition probabilities is the Poisson semigroup on $\mathbf{Z}$ and whose paths are all right continuous isotone functions $f: \mathbf{R}_+ \to \mathbf{Z}$ with jumps of size 1.

In analogy to Theorem 12.6.2, we now have:

12.7.2. Theorem. There is a Poisson process $(\Omega,\mathfrak{A},(P^x)_{x\in\mathbf{Z}},(X_t)_{t\in\mathbf{R}_+})$. In particular, Ω can be chosen equal to the set $W \subset \mathbf{Z}^{\mathbf{R}_+}$ of all right continuous isotone mappings of $\mathbf{R}_+$ into $\mathbf{Z}$ with jumps of size 1, and $\mathfrak{A}$ equal to the trace of $\mathfrak{P}(\mathbf{Z})^{\mathbf{R}_+}$ in W.

Proof. The idea of the proof is analogous to that of the proof of Theorem 12.6.2. By Theorem 12.4.5 there exists a Markov process $(\Omega,\mathfrak{A},(P^x)_{x\in\mathbf{Z}}, (X_t)_{t\in\mathbf{R}_+})$ with $\mathbf{Z}$ as state space and the Poisson semigroup $(P_t)_{t\in\mathbf{R}_+}$ as semigroup of transition probabilities. Here we can again assume that each of the processes $(\Omega,\mathfrak{A},P^x,(X_t)_{t\in\mathbf{R}_+})$ is canonical, and thus in particular $\Omega = \mathbf{Z}^{\mathbf{R}_+}$, $\mathfrak{A} = \mathfrak{P}(\mathbf{Z})^{\mathbf{R}_+}$, and $P^x = \underleftarrow{\lim}\, P_J^x$, where (P_J^x) is the projective family of probability measures associated with (P_t) and $x \in \mathbf{Z}$ in the sense of (12.4.12). The theorem is proved if W is essential relative to (P_J^x) for every $x \in \mathbf{Z}$. The verification will be accomplished by a modification of the reasoning used for Lemma 12.2.5 and Theorem 12.2.6. Thus, we let $(\pi_t)_{t\in\mathbf{R}_+}$ denote the Poisson convolution semigroup of probability measures on $\mathbf{R}$. Then, by Theorem 12.5.2, $P^x_{X_t-X_s} = \pi_{t-s}$ for arbitrary $x \in \mathbf{Z}$ and $s, t \in \mathbf{R}_+$ with $s \leqq t$. Let S denote the set of all dyadic numbers $\geqq 0$. For arbitrary given $x \in \mathbf{Z}$, we then show the following:

1. There is a set $\Omega_1 \in \mathfrak{A}$ with $P^x(\Omega_1) = 1$ such that

$$X_s(\omega) \leqq X_t(\omega) \qquad (s, t \in S,\ s \leqq t;\ \omega \in \Omega_1). \tag{12.7.1}$$

Indeed, for arbitrary $s, t \in \mathbf{R}_+$ with $s \leqq t$, we have

$$P^x\{X_t - X_s < 0\} = \pi_{t-s}(\{-1, -2, \ldots\}) = 0,$$

that is, $X_s \leqq X_t$ P^x-almost surely. Hence, the existence of Ω_1 follows, since the set of all pairs $(s,t) \in S \times S$ with $s \leqq t$ is countable.

2. There is a set $\Omega_2 \in \mathfrak{A}$ such that $\Omega_2 \subset \Omega_1$, $P^x(\Omega_2) = 1$ and such that

$$\Omega_2 \subset \bigcap_{m\in\mathbf{N}} \bigcup_{n=m}^{\infty} \bigcap_{i=1}^{n2^n-1} \{X_{(i+1)2^{-n}} - X_{(i-1)2^{-n}} \leqq 1\}. \tag{12.7.2}$$

In order to prove this we show that each of the events

$$B_m \equiv \bigcap_{n=m}^{\infty} \bigcup_{i=1}^{n2^n-1} \{X_{(i+1)2^{-n}} - X_{(i-1)2^{-n}} \geqq 2\} \qquad (m \in \mathbf{N})$$

and hence the event $\bigcup B_m$ has probability zero. Then, obviously, $\Omega_2 \equiv \Omega_1 \setminus \bigcup_{m=1}^{\infty} B_m = \Omega_1 \cap \bigcap_{m=1}^{\infty} \complement B_m$ has the desired properties. But

$$P^x(B_m) = 0 \qquad (m \in \mathbf{N}) \tag{12.7.3}$$

can be seen as follows: For all $m, n \in \mathbf{N}$ with $n \geqq m$, we have

$$\begin{aligned} P^x(B_m) &\leqq P^x\left(\bigcup_{i=1}^{n2^n-1} \{X_{(i+1)2^{-n}} - X_{(i-1)2^{-n}} \geqq 2\}\right) \\ &\leqq \sum_{i=1}^{n2^n-1} P^x\{X_{(i+1)2^{-n}} - X_{(i-1)2^{-n}} \geqq 2\} \\ &= \sum_{i=1}^{n2^n-1} \pi_{2^{-n+1}}(\{2,3, \ldots\}) = (n2^n - 1)e^{-2^{-n+1}} \sum_{k=2}^{\infty} \frac{2^{-(n-1)k}}{k!} \\ &= (n2^n - 1)e^{-2^{-n+1}} 2^{-2(n-1)} \sum_{k=0}^{\infty} \frac{2^{-(n-1)k}}{(k+2)!} \leqq (n2^n - 1)2^{-2(n-1)} \end{aligned}$$

Since $\lim_{n\to\infty} (n2^n - 1)2^{-2(n-1)} = 0$, this implies $P^x(B_m) = 0$.

3. We have

$$P^x\{X_t - X_s \geqq 1\} = 1 - e^{-(t-s)} \qquad (s, t \in \mathbf{R}_+, s \leqq t). \tag{12.7.4}$$

For arbitrary s, t with $0 \leqq s \leqq t$, this follows from

$$P^x\{X_t - X_s \geqq 1\} = \pi_{t-s}(\mathbf{N}) = e^{-(t-s)} \sum_{k=1}^{\infty} \frac{(t-s)^k}{k!}.$$

4. By 1, $s \to X_s(\omega)$ is an isotone mapping of S into $\mathbf{Z}$ for every $\omega \in \Omega_1$ and thus for every $\omega \in \Omega_2$. Therefore, for every $\omega \in \Omega_2$, there exists

$$\tilde{X}_t(\omega) \equiv \lim_{\substack{s\to t \\ s\in S, s>t}} X_s(\omega) \qquad \text{for arbitrary } t \in \mathbf{R}_+.^{21} \tag{12.7.5}$$

Hence, by 1, $t \to \tilde{X}_t(\omega)$ is a right continuous isotone mapping of $\mathbf{R}_+$ into $\mathbf{Z}$, for every $\omega \in \Omega_2$. It follows from 2 that $t \to \tilde{X}_t(\omega)$ only has jumps of size 1. Indeed, for $\omega \in \Omega_2$, let $s_t(\omega)$ be the jump at t of the function $\tau \to \tilde{X}_\tau(\omega)$. If $s_t(\omega) \geqq 2$ for some $t \in \mathbf{R}_+$, then, for every natural number $n > t$, there obviously exists a number $i \in \{1,2, \ldots, n2^n - 1\}$ such that

$$X_{(i+1)2^{-n}}(\omega) - X_{(i-1)2^{-n}}(\omega) \geqq 2.$$

[21] Since the discrete topology is induced in $\mathbf{Z}$ by $\mathbf{R}$, this means: For $\omega \in \Omega_2$ and $t \in \mathbf{R}_+$, there is a number $\delta > 0$ such that $X_s(\omega)$ is *constant* and equal to $\tilde{X}_t(\omega)$ for all $s \in S$ with $0 < s - t < \delta$.

Thus we have

$$\{\omega \in \Omega_2 \colon s_t(\omega) \geqq 2\} \subset \Omega_2 \cap \bigcup_{m=1}^{\infty} B_m.$$

Since, by the definition of Ω_2, the sets Ω_2 and $\bigcup B_m$ are disjoint, this proves $s_t(\omega) < 2$ for all $t \in \mathbf{R}_+$. Now $t \to X_t(\omega)$ and hence $t \to \tilde{X}_t(\omega)$ only take on entire values. Hence $s_t(\omega) = 0$ or 1, that is, $t \to \tilde{X}_t(\omega)$ has jumps of size 1.

5. For every ω from the P^x-null set $\complement\Omega_2$, we define

$$\tilde{X}_t(\omega) \equiv x \qquad (t \in \mathbf{R}_+). \tag{12.7.6}$$

Since S is countable, then obviously each of the functions $\tilde{X}_t \colon \Omega \to \mathbf{Z}$ is the pointwise limit of a sequence of random variables on the probability space $(\Omega,\mathfrak{A},P^x)$ with values in $\mathbf{Z}$, and hence $(\Omega,\mathfrak{A},P^x, (\tilde{X}_t)_{t\in\mathbf{R}_+})$ is a stochastic process, with $\mathbf{Z}$ as state space, whose paths are all right continuous isotone functions with jumps of size 1. Moreover, for every $t \in \mathbf{R}_+$,

$$\tilde{X}_t = X_t, \qquad P^x\text{-almost surely.} \tag{12.7.7}$$

In fact, by (12.7.5), for every t there exists an antitone sequence (s_n) in S with $\lim s_n = t$ such that (X_{s_n}) converges P^x-almost surely and hence stochastically (relative to P^x) to $\tilde{X}_t$. But by 3 the sequence (X_{s_n}) also converges stochastically (relative to P^x) to X_t. The almost sure equality of X_t and $\tilde{X}_t$ now follows from this and from Section 2.11, Remark 3.

6. Finally, we have to repeat the reasoning of the proof of Lemma 12.2.8 with W in place of C in order to see that, because of 3, the set W is essential relative to the family (P_J^x). ┘

Remarks. 1. Remark 2 of the preceding section is obviously valid here too. Therefore we speak of *the Poisson process.*

2. If $(\Omega,\mathfrak{A},(P^x)_{x\in\mathbf{Z}},(X_t)_{t\in\mathbf{R}_+})$ is the Poisson process and $\alpha > 0$ is a real number, then we set

$$Y_t \equiv X_{\alpha t} \qquad (t \in \mathbf{R}_+).$$

Thus, we are stretching the time scale by the factor α. Then Definition 12.4.4 immediately shows that $(\Omega,\mathfrak{A},(P^x)_{x\in\mathbf{Z}},(Y_t)_{t\in\mathbf{R}_+})$ is also a Markov process, with state space $\mathbf{Z}$, whose paths are all right continuous isotone functions $f \colon \mathbf{R}_+ \to \mathbf{Z}$ with jumps of size 1. For the associated semigroup $(Q_t)_{t\in\mathbf{R}_+}$ of transition probabilities, we then obviously have $Q_t(x,B) = \pi_{\alpha t}(B - x)$ for arbitrary $x \in \mathbf{Z}$ and $B \in \mathfrak{P}(\mathbf{Z})$. Here $\pi_{\alpha t}$ again denotes the Poisson distribution with parameter αt in the case $t > 0$ and the measure ϵ_0 in the case $t = 0$. Therefore, we call the new Markov process the *Poisson process with parameter* $\alpha > 0$. It obviously suffices to consider the case $\alpha = 1$ specified by Definition 12.7.1.

The natural question concerning the probability of continuous paths for the Poisson process can be answered immediately:

12.7.3. Theorem. Let $(\Omega,\mathfrak{A},(P^x)_{x\in\mathbf{Z}},(X_t)_{t\in\mathbf{R}_+})$ be the Poisson process. Then for every $x \in \mathbf{Z}$ we have:

(a) For each $t_0 \in \mathbf{R}_+$, every path $t \to X_t(\omega)$ is P^x-almost surely continuous at t_0.
(b) Every path $t \to X_t(\omega)$ has a point of discontinuity P^x-almost surely.

Proof. For (a): The set A of all $\omega \in \Omega$ for which $t \to X_t(\omega)$ is discontinuous at t_0 lies in $\mathfrak{A}$. Indeed, when $t_0 > 0$,

$$A = \bigcap_{\substack{0 \leqq s < t_0 \\ s \text{ rational}}} \{X_{t_0} - X_s \geqq 1\}.$$

From this we obtain

$$P^x(A) \leqq P^x\{X_{t_0} - X_s \geqq 1\} = 1 - e^{-t_0+s},$$

for all rational s with $0 \leqq s < t_0$. By the passage to the limit $s \to t_0$ we obtain $P^x(A) = 0$, and hence the assertion. At $t_0 = 0$ all paths are continuous because they are right continuous.

For (b): The set B_{t_0} of all $\omega \in \Omega$ for which $t \to X_t(\omega)$ is continuous on $[0,t_0]$ lies in $\mathfrak{A}$, since, due to the right continuity of all paths, $B_{t_0} = \{X_{t_0} = X_0\}$ $(t_0 \in \mathbf{R}_+)$. Hence, and from (12.7.4), we have

$$P^x(B_{t_0}) = P^x\{X_{t_0} - X_0 = 0\} = e^{-t_0}.$$

Then for the event of interest $B_\infty \equiv \bigcap_{n=1}^{\infty} B_n$, we obviously have

$$P^x(B_\infty) = \lim_{n\to\infty} P^x(B_n) = 0. \quad \lrcorner$$

The above theorem as well as the remarks below seem to justify the use of the Poisson process for describing so-called *signal processes*. Let $(\Omega,\mathfrak{A},(P^x)_{x\in\mathbf{Z}},(X_t)_{t\in\mathbf{R}_+})$ be the Poisson process. We are particularly interested in the stochastic process $(\Omega,\mathfrak{A},P^0,(X_t)_{t\in\mathbf{R}_+})$ which is obtained from the above family for $x = 0$. Since $P^0\{X_0 = 0\} = P_0(0,\{0\}) = 1$, we then have:

(a) Every path $t \to X_t(\omega)$ is a right continuous isotone mapping of $\mathbf{R}_+$ into $\mathbf{Z}$ with jumps of size 1. $X_t \geqq 0$ holds P^0-almost surely.
(b) The process has stationary and independent increments.

This last property (b) follows from Theorems 12.5.2 and 12.5.3.

We can now interpret X_t as the number of all "*signals*" sent randomly by a "transmitter" in the time interval $[0,t[$. Here the transmitter can be a radioactive material; every emission of an α-particle would then be interpreted as a signal. But the transmitter can also be considered as the set of subscribers connected with a telephone central. Every call by one of the subscribers would then be a signal. Now

$$\pi_{t-s}(\{i\}) = e^{-(t-s)}((t-s)^i/i!)$$

is the probability that exactly $i = 0, 1, \ldots$ signals occur in a time interval of length $t - s > 0$. Then obviously

$$\text{(c)} \qquad \lim_{\substack{t\to s\\ t>s}} \frac{1}{t-s}\pi_{t-s}(\{1\}) = 1$$

and

$$\lim_{\substack{t\to s\\ t>s}} \frac{1}{t-s}(1 - \pi_{t-s}(\{0\}) - \pi_{t-s}(\{1\})) = 0.$$

The probability of the occurrence of a single signal in a time interval of short length $t - s$ is thus asymptotically equal to this length, while the probability of occurrence of at least two signals in a small time interval $[s,t]$ is small compared to the length $t - s$. A signal process is defined by means of the properties (a) to (c). Here it is appropriate to replace the number 1 on the right side of the first equality in (c) by a number $\alpha > 0$. This amounts to the introduction of the Poisson process with parameter $\alpha > 0$ (Remark 2).

Remark. 3. Besides the Brownian and Poisson processes, the *stable symmetric processes* are also significant among Markov processes. These are derived from the stable symmetric semigroup of order $\alpha \in]0,2]$ with parameter $c > 0$ (from Section 12.3, Example 6) and have right continuous paths by definition. The interested reader will find the details in Blumenthal-Getoor [24] and Meyer [39].

APPENDIX: CONTINUOUS MAPPINGS INTO THE CIRCLE

Let E be a topological space and let

$$\mathbf{U} \equiv \{z \in \mathbf{C}: |z| = 1\}$$

be the unit circle. A continuous mapping $f: E \to \mathbf{U}$ is said to be *unessential* if there is a continuous real function $\varphi: E \to \mathbf{R}$ such that

$$f = e^{i\varphi}.\text{[1]} \tag{A.1}$$

Obviously, together with φ, $\varphi + 2\pi k$ is also a function of this kind for every $k \in \mathbf{Z}$. For a *connected* space E, these are all continuous real functions on E with property (A.1). Indeed, for another such function ψ, we have $e^{i(\varphi-\psi)} = 1$, and thus $\Phi \equiv (1/2\pi)(\psi - \varphi)$ is a continuous mapping of E into the discrete space $\mathbf{Z}$ of integers. Since E is connected, $\Phi(E)$ is also connected, and thus $\Phi(E) = \{k\}$ for a fixed $k \in \mathbf{Z}$. Hence $\psi = \varphi + 2\pi k$. For connected E, the continuous real function φ is *uniquely* determined by (A.1) and its value at a given point of E.

The principal aim of the Appendix is the proof of the following theorem and several of its consequences.[2]

A.1. Theorem. Every continuous mapping $f: \mathbf{R}^p \to \mathbf{U}$ is unessential ($p = 1, 2, \ldots$).

[1] If there are no such functions φ, then the mapping f is said to be *essential*.

[2] In the first part of the reasoning we follow J. Dieudonné, *Foundations of Modern Analysis*, Academic Press, Inc., New York: 1960.

We prepare the proof by some simple properties of unessential mappings. We let E denote a topological space, arbitrary at the moment.

Every continuous mapping $f\colon E \to \mathbf{U}$ with $f(E) \neq \mathbf{U}$ is unessential. (A.2)

In fact, let $e^{i\alpha} \in \mathbf{U} \setminus f(E)$ for a suitable $\alpha \in \mathbf{R}$. Then we know that $t \to e^{it}$ is a homeomorphism Φ of $]\alpha,\alpha + 2\pi[$ onto $\mathbf{U} \setminus \{e^{i\alpha}\}$. For the inverse homeomorphism Ψ we have $z = e^{i\Psi(z)}$ for all $z \in \mathbf{U} \setminus \{e^{i\alpha}\}$. Thus, $\varphi \equiv \Psi \circ f$ is a continuous real function on E with $f = e^{i\varphi}$. ⌟

If f_1 and f_2 are continuous mappings of E into $\mathbf{U}$ such that $f_1(x) \neq -f_2(x)$ for all $x \in E$, then together with f_1, f_2 is also unessential. (A.3)

Indeed, $f \equiv f_1/f_2$ is a continuous mapping of E into $\mathbf{U}$, which does not attain the value -1 and thus is unessential by (A.2). Hence there are continuous real functions φ and φ_1 on E such that $f = e^{i\varphi}$ and $f_1 = e^{i\varphi_1}$. Since $f_2 = e^{i(\varphi_1 - \varphi)}$, f_2 is also unessential. ⌟

If E is a compact space and $f\colon E \times [0,1] \to \mathbf{U}$ is a continuous mapping, then along with $x \to f(x,0)$, the mapping $x \to f(x,1)$ is also unessential.[3] (A.4)

Because of the uniform continuity of f on the compact space $E \times [0,1]$, there is a natural number n such that $x \in E$, $s, t \in [0,1]$ and $|s - t| \leqq 1/n$ imply $|f(x,s) - f(x,t)| \leqq 1$. Thus if, for every $j = 0, 1, \ldots, n$, we define

$$f_j(x) \equiv f\left(x, \frac{j}{n}\right) \qquad (x \in E),$$

then $|f_j(x) - f_{j+1}(x)| \leqq 1$ for all $x \in E$. If we had $f_j(x) = -f_{j+1}(x)$ for some $x \in E$, this would imply that $2|f_j(x)| \leqq 1$, which contradicts $|f_j(x)| = 1$. Thus $f_j(x) \neq -f_{j+1}(x)$ for all $x \in E$ and $j = 0, 1, \ldots, n - 1$. The n-fold application of (A.3) yields the assertion. ⌟

Now we carry out the proof of Theorem A.1.

Proof. Every continuous mapping $f_r\colon K_r \to \mathbf{U}$ of the compact ball $K_r \equiv \{x \in \mathbf{R}^p\colon |x| \leqq r\}$ in $\mathbf{R}^p$ of radius $r > 0$ into the unit circle $\mathbf{U}$ is unessential. Indeed, $g(x,t) \equiv f_r(tx)$ is a continuous mapping of $K_r \times [0,1]$ into $\mathbf{U}$ with $g(x,1) = f_r(x)$ and $g(x,0) = f_r(0)$ for all $x \in K_r$. Thus, the fact that f_r is unessential follows from (A.2) and (A.4).

For a continuous mapping $f\colon \mathbf{R}^p \to \mathbf{U}$, we consider its restriction f_n to

[3] In other words, (A.4) says: Every mapping $f_1\colon E \to \mathbf{U}$, which is *homotopic* to an unessential mapping $f_0\colon E \to \mathbf{U}$ is itself unessential for compact E.

K_n ($n = 1, 2, \ldots$). Then, by what has just been proved, for every $n \in \mathbf{N}$, there is a continuous function $\varphi_n: K_n \to \mathbf{R}$ with $f_n = e^{i\varphi_n}$. Let ψ_{n+1} be the restriction of φ_{n+1} to K_n. Since K_n is connected, then according to our observations following (A.1) there is an integer k_n such that $\psi_{n+1} = \varphi_n + 2\pi k_n$. Thus, if we set $\varphi_1^* \equiv \varphi_1$ and $\varphi_{n+1}^* \equiv \varphi_{n+1} - 2\pi(k_1 + \cdots + k_n)$ for each $n = 1, 2, \ldots$, every $\varphi_n^*: K_n \to \mathbf{R}$ is a continuous function with $f_n = e^{i\varphi_n^*}$ and moreover φ_n^* is the restriction of φ_{n+1}^* to K_n ($n = 1, 2, \ldots$). Then if $\varphi: \mathbf{R}^p \to \mathbf{R}$ denotes that function whose restriction to K_n coincides with φ_n^* ($n \in \mathbf{N}$), φ is continuous and $f = e^{i\varphi}$; hence, f is unessential. ⌟

A.2. Corollary 1. For every continuous mapping $f: \mathbf{R}^p \to \mathbf{C}^*$ of $\mathbf{R}^p$ into the punctured complex plane $\mathbf{C}^* \equiv \mathbf{C} \setminus \{0\}$ with $f(0) \in \mathbf{R}_+$, there exists exactly one continuous function $\varphi: \mathbf{R}^p \to \mathbf{R}$ such that $\varphi(0) = 0$ and

$$f = |f|e^{i\varphi}.$$

Proof. The mapping $g \equiv f/|f|$ is unessential by Theorem A.1, and thus there exists a continuous real function ψ on $\mathbf{R}^p$ with $g = e^{i\psi}$. Since $g(0) = 1$, we have $\psi(0) = 2\pi k$ with suitable $k \in \mathbf{Z}$. Therefore, $\varphi \equiv \psi - 2\pi k$ is a continuous real function on $\mathbf{R}^p$ with $\varphi(0) = 0$ and $f = |f|e^{i\varphi}$. It follows from the introductory discussion that φ is uniquely determined by the given properties. ⌟

A.3. Corollary 2. If the mapping f of Corollary A.2 is n times continuously differentiable, then the associated function φ is also n times continuously differentiable ($n = 1, 2, \ldots$).

Proof. Since f is n times continuously differentiable on $\mathbf{R}^p$, it follows that $f/|f|$ is too. Thus we can assume that $|f(x)| = 1$ for all $x \in \mathbf{R}^p$. For arbitrary given $x_0 \in \mathbf{R}^p$, we choose a number $r > 0$ such that $f(K) \subset D$, where $D \equiv \{z \in \mathbf{C}: |z - f(x_0)| < 1\}$ and where K is the open ball $\{x \in \mathbf{R}^p: |x - x_0| < r\}$ with center x_0 and radius r. There exists a (holomorphic) branch of the complex logarithm in D, say $z \to \log z$; hence, $\log f(x)$ is defined for all $x \in K$. Since $f(x) = e^{i\varphi(x)} = e^{\log f(x)}$ ($x \in K$), there is a function $k: K \to \mathbf{Z}$ such that $\varphi(x) = -i \log f(x) + 2\pi k(x)$ for all $x \in K$. Hence, the continuity of k follows; thus, the function k is constant because of the connectedness of K. Now $\varphi(x) = -i \log f(x) +$ const. for all $x \in K$ implies the assertion, since the branch $z \to \log z$ in D is holomorphic. ⌟

Finally, we show:

A.4. Theorem. Let $(\varphi_k)_{k \in \mathbf{N}}$ be a sequence of continuous real functions on a connected compact space E. If the sequence $(e^{i\varphi_k})_{k \in \mathbf{N}}$ on E

converges uniformly to the constant function 1 and if there is an element $a \in E$ with $\varphi_k(a) = 0$ for all $k = 1, 2, \ldots$, then the sequence (φ_k) on E converges uniformly to 0.

Proof.[4] We write the sequence being investigated in the form $(2\pi\varphi_k)$ without having to change anything in the hypotheses on the function φ_k. For every real number $\gamma \in \mathbf{R}$, let $\{\gamma\}$ denote an integer satisfying

$$|\gamma - \{\gamma\}| \leqq \tfrac{1}{2}.$$

This is uniquely determined as the nearest integer if γ is not of the form $n + \frac{1}{2}$ with $n \in \mathbf{Z}$. Then $\{\varphi_k\}$ is a function on E with integer values, that is, $e^{-2\pi i\{\varphi_k\}} = 1$. Hence $(e^{-2\pi i(\varphi_k - \{\varphi_k\})})$ converges uniformly to 1. Thus for every $\epsilon > 0$ there is a natural number k_ϵ with

$$|e^{2\pi i(\varphi_k(x) - \{\varphi_k(x)\})} - 1| < \epsilon, \qquad \text{for all } x \in E \text{ and } k \geqq k_\epsilon.$$

A simple calculation of the square of the above absolute value yields

$$|\sin \pi(\varphi_k(x) - \{\varphi_k(x)\})| < \frac{\epsilon}{2}, \qquad \text{for all } x \in E \text{ and } k \geqq k_\epsilon.$$

Now we have $|\varphi_k(x) - \{\varphi_k(x)\}| \leqq \frac{1}{2}$ and thus $|\pi(\varphi_k - \{\varphi_k\})| \leqq \pi/2$ for all $k \in \mathbf{N}$. From the inequality $(2/\pi)|\xi| \leqq |\sin \xi|$, valid for all $\xi \in \mathbf{R}$ with $|\xi| \leqq \pi/2$, it follows that

$$|\varphi_k(x) - \{\varphi_k(x)\}| < \frac{\epsilon}{4}, \qquad \text{for all } x \in E \text{ and } k \geqq k_\epsilon. \tag{A.5}$$

As a consequence of the uniform continuity of φ_k on the compact space E for every $k \geqq k_\epsilon$, there is a covering of E by finitely many nonempty open sets $U_1^{(k)}, \ldots, U_n^{(k)}$, depending on k, such that for arbitrary points x, x' in the same set $U_j^{(k)}$ $(j = 1, \ldots, n_k)$,

$$|\varphi_k(x) - \varphi_k(x')| < \frac{\epsilon}{4}.$$

Hence, and from (A.5), we now obviously have

$$|\{\varphi_k(x)\} - \{\varphi_k(x')\}| < \tfrac{3}{4}\epsilon. \tag{A.6}$$

If $\epsilon < \frac{4}{3}$, then (A.6) yields

$$|\{\varphi_k(x)\} - \{\varphi_k(x')\}| < 1$$

and thus

$$\{\varphi_k(x)\} = \{\varphi_k(x')\}, \tag{A.7}$$

for arbitrary $x, x' \in U_j^{(k)}$ and all $j = 1, \ldots, n_k$. If we decompose $\{1, \ldots, n_k\}$ into two disjoint nonempty sets J and K, then, because E

[4] This proof arose from a discussion with Professor E. Hlawka.

is connected, the set $\bigcup_{j \in J} U_j^{(k)}$ is not disjoint from $\bigcup_{j \in K} U_j^{(k)}$. Therefore, (A.6) implies that

$$\{\varphi_k(x)\} = \text{const.}, \qquad \text{for all } x \in E.$$

But since $\varphi_k(a) = 0$ by hypothesis and therefore $\{\varphi_k(a)\} = 0$, we have

$$\{\varphi_k(x)\} = 0 \qquad (x \in E,\ k \geqq k_\epsilon).$$

By (A.5) we finally obtain $|\varphi_k(x)| < \epsilon$ for all $x \in E$ and all $k \geqq k_\epsilon$, provided $\epsilon < \frac{4}{3}$. But this is the asserted uniform convergence on E of (φ_k) to zero. ┘

A.5. Corollary. Let $(\varphi_k)_{k \in \mathbf{N}}$ be a sequence of continuous real functions on $\mathbf{R}^p$. Then, if the sequence $(e^{i\varphi_k})_{k \in \mathbf{N}}$ converges on every compact subset of $\mathbf{R}^p$ uniformly to the constant 1 and if $\varphi_k(0) = 0$ for all $k = 1, 2, \ldots$, the sequence (φ_k) converges uniformly to 0 on every compact subset of $\mathbf{R}^p$.

Proof. We apply the above theorem to every compact ball $E = K_n$ of radius $n \in \mathbf{N}$ with center 0. ┘

BIBLIOGRAPHY

I TEXTBOOKS AND LECTURE NOTES

Only those works are given which, in the opinion of the author, are especially suited to extending and deepening the material presented here.

(a) Measure and integration theory

[1] N. Bourbaki, *Éléments de Mathématique*, Livre VI: *Intégration*. Chaps. 1–8. Paris: Hermann, 1956–1965.

[2] P. Courrège, *Théorie de la mesure*. Paris: Centre de Documentation Universitaire, 1962.

[3] P. R. Halmos, *Measure Theory*. Princeton, N.J.: D. Van Nostrand Company, Inc., 1950.

[4] O. Haupt G. Aumann, and C. Y. Pauc, *Differential- und Integralrechnung*, Bd. III (*Integralrechnung*). Berlin: Walter de Gruyter & Co. 1955, 2. Auflage.

[5] K. Jacobs, *Mass und Integral*. Erlangen, 1966. Lecture Notes.

[6] J. F. C. Kingman and S. J. Taylor, *Introduction to Measure and Probability*. London: Cambridge University Press, 1966.

[7] L. Nachbin, *The Haar Integral*. Princeton, N.J.: D. Van Nostrand Company, Inc., 1965.

[8] A. C. Zaanen, *Integration*. Amsterdam: North Holland Publ. Co., 1967.

(b) Probability theory

[9] L. Breiman, *Probability*. Reading, Mass.: Addison-Wesley Publishing Company, Inc., 1968.

[10] K. L. Chung, *A Course in Probability Theory*. New York: Harcourt, Brace and World, Inc., 1968.

[11] J. L. Doob, *Stochastic Processes*. New York: John Wiley & Sons, Inc., 1953.

[12] W. Feller, *An Introduction to Probability and Its Applications*, Vols. I–II. New York: John Wiley & Sons, Inc., 1960 and 1966.

[13] M. Fisz, *Probability Theory and Mathematical Statistics* (Translation from the Polish). New York: John Wiley & Sons, Inc., 1963.

[14] B. V. Gnedenko, *The Theory of Probability* (Translation from the Russian). New York: Chelsea Publ. Co., 1967.

[15] K. Krickeberg, *Probability Theory* (Translation from the German). Reading, Mass.: Addison-Wesley Publishing Company, Inc., 1965.

[16] J. Lamperty, *Probability*. New York: W. A. Benjamin, 1966.

[17] M. Loève, *Probability Theory*. Princeton, N.J.: D. Van Nostrand Company, Inc., 1963, 3rd edition.

[18] J. Neveu, *Mathematical Foundations of the Calculus of Probability* (Translation from the French). San Francisco: Holden-Day, 1965.

[19] A. Rényi, *Probability Theory* (Translation from the German). Amsterdam: North Holland Publ. Co., 1970.

[20] H. Richter, *Wahrscheinlichkeitstheorie*. Berlin: Springer-Verlag, 1966, 2. Auflage.

II FURTHER REFERENCES CITED IN TEXT

[21] L. V. Ahlfors and L. Sario, *Riemann Surfaces*. Princeton, N.J.: Princeton Univ. Press, 1960.

[22] E. S. Andersen and B. Jessen, *On the Introduction of Measures in Infinite Product Sets*. Danske Vid. Selsk.Mat.-Fys. Medd. 25, no. 4 (1948).

[23] G. Aumann, *Relle Funktionen*. Berlin: Springer-Verlag, 1954.

[24] R. M. Blumenthal and R. K. Getoor, *Markov Processes and Potential Theory*. New York: Academic Press, Inc., 1968.

[25] S. Bochner, *Harmonic Analysis and the Theory of Probability*. Berkeley, Calif.: University of California Press, 1955.

[26] N. Bourbaki, *Éléments de Mathématique*, Livre III: *Topologie Générale*. Chap. 9, Utilisation des nombres réels en topologie générale. Paris: Hermann, 1958, 2nd edition. [English translation contained in: N. Bourbaki, *Elements of Mathematics. General Topology*. Part 2. Reading, Mass.: Addison-Wesley Publishing Company, Inc., 1966.]

[27] N. Bourbaki, *Éléménts de Mathématique*, Livre III: *Topologie Générale*. Chap. 10, Espaces Fonctionnels. Paris: Hermann, 1961, 2nd edition. [English translation contained in: N. Bourbaki, *Elements of Mathematics. General Topology*. Part 2. Reading, Mass.: Addison-Wesley Publishing Company, Inc., 1966.]

[28] P. Courrège, *Le Processus Stochastique du Mouvement Brownien*. Paris: Centre de Documentation Universitaire, 1963.

[29] P. J. Daniell, *A General Form of Integral*. Annals of Mathematics, **19** (1917–1918), 279–294.

[30] J. Dieudonné, *Sur un Théoréme de Jessen*. Fund. Math. **37** (1950), 242–248.

[31] E. B. Dynkin, *Theory of Markov Processes* (Translation from the Russian). Englewood Cliffs, N.J.: Prentice-Hall, Inc., 1961.

[32] W. Franz, *General Topology* (Translation from the German). New York: Frederick Ungar Publishing Co., 1965.

[33] B. V. Gnedenko and A. N. Kolmogorov, *Limit Distributions for Sums of Independent Random Variables* (Translation from the Russian). Reading, Mass.: Addison-Wesley Publishing Company, Inc., 1954.

[34] S. Guber, *Zur Bewegungsinvarianz des Lebesgue-Masses.* Sitz. Ber. d. Bayer. Akad. d. Wiss., Math.-Naturw. Kl. (1964), 91–92.

[35] H. Hahn and A. Rosenthal, *Set Functions.* Albuquerque, New Mexico: The Univ. of New Mexico Press, 1948.

[36] E. Hewitt and K. A. Ross, *Abstract Harmonic Analysis,* Part I. Berlin: Springer-Verlag, 1963.

[37] S. Johansen, *An Application of Extreme Point Methods to the Representation of Infinitely Divisible Distributions.* Z. Wahrsch.theorie verw. Geb. **5** (1966), 304–316.

[38] A. N. Kolmogorov, *Grundbegriffe der Wahrscheinlichkeitsrechnung,* Ergebn. d. Math. 2, Heft 3. Berlin: Springer-Verlag, 1933. (English translation: *Foundations of the Theory of Probability.* New York: Chelsea Publ. Co., 1950.)

[39] P.-A. Meyer, *Probability and Potentials.* Waltham, Mass.: Blaisdell Publishing Co., 1966.

[40] P.-A. Meyer, *Processus de Markov.* Lecture Notes in Math. 26. Berlin: Springer-Verlag, 1967.

[41] Yu. V. Prohorov, *Convergence of Random Processes and Limit Theorems in Probability Theory* (Translation from the Russian). Theory Prob. Applications 1 (1956), 157–214.

[42] W. Rudin, *Fourier Analysis on Groups.* New York: Interscience Publishers, 1962.

[43] M. H. Stone, *Notes on Integration I-IV.* Proc. Nat. Acad. Sci. U.S. **34** (1948), 336–342, 447–455, 483–490; **35** (1949), 50–58.

LIST OF SYMBOLS

Page numbers are given for first occurrence of these symbols.

$\mathbf{R}, \bar{\mathbf{R}}, \mathbf{R}_+, \bar{\mathbf{R}}_+, \mathbf{N}$, 2, 3
$\mathbf{Z}$, 376
$\mathbf{C}$, 242
$\subset, \cup, \cap, \bigcup, \bigcap, \complement, \setminus$, 3
$f: A \to B, (a_n)_{n \in \mathbf{N}}$, 3
$\mathfrak{P}(\Omega)$, 7
$\Omega' \cap \mathfrak{A}$ ($\mathfrak{A}$ a σ-algebra), 8
T^{-1}, 8
$\mathfrak{A}(\mathfrak{E})$, 8, 9
$\mathfrak{A}(T_{ij}; i \in I)$, 35
$\mathfrak{A}(\mathfrak{F})$, 196
$\mathfrak{D}(\mathfrak{E})$, 11
ϵ_ω, 13, 23
$E_n \uparrow E, E_n \downarrow E$, 15
$\triangleleft$, 18
$[a,b[$, 18
$]a,b[$, 31
$]a,b], [a,b]$, 33
$\mathfrak{F}^p, \mathfrak{I}^p$, 18
λ^p, 22, 30
λ^p_C, 31
$\mathfrak{B}^p$, 30
$\mathfrak{B} = \mathfrak{B}(E), \mathfrak{B}_0 = \mathfrak{B}_0(E)$, ($E$ a topological space), 204
$\mathfrak{C}^p, \mathfrak{K}^p, \mathfrak{O}^p$, 31
$(\Omega, \mathfrak{A}), (\Omega, \mathfrak{A}, \mu)$, 34

$(\Omega, \mathfrak{A}, P)$, 131
$T: (\Omega, \mathfrak{A}) \to (\Omega, \mathfrak{A}')$, 34
$T(\mu)$, 36
$\bar{\mathfrak{B}}^1$, 44
1_A, 45
$\{f \leqq g\}, \{f < g\}, \{f = g\},$
  $\{f \geqq g\}, \{f > g\}$, 46
$\{X \in A'\}$, 140
$\mathcal{E} = \mathcal{E}(\mathfrak{A})$, 49
$\mathcal{E}^*$, 53
$\int f\, d\mu, \int f(\omega)\mu(d\omega), \int f(\omega)\, d\mu(\omega)$,
  50, 54, 58
$\int_A f\, d\mu$, 77
f^+, f^-, 58
$N_p(f)$, 66, 76
$\mathcal{L}^p(\mu)$, 68, 69
$L^p(\mu)$, 75, 76
$\limsup_{n \to \infty} A_n, \liminf_{n \to \infty} A_n$, 71
$\nu \ll \mu$, 84
$\mu\text{-}\lim_{n \to \infty}$, 93
$\prod_{i=1}^{n} \Omega_i = \Omega_1 \times \cdots \times \Omega_n$, 109
$\prod_{i \in I} \Omega_i$ (Ω_i sets), 163

$\bigotimes_{i=1}^{n} \mathfrak{A}_i = \mathfrak{A}_1 \otimes \cdots \otimes \mathfrak{A}_n$, 109
$\bigotimes_{i\in I} \mathfrak{A}_i$ ($\mathfrak{A}_i$ σ-algebras), 164
$\bigotimes_{i=1}^{n} \mu_i = \mu_1 \otimes \cdots \otimes \mu_n$ (μ_i measures), 113, 119
$\bigotimes_{i\in I} P_i$ (P_i probability measures), 168
$\bigotimes_{i=1}^{n} (\Omega_i,\mathfrak{A}_i,\mu_i)$, 120
$\bigotimes_{i\in I} (\Omega_i,\mathfrak{A}_i,P_i)$, 168
$\bigotimes_{i=1}^{n} X_i = X_1 \otimes \cdots \otimes X_n$, 156
$\bigotimes_{i\in I} X_i$ (X_i mappings, random variables), 168
$\mu_1 * \cdots * \mu_n$ (μ_i measures), 121
$f * \nu$ (f a function, ν a measure), 123
$f * g$ (f,g functions), 123, 244
$P(A \mid B)$, 136
$P\{X \in A'\}$, 140,
P_X, 140
$E(X)$, $V(X)$, $\sigma(X)$, 140, 141
Cov (X,Y), 160
ϵ_x, 19, 23, 142
β_n^p, π_α, g_{α,σ^2}, ν_{α,σ^2}, γ_α, 143–145
$\mathfrak{F}_+$, 194
$\mathfrak{F}_+^*$, 196
$\mathcal{C} = \mathcal{C}(E)$, $\mathcal{C}^b = \mathcal{C}^b(E)$, 204
$\mathcal{C}(E,\mathbf{C})$, $\mathcal{C}^b(E,\mathbf{C})$ (E a topological space), 244
$\mathcal{C}^c = \mathcal{C}^c(E)$, $\mathcal{C}^0 = \mathcal{C}^0(E)$, 212, 215
$\mathcal{C}^c(E,\mathbf{C})$, $\mathcal{C}^0(E,\mathbf{C})$ (E a locally compact space), 244
$\|f\|$, 215, 244
S_f, 212, 244
$\mathfrak{M} = \mathfrak{M}(E)$, $\mathfrak{M}^e = \mathfrak{M}^e(E)$, $\mathfrak{M}^1 = \mathfrak{M}^1(E)$, 226
$\mathcal{E}_\alpha$, 239
$\mathcal{R}f$, $\mathcal{I}f$, 242
$\langle x,y \rangle$, $|x|$, 244
$\hat{\mu}$, $\hat{f}$, 245, 249
φ_X, 245
$M_{k_1,\ldots,k_p}$, 263
$E^{\mathfrak{Z}}(X) = E(X \mid \mathfrak{Z})$, $E^{(Y_i)_{i\in I}}(X) = E(X \mid Y_i, i \in I)$, $E^{Y_1,\ldots,Y_n}(X) = E(X \mid Y_1, \ldots, Y_n)$, 304
$P^{\mathfrak{Z}} = P(A \mid \mathfrak{Z})$, 309
$E^{Y=y}(X) = E(X \mid Y = y)$, 313
$E_B(X)$, P_B (B an event with $P(B) > 0$), 302
μK, Kf'(K a kernel), 316
$P_{X|\mathfrak{Z}}$, 318
$\mathfrak{A}_T$, X_T (T a stopping time), 334
$\bar{U}_{[a,b]}$, $\underline{U}_{[a,b]}$, 338
$(\Omega,\mathfrak{A},P,(X_t)_{t\in I})$, 358
E^I, $\mathfrak{B}^I$, 358
$\{X_1 \in B_1, \ldots, X_n \in B_n\}$, 359
$\underset{J\in\mathfrak{H}(I)}{\overleftarrow{\lim}} P_J = \overleftarrow{\lim} P_J$, 360
KL (K,L kernels), 373
P^μ, E^μ, $P^x = P^{\epsilon_x}$, $E^x = E^{\epsilon_x}$, 379, 386
$(\Omega,\mathfrak{A},(P^x)_{x\in E},(X_t)_{t\in\mathbf{R}_+})$, 385

INDEX